American Welding Society.

Welding handbook

DATE			

Welding Handbook

Seventh Edition, Volume 5

Engineering, Costs, Quality, and Safety

The Five Volumes of the Welding Handbook, Seventh Edition

1 Fundamentals of Welding

2 Welding Processes—
Arc and Gas Welding and Cutting, Brazing, and Soldering

3 Welding Processes—
Resistance and Solid-State Welding and Other Joining Processes

4 Metals and Their Weldability

5 Engineering, Costs, Quality, and Safety

Welding Handbook

Seventh Edition, Volume 5

Engineering, Costs, Quality, and Safety

W. H. Kearns, Editor

AMERICAN WELDING SOCIETY
550 N.W. LeJeune Road
P.O. Box 351040
Miami, FL 33135

Library of Congress Number: 75-28707
International Standard Book Number: 0-87171-239-3

American Welding Society, 550 N.W. LeJeune Rd., Miami, FL 33126

The Welding Handbook is a collective effort of many volunteer technical specialists to provide information to assist with the design and application of welding and allied processes.

Reasonable care is taken in the compilation and publication of the Welding Handbook to insure authenticity of the contents. No representation or warranty is made as to the accuracy or reliability of this information.

The information contained in the Welding Handbook shall not be construed as a grant of any right of manufacture, sale, use, or reproduction in connection with any method, process, apparatus, product, composition or system, which is covered by patent, copyright, or trademark. Also, it shall not be construed as a defense against any liability for such infringement. No effort has been made to determine whether any information in the Handbook is covered by any patent, copyright, or trademark, and no research has been conducted to determine whether an infringement would occur.

Printed in the United States of America

Contents

Welding Handbook Committee ix

Preface xi

Chapter 1, Design for Welding 2
- General Considerations 2
- Properties of Metals 3
- Design Program 13
- Design Considerations 17
- Design of Welded Joints 32
- Sizing of Steel Welds 45
- Structural Tubular Connections 68
- Aluminum Structures 81
- Supplementary Reading List 91

Chapter 2, Symbols for Welding, Brazing, and Nondestructive Examination 94
- Purpose 94
- Welding Symbols 94
- Brazing Symbols 113
- Nondestructive Examination Symbols 116

Chapter 3, Economics and Cost Estimating 122
- The Cost Estimate 122
- Welding Costs 123
- Economics of Arc Welding 134
- Economics of Welding Automation and Robotics 135
- Economics of New Joining Processes 139
- Economics of Brazing and Soldering 142
- Economics of Thermal Cutting 143
- Supplementary Reading List 148

Chapter 4, Fixtures and Positioners 150
- Fixtures 150
- Positioners 159
- Supplementary Reading List 175

Chapter 5, Automation and Control 178
Introduction 178
Fundamentals of Welding Automation 178
Extent of Automation 186
Planning for Automation 187
Resistance Welding Automation 193
Arc Welding Automation 200
Brazing Automation 208
Problems of Automation 210
Supplementary Reading List 213

Chapter 6, Codes and Standards 216
Definitions 216
Sources 216
Applications 217
American Association of State Highway and Transportation Officials 219
American Bureau of Shipping 219
American Institute of Steel Construction 221
American National Standards Institute 221
American Petroleum Institute 222
American Railway Engineering Association 223
American Society of Mechanical Engineers 223
American Society for Testing and Materials 225
American Waterworks Association 226
American Welding Society 226
Association of American Railroads 231
Canadian Standards Association 231
Compressed Gas Association 232
Federal Government 232
International Organization for Standardization 237
National Board of Boiler and Pressure Vessel Inspectors 237
National Fire Protection Association 238
Pipe Fabrication Institute 238
Society of Automotive Engineers 239
Manufacturers Association 244

Chapter 7, Qualification and Certification 246
Introduction 246
Procedure Specifications 248
Qualification of Procedure Specifications 256
Performance Qualification 261
Standardization of Qualification Requirements 273
Supplementary Reading List 274

Chapter 8, Weld Quality 276
Introduction 276
Terminology 278
Discontinuities in Fusion Welded Joints 278
Causes and Remedies for Fusion Weld Discontinuities 293
Discontinuities in Resistance and Solid State Welds 303
Discontinuities in Brazed and Soldered Joints 306
Significance of Weld Discontinuities 308
Supplementary Reading List 316

Chapter 9, Inspection 318
Welding Inspectors 318
Inspection Plan 319
Nondestructive Testing 321
Destructive Testing 370
Proof Testing 374
Brazed Joints 376
Supplementary Reading List 377

Chapter 10, Safe Practices 380
General Welding Safety 380
Fumes and Gases 387
Handling of Compressed Gases 397
Electrical Safety 402
Processes 405
Supplementary Reading List 415

Index of Major Subjects 417
Index 431

Welding Handbook Committee

J.R. Hannahs, Chairman	Midwest Testing Laboratories
M.J. Tomsic, Vice Chairman	Hobart Brothers Company
W.H. Kearns, Secretary	American Welding Society
D.R. Amos	Westinghouse Electric Corporation
C.W. Case	Inco Alloys International
J.R. Condra	E.I. du Pont de Nemours and Company
E.H. Daggett	Babcock and Wilcox
R.L. Frohlich[a]	Westinghouse Electric Corporation
A.F. Manz	Union Carbide Corporation
L.J. Privoznik	Westinghouse Electric Corporation
E.G. Signes	Bethlehem Steel Corporation

a. Resigned April 6, 1983

Preface

This is the fifth and final volume of the 7th Edition of the *Welding Handbook*. This volume contains revisions to several chapters that last appeared in Volume (Section) 1 of the 6th Edition of the Handbook. These revisions include symbols for welding, brazing, and nondestructive testing; economics and cost estimating; codes and standards; inspection; and safe practices. The volume also includes new chapters on the design for welding, fixtures and positioners, automation and control, qualification and certification, and weld quality. The information in this volume will be particularly useful to engineering as well as manufacturing and inspection personnel.

The chapter on symbols is a guide to the use of symbols, not the standard for symbols that is published separately (AWS A2.4, Symbols for Welding and Nondestructive Testing). The information on codes and standards includes currently published documents. The reader should verify the existence and latest edition of a code or standard before applying it to a product. Safe practices, the last but perhaps the most important chapter, should be referred to when planning an installation as well as during equipment installation and operation.

An index of the major subjects in all five volumes of the *Welding Handbook* preceeds the volume index. It enables the reader to quickly determine the appropriate volume and chapter in which the desired information can be found.

This volume, like the others, was a voluntary effort by the Welding Handbook Committee and the Chapter Committees. The Chapter Committee Members and the Handbook Committee Member responsible for a chapter are recognized on the title page of that chapter. Other individuals also contributed in a variety of ways, particularly in chapter reviews. All participants contributed generously of their time and talent, and the American Welding Society expresses herewith its appreciation to them and to their employers for supporting the work.

The Welding Handbook Committee expresses its appreciation to Richard French, Deborah Givens, Hallock Campbell, and other AWS Staff Members for their assistance in the production of this volume.

The Welding Handbook Committee welcomes your comments on the Handbook. Please address them to the Editor, Welding Handbook, American Welding Society, 550 N.W. LeJeune Road, P.O. Box 351040, Miami, FL 33135.

J.R. Condra, *Chairman*
Welding Handbook Committee
1981-1984

W.H. Kearns, *Editor*
Welding Handbook

1
Design for Welding

General Considerations *2*
Properties of Metals *3*
Design Program *13*
Design Considerations *17*
Design of Welded Joints *32*
Sizing of Steel Welds *45*
Structural Tubular Connections *68*
Aluminum Structures *81*
Supplementary Reading List *91*

Prepared by

O.W. BLODGETT
Lincoln Electric Company

L.J. PRIVOZNIK
Westinghouse Electric Corporation

Reviewed by

W.A. MILEK
Consulting Engineer

D.D. RAGER
Reynolds Metals Company

W.H. MUNSE
University of Illinois

F.H. RAY
Consultant

C.L. TSAI
Ohio State University

Welding Handbook Committee Member

L.J. PRIVOZNIK
Westinghouse Electric Corporation

1
Design for Welding

GENERAL CONSIDERATIONS

OBJECTIVES

A weldment is an assembly that has the component parts joined by welding. It may be a bridge, a building frame, an automobile or truck body, a trailer hitch, a piece of machinery, or an offshore tubular structure.

The basic objectives of weldment design[1] are ideally to provide a weldment that

(1) Will perform its intended functions.

(2) Will have the required reliability and safety.

(3) Is capable of being fabricated, inspected, transported, and placed in service at minimum total cost.

Total cost includes costs of the following:

(1) Design
(2) Materials
(3) Fabrication
(4) Erection
(5) Inspection
(6) Operation
(7) Repair
(8) Maintenance

THE DESIGNER

Designers of weldments should have some knowledge and experience in the following processes and procedures in addition to the basic design concepts for the product or structure:

(1) Cutting and shaping of metals

(2) Assembly of components

(3) Preparation and fabrication of welded joints

(4) Weld acceptance criteria, inspection, mechanical testing, and evaluation

Designers also need a general knowledge of the following subjects[2] and their effects on the design of weldments:

(1) Mechanical and physical properties of metals and weldments

(2) Weldability of metals

(3) Welding processes, costs, and variations in welding procedures

(4) Filler metals and properties of weld metals

(5) Thermal effects of welding

(6) Effects of restraint and stress concentrations

(7) Control of distortion

(8) Communication of weldment design to the shop, including the use of welding symbols

(9) Applicable welding and safety codes and standards

Engineers who are responsible for designing welded machine parts and structures need knowl-

1. There is similarity between the design of weldments and brazements except for joint designs and joining processes. Much information presented here can be applied to brazement design. Also, see the *Welding Handbook,* Vol. 2, 7th Ed., 1978: 370-438.

2. The listed subjects are covered in this and the other four volumes of the *Welding Handbook,* 7th Edition.

edge of several areas related to weldments including the following:

(1) Efficient use of steel, aluminum, and other metals in weldments.

(2) Design for appropriate stiffness or flexibility in welded beams and other structural members.

(3) Design for torsional resistance.

(4) Effects of thermal strains induced by welding in the presence of restraints.

(5) Effects of stresses induced by welding in combination with design stresses.

(6) Practical considerations of welding and selection of proper joint designs for the application.

One common mistake in the design of weldments is copying the over-all shape and appearance of a casting or other form that is to be replaced by a weldment. The designer should be aware that weldments are different from castings or forgings and can look different. There is no advantage in shaping a weldment with protrusions, separate legs, brackets, and housings that are used on castings. A weldment should have a shape appropriate for the intended function.

When a change is made from a casting to a weldment, both appearance and function may be improved because weldment design generally involves a more conservative and strategic use of materials. The motivating force for such a change is a desire to decrease production cost by fabricating the machine, part, or structure more economically. Cost, therefore, must be taken into consideration at every step in the design program. The designer must consider not only the obvious elements entering into production costs but also all incidentals from the selection of materials and methods of fabrication to the final inspection of the finished product and preparation for shipment.

PROPERTIES OF METALS

STRUCTURE SENSITIVITY OF PROPERTIES

The properties of metals can be divided into three general groups, (1) mechanical properties, (2) physical properties, and (3) corrosion properties. Other groups could be added, such as optical properties and nuclear properties. Table 1.1 provides a list of the properties of metals, but not all are discussed herein. It should be noted that the specific properties listed under each general heading are divided into *structure-insensitive* properties and *structure-sensitive* properties. This separation in properties is commonly made by most textbooks on metals to emphasize the consideration that should be given to reported property values.

Structure-insensitive properties are well defined properties of a metal. They do not vary from one piece of metal to another of the same kind. This is true, at least for most engineering purposes, and is verified by the data obtained from standard engineering tests. The structure-insensitive properties often can be calculated or rationalized by consideration of the chemical composition and the crystallographic structure of the metal.[3] These properties (Table 1.1) commonly are listed in handbooks as constants for the particular metals. Seldom is any mention made of the size or condition of the sample upon which the determination was made.

Structure-sensitive properties, on the other hand, are dependent upon not only chemical composition and crystallographic structure, but also the microstructural details that are affected in subtle ways by the manufacturing and processing history of the metal. Even the size of the sample can influence the test results obtained for a structure-sensitive property. Therefore, different

3. General and welding metallurgy of metals is covered in the *Welding Handbook,* Vol. 1, 7th Ed., 1976: 100-151.

Table 1.1
Properties of metals

General groups	Structure-insensitive properties	Structure-sensitive properties
Mechanical	Elastic moduli	Ultimate strength Yield strength Fatigue strength Impact strength Hardness Ductility Elastic limit Damping capacity Creep strength Rupture strength
Physical	Thermal expansion Thermal conductivity Melting point Specific heat Emissivity Thermal evaporation rate Density Vapor pressure Electrical conductivity Thermoelectric properties Magnetic properties Thermionic emission	Ferromagnetic properties
Corrosion	Electrochemical potential Oxidation resistance	
Optical	Color Reflectivity	
Nuclear	Radiation absorbtivity Nuclear cross section Wavelength of characteristic x-rays	

samples of the same kind of metal or alloy will have essentially identical structure-insensitive properties, but the structure-sensitive properties are likely to vary to some degree because of differences in the treatment and preparation of the samples.

All of the mechanical properties of metals, with the exception of elastic moduli, are structure-sensitive, as listed in Table 1.1. This suggests that single values, or even a narrow range of values, published in a handbook for the strength of a particular metal or alloy must be accepted with reservation. It is not uncommon to find that plates or bars of a metal, which represent unusual sizes or conditions of treatment, may display significant deviations in mechanical properties from those published for the metal. Even the direction in which wrought metal is tested (longitudinal, transverse, or through-thickness) may give significantly different values for strength and ductility. Although the physical and corrosion properties of metals are indicated to be structure-insensitive for the most part, it is becoming more evident that some of the values established for these properties apply only to common multi-grained metals.

MECHANICAL PROPERTIES

Metals are the most useful material for construction because they are generally strong, tough, and ductile. This combination of properties is not often found in nonmetallic materials. Furthermore, in metals, the relative degrees of strength, toughness, and ductility can be varied individually by alloy selection or by heat treatment. Joining of metals by welding or brazing also affects their mechanical properties. The applied heat, cooling rates, filler metal addition, and metallurgical structure of the joint are some of the factors that affect mechanical properties.[4]

Metals not only offer many useful properties and characteristics in their mechanical behavior, but they can also develop a large number of combinations of these properties. The versatility of metals with respect to mechanical properties has encouraged the selection of the best combination of properties to facilitate fabrication and to insure good service performance. Some applications require considerable thought about base and filler metal selections and treatment, particularly where fabricating properties differ from the required service properties. To solve these selection problems, or to effect a compromise, it is necessary to examine first the governing properties and then consider their combined effect upon the design and service behavior of the weldment.

Modulus of Elasticity

A convenient way of appraising the ability of a metal to resist elastic stretching (strain) under stress is by the ratio E between the stress and the corresponding strain. This ratio is known as *Young's modulus* or the *modulus of elasticity*. It is commonly expressed by the following formula:

$$E = \frac{\sigma}{\varepsilon}$$

where

σ = the unit stress, psi

ε = the unit strain, in./in.

4. The testing and evaulation of welded joints are discussed in the *Welding Handbook,* Vol. 1, 7th Ed., 1976: 154-219.

This modulus is a constant characteristic of a metal as measured in polycrystalline metals during standard tensile, compression, bending, and other engineering tests in which stress and strain are correlated. The elastic modulus is a structure-insensitive property; it is virtually unaffected by grain size, cleanliness, or condition of heat treatment. In fact, the modulus of elasticity often remains unchanged for a metal even after substantial alloying additions have been made. Table 1.2 lists the modulus of elasticity for a number of metals.

Table 1.2
Modulus of elasticity of metals

Metal	Modulus of elasticity, psi
Aluminum	9.0×10^6
Beryllium	42.0
Columbium	15.0
Copper	16.0
Iron	28.5
Lead	2.0
Molybdenum	46.0
Nickel	30.0
Steel, carbon & alloy	29.0
Tantalum	27.0
Titanium	16.8
Tungsten	59.0

The elastic modulus can be used practically to indicate how much a tie rod will stretch elastically when a load is placed upon it, or how much a beam will deflect elastically when under load. A frequent use of the elastic modulus in welding, however, is to determine the level of stress created in a piece of metal when it is forced to stretch elastically for a specified amount. The stress can be determined by multiplying the strain by the elastic modulus. It is important to point out that the modulus of elasticity decreases with increasing temperature, and that the change in elastic modulus with temperature varies with different metals.

Elastic Limit

The elastic behavior of metal reaches a limit at a level of stress called the *elastic limit*. This term is more a definition rather than an exact stress level because the value determined for a metal is quite structure-sensitive and dependent on stain rate. The elastic limit is the upper bound of stress where the member will return to its original dimensions when the load is released. When the elastic limit is exceeded, permanent deformation will take place in the member.

Because an engineer usually is interested in knowing the capability of a metal to carry loads without plastic deformation, several properties closely related to the elastic limit have been defined for guidance. These properties can be determined easily from a stress-strain diagram, as is commonly plotted for a tensile test, Fig. 1.1. The curve on a stress-strain diagram first proceeds as a straight line, A-A′. The slope of this line conforms to the modulus of elasticity for the metal under test. As the line proceeds upward, a point is reached where the strain exceeds the amount predicted by the earlier straight line relationship. It is difficult to state exactly where the proportionality ends between stress and strain because the clarity and interpretation of the curve may vary. The proportional limit on the stress-strain curve in Fig. 1.1 is about 28 ksi. This is the maximum stress at which the strain remained directly proportional to stress. Some additional strain may be applied beyond the proportional limit where the metal still behaves elastically. However, the upper limit of elastic behavior is the elastic limit.

Any strain in a metal below the elastic limit is recoverable upon removal of the load. When metal is stressed beyond the elastic limit, the additional strain is plastic in nature and represents permanent deformation. As an example, if the tensile specimen depicted in Fig. 1.1 were loaded to 32 ksi (S_2), the specimen would elongate 0.00125 in./in. of length. Upon removal of the load, the specimen would not return to its original length, but would display a permanent stretch of about 0.00015 in./in., line B-B′.

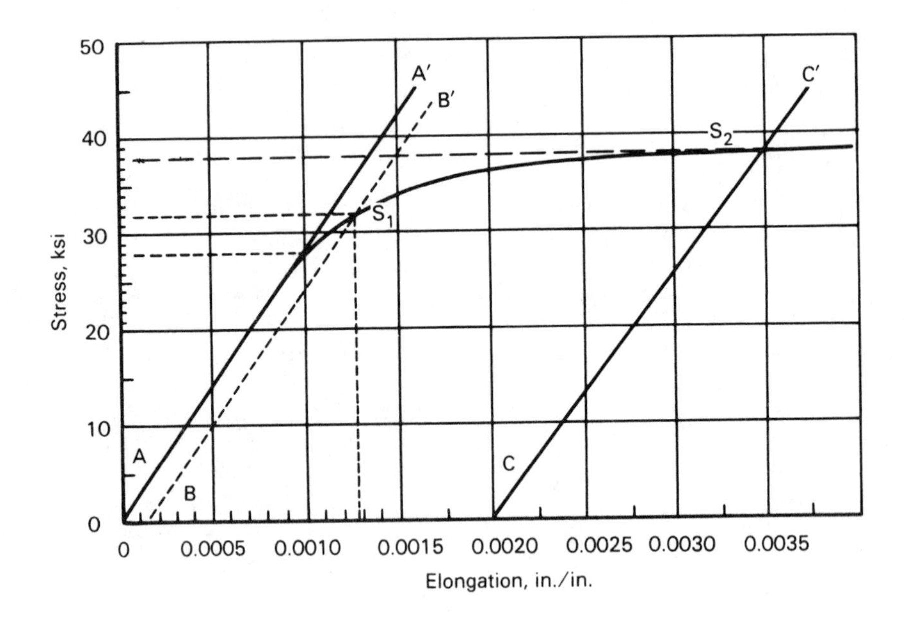

Fig. 1.1—Typical tensile stress-strain diagram for a metal stressed beyond the limit of elastic behavior

Yield Strength

The *yield strength* of a metal is the stress level at which the metal exhibits a specified deviation from proportionality of stress and strain. A practicable method of determining the yield strength of a metal is illustrated in Fig. 1.1. The line C-C′ is drawn parallel to the elastic line A-A′ from a point on the abscissa representing 0.2 percent (0.0020 in./in.) elongation. The line C-C′ will intersect the stress-strain curve at S_2 where the stress level is about 38 ksi. This stress is the yield strength of the tested metal.

While a 0.2 percent offset yield strength is commonly used in engineering design, offsets of 0.1 and 0.5 percent are sometimes employed in the same manner for some metals.

Certain metals, such as low carbon steel, exhibit a *yield point* which is the stress just above the elastic limit where an increase in strain occurs without an increase in stress. The yield point is generally mentioned only for low strength steels. Offset yield strength is the more commonly used index of load-carrying ability without excessive plastic strain.

Tensile Strength

The ratio of the maximum load sustained by a tensile test specimen to the original cross-sectional area is called the *ultimate tensile strength* (UTS), which is the value regularly listed for the strength of a metal. The ultimate tensile strength represents a convenient value calculated from a standard engineering test. The "true" tensile strength of the metal usually is substantially higher than the reported tensile strength.

After the ultimate strength of a tensile test specimen is exceeded, the loss in load-carrying ability is caused by a marked reduction in the cross section as plastic deformation increases. It is standard practice to calculate the tensile stress on the basis of the original cross section area. Therefore, the metal appears to decrease in strength as it nears the breaking point. Actually, the strength of the metal increases steadily as plastic strain takes place because of the continuing effect of cold work, and a maximum strength is reached just prior to fracture.

Tensile strength values obtained for metals are influenced by many factors. Tensile strength is a structure-sensitive property, and is dependent upon chemistry, microstructure, orientation, grain size, strain history, and other factors. The size and shape of the specimen and the rate of loading can also affect the result. For these reasons, the ultimate tensile strengths of the weld metal and the heat-affected zone may be different from that of the unaffected base metal.

Fatigue Strength

The effect of cyclic loading is an important aspect of the strength of metals and welded joints. Fracture can take place in a metal at a stress level well below the tensile strength when a load is applied repeatedly. Fatigue fractures develop as each application of the applied tensile stress causes the tip of a crack to advance a minute amount (stable crack growth). The rate of advance increases as the area of section ahead of the crack decreases with each application of applied load until the stress at the crack tip causes unstable crack growth. Then, sudden, complete failure occurs. Crack growth does not occur when the net stress at the crack tip is compressive; thus, while a crack may initiate as a result of tensile residual stress in combination with fluctuating compressive stress, the small crack relieves the residual welding stress and crack growth ceases.

The stress that a metal can endure without fracture decreases as the number of repeated stress applications increases. The *fatigue strength* is generally defined as the maximum stress that can be sustained for a stated number of cycles without failure. Unless otherwise stated, the stress is completely reversed within each cycle. As the desired number of stress repetitions is increased, the corresponding fatigue strength becomes smaller. For steel, the fatigue strength is usually almost constant beyond about two million cycles. Many million additional cycles are required to cause a significant reduction in fatigue strength. The *fatigue limit*, therefore, practically becomes the maximum stress or stress range which the metal will bear for an infinite number of cycles without fracture. Such a limiting stress level is often called the *endurance limit*. The *fatigue life*, accordingly, is the number of cycles of stress that can be sustained by a metal under stipulated conditions.

The endurance limits reported for metals in engineering handbooks usually have been determined with polished round specimens tested in air. While such data are valid and useful for design of real members represented by the specimen used in the tests, such as finished shafts in rotating machinery, it may have little relevance in the design of as-welded weldments. Weldments are characterized by abrupt changes in cross section, geometrical and metallurgical discontinuities, and residual stresses.

The life (cycle life) of a welded structural member subject to repeated variation of tensile or alternately tensile and compressive stress primarily within the elastic range of the material is principally dependent on the stress range and joint geometry. *Life* is defined as the number of times a member can be subjected to a specific load prior to the initiation and growth of a fatigue crack to size where either failure of the structural component or collapse of the structure takes place.

The *stress range* is the absolute magnitude of stress variation caused by the application and removal of loads. For vehicular bridges, this load is usually a truck or train traversing the structure. Structural details and joint geometry include the type of joint, the type of weld, surface finish, and structural details that affect stress amplification caused by mechanical notches. The differences in stress concentration effects of joints and structural details are largely responsible for the variation in life obtained from details and members.

When designing welded built-up members and welded connections for structures subject to fatigue loading, the applicable code or standard governing the subject structure must be followed. In the absence of a specific code, an appropriate code or standard for similar structures should be used as a design guide.

Localized stresses within a structure may result entirely from external loading, or there may be a combination of applied and residual stresses. Residual stresses per se do not cycle, of course, but they may augment or detract from applied stresses, depending upon their respective signs. For this reason, it may be advantageous to induce, if possible, compressive residual stress in critical areas of a weldment where cyclic tension stresses are expected. This may be accomplished by a welding sequence that controls the residual stresses from welding, or by a localized treatment that acts to place the surface in compression.

Thermal stresses also must be considered in the same light as an applied stress because thermal cycling can lead to fatigue failure if the thermal gradients are steep or if the thermal stresses are concentrated by a stress raiser.

The rate of repetition of loading is important because it determines the time required for the number of cycles that will cause a crack to initiate and propagate to a critical length. Weldments in rotating equipment are particularly prone to fatigue failure. Pressure vessels can fail by fatigue also when pressurization is cyclic and stress above the fatigue strength is concentrated at some point.

Designers of weldments need to thoroughly understand the fatigue characteristics of metals, particularly in weldments. The most common cause of fracture in weldments is fatigue. One of the reasons for this is the frequent presence of stress raisers (changes in cross section, discontinuities, etc.) that concentrate imposed cyclic stresses to levels above the fatigue limit of the metal for the existing conditions.

Ductility

The amount of plastic deformation that an unwelded or welded specimen undergoes in a mechanical test carried to fracture is considered a measure of the ductility of the metal or the weld. Values expressing ductility in various mechanical tests do not measure any fundamental characteristic, but merely provide relative values for comparison of ductilities of metals subjected to the identical test conditions. The plasticity exhibited by a specimen is simply the deformation accomplished during the yielding process.

Ductility, regardless of the method of measurement, is a structure-sensitive property, and is affected by many of the conditions of testing. The size and shape of the specimen, the ambient temperature, the rate of straining, metallurgical structure, and surface conditions all influence the amount and location of plastic deformation prior to fracture.

Ductility values obtained from precise or elaborate tests are not used directly in design. Most structures are designed to operate at stresses below the yield strength, and any significant deformation usually makes the unit or article unfit for service. Ductility values give an indication of the ability of a metal to yield and relieve high secondary stresses. These values also may give some idea of the reserve of plasticity available to insure against sudden fracture under unexpected overloading. However, ductility values do not necessarily indicate the amount of plastic deformation that will take place under all conditions of loading. Most structures are sensitive to both loading rate and ambient temperature.

Fracture Toughness

A metal that is judged *ductile* by a standard tensile test or slow bend test may perform in a *brittle* manner when exposed to different conditions in another type of test. The only forecast that can be made with reasonable certainty from tensile or bend test results is that a metal with very little ductility is not likely to behave in a ductile fashion in any other type of mechanical test carried to fracture. A metal that displays good ductility in a tensile or bend test may or may not behave in a ductile manner in all other kinds of mechanical tests. In fact, there have been numerous cases where ductile metals (as judged by tensile and bend tests) have fractured in service with little or no plastic deformation. The lack of deformation and other aspects of such failures usually show that little energy was required to produce the fracture. This general experience prompts the metallurgist to speak of the toughness of metal as a property distinct from ductility.[5]

Toughness can be described as the ability of metal containing an existing small crack or other stress raiser to resist fracture while being loaded under conditions that are unfavorable for energy absorption and plastic deformation. Three conditions markedly influence the behavior of a metal; namely, (1) the rate of straining, (2) the nature of the load, that is, whether the imposed stresses are uniaxial or multiaxial, and (3) the temperature of the metal.

While many metals can absorb energy and deform plastically under the simple circumstances represented in tensile or bend tests (and therefore would be judged ductile), a lesser number of these metals are considered to have good toughness when tested under conditions of high stress concentration. Briefly, the toughness displayed by a metal tends to decrease as (1) the rate of straining increases, (2) the stresses become multiaxial, and (3) the temperature of the metal is lowered. Weld metal in service easily may be exposed to one or more of these conditions. Consequently, there is good reason to be concerned about the toughness of weld metal.

In designing with ductile metals, the fail-safe load carrying capability of an engineering structure, including the welds, is normally based on a stress analysis to assure that the nominal stresses are below the yield strength. Failures that occur at stresses below the yield strength are broadly classified as brittle factures. These failures can result from the effects of critical size discontinuities or crack-like defects in welds or base metal that do not greatly alter the nominal stress distribution and are customarily neglected in the stress analysis.

Conditions of high lateral restraint could cause brittle fracture in a structure because a discontinuity can greatly decrease the ductility that may have been predicted on the basis of smooth tensile specimens. Such lateral restraint may be caused by relatively large material thickness; a detail design that has been proven satisfactory in service may be unsatisfactory if the proportions are simply extrapolated upward.

It is evident that a complete fracture-safe analysis requires proper attention to the role of a discontinuity. For many classes of structures, such as ships, bridges, and pressure vessels, experience with specific designs, materials, and fabrication procedures has established a satisfactory correlation between notch test standards for base

---

5. For information on fracture toughness, see the *Welding Handbook,* Vol. 1, 7th Ed., 1976: 170-185; and Rolfe, S., et al, *Fracture and Fatigue Control in Structures: Applications of Fracture Mechanics,* Englewood Cliffs, NJ: Prentice-Hall, 1977.

and weld metals and acceptable service. The problem is to insure the soundness and integrity of a new design.

One of the motivations for the application of fracture mechanics concepts and tests to welded joints is the possibility of designing safely against the effects of common weld discontinuities. It is widely recognized that welded joints almost always contain some discontinuities, and this places the designer using welded joints in a dilemma. The designer likes to think of joints that are entirely free of flaws, but this is not realistic. The practical approach is to recognize that discontinuities are present, and to place a reasonable limit on their existence. The problem is how to determine the types and extent of discontinuities that are acceptable. While conventional toughness testing procedures cannot deal directly with this problem, fracture mechanics tests, where applicable, specifically define a relationship between flaw size and fracture stress for a given base metal or weld joint. Thus, the tests permit a direct estimate of allowable flaw sizes for different geometrical configurations and operating conditions.

Where structural elements are subject to cyclical stresses or where other flaw extension mechanisms are at work, the allowable discontinuity size is further complicated. Fracture mechanics may be able to establish the safe maximum flaw size, but this size may not provide a useful life for elements subject to fatigue or corrosion, or both. For these elements, information relative to flaw growth rates is essential.

The specification of allowable flaw sizes and inspection procedures for flaw detection can be referred to a rational and logical procedure rather than one based solely on experience or opinion. As weldments of more complicated design and higher strength requirements are introduced, the designer will need to take an analytical approach to the problem of weld discontinuities.

Low Temperature Properties

Lowering the temperature of a metal profoundly affects fracture characteristics, particularly if the metal possesses a body-centered cubic crystalline structure (carbon steel is an example). Strength, ductility, and other properties are changed in all metals and alloys as the temperature decreases to near absolute zero.

The properties of metals at very low temperatures are of more than casual interest. Welded pressure vessels and other pieces of equipment sometimes are expected to operate satisfactorily at temperatures far below room temperature. Very low temperatures are involved in so-called cryogenic service that entails the storage and use of liquefied industrial gases, such as oxygen and nitrogen.

As the temperature is lowered, a number of changes in properties take place in metals. The elastic modulus, for example, increases. In general, the tensile and yield strengths of all metals and alloys increase as the temperature is lowered.

The ductility of most metals and alloys tends to decrease as temperature is lowered. However, some metals and alloys have considerable ductility at very low temperatures. Because of the notch-toughness transition behavior of carbon and low alloy steels and certain other metals, the suitability of metals for low temperature service is judged by tests that evaluate propensity to brittle fracture rather than simple tensile or bend ductility. The favorite specimens for low-temperature testing are the notched tensile specimen and the notched-bar impact specimen. The principal factors that determine the low temperature behavior of a metal during mechanical testing are (1) crystal structure, (2) chemical composition, (3) size and shape of the test specimen, (4) conditions of manufacture and heat treatment, and (5) rate of loading. The notched-bar impact strengths for five common metals are listed in Table 1.3. Iron and steel suffer a marked reduction in impact strength at low temperatures. The addition of alloying elements to steel, especially nickel and manganese, can markedly improve notch toughness at low temperatures, while increased carbon and phosphorus can greatly decrease notch toughness.

Elevated Temperature Properties

The mechanical properties of metals at elevated temperatures are of concern for three reasons. First, many metal sections, employed in weldments are produced or formed by working

Table 1.3
Notch-bar impact strengths of metals at low temperatures

Temperature, °F	Charpy V-notch impact strength, ft · lb				
	Aluminum[a]	Copper[a]	Nickel[a]	Iron[b]	Titanium[c]
75 (room temperature)	20	40	90	75	15
0	20	42	92	30	13
-100	22	44	93	2	11
-200	24	46	94	1	9
-320	27	50	95	1	7

a. Face-centered cubic crystal structure.
b. Body-centered cubic crystal structure.
c. Hexangonal close-packed crystal structure.

Source: Linnert, G.E., *Welding Metallurgy*, Vol. 1, 3rd Ed., Miami: American Welding Society, 1965.

the metal at elevated temperatures. If the metal does not possess suitable mechanical properties for these hot working operations, flaws can be created in the section that may later interfere with the making of sound welds. Second, most welding operations involve the application of heat. Therefore, information may be needed on the changes in mechanical properties of the base metal as the temperature rises and falls to predict behavior under induced strain during welding. Even the strength and ductility of the weld metal as it cools from the solidification point to room temperature may be useful information. Finally, the properties of metals at elevated temperatures are important because many weldments are exposed to high temperatures either during heat treatments or while in service.

The strength of a metal decreases as the temperature is raised. The elastic modulus is reduced, and plastic deformation mechanisms are able to operate more freely. However, metals vary considerably in the way that their strengths and ductilities change with temperature. A metal that is strong at one temperature may be weaker than other metals at another temperature.

During welding, the temperature of the metal at the joint rises to the melting point and falls when the heat source is removed. If the parts are restrained, the hot weld metal and heat-affected zone are subjected to complex thermal strains. Whether the metal fails (cracks) or distorts badly from shrinkage stresses depends upon the various properties of the metal at the varying temperatures encountered.

Performance of a metal in service at an elevated temperature is governed by other factors in addition to strength and ductility. Time becomes a factor because at high temperatures metal will creep; that is, the section under stress will continue to deform even if the load is maintained constant. The rate at which a metal creeps under load increases rapidly with increasing temperature. Consequently, the time over which a metal under load will deform too much to be usable can vary from many years at a slightly elevated temperature to a few minutes at a temperature near the melting point.

As would be expected, the creep rates of metals and alloys differ considerably. If the temperature and stress are sufficiently high, the metal will creep until rupture occurs. The term *creep-rupture* is used to identify the mechanics of deformation and failure of metals under stress at elevated temperatures.

PHYSICAL PROPERTIES

The physical properties of metals seldom receive the attention regularly given to mechanical properties in a general treatise on welding. Nevertheless, the physical properties are an important aspect of metal characteristics. Oftentimes, the welding engineer (and welder) may be unaware that the success of the joining operation depends heavily on a particular physical property. The physical properties of regular polycrystalline metals are not so structure-sensitive as the mechanical properties. Constant values usually are provided for metals and alloys, and they serve satisfactorily for most engineering purposes. Only the physical proper ties that may require some consideration in designing or fabricating a weldment are discussed here.[6]

Thermal Conductivity

The rate at which heat is transmitted through a material by conduction is called *thermal conductivity* or *thermal transmittance*. Metals are better heat conductors than nonmetals, and metals with high electrical conductivity generally have high thermal conductivity.

Metals differ considerably in their thermal conductivities. Copper and aluminum are excellent conductors, which accounts for the difficulty in attempting to weld these metals using a relatively low-temperature heat source, like an oxyacetylene flame. Conversely, the good conductivity of copper makes it a good heat sink when employed as a hold down or backing bar. Steel is a relatively poor conductor; this partly accounts for the ease with which it can be welded and thermally cut.

Melting Temperature

The higher the melting point or range, the larger is the amount of heat needed to melt a given volume of metal. Hence, the temperature of the heat source in welding must be well above the melting range of the metal. However, two pieces of a metal, such as iron, may be joined with a metal of lower melting point, such as bronze. The brazing filler metal wets and adheres to the steel faces to which it is applied. Welding of two metals of dissimilar composition becomes increasingly difficult as the difference in melting ranges widens.

Coefficient of Thermal Expansion

Most metals increase in volume when they are heated. The coefficient of thermal expansion is the unit change in linear dimensions of a body when its temperature is changed by one degree. The coefficient also serves to indicate contraction when the temperature is decreased. Engineers usually are concerned with changes in length in metal components, and most handbooks provide a linear coefficient of thermal expansion rather than a coefficient for volume change.

Metals change in volume when they are heated and cooled during welding. The greater the increase in volume and localized upsetting during heating, the more pronounced will be the distortion and shrinkage from welding. Changes in dimensions from welding must be considered during weldment design when setting part dimensions and tolerances.

Electrical Conductivity

Metals are relatively good conductors of electricity. Increasing the temperature of a metal interferes with electron flow; consequently, electrical conductivity decreases. Adding alloying elements to a metal or cold working also decreases conductivity. These characteristics are important variables affecting resistance welding processes.

CORROSION PROPERTIES

The corrosion properties of a metal determine its mode and rate of deterioration by chemical or electrochemical reaction with the surrounding environment. Metals and alloys differ greatly in their corrosion resistance. Corrosion resistance often is an important consideration in planning and fabricating a weldment for a particular service. Therefore, the designer should

6. Physical properties of metals are discussed in the *Welding Handbook*, Vol. 1, 7th Ed., 1976: 71-76.

know something about the behavior of weld joints under corrosive conditions.[7]

Many times, weld joints display corrosion properties that differ from the remainder of the weldment. These differences may be observed between the weld metal and the base metal, and sometimes between the heat-affected zone and the unaffected base metal. Even the surface effects produced by welding, like heat tint formation or oxidation, fluxing action of slag, and moisture absorption by slag particles can be important factors in the corrosion behavior of the weld metal. These considerations are particularly important in the design and fabrication of weldments of unpainted weathering steels, including the selection of filler metals.

Welds made between dissimilar metals or with a dissimilar filler metal may be subject to electrochemical corrosion. Brazed joints can be particularly vulnerable. Appropriate protective coatings are required to avoid corrosion in sensitive environments.

DESIGN PROGRAM

A weldment design program starts with a recognition of a need. The need may be for improving an existing machine or for building an entirely new product or structure using advanced design and fabrication techniques. In any event, many factors must be taken into account before a design is finalized. These considerations involve numerous questions and considerable research into the various areas of marketing, engineering, and production.

ANALYSIS OF EXISTING DESIGN

When designing an entirely new machine or structure, an attempt should be made to obtain information on similar units including those of other manufacturers or builders. If a new design is to replace a current one, the good and bad points of the latter should be understood. The following questions can help in determining these points:

(1) What are the opinions of customers and the sales force about the current design?

(2) What has been its history of failures?

(3) What features should be retained, discarded, or added?

(4) What suggestions for improvements have been made?

(5) Is it over or under designed?

DETERMINATION OF LOAD CONDITIONS

The service requirements of a weldment and the conditions of service that might result in overloading should be ascertained. From such information, the load on individual members can be determined. As a starting point for calculating loading, a designer may find one or more of the following methods useful:

(1) From the motor horsepower and speed, determine the torque on a shaft or revolving part.

(2) Calculate the forces on members caused by the dead weight of parts.

(3) Determine the load on members of a crane hoist, shovel, lift truck, or similar material handling equipment from the load required to tilt the machine.

(4) Use the maximum strength of critical cables on such equipment to determine the maximum loads on machine members.

7. Types of corrosion and corrosion testing of welds are discussed in the *Welding Handbook,* Vol. 1, 7th Ed., 1976: 200-207.

(5) Consider the force required to shear a critical pin as an indication of maximum loading on a member.

If a satisfactory starting point cannot be found, design for an assumed load, and adjust from experience and tests.

MAJOR DESIGN FACTORS

In developing a design, the designer should consider how decisions will affect production operations, manufacturing costs, product performance, appearance, and customer acceptance. Many factors far removed from engineering considerations per se become major design factors. Some of these, along with other relevant rules, are as follows:

(1) The design should satisfy strength and stiffness requirements. Overdesigning wastes materials and increases production and shipping costs.

(2) The safety factor should not be unrealistically high.

(3) Good appearance may be necessary, but only in areas that are exposed to view. The drawing or specifications should specify those welds that are critical in respect to appearance.

(4) Deep, symmetrical sections should be used to efficiently resist bending.

(5) Welding the ends of beams rigidly to supports increases strength and stiffness.

(6) Rigidity may be provided with welded stiffeners to minimize the weight of material.

(7) Tubular sections or diagonal bracing should be used for torsion loading. A closed tubular section is significantly more effective in resisting torsion than an open section of similar weight and proportions.

(8) Standard rolled sections, plate, and bar should be used for economy and availability.

(9) Accessibility for maintenance must be considered during the design phase.

(10) Standard, commercially available components should be specified when they will serve the purpose. Examples are index tables, way units, heads, and columns.

DESIGNING THE WELDMENT

To a designer familiar only with castings or forgings, the design of a weldment may seem complex because of the many possible choices. Variety in the possibilities for layout, however, is one of the advantages of welded design; opportunities for saving are presented. Certain general pointers for effective design are as follows:

(1) Design for easy handling of materials, for inexpensive tooling, and for accessibility of the joints for reliable welding.

(2) Check with the shop for ideas that can contribute to cost savings.

(3) Establish realistic tolerances based on end use and suitability for service. Check the tolerances with the shop. Excessively close tolerances and fits may serve no useful purpose, and may be beyond the ability of the shop to produce them economically.

(4) Plan the design to minimize the number of pieces. This will reduce assembly time and the amount of welding.

Part Preparation

Flame cutting, shearing, sawing, blanking, nibbling, and machining are methods for cutting blanks from stock material. Selection of the appropriate method depends on the available material and equipment and the relative costs. The quality of edge needed for good fit-up and the type of edge preparation for groove welds must also be kept in mind. The following points should be considered:

(1) Dimensioning of a blank may require stock allowance for subsequent edge preparation.

(2) The extent of welding must be considered when proposing to cut the blank and prepare the edge for welding simultaneously.

(3) Weld metal costs can be reduced for thick plate by specifying J- or U-groove preparations.

(4) Consider air carbon arc gouging, flame gouging, or chipping for back weld preparation.

Forming

Forming of parts can sometimes reduce the cost of a weldment by avoiding joints and machining operations. The base metal composition, part thickness, over-all dimensions, production volume, tolerances, and cost influence the choice of forming methods. The following suggestions may be helpful in making decisions in this area:

(1) Create a corner by bending or forming rather than by welding two pieces together.

(2) Bend flanges on plate rather than welding flanges to it.

(3) Use a casting or forging in place of a complex weldment to simplify design and reduce manufacturing costs.

(4) Use a surfacing weld on an inexpensive component to provide wear resistance or other properties in place of an expensive alloy component.

With some alloys, cold forming may increase the likelihood of brittle fracture in service, and heat from welding near the formed area may also contribute to the problem. The matter should be investigated before cold forming and welding of susceptible materials is specified in a design.

Weld Joint Design

The joint design should be selected primarily on the basis of load requirements. However, variables in design and layout can substantially affect costs. Generally, the following rules apply:

(1) Select the joint design that requires the least amount of weld metal.

(2) Where possible, use a square-groove joint together with a welding process capable of deep joint penetration or of welding within the groove (electrogas, electroslag, or narrow-gap welding).

(3) Use the smallest practical root opening and groove angle to minimize the amount of filler metal required.

(4) On thick plates, use double instead of single V- or U-groove welds to reduce the amount of weld metal and to control distortion.

(5) For corner joints in steel, specify a depth of preparation of the groove to intersect the full thickness of the material to minimize lamellar tearing, especially in thicknesses of 2 in. and over.

(6) For T-joints where fatigue is not a primary concern, use double fillet welds to minimize through-thickness shrinkage strains and to spread the transfer of applied forces.

(7) Use one weld in place of two welds, where possible, to joint three parts at one location.

(8) Avoid joints where it is difficult to obtain fusion at the root. For example, welds joining two surfaces at less than 30 degrees to each other are unreliable.

(9) Design the assembly and the joints for good accessibility for welding.

Size and Amount of Weld

Overdesign is a common error, as is overwelding in production. Control of weld size begins with design, but it must be maintained during the assembly and welding operations. The following are basic guides:

(1) Specify welds of minimum size and length but adequate for the forces to be transferred. Oversize welds may cause increased distortion and higher residual stress without improving suitability for service. (They also contribute to increased costs.)

(2) The size of a fillet weld is especially important because the amount of weld required increases as the square of the increase in leg length (equal legs).

(3) For equivalent strength, a continuous fillet weld of a given size is usually less costly than a larger sized intermittent fillet weld. Also, there are fewer weld terminations that are potential sites of discontinuities.

(4) An intermittent fillet weld can be used in place of a continuous fillet weld of minimum size when static load conditions do not require a continuous weld. An intermittent fillet weld should not be used under cyclic loading conditions.

(5) To derive maximum advantage of automatic welding, it may be better to use one continuous weld rather than several short welds.

(6) The weld should be placed in the section of least thickness, and the weld size should be based on the load or other requirements of that section.

(7) Welding of stiffeners or diaphragms should be limited to that required to carry the load, and should be based on expected out-of-plane distortion of the supported components under service loads as well as during shipment and handling.

(8) The amount of welding should be kept to a minimum to limit distortion and internal stresses and, thus, the need and cost for stress-relieving and straightening.

Subassemblies

In visualizing assembly procedures, the designer should break the weldment into subassemblies in several ways to determine the arrangement that offers the greatest cost savings. The following are advantages of subassemblies:

(1) Two or more subassemblies can be worked on simultaneously.

(2) Subassemblies usually provide better access for welding, and may permit automatic welding.

(3) Distortion in the finished weldment may be easier to control.

(4) Large size welds may be deposited under lesser restraint in subassemblies, which, in turn, helps to minimize residual stresses in the completed weldment.

(5) Machining of subassemblies to close tolerances can be done before final assembly. If necessary, stress relief of certain sections can be performed before final assembly.

(6) Chamber compartments can be leak tested and painted before final assembly.

(7) In-process inspection and repair is facilitated.

(8) Handling costs may be much lower.

When possible, it is desirable to construct the weldment from standard sections, so that the welding of each can be balanced about the neutral axis. Welding the more flexible sections first facilitates any straightening that might be required before final assembly.

FIXTURES, POSITIONERS, AND ROBOTS

Welding fixtures, positioners, and robots should be used to minimize fabrication time. In the planning of assemblies and subassemblies, the designer should decide if the fixture is to be used simply to aid in assembly and tacking or if the entire welding operation is to be done in the fixture. Welding fixtures and positioners are discussed in Chapter 4; automatic welding machines and robotic equipment are covered in Chapter 5.

WELDING PROCEDURES

Although designers may have little control of welding procedures, they can influence which procedures are used in production. The following guidelines can help to effect the ultimate success of weldment design:

(1) Backing strips increase the speed of welding when making the first pass in groove welds.

(2) The use of low-hydrogen steel electrodes or welding processes eliminates or reduces preheat requirements for steel.

(3) If plates are not too thick, consider a joint design requiring welding only from one side to avoid manipulation or overhead welding.

(4) Excessive reinforcement of a weld is generally unnecessary to obtain a full-strength joint.

(5) With T-joints in thick plate subject to tensile loading in the through-thickness direction, the surface of the plate should be buttered with weld metal for some distance beyond the intended weld terminations to discourage lamellar tearing in susceptible steels.

(6) Joints in thick sections should be welded under conditions of least restraint; for example, prior to installation of stiffeners.

RESIDUAL STRESSES AND DISTORTION

The magnitude and extent of residual stresses and the resulting distortion during the fabrication of a weldment can cause problems with fit-up and welding operations. Such problems may offset the quality and soundness of welded joints and ultimately, the serviceability of the weldment.

Weldment design, welding procedures, and assembly methods can be beneficial in controlling residual stresses and distortion in a completed structure. Procedures for reducing residual welding stresses and for controlling distortion in a weldment are discussed in the *Welding Handbook*, Vol. 1, 7th Ed., 1972: 265-68.

CLEANING AND INSPECTION

Design specifications can have some effect on cleaning and inspection costs. The design factor of safety determines the type and amount of inspection required. The following shop practices also affect these costs:

(1) As-welded joints that have uniform appearance are acceptable for many applications. Therefore, the surface of a weld need not be

machined smooth or flush unless that is required for another reason. Smoothing a weld is an expensive operation.

(2) Undesirable overwelding should be noted during inspection because it can be costly and also contributes to distortion. Corrective action should be directed at work in progress rather than at completed weldments.

(3) The type of nondestructive inspection to be used on weldments must be capable of detecting the types and sizes of weld discontinuities that are to be evaluated for acceptability.

DESIGN CONSIDERATIONS

The performance of any member of a structure depends on the properties of the material and the characteristics of section. If a design is based on the efficient use of these properties, the weldment should be functionally good and conservative of materials.

Engineers assigned to design welded steel members need to know (1) how to select the most efficient structural section and determine required dimensions, and (2) when to use stiffeners and how to size and place them.

The mathematical formulas for calculating forces and their effects on sections, and for determining the sections needed to resist such forces appear quite forbidding to the novice. With the proper approach, however, it is possible to simplify the design analysis and the use of those formulas. In fact, as will be explained later, it is often possible to make correct design decisions merely by examining one or two factors in an equation, without using tedious calculations. On the whole, the mathematics of weld design is no more complex than in other engineering fields.

THE DESIGN APPROACH

Considerations other than the engineer's wishes may prevail; for example, when a machine is to be converted from a cast to a welded design. Management may favor the redesign of one or more components to weldments, and conversion of the design over a period of years to an all-welded product. Gradual conversion avoids the obsolescence of facilities and skills, and limits the requirement for new equipment. Available capital and personnel considerations often limit the ability of a company when changing to welded design. Supplementing these considerations is the need to maintain a smooth production flow, and to test the production and market value of the conversion as it is made step by step.

From the standpoint of performance and ultimate production economics, redesign of the machine or structure as a whole is preferable. The designer then is unrestricted by the previous design, and in many cases is able to reduce the number of pieces, the amount of material used, and the labor for assembly. A better, lower-cost product is realized immediately. When the adjustment to changes in production procedures is complete, the company is in a position to benefit more fully from welded design technology.

STRUCTURAL SAFETY

A safe welded structure depends upon a combination of good design practices, good workmanship during fabrication, and good construction methods. In design, the selection of a safe load factor or safety factor, whichever applies, and correct analytical procedures requires experience and sound engineering judgement. Deterioration from corrosion or other service conditions during the life of the structure, variations in material properties, potential imperfections in materials and welded joints, and many other factors also need consideration in design.

A rational approach to structural safety is a statistical evaluation of the random nature of all the variables that determine the strength of a structure and also the variables that may cause it to fail. From these data, the risk of failure may be evaluated and the probability of occurrence kept at a safe level for the application considering the risk of injury, death, or extensive property damage.

The choice of materials and safe stress levels in members may not produce the most economical structure. However, safety must take precedence over cost savings when there is a question of which should govern. Great skill, care, and detailed stress analyses are needed when the designer attempts a new design or structural concept. Laboratory tests of models or sections of prototype structures should be used to verify the design.

SELECTING A BASIS FOR WELDED DESIGN

A redesign of a product may be based on the previous design or on loading considerations solely. Following a previous design has the advantages of offering a "safe" starting point; the old design is known to perform satisfactorily. This approach, however, has disadvantages in that it stifles creative thinking toward developing an entirely new concept to solve the basic problem.

Little demand is made on the ingenuity of the designer when the welded design is modeled on the previous product. Tables of equivalent sections or nomographs can be used to determine required dimensions for strength and rigidity.

A design based on the loading requires designers to analyze what is wanted and come up with configurations and materials that best satisfy the need. They must know or determine the type and amount of load, and the values for stress allowables in a strength design, or deflection allowables in a rigidity design.

DESIGNING FOR STRENGTH AND RIGIDITY

A design may require strength only, or strength and rigidity to support the load. All designs must have sufficient strength so that the members will not fail by breaking or yielding when subjected to normal operating loads or reasonable overloads. Strength designs are common in road machinery, farm implements, motor brackets, and various types of structures. If a weldment design is based on calculated loading, design formulas for strength are used to dimension the members.

In certain weldments, such as machine tools, rigidity as well as strength is important, because excessive deflection under load would result in lack of precision in the product. A design based on rigidity also requires the use of design formulas for sizing members.

Some parts of a weldment serve their design function without being subjected to loadings much greater than their own weight (dead load). Typical members are fenders and dust shields, safety guards, cover plates for access holes, and enclosures for esthetic purposes. Only casual attention to strength and rigidity is required in sizing such members.

DESIGN FORMULAS

The design formulas for strength and rigidity always contain terms representing the load, the member, the stress, and the strain or deformation. If any two of the first three terms are known, the others can be calculated. All problems of design thus resolve into one of the following:

(1) Finding the internal stress or the deformation caused by an external load on a given member.

(2) Finding the external load that may be placed on a given member for any allowable stress, or deformation.

(3) Selecting a member to carry a given load without exceeding a specified stress or deformation.

A load is a force that stresses a member. The result is a deformation strain measured by relative displacements in the member as elongation, contraction, deflection, or angular twist. A useful member must be designed to carry a certain type of load within an allowable stress or deformation. In designing within the allowable limits, the designer should generally select the most efficient material and the most efficient section size and shape. The properties of the material and those of the section determine the ability of a member to carry a given load.

The design formulas in use, developed for various conditions and member types, are much too numerous for inclusion here. However, some are used to illustrate specific design problems. Table 1.4 summarizes the components of design formulas.[8] These components are terms that describe the three basic factors: load, member, and stress or deformation. The symbols for values and properties normally used in design formulas are also given in the table.

The use of design formulas may be illustrated by the problem of obtaining adequate stiffness in a cantilever beam. The amount of vertical deflection at the end of the beam under a concentrated load, Fig. 1.2, can be determined using the following deflection formula:

$$\Delta = (FL^3)/(3EI)$$

where

Δ = the deflection

F = the applied force or load

L = the length of the beam

E = the modulus of elasticity of the metal

I = the area moment of inertia of the beam section

8. The properties of sections for standard structural shapes are given in the *Manual of Steel Cosntruction,* Chicago: American Institute of Steel Construction, (latest edition), and *Engineering Data for Aluminum Structures* (ED-33), Washington, D.C.: The Aluminum Association (latest edition).

Table 1.4
Components of design formulas

(A) Load Components

Methods of application	Types of load	Concept	Symbol
1. Steady	1. Tension	Force	F
2. Variable	2. Compression		
3. Cyclic	3. Shear		
4. Impact	4. Bending	Moment	M
	5. Torsion	Torque	T

(B) Member Components

Property of material	Symbol	Property of section	Symbol
1. Tensile strength	σ_t	1. Area	A
2. Compressive strength	σ_c	2. Length	L, l
3. Yield strength	σ_y	3. Moment of inertia	I
4. Shear strength	τ	4. Section modulus, elastic	S
5. Fatigue strength	FS	5. Torsional resistance	R
6. Modulus of elasticity (Young's modulus)	E	6. Radius of gyration	r, k
7. Modulus of rigidity in shear	G		

Table 1.4 (Continued)

Stress	Symbol	Strain or deformation	Symbol
1. Tensile	σ_t	1. Unit linear	ε
2. Compressive	σ_c	2. Total	Δ
3. Shear	τ	3. Angular rotation	θ

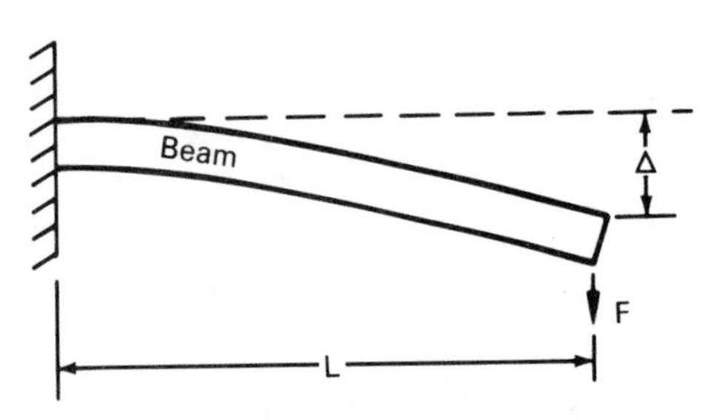

Fig. 1.2—Deflection, Δ, of a cantilever beam under a concentrated load F

It is normally desirable to have the least amount of deflection. Therefore, *E* and *I* values should be as large as possible. The common structural metal having the highest modulus of elasticity (*E*) is steel, with a value of about 30×10^6 psi. The other factor requiring a decision relative to deflection is *I*, the area moment of inertia. This is a *member* (geometrical) property of the cross section (Table 1.4). The beam must have a cross section with a moment of inertia about the horizontal axis large enough to limit the deflection to a permissible value. A section with adequate in-plane moment of inertia will satisfy the vertical deflection requirement, whatever the shape of the section. However, the out-of-plane stability of the beam may also need consideration if forces transverse to the principle axis or torsion are involved. The designer must then decide the shape to use for the best design at lowest cost.

TYPES OF LOADING

As indicated in Table 1.4, there are five basic types of loading: tension, compression, bending, shear, and torsion. When one or more of these types of loading are applied to a member, they induce stress in addition to any residual stresses. The stresses result in strains or movements within the member, the magnitudes of which are governed by the modulus of elasticity (*E*) of the metal. Some movement always takes place in a member when the load is applied because of the stress and strain produced.

Tension

Tension is the simplest type of load. It subjects the member to tensile stresses. A straight bar under a simple axial tension force has no tendency to bend. The axial tensile load causes axial strains and elongation of the bar. The only requirement is adequate cross-sectional area to carry the load.

Compression

A compressive load may require designing to prevent buckling. Few compression members fail by crushing or by exceeding the ultimate compressive strength of the material. If a straight compression member, such as the column in Fig. 1.3 (A), is loaded through its center of gravity, the resulting stresses are simple axial compressive stresses. In the case of a slender column, it will start to bow laterally, as the load is increased, at a stress lower than the yield strength. This movement is shown in Fig. 1.3 (B). As a result of bowing, the central portion of the column becomes increasingly eccentric to the axis of the force and causes a bending moment on the column, as shown in Fig. 1.3 (C). Under a steady load, the column will remain stable under the combined effect of the axial stress and the bending moment. However, with increasing load and associated curvature, Fig. 1.3 (D), a critical point will be reached where the column will buckle and fail. As a column deflects under load, a bending moment can develop in semirigid or rigid welded end connections.

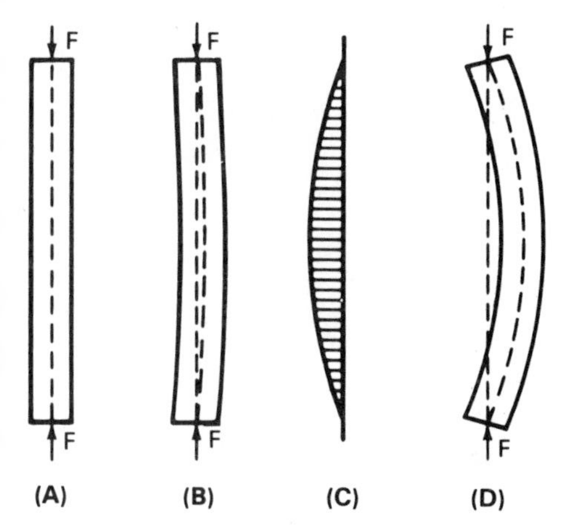

Fig. 1.3—(A) Straight column with a compressive load; (B) column starts to deflect laterally with increasing load; (C) bending moment diagram; (D) increased deflection with higher loading

Two properties of a column are important to compressive strength; cross-sectional area and radius of gyration. The area is multiplied by the allowable compressive stress to arrive at the compressive load that the column can support without buckling. The radius of gyration indicates, to a certain extent, the ability of the column to resist buckling. It is the distance from the neutral axis of the section to an imaginary line in the cross section about which the entire area of the section could be concentrated and still have the same moment of inertia about the neutral axis of the section. The formula for the radius of gyration is as follows:

$$r = (I/A)^{1/2}.$$

where

r = radius of gyration

I = moment of inertia about the neutral axis

A = cross-sectional area of the member

The worst condition is of concern in design work and, therefore, it is necessary to use the least radius of gyration relative to the unbraced length. The smaller of the two moments of inertia about the x-x and y-y axes is used to obtain the least radius of gyration. The slenderness ratio of a column is the ratio of its length, l, to its radius of gyration, r.

The design of a compression member is by trial and error. A trial section is selected, and the cross-sectional area and the least radius of gyration are determined. A suitable column table (*AISC Manual of Steel Construction, Appendix A*) gives the allowable compressive stress for the column length and the radius of gyration (l/r ratio). This allowable stress is then multiplied by the cross-sectional area, A, to give the allowable total compressive load that may be placed on the column. If this value is less than the load to be applied, the design must be changed to a larger section and tried again. Table 1.5 gives the AISC column formulas, and Fig. 1.4 gives allowable compressive stresses with various slenderness ratios.

Table 1.5
Allowable axial compressive stresses (AISC)

(1) When the effective slenderness ratio, Kl/r, is less than C_c:

$$\sigma_c = \frac{\left(1 - \dfrac{(Kl/r)^2}{2\,C_c^2}\right)\sigma_y}{\dfrac{5}{3} + \dfrac{3\,(Kl/r)}{8\,C_c} - \dfrac{(Kl/r)^3}{8\,C_c^3}}$$

(2) When the effective slenderness ratio, Kl/r, exceeds C_c:

$$\sigma_c = \frac{12\,\pi^2\,E}{23\,(Kl/r)^2}$$

where:

σ_c = allowable compressive stress, psi

K = effective length factor (ratio of the length of an equivalent pinned-end member to the length of the actual member)

l = unbraced length of the member, in.

r = radius of gyration, in.

$C_c = (2\,\pi^2\,E/\sigma_y)^{1/2}$

E = modulus of elasticity, psi

σ_y = minimum yield strength, psi

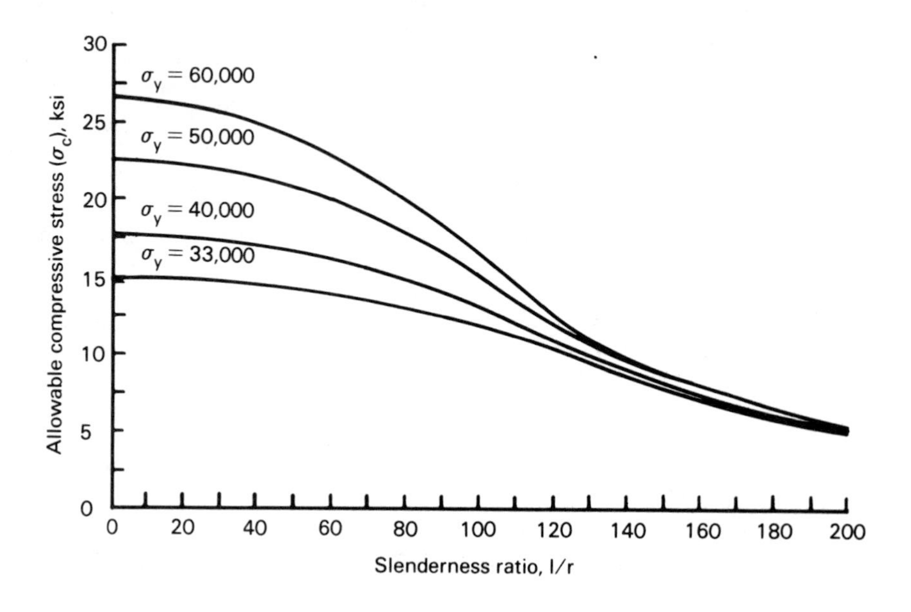

Fig. 1.4—Allowable compressive stress for steel columns of various yield strengths and slenderness ratios

Bending

Figure 1.5 illustrates bending of a member under uniform loading. Loads may also be non-uniform or concentrated at specific locations on the beam. When a member is loaded in bending within the elastic range, the bending stresses are zero along the neutral axis and increase linearly to a maximum value at the outer fibers. The bending stress at any distance, d, from the neutral axis in the cross section of a straight beam, Fig. 1.6, may be found by the formula:

$$\sigma = \frac{Md}{I}$$

where

σ = bending stress, tension, or compression, psi

M = bending moment at the point of interest, lbf · in.

d = distance from neutral axis of bending to point d, in.

I = moment of inertia about the neutral axis of bending, in.^{4}

In most cases, the maximum bending stress is of greatest interest, in which case the formula becomes:

$$\sigma_m = \frac{Mc}{I} = \frac{M}{S}$$

where

σ_m = maximum bending stress, psi

M = bending moment, lbf · in.

c = distance from the neutral axis of bending to the extreme fibers, in.

I = moment of inertia, in.^{4}

S = section modulus (I/c), in.^{3}

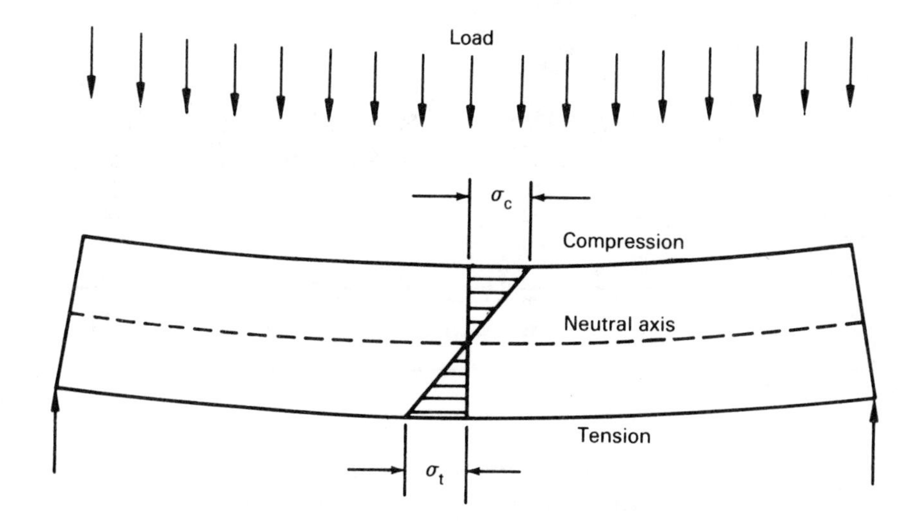

Fig. 1.5—Bending of a beam with uniform loading

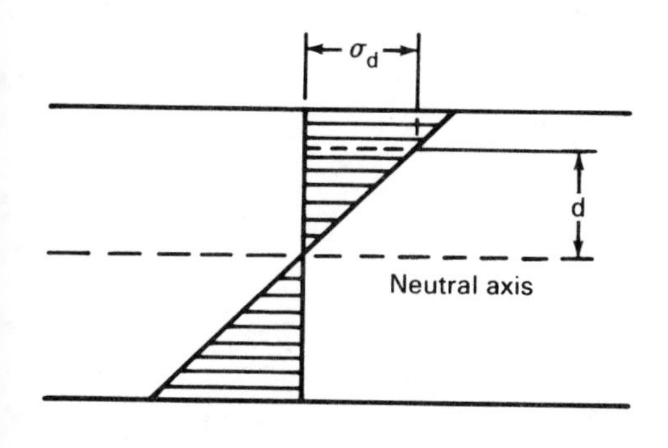

Fig. 1.6—Bending stress at any point d in the cross section of a straight beam

As the bending moment decreases along the length of a simply supported beam toward the ends, the bending stresses (tension and compression) in the beam also decrease. If a beam has the shape of an I-section, the bending stress in the flange decreases as the end of the beam is approached. If a short length of the tension flange within the beam is considered, a difference exists in the the tensile forces F_1 and F_2 at the two locations in the flange, as shown enlarged above the beam in Fig. 1.7. (The tensile force, *F*, is the product of the tensile stress, σ_t, and the flange cross-sectional area, *A*.)

The decrease in the tensile force in the flange results in a corresponding shearing force between the flange and the web. This shearing force must be transmitted by the fillet welds joining the two together. The same reaction takes place in the upper flange, which is in compression. The change in tensile force in the lower flange transfers as shear through the web to the upper flange, and is equal to the change in compression in that flange.

A common bending problem in machinery design involves the deflection of beams. Beam formulas found in many engineering handbooks are useful for quick approximations of deflections with common types of beams where the span is large compared to the beam depth. An example of a typical beam with applicable formulas is shown in Fig. 1.8. Beams that are supported or loaded differently have other applicable design formulas. To meet stiffness requirements, beam depth should be as large as practical.

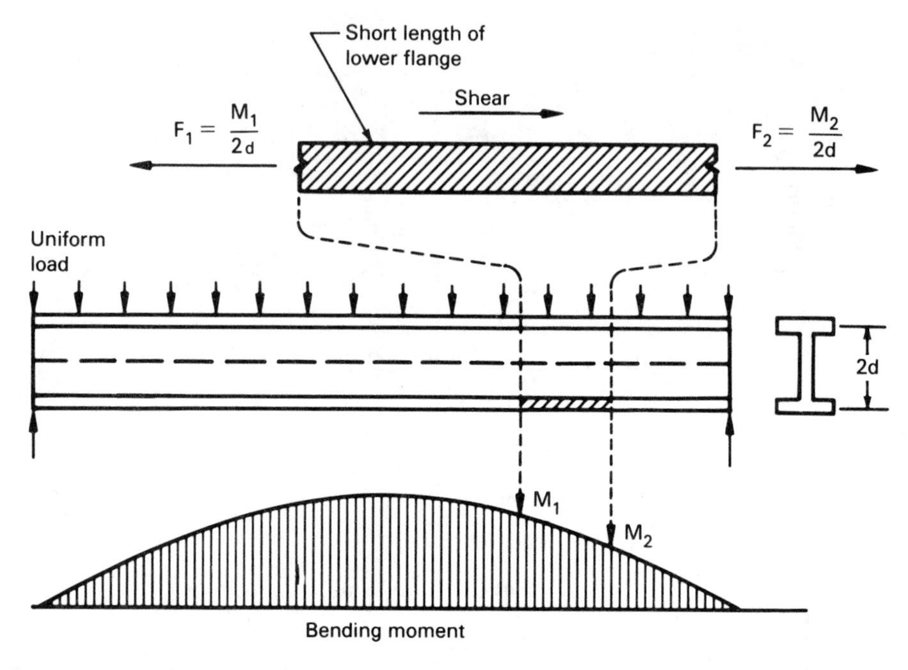

Fig. 1.7—Approximate tensile forces on a section of the lower flange of a loaded beam

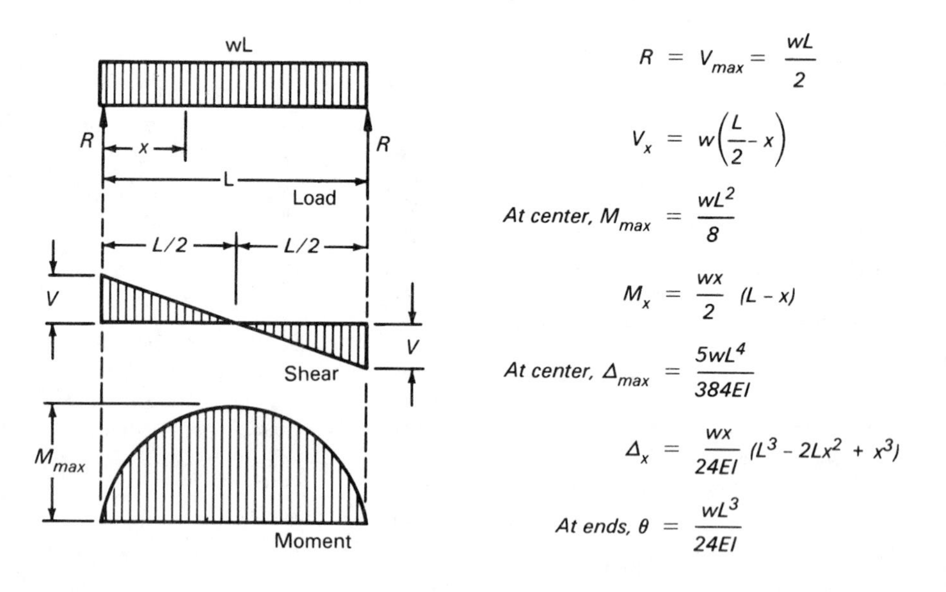

Fig. 1.8—Formulae for a simply supported beam with a uniformly distributed load on the span

Information concerning the design of a compression member or column also may apply to the compression flange of a beam. The lateral buckling resistance of the compression flange must also be considered. It should have adequate width and thickness to resist local buckling, be properly supported to prevent twisting or lateral movement, and be subjected to compressive stresses within allowable limits.

Shear

Figure 1.9 illustrates the shear forces in the web of a beam under load. They are both horizontal and vertical, and create diagonal tensile and diagonal compressive stresses.

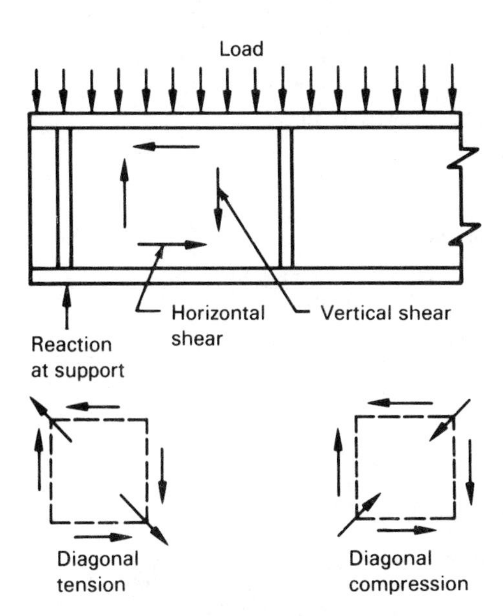

Fig. 1.9—Shear forces in the web of a beam under load

The shear capacity of a beam of either an I or a box cross section is dependent upon the slenderness proportions of the web or webs. For virtually all hot rolled beams and welded beams of similar proportions where the web slenderness ratio (h/t) is less than 260, vertical shear load is resisted by pure beam shear without lateral buckling to a level of loading well above that where unacceptable deflections will develop. Thus, design is based upon keeping the shear stress on the gross area of the web below the allowable value of 0.4 σ_y to prevent shear yielding.

In plate girders with slender webs, shear is resisted by plane beam shear up to a level of stress which will cause shear buckling. However, webs subject to shear stress have considerable post-buckling strength, and current design specifications take this strength into account. After buckling occurs, the web resists larger shear loads by a combination of beam shear and diagonal tension. If the shear span has a length-to-depth ratio greater than approximately three, the direction of the diagonal tension becomes too near to the horizontal for it to be effective, and transverse stiffeners must be provided. The stiffeners act in a manner analogous to the compression vertical members of a Pratt truss.

The onset of diagonal compression buckling in a web panel is of negligible structural significance because the magnitude of the buckle is limited by the crossing diagonal tension field. With properly proportioned transverse stiffeners, the web continues to resist higher levels of shear loading up to the point where diagonal tension yielding occurs. The true maximum shear strength would exist at this point.[9]

There are a few practical considerations independent of maximum strength that may govern a design. For architectural reasons, especially in exposed fascia girders, the waviness of the web caused by controlled compression buckles may be deemed unsightly. In plate girders subject to cyclic loading to a level which would initiate web shear buckling, each application of the critical load will cause an "oil canning" or breathing action of the web panels. This action causes out-of-plane bending stresses at the toes of web-to-flange fillet welds and the stiffener welds. These cyclic stresses will eventually initiate fatigue cracking. For these cases, web stresses should be limited to the values where shear buckling is precluded.

---

9. Reliable criteria for the design of plate and box girder webs to achieve full maximum strength from beam shear and tension field action may be found in the *Manual of Steel Construction*, Section 5, latest edition, published by the American Institute of Steel Construction.

In the case of a beam fabricated by welding, the unit shear load on the welds joining the flanges of the beam to the web can be calculated by the formula:

$$W_s = (Vay)/(In)$$

where

W_s = load per unit length of weld.

V = external shear force on the member at this location.

a = cross-sectional area of the flange.

y = distance between the center of gravity of the flange and the neutral axis of bending of the whole section.

I = moment of inertia of the whole section about the neutral axis of bending.

n = number of welds used to attach the web to the flange.

Torsion

Torsion creates greater design problems for bases and frames than for other machine members. A machine with a rotating unit may subject the base to torsional loading. This becomes apparent by the lifting of one corner of the base, if the base is not anchored to the floor.

If torsion is a problem, closed tubular sections or diagonal bracing should be used, as shown in Fig. 1.10. Closed tubular sections are significantly better for resisting torsion than comparable open sections. Closed members can easily be made from channel or I-sections by intermittently welding flat plate to the toes of the rolled sections. The torsion effect on the perimeter of an existing frame may be eliminated or the frame stiffened for torsion by welding in cross bracing. Torsion problems in structures can be avoided by judicious arrangement of members to transmit loads by direct stresses or bending moments.

Torsional Resistance. The torsional resistance of a flat strip or open section (I-beam or channel) is very low. The torsional resistance of a solid rectangular section having a width of several times the thickness may be approximated by the following formula:

$$R = \frac{bt^3}{3}$$

where

R = torsional resistance, in.^{4}

b = width of the section, in.

t = thickness of the section, in.

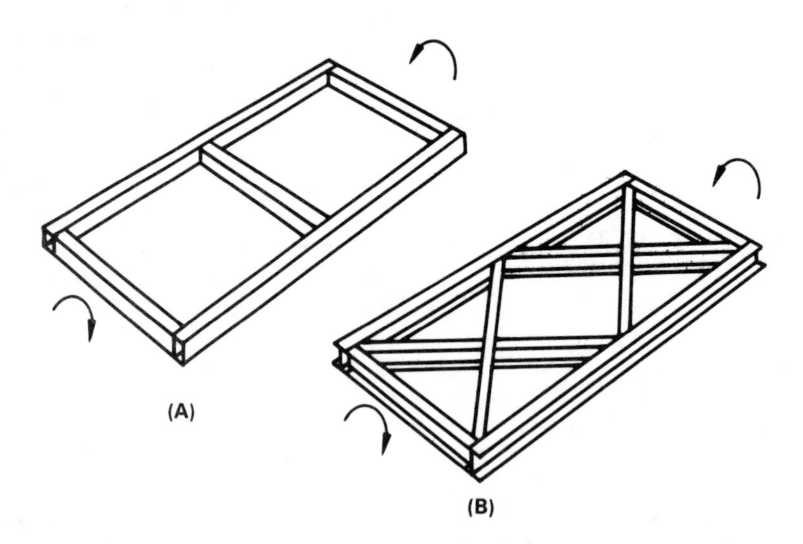

Fig. 1.10—Application of (A) closed tubular sections or (B) diagonal bracing to resist torsion

The total angular twist (rotation) of a member can be estimated by the following formula:

$$\theta = Tl/GR$$

where

θ = angle of twist, radians

T = torque, lbf · in.

l = length of member, in.

G = modulus of elasticity in shear, psi (12×10^6 for steel)

R = total torsional resistance, in.^{4}

The unit angular twist, ϕ, is equal to the total angular twist, θ divided by the length, l, of the member.

The torsional resistance of a structural member, such as an I-beam or a channel, is approximately equal to the sum of the torsional resistances of the individual flat sections into which the member can be divided. This is illustrated in Table 1.6, listing the actual and calculated angle of twist of a flat strip and an I-shape made up of three of the flat strips. The applied torque was the same for both sections.

Table 1.6
Calculated and actual angle of twist

	Angle of twist, degrees	
	Strip[a]	I-section[b]
Calculated using torsional resistance	21.8	7.3
Actual twist	22	9.5

a. 0.055 in. by 2 in.
b. Made of three of the strips.

Torsional resistance increases markedly with closed cross sections, such as circular or rectangular tubing. As a result, the angular twist is greatly reduced because it varies inversely with torsional resistance. The torsional resistance R of any closed box shape enclosing only one cell can be estimated by the following procedure. Draw a dotted line through the mid-thickness around the section, as shown in Fig. 1.11. The area enclosed by the dotted lines is A. Divide the cross section of the member into convenient lengths, S_x having thicknesses t_x. Determine the ratios of these individual lengths to their corresponding thicknesses and total them. Torsional resistance is then obtained from the following formula:

$$R = \frac{4A^2}{\Sigma(S_x/t_x)}$$

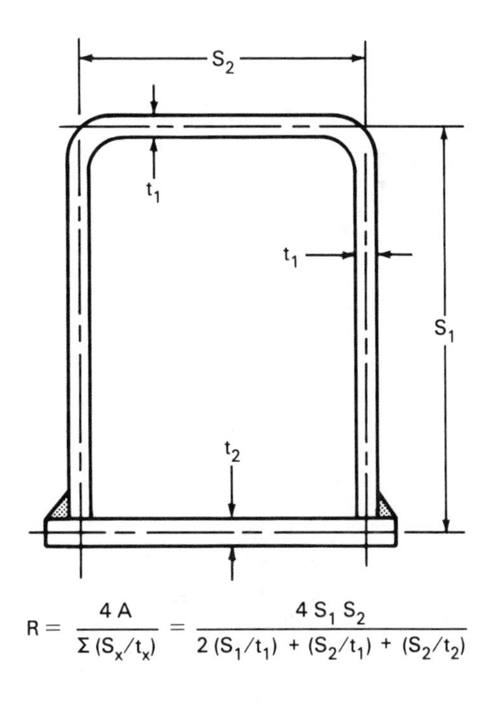

Fig. 1.11—Torsional resistance of a closed box section

The maximum shear stress in a rectangular section in torsion is on the surface at the center of the long side. Knowing the unit angular twist, the following formula will give the maximum shear stress at the surface of a rectangular part.

$$\tau = \phi tG = \frac{Tt}{R}$$

where

τ = maximum unit shear stress, psi

ϕ = unit angular twist, radians/in.

G = modulus of elasticity in shear, psi

T = applied torque, lbf · in.

t = thickness of the section, in.

R = torsional resistance, in.^{4}

This formula can be applied to a flat plate or a rectangular area of an open structural shape (channel, angle, I-beam). In the latter case, R is the torsional resistance of the whole structural shape.

Diagonal Bracing. Diagonal bracing is very effective in preventing the twisting of frames. A simple explanation of the effectiveness of diagonal bracing involves an understanding of the directions of the forces involved.

A flat bar of steel has little resistance to twisting but has exceptional resistance to bending (stiffness) parallel to the width of the bar. Transverse bars or open sections at 90° to the main members are not effective for increasing the torsional resistance of a frame because, as shown in Fig. 1.12(A), they contribute only relatively low torsional resistance. However, if the bars are oriented diagonally across the frame, as in Fig. 1.12(B), twisting of the frame is resisted by the stiffness of the bars. To be effective, the diagonal braces must have good stiffness perpendicular to the plane of the frame.

TRANSFER OF FORCES

Loads create forces that must be transmitted through the structure to suitable places for counteraction. The designer needs to know how to provide efficient pathways.

One of the basic rules is that a force applied transversely to a member will ultimately enter that portion of the section that lies parallel to the applied force. Figure 1.13, for example, shows a lug welded parallel to the length of a beam. The portion of the beam that is parallel to the applied force is the web. The force in the lug is easily transferred through the connecting welds into the web. No additional stiffeners or attaching plates are required.

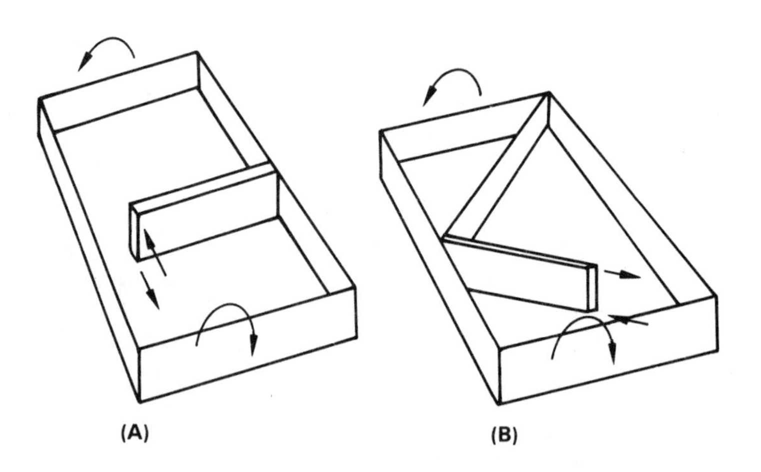

Fig. 1.12—Frames subjected to torsion with (A) transverse rib bracing and (B) diagonal bracing

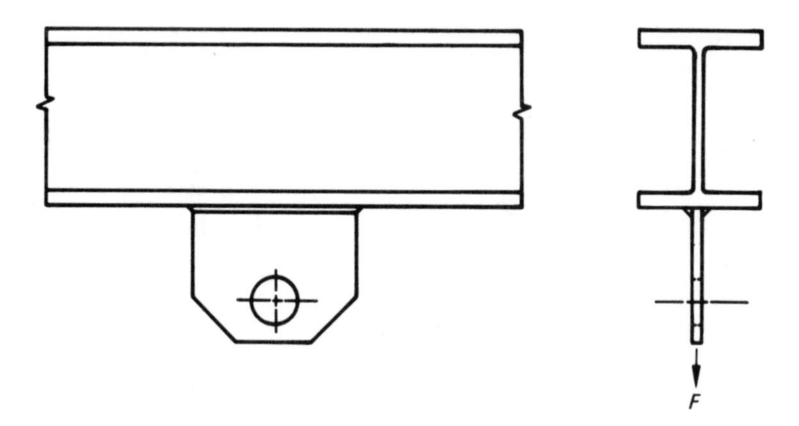

Fig. 1.13—Lug welded parallel to the length of a beam

Suppose, however, that the lug was welded to the beam flange at right angles to the length of the beam, Fig. 1.14(A). The outer edges of the flange tend to deflect rather than support much load. This forces a small portion of the weld in line with the web to carry a disproportionate share of the load, Fig. 1.14(B). To uniformly distribute the load on the attachment weld, two stiffeners can be aligned with the lug and then welded to the web and to the adjacent flange of the beam, as shown in Fig. 1.15. The stiffeners reinforce the bottom flange and also transmit part of the load to the web of the beam. The welds labeled *A, B,* and *C* in Fig. 1.15 must all be designed to carry the applied force, *F*.

If a force is to be applied parallel to the flanges, it would not be acceptable to simply weld an attachment plate to the web of the beam. The applied force on the attachment could cause excessive bending of the web, as shown in Fig. 1.16. In such a situation, the plate can be welded to the flanges of the beam, as in Fig. 1.17 for low forces, and no force is transferred to the web. For large forces, it might be necessary to place a stiffener on the opposite side of the web, Fig. 1.18. In this case, both the plate and the stiffener should be welded to the web as well as to both flanges. The stiffener is loaded only through the welds along the web. The fillet welds joining the plate to the beam flanges must not extend around the plate along the edges of the flanges. The abrupt changes in weld direction can intensify stress concentrations.

When a force in a structure changes direction, a force component is involved. This is illustrated in Fig. 1.19, a knee of a rigid frame subjected to a bending moment. The compressive force in the interior flanges must change direction at the knee. To prevent bending of the web and

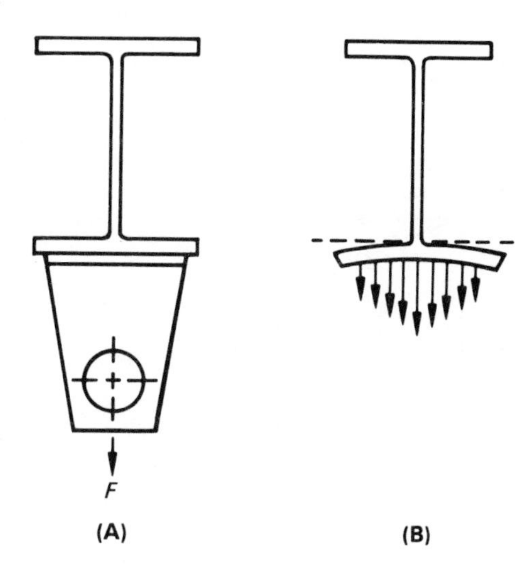

Fig. 1.14—(A) Lug welded to a flange transverse to the beam, (B) Load distribution with unreinforced flange

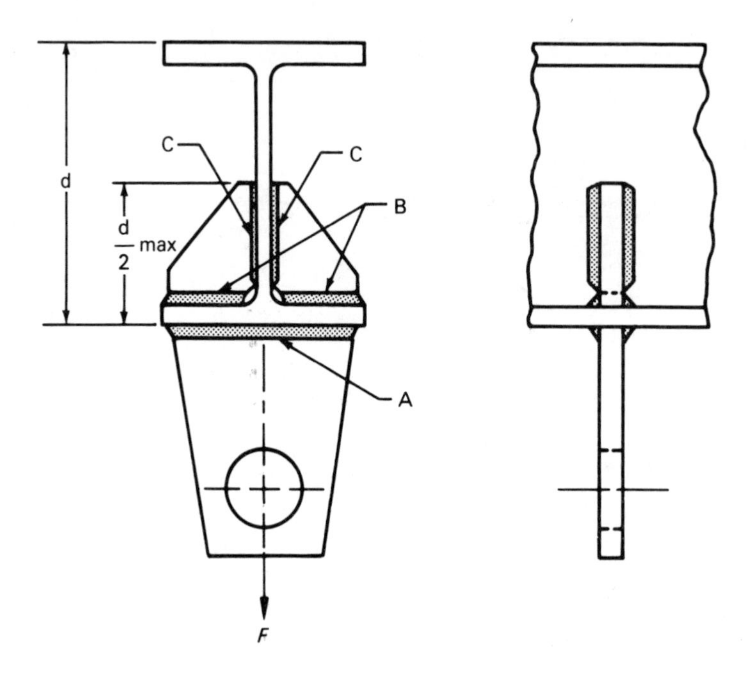

Fig. 1.15—Stiffeners transmit a portion of the load from the flange to the web of the beam through welds A, B, and C

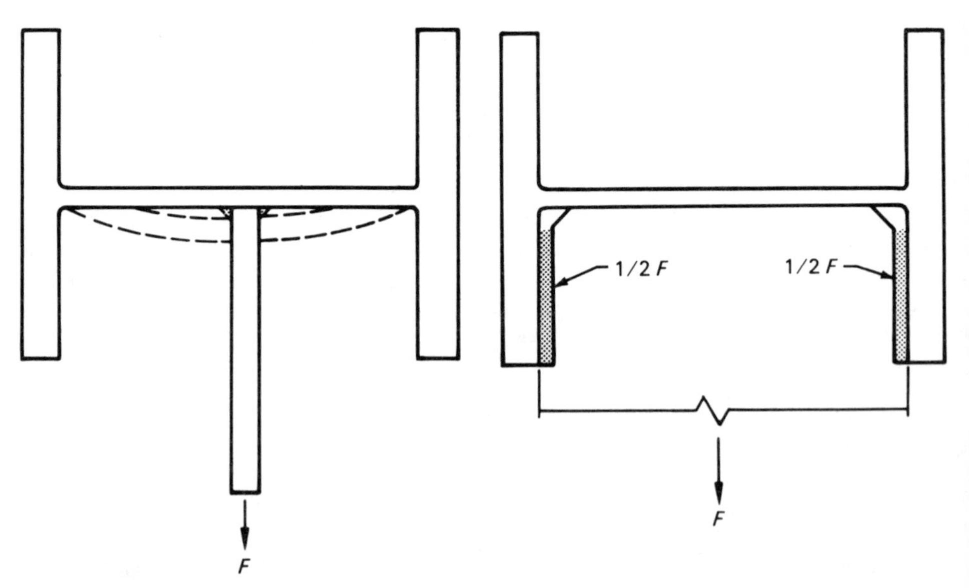

Fig. 1.16—Deflection of beam web from loaded attachment

Fig. 1.17—Attachment welded to beam flanges to avoid web defection

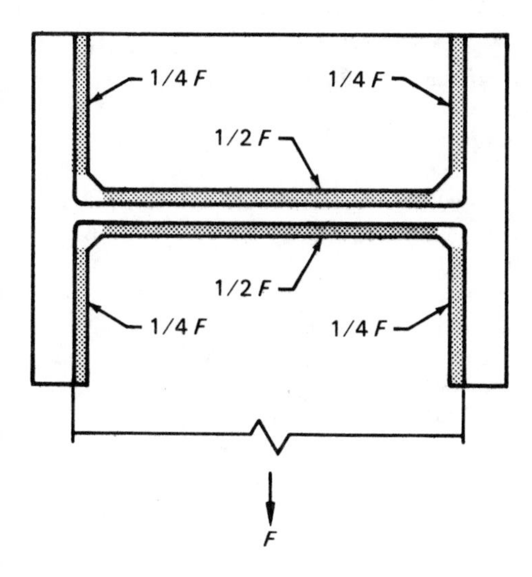

Fig. 1.18—Stiffener on opposite side of attachment, welded as shown, for large loads

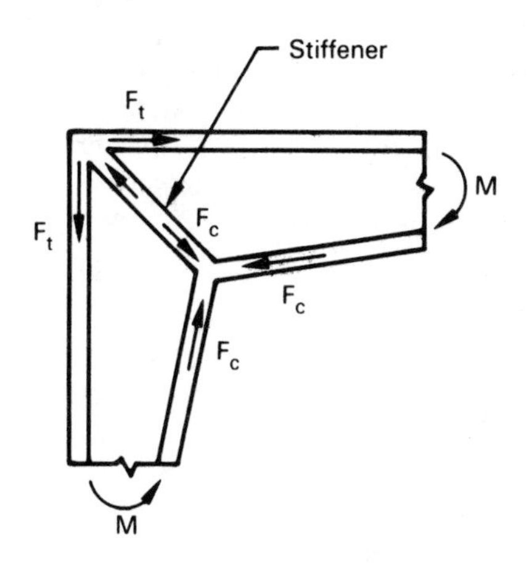

Fig. 1.19—Knee of rigid frame

failure of the structure, diagonal stiffeners are placed on both sides of the web at the intersections of the two flanges. The compressive force component in the web and stiffeners balances the change in direction of the tensile force in the outer flanges.

INFLUENCE OF WELDING PROCESS

Most designs for welding are based on the use of arc welding processes where filler metal is added in multiple passes, either in a groove to obtain required penetration or in a fillet weld on one or both sides of a member.[10] Adding filler metal in a groove produces lateral and longitudinal shrinkage between components that can result in angular distortion about the axis of the weld and longitudinal bowing of the joints. Fillet welds primarily cause angular distortion. The amount of shrinkage and distortion depends on the size of the weld in proportion to the thickness of the components and the degree of restraint imposed on the joint during welding and cooling.

When components are restrained from shrinkage or distortion, high residual stresses remain in the assembly. Depending on the size of the weld and the level of stresses, additional distortion can occur during subsequent thermal stress relieving.

When welding quenched-and-tempered steels, the heat input must be controlled to maintain the properties of the weld and heat-affected zone of the base metal. This usually dictates the use of small weld beads using relatively low welding currents.

Most welding processes require the weld joint to be readily accessible to the welder and the welding equipment. Several high-energy welding processes sometimes used in production are unique. When applicable, they offer some added flexibility in the design that may not be available with conventional arc welding processes.

The electron beam welding (EBW) process can produce narrow, deep-penetration welds with minimal lateral shrinkage and angular distortion.[11] Electron beam welds made in restrained sections have lower residual stresses than welds made with most other welding processes. The heat-affected zones of electron beam welds are normally narrower than those of welds made with arc welding processes. Quenched-and-

---

10. Arc welding processes are discussed in the *Welding Handbook*, Vol. 2, 7th Ed., 1978.

11. Electron beam welding is covered in the *Welding Handbook*, Vol. 3, 7th Ed., 1980: 169-215.

tempered steels can be welded with minimal degradation of properties. With some designs, welds can be made with the components in the finish-machined state. In contrast, components that are welded with conventional arc welding processes require excess stock allowance for machining the weldment in those areas that must meet close dimensional tolerances.

A high-power laser beam can produce welds with characteristics similar to those made with the nonvacuum electron beam process.[12] However, equipment costs with both processes must be considered when specifying these processes.

The joint design for electron beam or laser beam welding consists of a simple square groove. Fit-up for these processes is critical, and root openings must be held to a minimum (preferably metal to metal contact). To obtain the proper fit-up for welding with these processes, weld joints must be prepared by machining. Fortunately, welds made with the electron beam or laser beam normally do not require filler metal addition. However, with either process, the joint must be positioned under the welding head, and the work or head moved automatically while maintaining the beam on the joint.

The electroslag welding process is unique from a design viewpoint.[13] It is a high-deposition, efficient process, but it imposes certain limitations or conditions on the designer. However, the process has some definite advantages where it can be applied.

The electroslag welding process deposits a continuous, parallel-sided weld vertically from one end of the joint to the other end. The weld has predictable lateral shrinkage, generally less than that of single or double groove welds, and essentially no angular distortion. Because of the high energy input, the weld is considered essentially self-stress relieving.

An electroslag weld is basically a continuous casting. Consequently, the weld generally has a coarse, columnar grain structure, and a wide heat-affected zone. Impact properties and fatigue strengths of welded joints in the as-welded and the stress-relieved conditions may be lower than with other welding processes. These properties can be improved with normalizing and tempering heat treatments. This process cannot be used with quenched-and-tempered steels unless the weld and heat-affected zone receive a full heat treatment after welding.

Joint preparation for electroslag welding consists of square flame-cut edges. Weld joint length is not critical, but the joint must be aligned and properly fixtured to permit continuous welding. In general, it is not practical to weld plate thicknesses less than 3/4 in. with this process. Two variations of the electroslag welding process are generally used, conventional and consumable guide. The specific variation determines joint accessibility requirements.

DESIGN OF WELDED JOINTS

TYPES OF JOINTS

The loads in a welded structure are transferred from one member to another through welds placed in the joints. The types of joints used in welded construction and the applicable welds are shown in Fig. 1.20. The configurations of the various welds are illustrated in Figs. 1.21, 1.22, and 1.23. Combinations of welds may be used to make a joint, depending on the strength requirements and load conditions. For example, fillet and groove welds are frequently combined in corner and T-joints.

Welded joints are designed primarily for the strength and safety required by the service conditions under which they must perform. The manner in which the stress will be applied in

12. Laser beam welding is discussed in the *Welding Handbook*, Vol. 3, 7th Ed., 1980: 218-32.

13. Electroslag welding is covered in the *Welding Handbook*, Vol. 2, 7th Ed., 1978: 225-48.

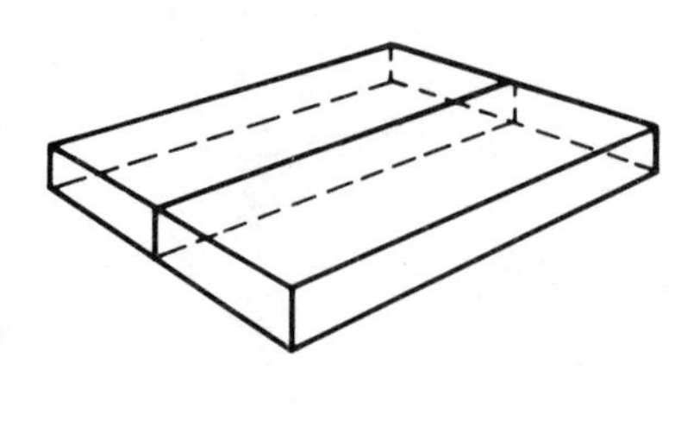

Applicable welds

- Square-groove
- V-groove
- Bevel-groove
- U-groove
- J-groove
- Flare-V-groove
- Flare-bevel-groove
- Edge-flange
- Braze

(A) Butt joint

Applicable welds

- Fillet
- Square-groove
- V-groove
- Bevel-groove
- U-groove
- J-groove
- Flare-V-groove
- Flare-bevel-groove
- Edge-flange
- Corner-flange
- Spot
- Projection
- Seam
- Braze

(B) Corner joint

Applicable welds

- Fillet
- Plug
- Slot
- Square-groove
- Bevel-groove
- J-groove
- Flare-bevel-groove
- Spot
- Projection
- Seam
- Braze

(C) T-joint

Applicable welds

- Fillet
- Plug
- Slot
- Bevel-groove
- J-groove
- Flare-bevel-groove
- Spot
- Projection
- Seam
- Braze

(D) Lap joint

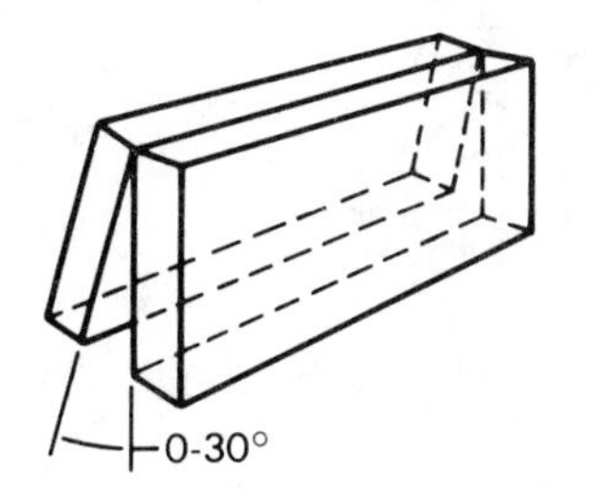

Applicable welds

- Square-groove
- Bevel-groove
- V-groove
- U-groove
- J-groove
- Edge-flange
- Corner-flange
- Seam
- Edge

(E) Edge joint

Fig. 1.20—Types of joints

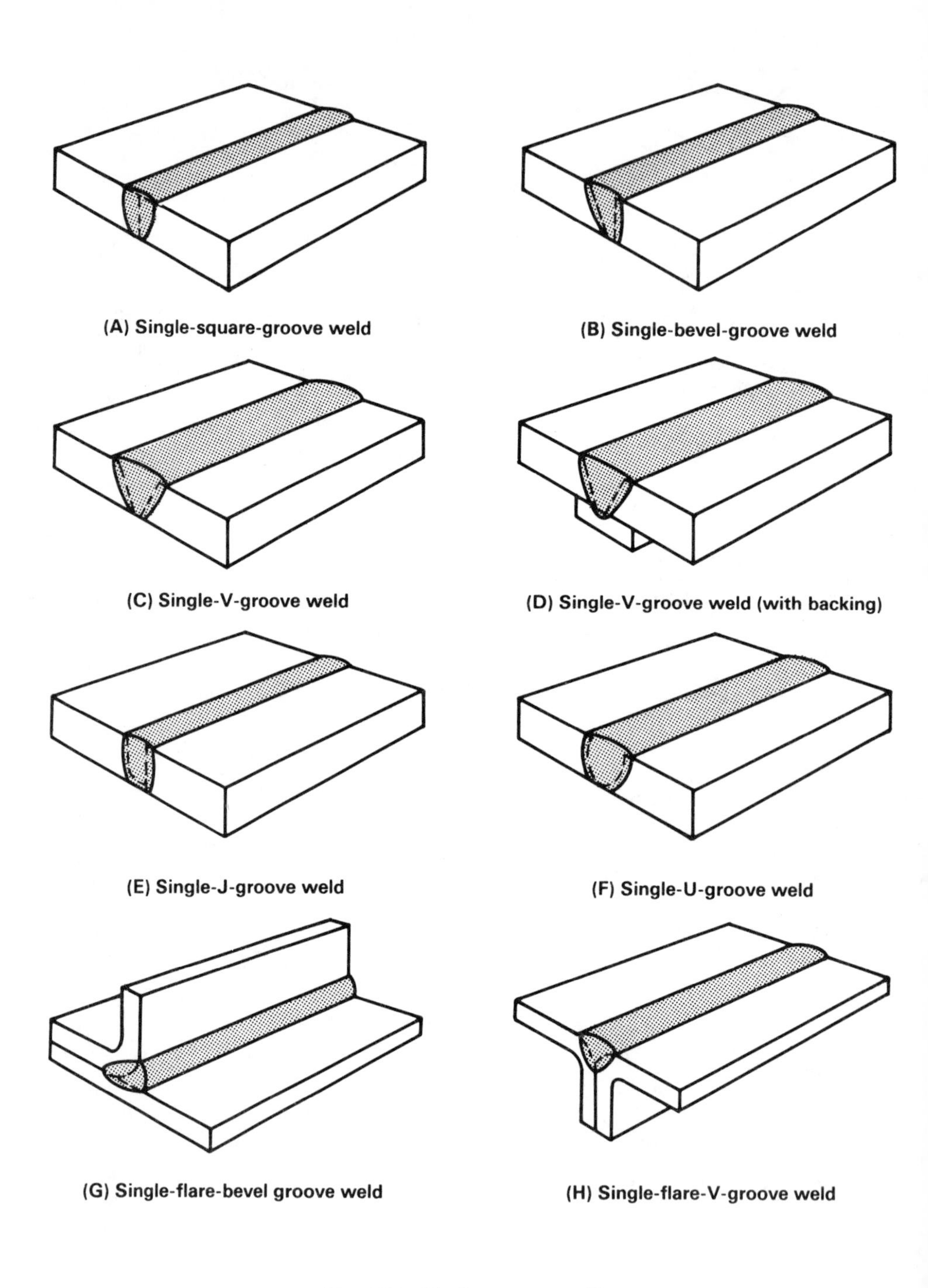

Fig. 1.21—Single-groove welds

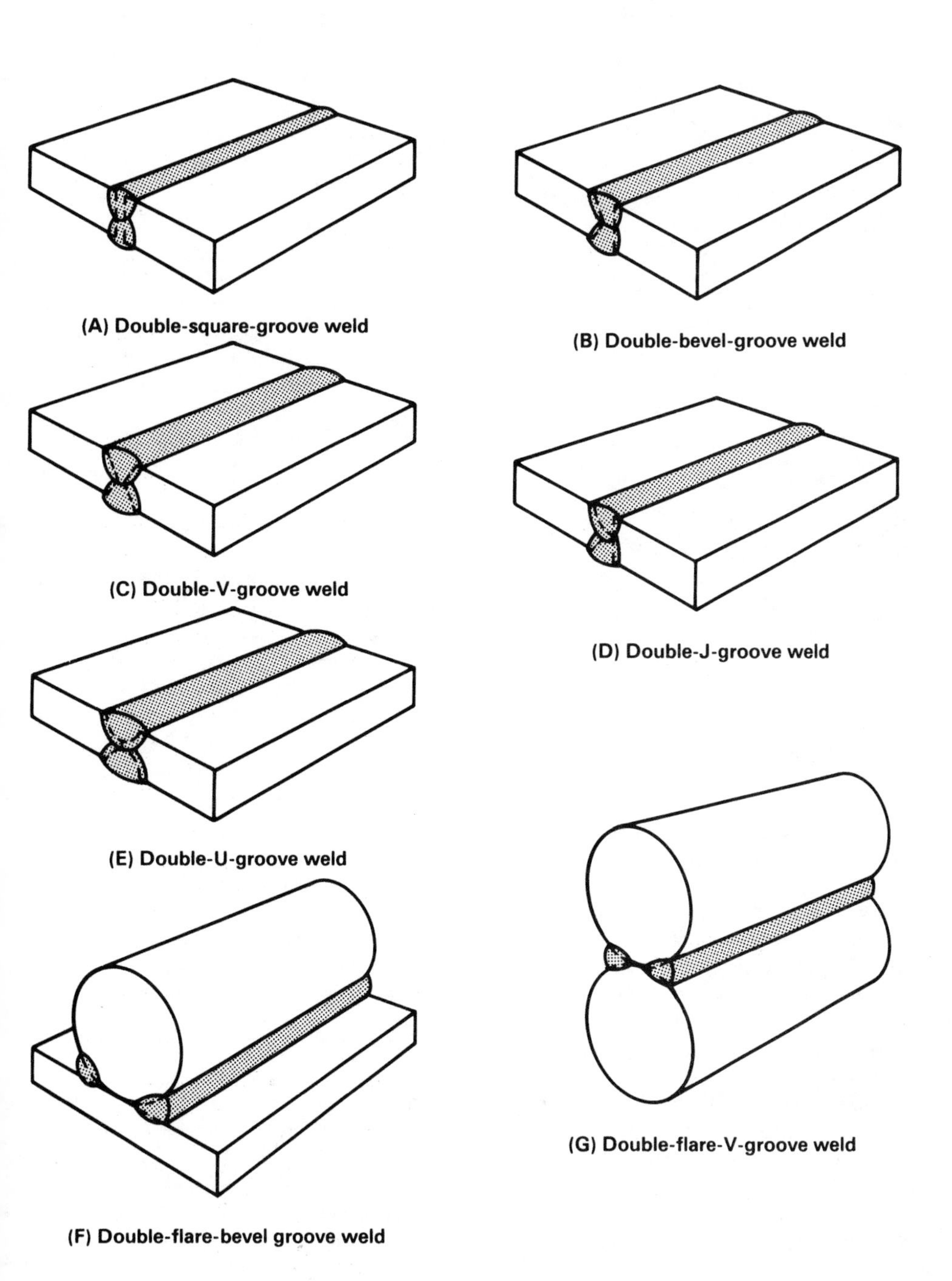

Fig. 1.22—Double-groove welds

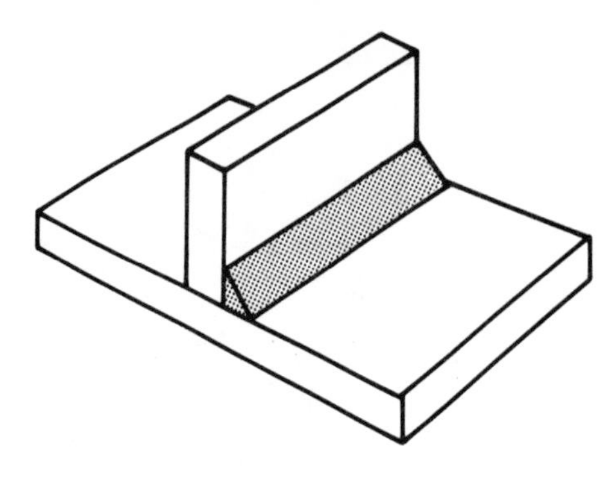

(A) Single fillet weld

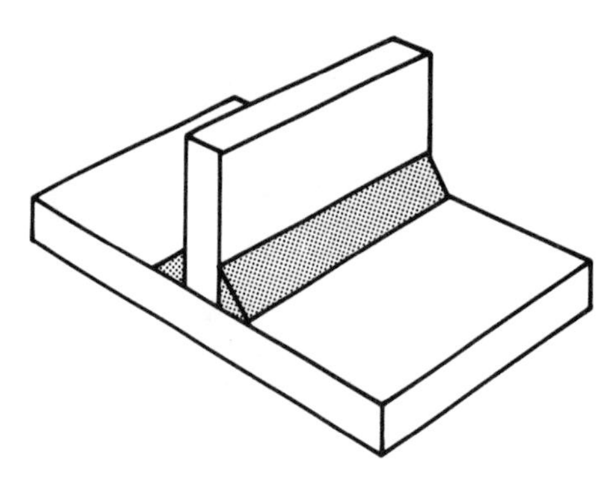

(B) Double fillet weld

Fig. 1.23—Fillet welds

service, whether tension, shear, bending, or torsion, must be considered. Different joint designs may be required depending on whether the loading is static or dynamic where fatigue must be considered. Joints may be designed to avoid stress-raisers and to obtain an optimum pattern of residual stresses. Conditions of corrosion or erosion require joints that are free of irregularities, crevices, and other areas that make them susceptible to such forms of attack. The design must reflect consideration of joint efficiency, which is the ratio of the joint strength to the base metal strength, generally expressed as a percentage.

In addition to the above, joints are detailed for economy and accessibility during construction and inspection. Among the factors involved in construction are control of distortion and cracking, facilitation of good workmanship, and production of sound welds. Accessibility during construction not only contributes to low costs, but also provides an opportunity for better workmanship, fewer flaws, and control of distortion and residual stresses.

GROOVE WELDS

Groove welds of different types are used in many combinations; the selection of which is influenced by accessibility, economy, adaptation to the particular design of the structure being fabricated, expected distortion, and the type of welding process to be used.

A square-groove weld is economical to use, provided satisfactory soundness and strength can be obtained, because it only requires a square edge on each member for welding. Its use is limited by the thickness of the joint.

For thick joints, the edge of each member must be prepared to a particular geometry to provide accessibility for welding and ensure desired soundness and strength. In the interest of economy, those joint designs should be selected with root openings and groove angles that require the smallest amount of weld metal and still give sufficient accessibility for sound welds. The selection of root opening and groove angle also is greatly influenced by the metals to be joined, the location of the joint in the weldment, and the required performance.

J- and U-groove welds may be used to minimize weld metal requirements when the savings are sufficient to justify the more costly machining of the edge preparation. These joints are particularly useful in the welding of thick sections. One disadvantage of J- and bevel-groove welds is that they are difficult to weld soundly when they are positioned to have a vertical side.

The most important criterion of the strength of a groove weld is the amount of joint penetration. Welded joints are usually designed so that they will be equal in strength to the base metal. To accomplish this, designs that permit penetration completely through the members being joined are most commonly used. One of the principles of welded design is the selection of joint designs that provide the desired degree of joint penetration.

The details of welding grooves (groove angle, root face, root opening, etc.) depend upon the welding process and type of power to be used, and the physical properties of the base metal(s) being joined.[14] Some welding processes characteristically provide deeper joint penetration than others. Some metals have relatively high thermal conductivities or specific heats, or both. Those metals require greater heat input for welding than other metals with lower heat absorption. Examples are aluminum and copper.

Use of a double-welded joint in preference to a single-welded joint can reduce the amount of welding by 50 percent or more, as indicated by Fig. 1.24. This also limits distortion when equal numbers of weld passes are placed alternately on each side of the joint.

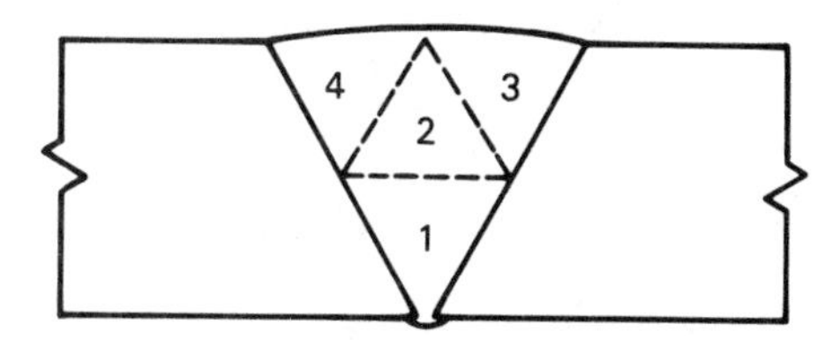

(A) Single-welded joint

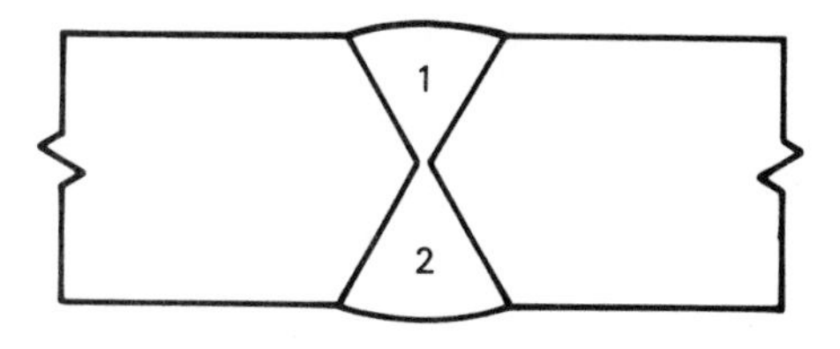

(B) Double-welded joint

Fig. 1.24—Comparison or relative volumes of filler metal for single- and double-V-groove welds of equal thickness and length

14. Groove-weld joint preparations used with various welding processes and base metals are discussed in the *Welding Handbook*, Vols. 2, 3, and 4, 7th Ed., 1978, 1980, and 1982.

The various types of groove welds have certain advantages and limitations with respect to their applications. In the following discussion, comments on joint penetration or effective throat apply to joining of carbon steel by shielded metal arc (SMAW), gas metal arc (GMAW), flux cored arc (FCAW), and submerged arc welding (SAW). Joint penetration with other processes and base metals may be different when their physical properties vary significantly from those of carbon steel.

Complete Joint Penetration

Groove welds with complete joint penetration are suitable for all types of loading provided they meet the acceptance criteria for the application. In many cases, a single-groove weld is not considered to have complete penetration, unless it was made using a suitable backing strip, or a back weld was applied.[15] In such cases, to ensure complete penetration with double-groove welds and single-groove welds without a backing strip, the root of the first weld must be back gouged to sound metal before making a weld pass on the other side.

Partial Joint Penetration

A partial-joint-penetration groove weld has an unwelded portion at the root of the weld. This condition may also exist in joints welded from one side without backing.

The unwelded portions of groove welded joints constitute a stress raiser having significance when cyclic loads are applied transversely to the joint. This fact is reflected in applicable fatigue criteria. However, when the load is applied longitudinally, there is no appreciable reduction in fatigue strength.

Regardless of the rules governing the service application of partial-joint-penetration welds, the eccentricity of shrinkage forces in relation to

15. Welders with good technique, welding on metals that have high surface tension when molten, are able to obtain complete joint penetration without the aid of backing in single-groove welds. An appropriate joint design and welding process must be used to ensure complete root penetration and a smooth root surface.

the center of gravity of the section can result in angular distortion on cooling after welding. This same eccentricity also tends to cause rotation of a transverse axial load across the joint. Therefore, means must be applied to restrain or preclude such rotation both during fabrication and in service.

For static loading, the allowable stresses in partial-joint-penetration groove welds depend upon the type and direction of loading and applicable code requirements. For steel, the allowable stress in tension and shear should be limited to about 30 percent of the nominal tensile strength of the weld metal. For tension or compression parallel to the weld axis, the allowable stress can be the same as that for the base metal. In no case should the actual tensile or shear stress in the base metal next to the weld exceed the allowable for the base metal.

Square-Groove Welds

Square-groove welds are economical in preparation and welding. Joint strength depends upon the amount of joint penetration. Adequate control of welding procedures and proper inspection are required to obtain the specified joint penetration and strength.

With most welding processes when welding from one side only, complete joint penetration can be reliably obtained only in thin sections unless a root opening and backing strip are used. With a suitable root opening and backing strip, joints up to 1/4 in. thickness can be welded by SMAW, and up to 3/8 in. thickness by GMAW, FCAW, and SAW. This type of joint should not be loaded in bending with the root of the weld in tension, nor subjected to transverse fatigue or impact loading. Also, it should not be used at low temperatures or in corrosive applications.

When welded from both sides, complete joint penetration can be obtained in sections up to 1/4-in. thickness by shielded metal arc welding, 3/8-in. thickness by gas metal arc and flux cored arc welding, and 1/2-in. thickness by submerged arc welding.

Single-V-Groove Welds

Single-V-groove welds are economical provided the depth of the groove does not exceed 3/4 in. Groove preparations and welding are relatively inexpensive.

The strength of the joint depends on the depth of penetration (effective throat). Joints with complete penetration are recommended for optimum mechanical properties.

Joints welded from one side only should not be used in bending with the root in tension. Also, they should not be subjected to transverse fatigue nor impact loading unless (1) complete joint penetration is obtained, (2) the backing strip, if used, is removed, and (3) the root face is machined smooth. Partially penetrated joints should not be exposed to corrosive conditions. The effective throat of such joints is never greater than the depth of joint preparation for design purposes.

Double-V-Groove Welds

The strength of a double-V-groove weld depends upon the total joint penetration (effective throat). For all types of loading, full strength joints can be obtained with complete joint penetration provided the root of the first weld is back gouged to sound metal before commencing welding on the other side. Partial penetration welds are recommended only for static loads transverse to the weld axis and for other types of loads parallel to the weld axis. This type of weld is economical when the depth of the grooves does not exceed 3/4 in. Double-V-groove welds with complete joint penetration may be costly when the joint thickness exceeds 1-1/2 in.; double-U-groove welds may be more economical.

Single-Bevel-Groove Welds

Single-bevel-groove welds have characteristics similar to single-V-groove welds with respect to properties and applications. The bevel type requires less joint preparation and weld metal; therefore, it is more economical.

One disadvantage of this type of weld is that proper welding procedures are required to obtain complete fusion with the perpendicular face of the joint. In the horizontal position, that face should be placed on the lower side of the joint to obtain good fusion.

Double-Bevel-Groove Welds

Double-bevel-groove welds have the same characteristics as double-V-groove welds. The perpendicular joint faces make compelte fusion more difficult. Also, back gouging the root of the first weld may be harder to accomplish.

A double-bevel joint design is economical when the depth of the groove does not exceed about 3/4 in., and the joint thickness is 1-1/2 in. or less for complete penetration.

Single- and Double-J-Groove Welds

J-groove welds have the same characteristics as similar bevel-groove welds. However, they may be more economical for thicker sections because less weld metal is required. Their use may be limited to the horizontal position in some applications.

Single- and Double-U-Groove Welds

U-groove welds and J-groove welds are used for similar applications. However, complete fusion is easier to obtain, and root gouging is more readily accomplished with U-grooves.

FILLET WELDS

Where the design permits, fillet welds are used in preference to groove welds for reason of economy. Fillet welded joints are very simple to prepare from the standpoint of edge preparation and fit-up, although groove welded joints sometimes require less welding.

When the smallest continuous fillet weld that is practicable to make results in a joint efficiency greater than that required, intermittent fillet welding may be used to avoid overwelding unless continuous welding is required by the service conditions.

Fillet weld size is measured by the length of the legs of the largest right triangle that may be inscribed within the fillet weld cross section as shown in Fig. 1.25. The effective throat, a better indication of weld shear strength, is the shortest distance between the root of the weld and the diagrammatical weld face. With a concave weld face, the size of a fillet weld is less than the actual leg of the weld. With a convex weld face, the actual throat may be larger than the effective throat of the weld. In any case, the strength of a fillet weld is based on the effective throat and the length of the weld (effective area of the weld).

It should be noted that the actual throat may be larger than the theoretical throat by virtue of joint penetration beyond the root of the weld. When deep root penetration can be reliably assured by the use of automatic submerged arc or flux cored arc welding, some codes permit the effective throat to be taken as the size of the weld for 3/8 in. and smaller fillet welds, and as the theoretical throat plus 0.11 in. for fillet welds over 3/8 in.

Applications

Fillet welds are used to join corner, T-, and lap joints because they are economical. Edge preparation is not required, but surface cleaning may be needed. Fillet welds are generally applicable where stresses are low, and the required weld size is less than about 5/8 in. If the load would require a fillet weld of 5/8 in. or larger, a groove weld should be used possibily in combination with a fillet weld to provide the required effective throat. Fillet welds may be used in skewed T- or corner joints having a dihedral angle between 60 and 135 degrees. Beyond these limits, a groove weld should be used.

Fillet welds are always designed on the basis of shear stress on the throat. The maximum shear stress appears on the effective area of the weld. In the case of steel, the maximum shear stress is normally limited to about 30 percent of the nominal (classification) tensile strength of the weld metal.

Weld Size

Fillet welds must be large enough to carry the applied load and to accommodate shrinkage of the weld metal during cooling if cracking is to be avoided, particularly with highly restrained, thick sections. However, the specified fillet weld size should not be excessive to minimize distortion and welding costs. Welds in lap joints cannot exceed in size the thickness of the exposed edge, which should be visible after welding.

Fillet welds may be designed with unequal leg sizes to provide the required effective throat or the needed heat balance for complete fusion with unequal base metal thicknesses.

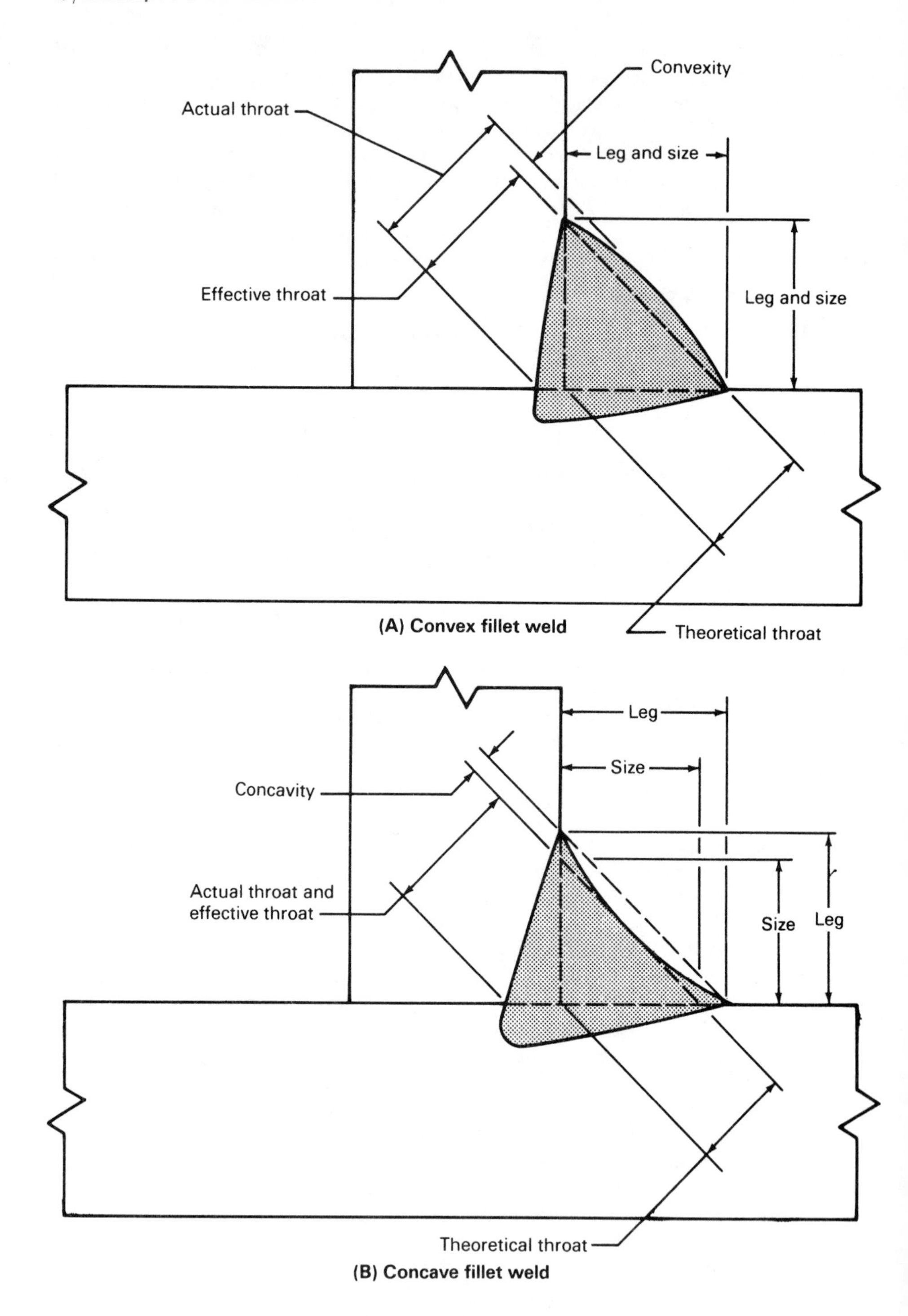

Fig. 1.25—Fillet weld size

Single Fillet Welds

Single fillet welds are limited to low loads. Bending moments that result in tension stresses in the root of a fillet weld should not be permitted because of the notch condition. For this reason, single fillet welds should not be used with lap joints that can rotate under load. The welds should not be subjected to impact loads. When used with fatigue loading, the allowable stress range must be subject to stringent limitations.

Double Fillet Welds

Smaller double fillet welds are preferred to a large single fillet weld. Full plate strength can be obtained with double fillet welds under static loading. Double fillet welding of corner and T-joints limits rotation of the members about the longitudinal axis of the joint, and minimizes tension stresses at the root of the welds. These types of joints can be cyclically loaded parallel to the weld axes.

Double-welded lap joints should have a minimum overlap of about five times the base metal thicknesses to limit joint rotation under load.

COMBINED WELDS

Combined partial-joint-penetration groove and fillet welds, as shown in Fig. 1.26, are useful for many joints. Occasionally, this practice is questioned because of differences in recommended welding electrodes for the two types of welds. The practice, however, is sound.

SELECTION OF WELD TYPE

The designer is frequently faced with the question of whether to use fillet or groove welds. Cost is a major consideration. Double fillet welds, Fig. 1.27(A), are easy to apply and require no special edge preparation. They can be made using large diameter electrodes with high welding currents for high deposition rates.

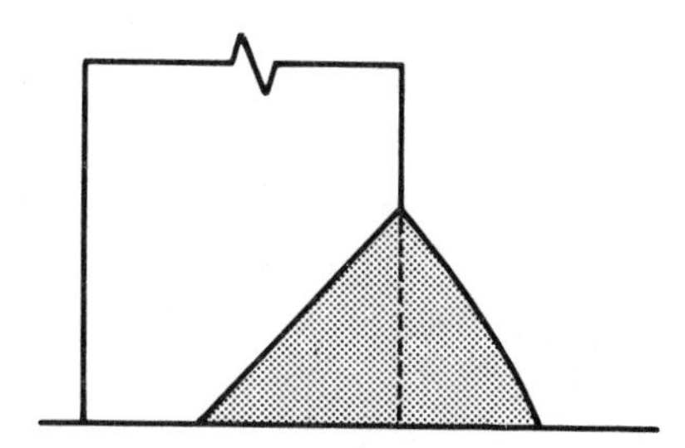

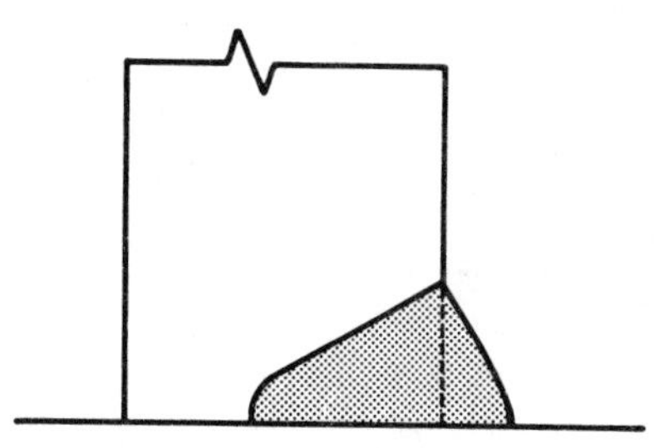

Fig. 1.26—Combined groove- and fillet-welded joints

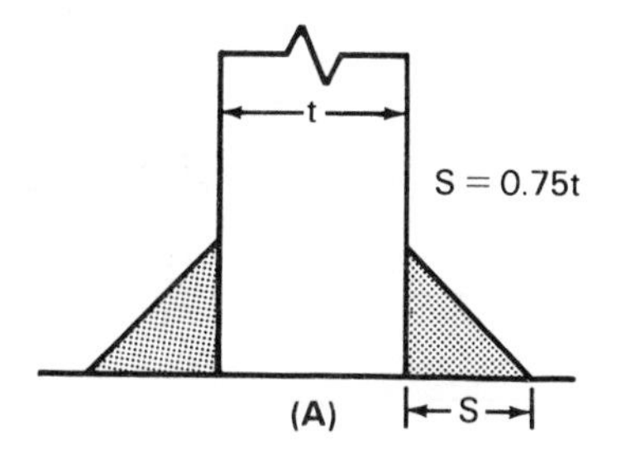

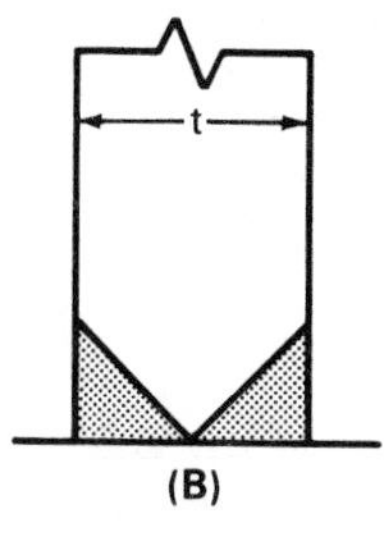

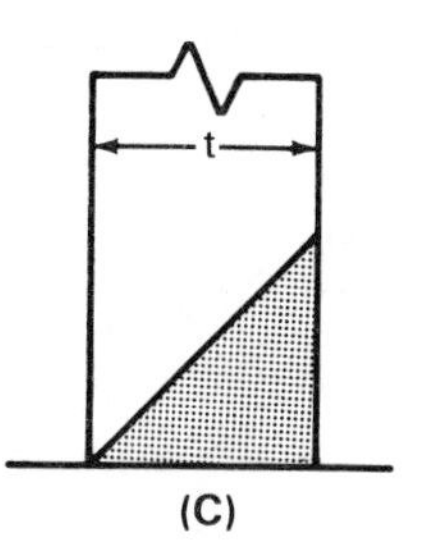

Fig. 1.27—Comparison of fillet welds and groove welds

In comparison, the double-bevel groove weld, Fig. 1.27(B), has about one half the cross-sectional area of the fillet welds. However, it requires edge preparation and the use of small diameter electrodes to make the root pass.

Referring to Fig. 1.27(C), a single-bevel-groove weld requires about the same amount of weld metal as the double fillet weld, Fig. 1.27(A). Thus, it has no apparent economic advantage. There are some disadvantages. The single-bevel weld requires edge preparation and a low-deposition root pass. From a design standpoint, it does offer direct transfer of force through the joint, which means that it is probably better than fillet welds under cyclic loading.

Double fillet welds having a leg size equal to 75 percent of the plate thickness would be sufficient for full strength. However, some codes have lower allowable stress limits than other codes for fillet welds, and may require a leg size equal to the plate thickness. The cost of a double fillet welded joint may exceed the cost of a single-bevel-groove weld in thick plates. Also, if the joint can be positioned so that the weld can be made in the flat position, a single-bevel-groove weld would be less expensive than a double fillet weld.

The construction of curves based on the best determination of the actual cost of joint preparation, positioning, and welding, such as those illustrated in Fig. 1.28, is a technique for determining the plate thicknesses where a double-bevel-groove weld becomes less costly. The intersection of the fillet weld curve with the groove weld curve is the point of interest. The validity of the information is dependent on the accuracy of the cost data used in constructing the curves.

The combined double-bevel-groove and fillet weld joint, shown in Fig. 1.29, is theoretically a full strength weld. The plate edge is beveled to 60 degrees on both sides to a depth of 30 percent of the thickness of the plate. After the groove on each side is welded, it is reinforced with a fillet weld of equal area and shape. The total effective throat of weld is equal to the plate thickness. This partial-joint-penetration weld has only about 60 percent of the weld metal in a full-strength,

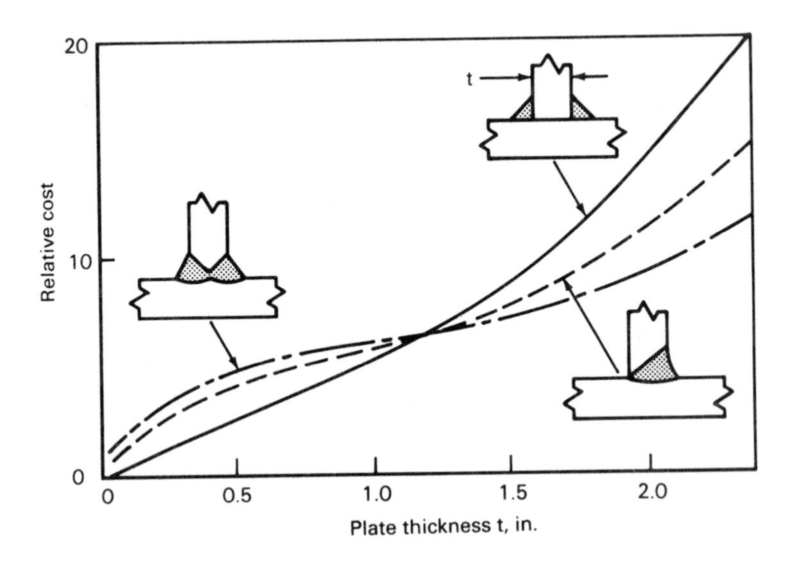

Fig. 1.28—Relative costs of full strength welds in plate

double fillet weld. It does require joint preparation, but the wide root face permits the use of large electrodes and high welding currents. It is recommended for submerged arc welding to achieve deep joint penetration.

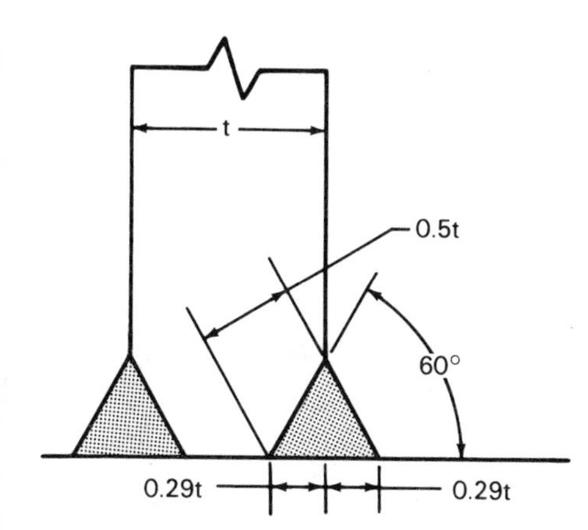

Fig. 1.29—Combined groove and fillet welds with partial joint penetration but capable of full strength

Full strength welds are not always required in a weldment, and economies can often be achieved by using smaller welds where applicable and permissible. With equal effective throats, a fillet weld, Fig. 1.30(A), requires twice the weld metal needed for a 45 degree partial-joint-penetration, single-bevel-groove weld, Fig. 1.30(B). The latter weld may not be as economical as a fillet weld, however, because of the cost of edge preparation. Also, some welding codes limit the effective throat of this type of weld to less than the depth of the bevel with certain welding processes because of incomplete root penetration.

If a single-bevel-groove weld is combined with a 45 degree fillet weld, Fig. 1.30(C), the cross-sectional area for the same effective throat is also about 50 percent of the area of the fillet weld in Fig. 1.30(A). Here, the bevel depth is smaller than it is with the single-bevel-groove weld in Fig. 1.30(B). A similar weld with a 60 degree groove angle and an unequal leg fillet, but with the same effective throat, Fig. 1.30(D), also requires less weld metal than a fillet weld alone. This joint allows the use of higher welding currents and larger electrodes to obtain deep root penetration.

The desired effective throat of combined groove and fillet welds can be obtained by adjustment of the groove dimensions and the fillet weld leg lengths. However, consideration must be given to the accessibility of the root of the joint for welding, and to stress concentrations at the toes of the fillet weld. When a partial-joint-penetration groove weld is reinforced with a fillet weld, the minimum effective throat is used for design purposes. The effective throat of the combined welds is not the sum of the effective throats of each weld. The combination is treated as a single weld when determining the effective throat.

CORNER JOINTS

Corner joints are widely used in machine design. Typical corner joint designs are illustrated in Fig. 1.31. The corner-to-corner joint, Fig. 1.31(A), is difficult to position and usually requires fixturing. Small electrodes with low welding currents must be used for the first weld pass to avoid excessive melt-thru. Also, the joint requires a large amount of weld metal.

The corner joints shown in Fig. 1.31(B) is easy to assemble, does not require backing, and needs only about half of the weld metal required to make the joint shown in Fig. 1.31(A). However, the joint has lower strength because the effective throat of the weld is smaller. Two fillet welds, one outside and the other inside, as in Fig. 1.31(C), can provide the same total effective throat as with the first design but with half the weld metal.

With thick sections, a partial-joint-penetration, single-V-groove weld, Fig. 1.31(D), is often used. It requires joint preparation. For deeper joint penetration, a J-groove, Fig. 1.31(E), or a U-groove may be used in preference to a bevel groove. A fillet weld on the inside corner, Fig. 1.31(F), makes a neat and economical corner. Such a fillet weld can be combined with a groove

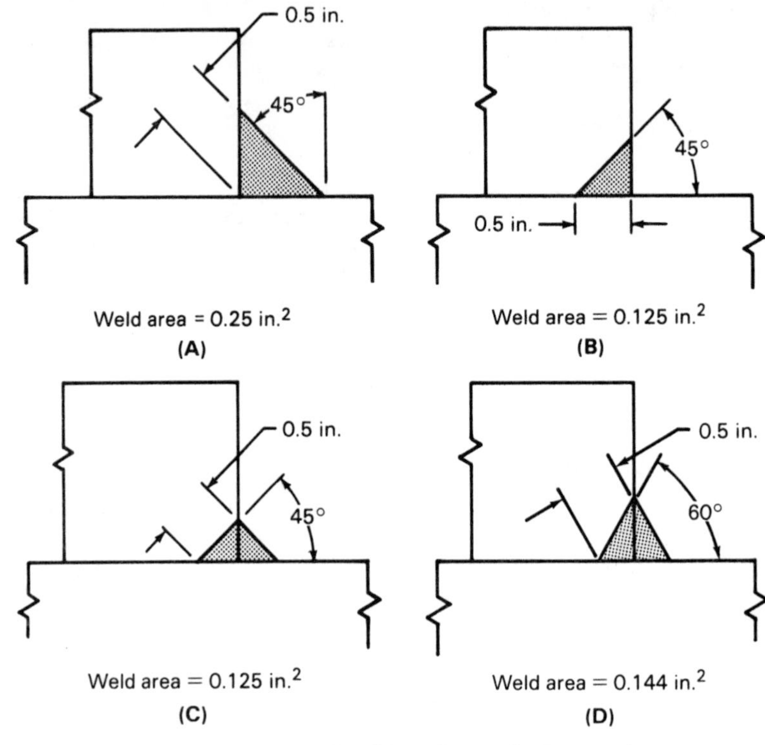

Fig. 1.30—Comparison of welds with equal effective throats

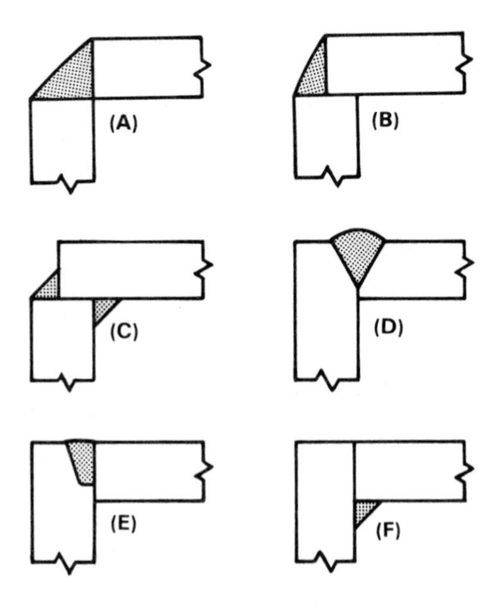

Fig. 1.31—Typical corner joint designs

weld on the outside of the joint when access to the inside is limited.

The size of the weld should always be designed with reference to the thickness of the thinner member. The joint cannot be stronger than the thinner member, and the weld metal requirements are minimized for low cost.

Lamellar tearing at the exposed edges of corner joints in thick steel plates must always be considered during the design phase. Figs. 1.31(D) and (E) are examples of susceptible designs. Through-thickness stresses in the base metal resulting from weld shrinkage are the cause of lamellar tearing. Recommendations to avoid lamellar tearing are discussed later under *Structural Tubular Connections*, page 67.

SIZING OF STEEL WELDS

Welds are sized for their ability to withstand static or cyclic loading. Allowable stresses in welds for various types of loading are normally specified by the applicable code or specification for the job. They are usually based on a percentage of the tensile or the yield strength of the metal (factor of safety) to make sure that a soundly welded joint can support the applied load for the expected service life. Allowable stresses or stress ranges are specified for various types of welds under static and cyclic loads. The allowable stress ranges for welded joints subjected to cyclic loading should be based on testing of representative full-size welded joints in actual or mockup structures.[16]

STATIC LOADING

Table 1.7 is an example of allowable static stresses for steel weld metal. Figure 1.32 illustrates the various types of loading for the welds in Table 1.7.

Complete-joint-penetration groove welds, illustrated in Figs. 1.32(A), (B), (C), and (D), are considered full-strength welds because they are capable of transferring the full strength of the connected members.

The allowable stresses in such welds are the same as those in the base metal, provided weld metal of the proper strength is used. In complete-joint-penetration welds, the mechanical properties of the weld metal must at least match those of the base metal. If two base metals of different strengths are welded together, the weld metal must at least match the strength of the weaker base metal.

Partial-joint-penetration groove welds, illustrated in Figs. 1.32(B), (C), (E), and (F), are widely used for economical welding of thick sections. Such welds can provide required joint strength and, in addition, accomplish savings in weld metal and welding time. The fast cooling rate and increased restraint imposed by thick sections justify minimum effective throat requirements for various thickness ranges. Table 1.8 gives suggested minimum effective throat sizes for partial-joint-penetration groove welds in steel.

Various factors must be considered in determining the allowable stresses on the throat of partial-joint-penetration groove welds. Joint configuration is one factor. If a V-, J-, or U-groove is specified, it is assumed that the root of the groove is accessible for welding, and that the effective throat equals the depth of the groove. If a bevel groove with a groove angle of 45 degrees or less is specified for shielded metal arc, gas metal arc, or flux cored arc welding, the effective throat is considered to be 1/8 in. less than the depth of the groove because of the difficulty in obtaining root penetration with these processes. This does not apply to the submerged arc welding process because of its deeper penetration characteristics.

---

16. Allowable unit stresses in welds for building construction from AWS D1.1-84, *Structural Welding Code — Steel*, American Welding Society, 1984.

Table 1.7
Allowable stress and strength levels for steel welds in building construction

Type of weld	Stress in weld[a]		Allowable stress	Required weld strength level
Complete joint penetration groove welds	Tension normal to the effective area.		Same as base metal.	Matching weld metal must be used.
	Compression normal to the effective area.		Same as base metal.	Weld metal with a strength level equal to or one classification (10 ksi) less than matching weld metal may be used.
	Tension or compression parallel to the axis of the weld.		Same as base metal.	Weld metal with a strength level equal to or less than matching weld metal may be used.
	Shear on the effective area.		0.30 nominal tensile strength of weld metal (ksi), except shear stress on base metal shall not exceed 0.40 yield strength of base metal.	
Partial joint penetration groove welds	Compression normal to the effective area.	Joint not designed to bear compression.	0.50 nominal tensile strength of weld metal (ksi), except stress on base metal shall not exceed 0.60 yield strength of of base metal.	Weld metal with a strength level equal to or less than matching weld metal may be used.
		Joint designed to bear compression.	Same as base metal.	
	Tension or compression parallel to the axis of the weld.		Same as base metal.	
	Shear parallel to the axis of the weld.		0.30 nominal tensile strength of weld metal (ksi), except shear stress on base metal shall not exceed 0.40 yield strength of base metal.	
	Tension normal to the effective area.		0.30 nominal tensile strength of weld metal (ksi), except tensile stress on base metal shall not exceed 0.60 yield strength of base metal	

Table 1.7 (Continued)

Type of weld	Stress in weld[a]	Allowable stress	Required weld strength level
Fillet welds	Shear on the effective area.	0.30 nominal tensile strength of weld metal (ksi), except shear stress on base metal shall not exceed 0.40 yield strength of base metal.	Weld metal with a strength level equal to or less than matching weld metal may be used.
	Tension or compression parallel to the axis of weld.	Same as base metal.	
Plug and slot welds	Shear parallel to the faying surfaces (on effective area).	0.30 nominal tensile strength of weld metal (ksi), except shear stress on base metal shall not exceed 0.40 yield strength of base metal.	Weld metal with a strength level equal to or less than matching weld metal may be used.

a. The effective weld area is the effective weld length multiplied by the effective throat.

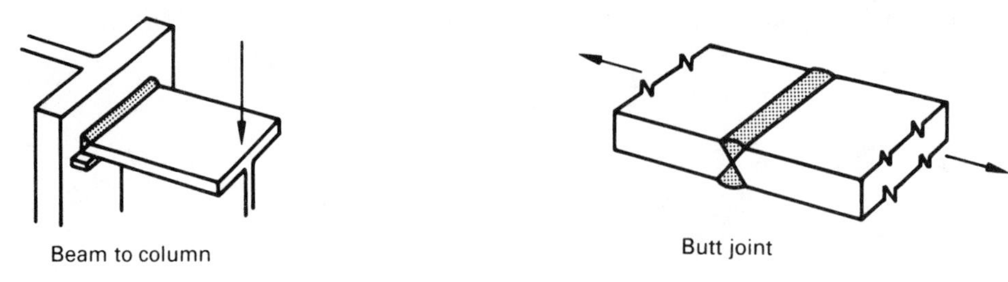

(A) Complete-joint-penetration groove weld in tension

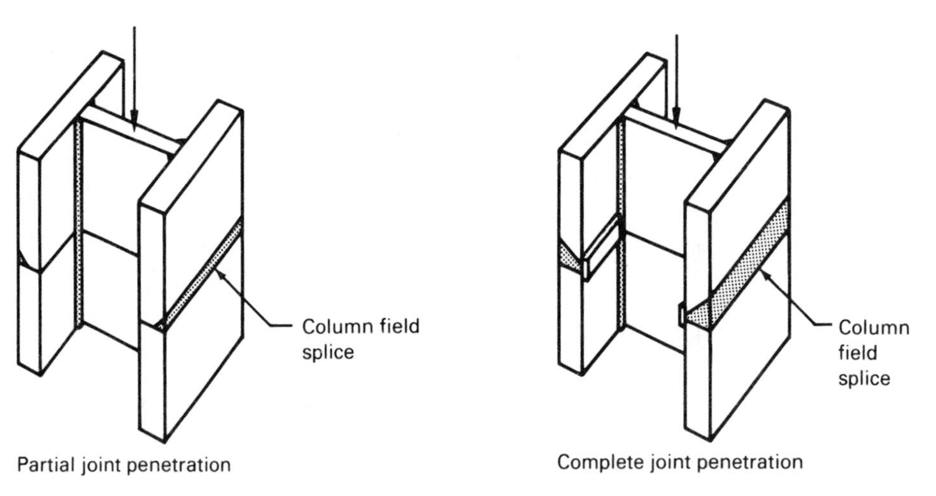

(B) Compression normal to axis of weld

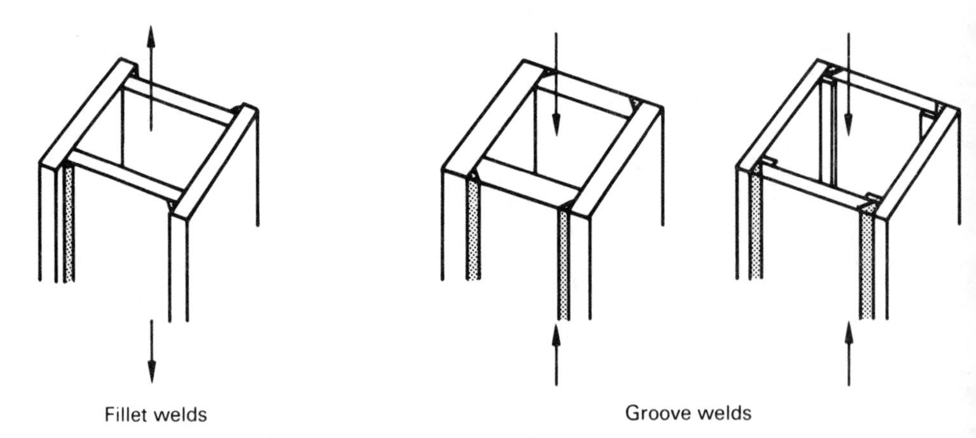

(C) Tension or compression parallel to weld axis

Fig. 1.32—Examples of welds with various types of loading

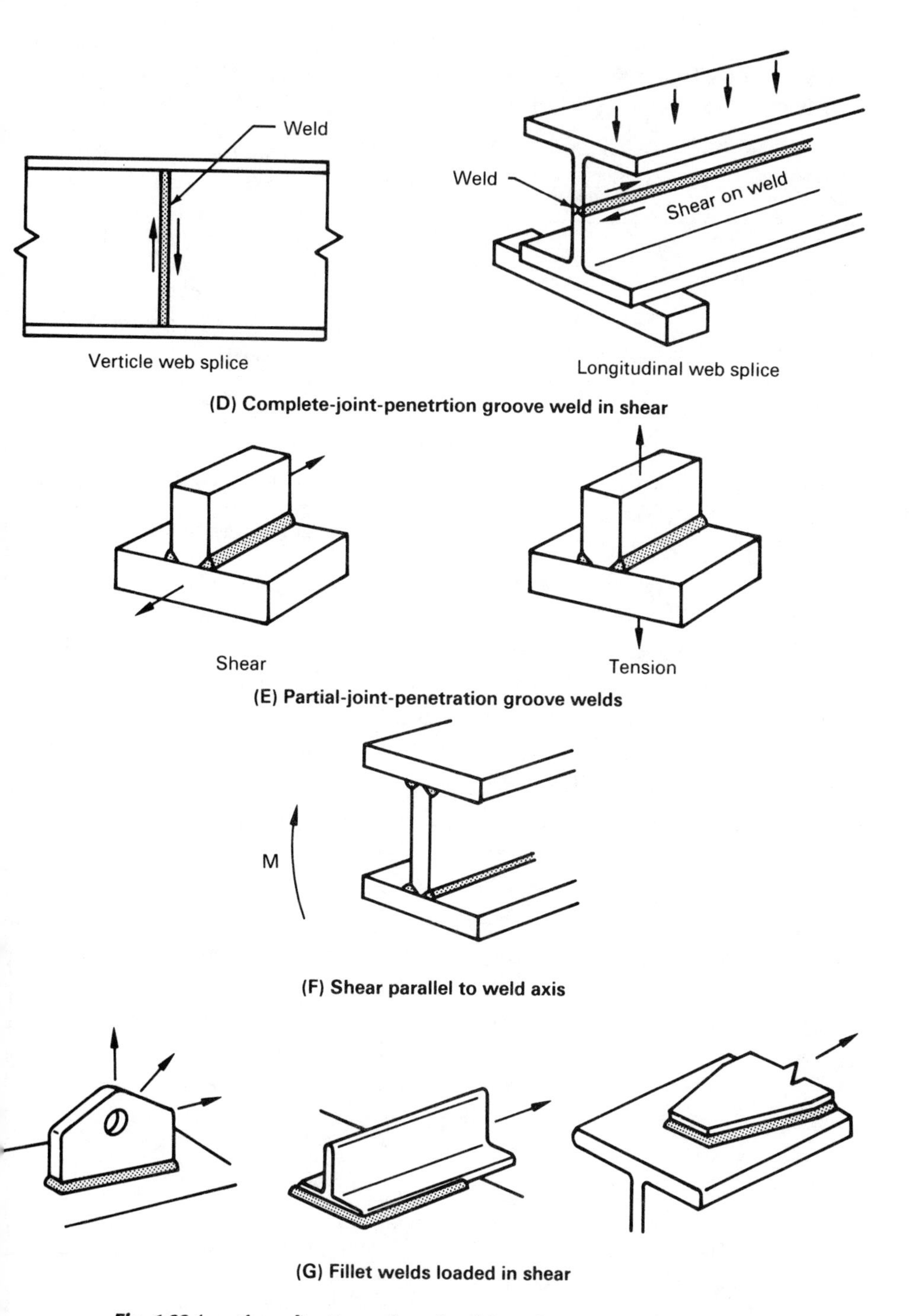

Fig. 1.32 (continued)—Examples of welds with various types of loading

Table 1.8
Minimum effective throat for partial-joint-penetration groove welds in steel

Thickness of base metal,[a] in.	Minimum effective throat[b], in.
1/8 to 3/16	1/16
Over 3/16 to 1/4	1/8
Over 1/4 to 1/2	3/16
Over 1/2 to 3/4	1/4
Over 3/4 to 1-1/2	5/16
Over 1-1/2 to 2-1/4	3/8
Over 2-1/4 to 6	1/2
Over 6	5/8

a. Thickness of thicker section with unequal thicknesses.

b. The effective throat need not exceed the thickness of the thinner part.

The allowable shear stress in steel weld metal in groove and fillet welds is about 30 percent of the nominal tensile strength of the weld metal.[17] Table 1.9 gives the allowable unit loads on various sizes of steel fillet welds of several strength levels. These values are for equal-leg fillet welds where the effective throat is 70.7 percent of the weld size. For example, the allowable unit force on a 1/2-in. fillet weld made with an electrode that deposits weld metal of 70,000 psi minimum tensile strength is determined as follows:

Allowable shear stress, $\gamma = 0.30\ (70{,}000) = 21{,}000$ psi

Unit force, $f = 0.707\ (1/2)\ (21{,}000) = 7420$ lb/in. of weld

17. The validity of the ratio was established by a series of fillet weld tests conducted jointly by the American Institute of Steel Construction and the American Welding Society.

Table 1.9
Allowable unit loads on steel fillet welds

	Strength level of weld metal, ksi						
	60[a]	70[a]	80	90[a]	100	110[a]	120
Weld size, in.	Allowable unit load, 10^3 lb/in.						
1/16	0.795	0.930	1.06	1.19	1.33	1.46	1.59
1/8	1.59	1.86	2.12	2.39	2.65	2.92	3.18
3/16	2.39	2.78	3.18	3.58	3.98	4.38	4.77
1/4	3.18	3.71	4.24	4.77	5.30	5.83	6.36
5/16	3.98	4.64	5.30	5.97	6.63	7.29	7.95
3/8	4.77	5.57	6.36	7.16	7.95	8.75	9.54
7/16	5.57	6.50	7.42	8.35	9.28	10.21	11.14
1/2	6.37	7.42	8.48	9.54	10.61	11.67	12.73
5/8	7.96	9.28	10.61	11.93	13.27	14.58	15.91
3/4	9.55	11.14	12.73	14.32	15.92	17.50	19.09
7/8	11.14	12.99	14.85	16.70	18.57	20.41	22.27
1	12.73	14.85	16.97	19.09	21.21	23.33	25.45

a. Fillet welds actually tested by the joint AISC-AWS Task Committee

The minimum allowable sizes of the first passes for steel fillet welds are given in Table 1.10. Where sections of different thickness are being joined, the minimum fillet weld size is governed by the thicker section. However, the weld size does not need to exceed the thickness of the thinner section unless a larger size is required by the load conditions.

Table 1.10
Recommended minimum fillet weld sizes for steel

Section thickness, in.[a]	Minimum fillet weld size, in.[b]
Over 1/8 to 1/4	1/8
Over 1/4 to 1/2	3/16
Over 1/2 to 3/4	1/4
Over 3/4	5/16

a. Thickness of thicker section when sections of unequal thicknesses are joined.

b. Single-pass welds or first pass of multiple pass welds. The size of the weld need not exceed the thickness of the thinner section being joined, provided sufficient preheat is used to ensure weld soundness.

The minimum fillet weld size is intended to ensure sufficient heat input to reduce the possibility of cracking in either the heat-affected zone or weld metal, especially in a restrained joint. The minimum size applies if it is greater than the size required to carry design stresses.

CYCLIC LOADING

When metals are subjected to cyclic tensile or alternating tensile-compressive stress, they may fail by fatigue. The performance of a weld under cyclic stress is an important consideration in structures and machinery. The specifications relating fatigue in steel structures have been developed by the American Institute of Steel Construction (AISC), the American Association of State Highway and Transportation Officials (AASHTO), and the American Railway Engineering Association (AREA). The applicable specifications of these organizations should be referred to for specific information.

Although sound weld metal may have about the same fatigue strength as the base metal, any change in cross section at a weld lowers the fatigue strength of the member. In the case of a complete-penetration groove weld, any reinforcement, undercut, incomplete joint penetration, or cracking acts as a notch or stress raiser. Each of these conditions is detrimental to fatigue life. The very nature of a fillet weld provides an abrupt change in section that limits fatigue life.

The AISC fatigue specifications cover a wide range of welded joints, and also address the fatigue strength of members attached by welds. Figure 1.33 shows various types of joints and the fatigue category of each type. Table 1.11 tabulates the allowable stress ranges for the six stress categories in Fig. 1.33 for four fatigue life ranges. A *stress range* is the magnitude of the change in stress that occurs with the application or removal of the cyclic load that causes tensile stress or a reversal of stress. Loads that cause only changes in magnitude of compressive stress do not cause fatigue.

The data in Table 1.11 applies to steels having yield strengths from 36 ksi (ASTM A36) to 100 ksi (ASTM A514), and to steel weld metal with tensile strengths of 60 to 120 ksi.

In the joint illustration, (Fig. 1.33), *M* and *W* indicate whether the allowable stress range applies to the base metal or the weld metal, or both.

In every case, the allowable stress must not exceed the allowable static stress for the base metal and weld metal. The fatigue formulas are used to reduce the allowable stress because of the fluctuating tensile stress conditions. When the value of *K* is high, the maximum calculated stress may exceed the allowable static stress; if it does, the design is controlled by the allowable static stress.

In Fig. 1.33(B), the allowable fatigue stress (Category B) in the base metal of a welded member in bending is really determined by the fillet welds parallel to the direction of the applied stress. The reason for this is that discontinuities or ripples in the weld face would be the initiation site for fatigue cracks. If stiffeners are used to

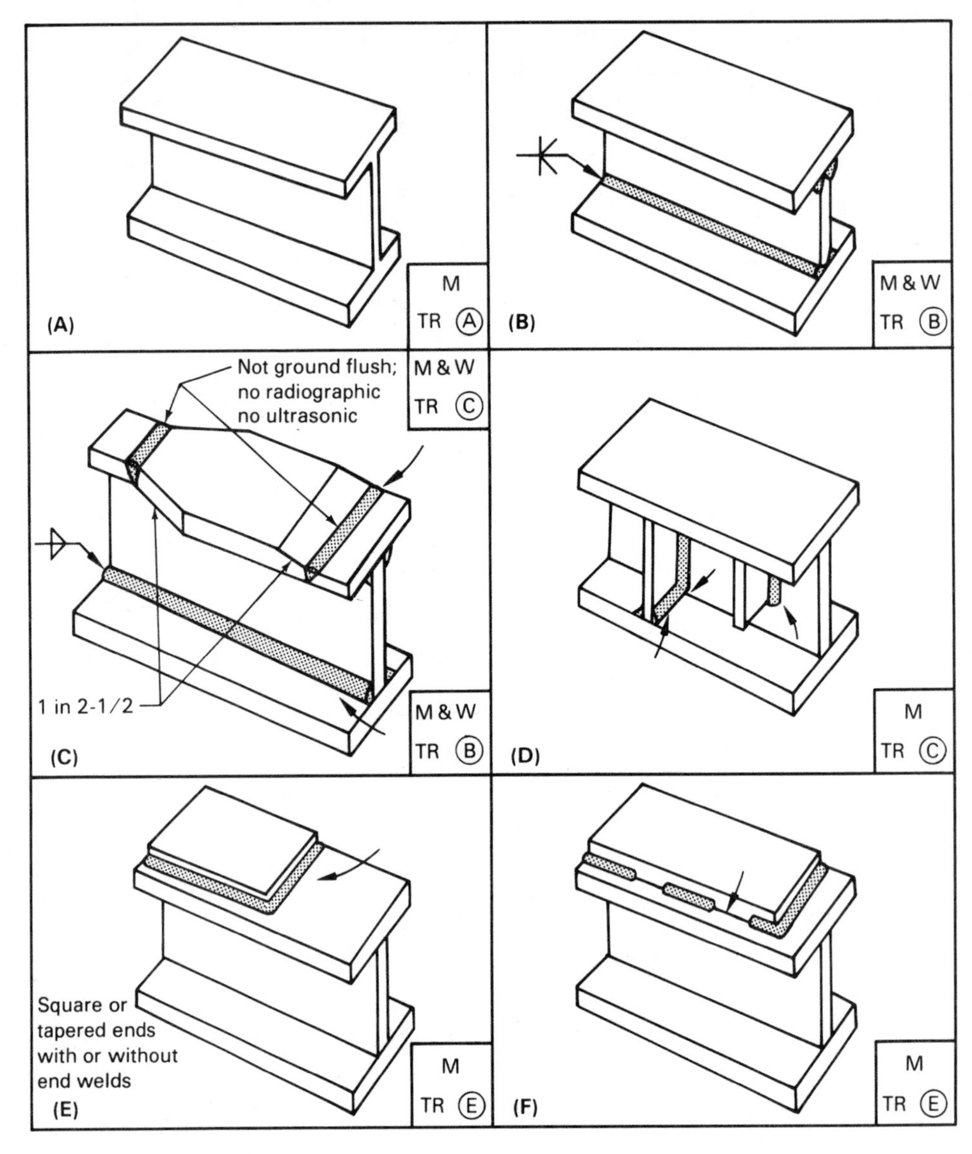

Notes:

1. Letter at lower left of each sketch identifies detail (A, ... HH).
2. Box at right indicates the following:

   M = stress in metal
   W = stress in weld
   S = shear
   T = tension
   R = reversal

3. Letter in the circle following TR or S is the stress category (A through F, see Table 1.11).
4. Curved arrows indicate region of application of fatigue allowables.
5. Straight arrows indicate applied forces.
6. Grind in the direction of stressing only.
7. When slope is mentioned, i.e., 1 in 2-1/2, it is always the maximum value. Less slope is permissible.

Fig. 1.33—Examples of various fatigue categories

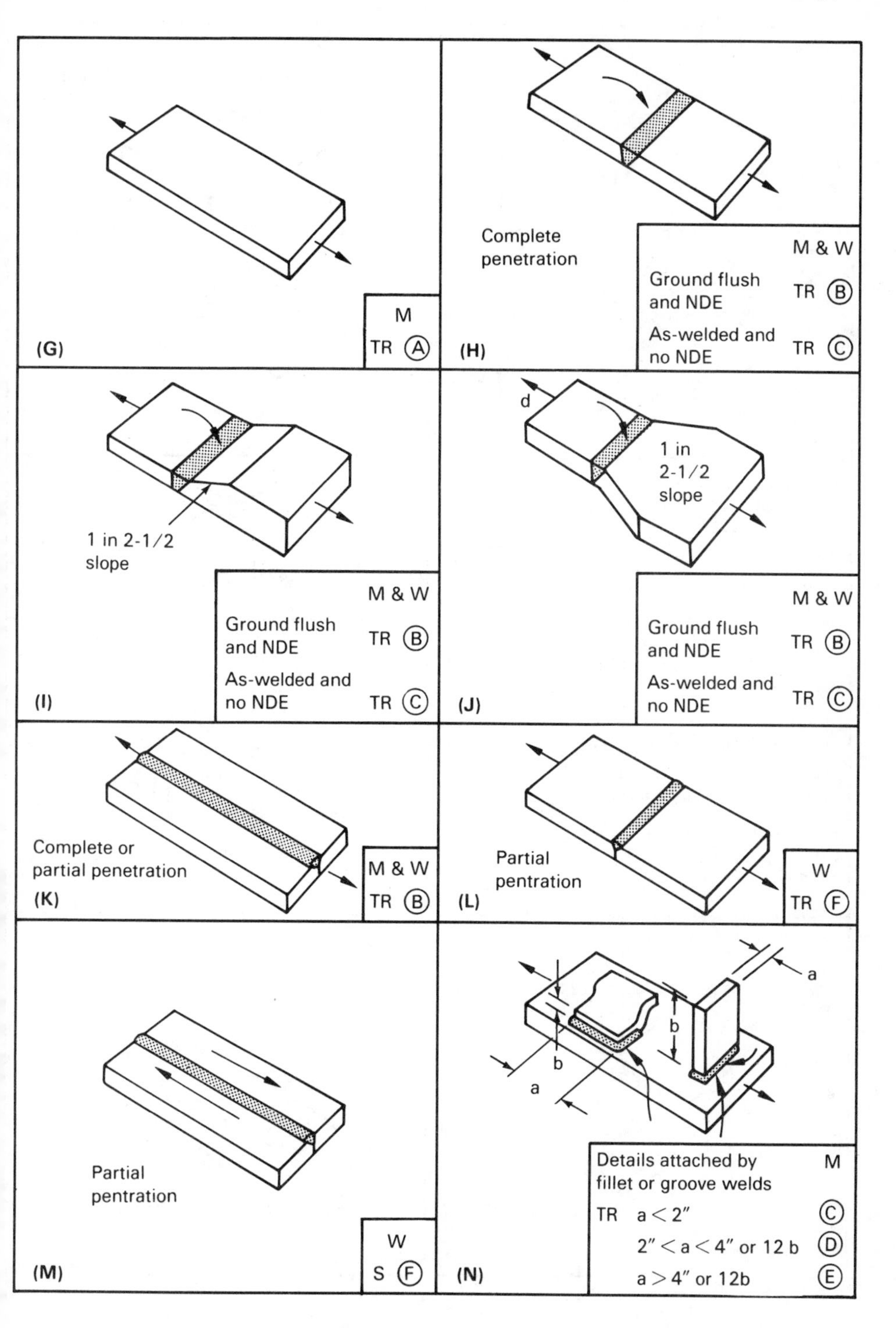

Fig. 1.33 (cont.)—Examples of various fatigue categories

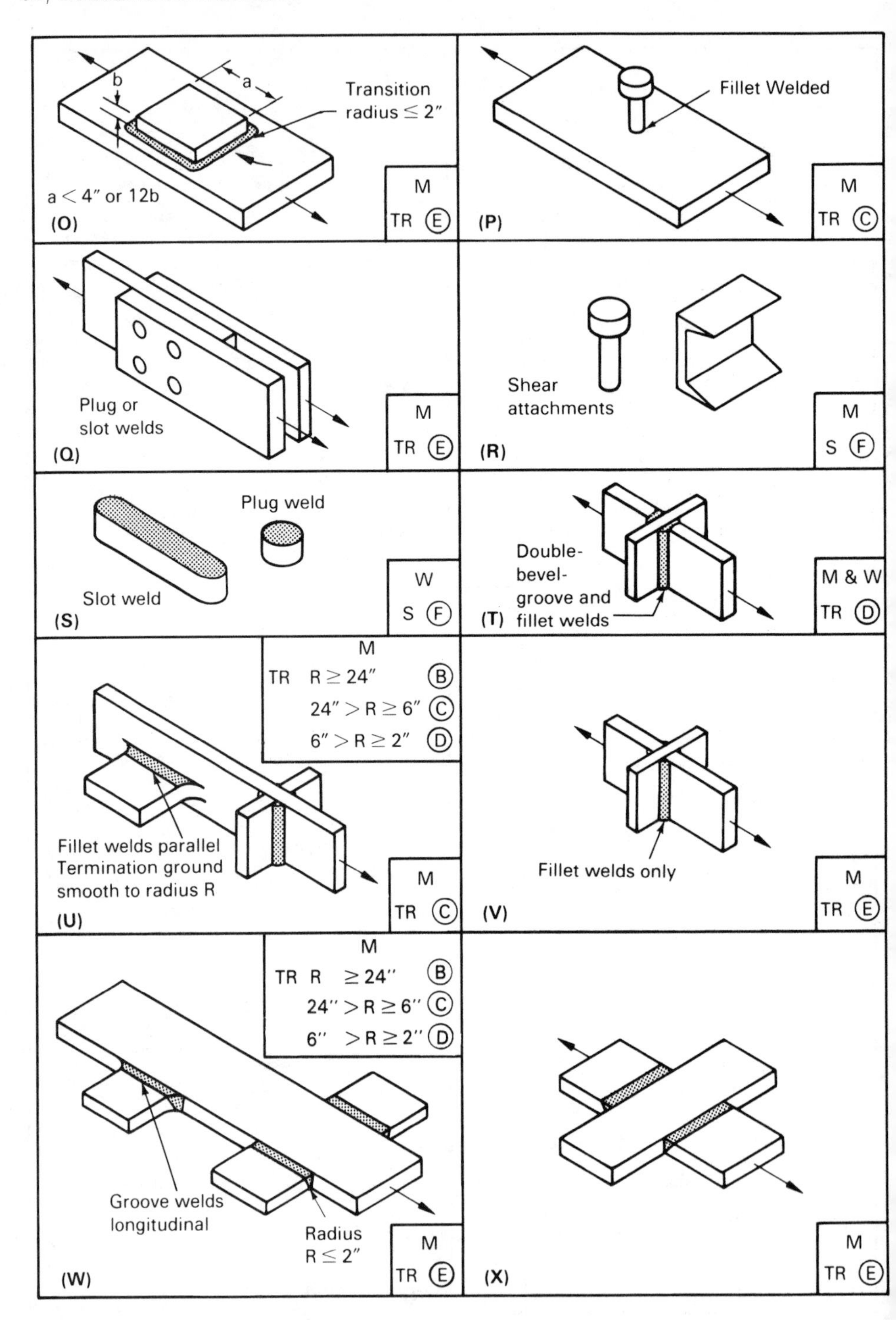

Fig. 1.33 (cont.)—Examples of various fatigue categories

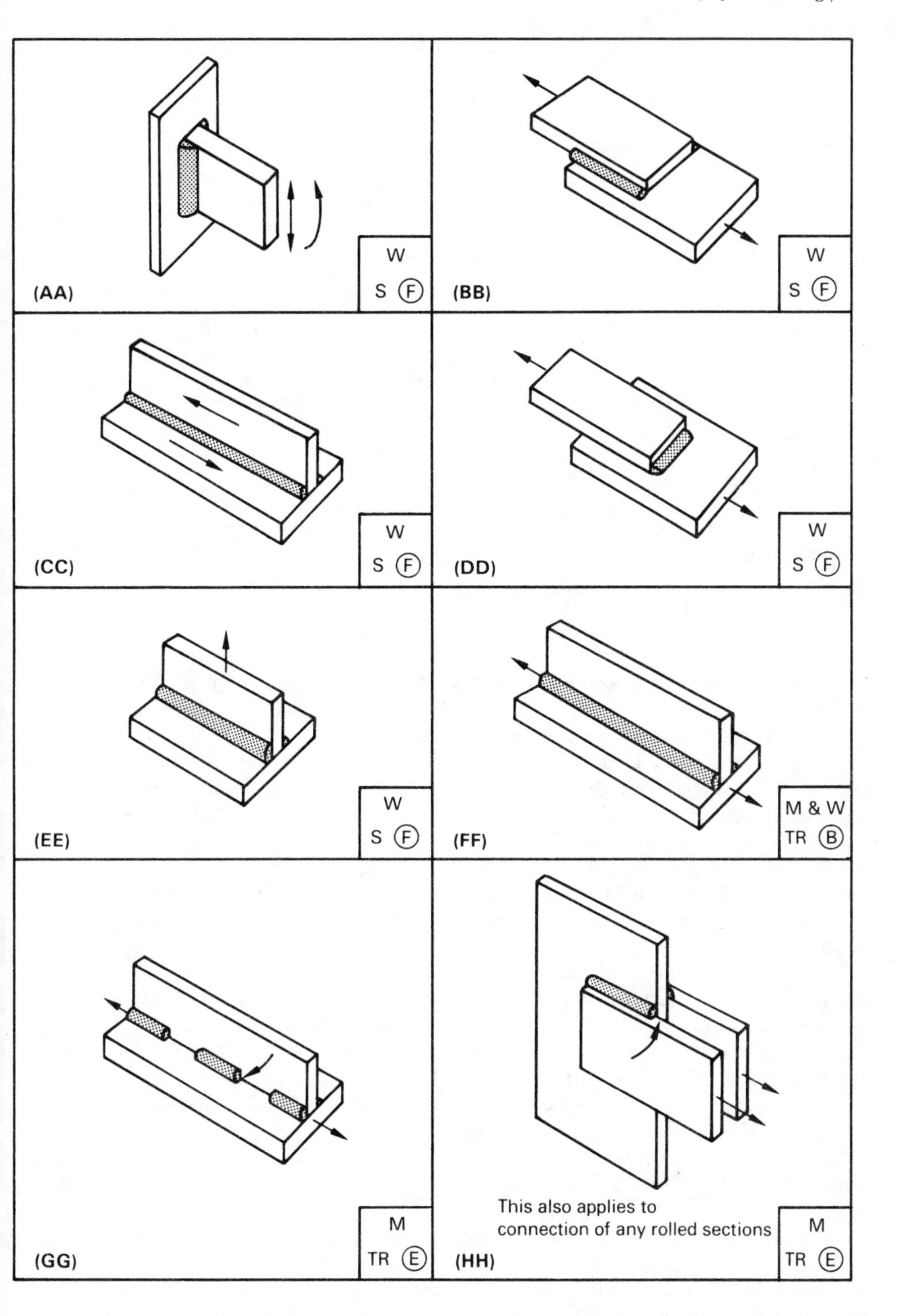

Fig. 1.33 (cont.)—Examples of various fatigue categories

Table 1.11
Suggested allowable fatigue stress range in steel weldments

Stress category[a]	Fatigue life range: 20,000 to 100,000	100,000 to 500,000	500,000 to 2,000,000	Over 2,000,000	Type of stress
	Stress range, ksi				
A	60	36	24	24	Normal, σ_{sr}
B	45	27.5	18	16	Normal, σ_{sr}
C	32	19	13	10 12[b]	Normal, σ_{sr}
D	27	16	10	7	Normal, σ_{sr}
E	21	12.5	8	5	Normal, σ_{sr}
F	15	12	9	8	Shear, τ_{sr}

a. Refer to Fig. 1.33, Note 6.

b. At toe of welds on girder webs or flanges.

reinforce a structural shape, as in Fig. 1.33(D), the allowable fatigue stress in the connected material is the calculated stress at the termination of the weld or adjacent to it. This stress is assigned to Category C because of the stress raisers at the weld terminations.

If intermittent fillet welds are used parallel to the direction of stress, the allowable fatigue stress range in the plate adjacent to the termination of the weld is Category E, as shown in Figs. 1.33(F) and (GG).

The allowable fatigue conditions for partial-joint-penetration groove welds may be determined by reference to Figs. 1.33(K), (L), and (M).

According to Figs. 1.33(N) and (O), the allowable fatigue stress range for a member with a transverse attachment increases as the length of the attachment decreases, measured parallel to the axis of the load. Although there may be a similar geometrical notch effect or abrupt change in section in both, it is the stress raiser that is important. The dimension of the transverse bar in Fig. 1.33(N) is parallel to the axis of the member, and the load is so short that very little load is transferred through the fillet welds into the bar. Consequently, the fillet welds are not severe stress raisers. A longer bar attachment shown in Fig. 1.33(O), however, is sufficiently long to carry part of the load through the connecting fillet welds. The fillet welds are higher stress raisers and, as a result, the allowable fatigue strength of the member is lower.

The joint in Fig. 1.33(EE) should not be confused with that shown in Fig. 1.33(FF). Both depict transverse fillet welds, but the former provides an allowable fatigue stress in shear for the throat of the fillet weld. The latter provides an allowable fatigue stress for the base metal adjacent to the fillet weld.

The static strength of a fillet weld loaded transversely is about one third stronger than a fillet weld loaded axially. However, the allowable fatigue stress range for parallel and transversely loaded fillet welds is the same, namely Category F. [See Figs. 1.33(BB) through (EE)]. The actual fatigue strength of a transverse fillet weld, Fig. 1.33(DD), is only slightly higher than that of a parallel fillet weld, Fig. 1.33(BB). However, both have been placed in Category F because natural scatter in all fatigue testing is probably larger than the difference between the two cases. Furthermore, unwarranted complexities in design would result from the cases of obliquely loaded welds. Although a transverse fillet-welded T-joint, Fig. 1.33(EE), might have a slightly lower fatigue strength than a transverse fillet-welded lap joint, Fig. 1.33(DD), both are Category F because of the stress concentration at the root of the welds.

Figure 1.34 is a modified Goodman design stress diagram[18] for a complete-joint-penetration, single-V-groove weld, as-welded without nondestructive examination, Fig. 1.33(H). The fatigue stress category is *C*, and the diagram for fatigue range is 500,000 to 2,000,000 cycles (Table 1.11). The vertical axis represents the maximum stress, σ_{max}, and the horizontal axis represents the minimum stress, σ_{min}, either positive or negative. A static load is represented by the 45 degree line to the right (K = + 1.0) and a complete reversal by the 45 degree line to the left (K = – 1.0). The region to the right of the vertical line (K = 0) represents tensile minimum stress. The region to the left of this line represents compressive minimum stress. The allowable cyclic stress levels for weld metals of four strength levels (60, 70, 100, and 110 psi) are shown in Fig. 1.34.

18. For additional information, refer to Munse, W.H., *Fatigue of Welded Steel Structures*, New York: Welding Research Council, 1964.

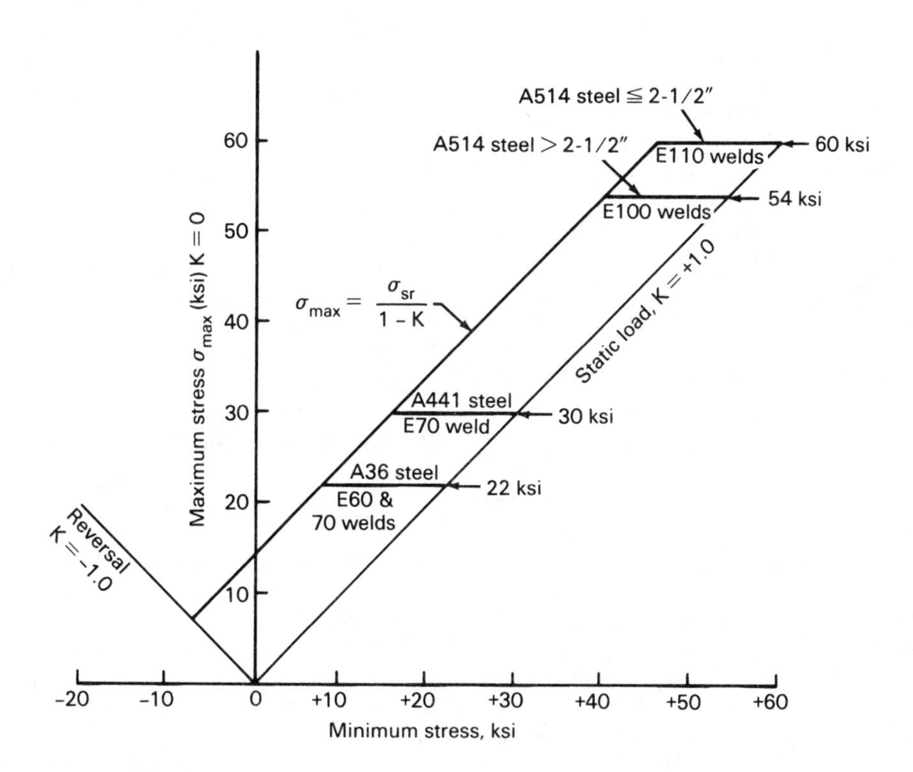

Fig. 1.34—Modified Goodman design stress diagram for complete-joint-penetration, single-V-groove welds, as welded, 500,000 to 2,000,000 cycles

The horizontal axis of Figs. 1.35(A) and (B), which are modifications of Fig. 1.34, represents the stress ratio, K, of the stresses. Here, two additional strength levels of weld metal have been added, E80 and E90, along with equivalent strength levels of steel. Note that in Fig. 1.35(A) for a stress ratio, K, greater than +0.35, higher strength steels and weld metals have higher allowable maximum fatigue stresses. Below a stress ratio, K, of about +0.35, steel welds and base metals have the same allowable maximum fatigue stress. However, throughout the entire stress ratio range, the allowable stress range is constant.

Figure 1.35(B) represents the same type of welded joint but with a lower life of 20,000 to 100,000 cycles. Here, the higher strength steel welds and base metals have higher allowable fatigue stresses over a wider range of fatigue stress ratios.

The fatigue formulas for maximum normal or shear stress generally reduce the allowable stress when fatigue conditions are encountered. Because the resulting allowable maximum fatigue stress should not exceed the usual static allowable stress, the allowable maximum fatigue stresses are terminated with horizontal lines representing the permissible static stress for the particular steel and weld metal use.

These graphs illustrate that when the stress ratio is lower, the usefulness of a high strength steel is reduced. When there is a reversal of stress there is generally no advantage in using a high strength steel for fatigue applications.

RIGID STRUCTURES

In machine design work, the primary design requirement for some members is rigidity. Such members are often heavy sections so that the movement under load will be within close tolerances. The resulting stresses in the members are very low. Often, the allowable stress in tension for mild steel is 20,000 psi, but a welded machine base or frame may have a working stress of only 2000 to 4000 psi. In this case, the weld sizes need to be designed for rigidity rather than load conditions.

A very practical method is to design the weld size to carry one-third to one-half of the load capacity of the thinner member being joined. This means that if the base metal is stressed to one third to one one half of the normal allowable stress, the weld would be strong enough to carry the load. Most rigid designs are stressed below these values. However, any reduction in weld size below one third of the normal full-strength size would give a weld that is too small in appearance for general acceptance.

PRIMARY AND SECONDARY WELDS

Welds may be classified as primary and secondary types. A primary weld transfers the entire load at the point where it is located. That weld must have the same strength properties as the members. If the weld fails, the structure fails. Secondary welds are those that simply hold the parts together to form a built-up member. In most cases, the forces on these welds are low.

When a full-strength, primary weld is required, a weld metal with mechanical properties that match those of the base metal must be used. Generally it is unnecessary for the weld metal and the base metal compositions to be exactly alike but they should be similar. An exception is when the weldment is to be heat treated and the weld metal must have the same mechanical properties as the base metal in service. Although quenched-and-tempered steel plate is cooled much faster during manufacture than hot-rolled plate, the cooling rate is limited somewhat by the thickness. A weld, on the other hand, is quenched at a faster rate by the cold adjacent base metal during welding than is the base metal during manufacture. Preheating or high welding heat input, or both, decrease the cooling rate of the weld metal by decreasing the temperature gradient.

In welding high strength steels, full-strength welds should not be used unless they are required. High strength steel may require preheat and special welding procedures because of its tendency for weld cracking, especially if the joint is restrained.

Secondary welds in high strength steels can be made with the weld metal of lower strength than the base metal. Low-hydrogen weld metal with 70,000 to 90,000 psi minimum tensile strength is preferred because the likelihood of cracking is lower than with matching weld metal. In any case, the weld must be sized to provide a joint of sufficient strength.

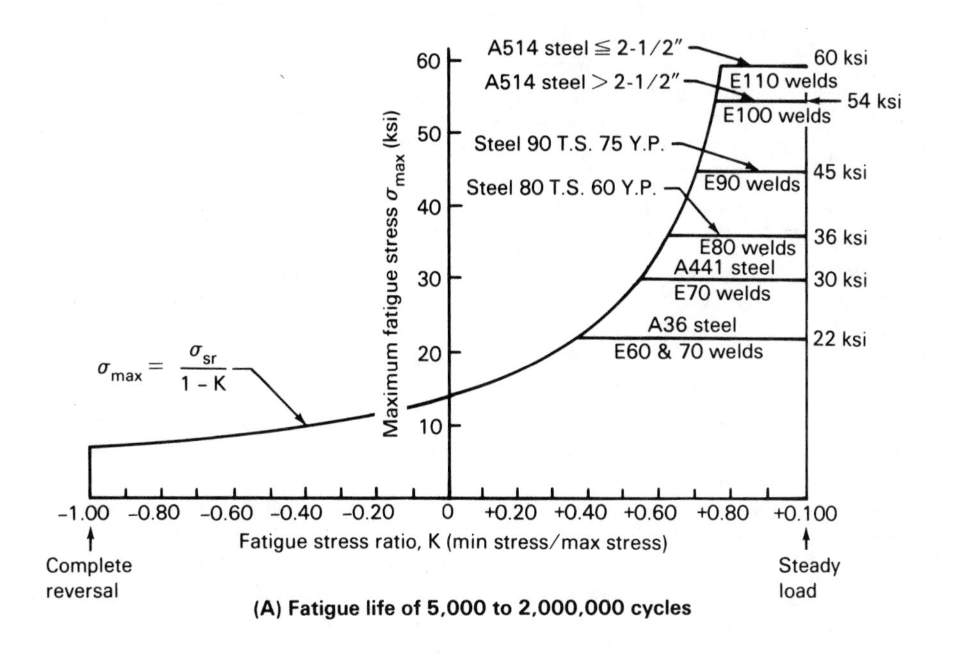

(A) Fatigue life of 5,000 to 2,000,000 cycles

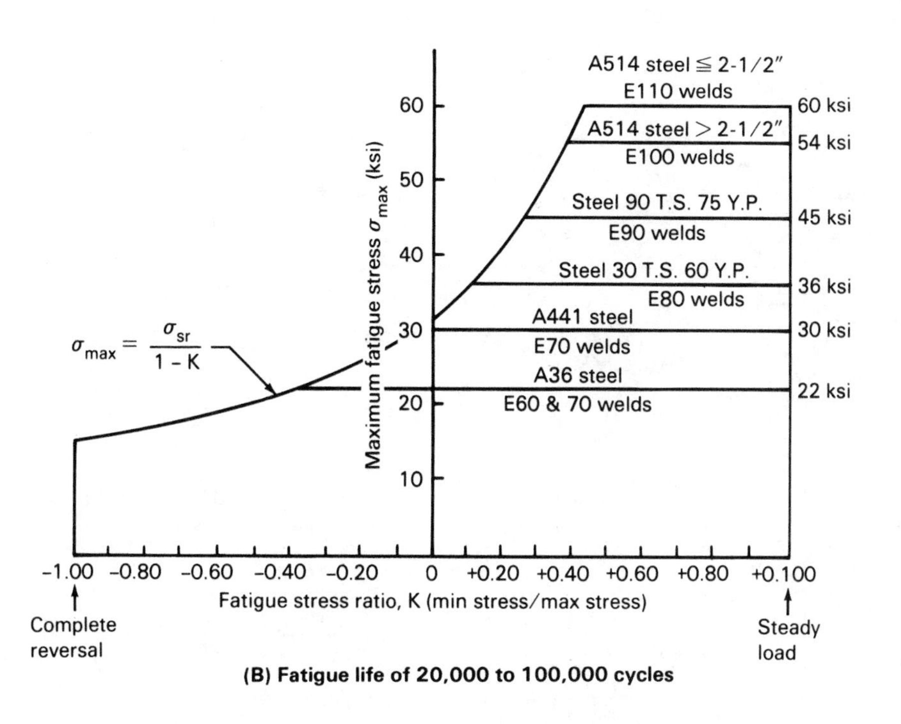

(B) Fatigue life of 20,000 to 100,000 cycles

Fig. 1.35—Maximum fatigue stress for complete-joint-penetration, single-V-groove welds, as-welded

A comparison of behaviors of full-strength and partial-strength welds in quenched-and-tempered ASTM A514 steel is shown in Fig. 1.36. The former weld is transverse to and the latter weld is parallel to the tensile load and the weldment. The plate has a tensile strength of a 110,000 psi, and it is welded with an E11018 covered electrode to provide a full strength weld, Fig. 1.36(A). When the stress is parallel to the weld axis, Fig. 1.36(B), a weld made with an E7018 covered electrode (70,000 psi minimum tensile strength) is adequate so long as there is sufficient weld to transmit any shear load from one member to the other.

In the full-strength welded joint, both the plate and the weld metal have equivalent strengths and their behavior under load is shown by the stress strain curve shown in Fig. 1.36(A). If a transversely loaded test weld were pulled in tension, it is likely that the plate would neck down and fail first. The weld would be stronger because of the reinforcement and the slightly higher strength of the weld metal and heat-affected zone as a result of rapid cooling following welding.

In the partial-strength weld loaded axially, Fig. 1.36(B), both the plate and the weld would be strained together. As the member is loaded, the strain increases from *1* to *2* on the stress-strain plot with a corresponding increase in the stress in both the plate and weld from *1* to *3*. At this point, the E7018 weld metal has reached its yield strength. On further loading, the strain is increased to *4*. The weld metal is stressed beyond its yield strength at *5*, and flows plastically. The stress in the plate, however, is still below its yield strength at *6*. With still further loading, the strain will reach *7* where the ductility of the plate will be exhausted. The plate will fail first because the weld metal has greater ductility. The weld will not fail until its unit strain reaches *8*.

It is obvious in the example that the 70,000 psi weld metal has sufficient strength to carry axial load because it carries only a small portion of the total axial load on the weldment. When a weld must transmit the total load, it has to be as strong as the base metal.

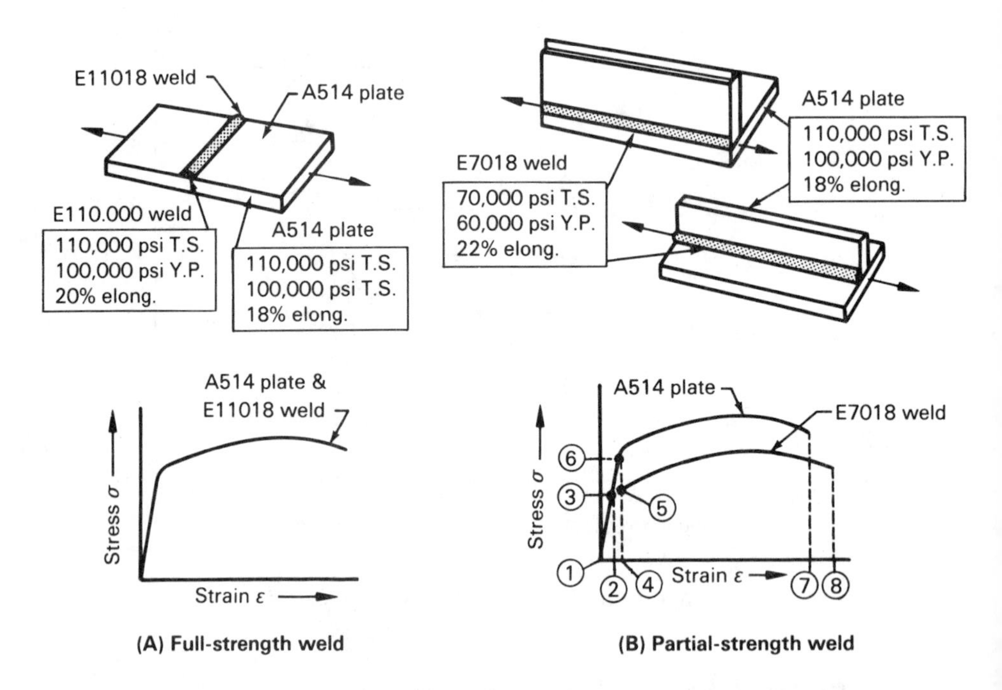

Fig. 1.36—Stress-strain characteristics of full and partial strength welds

SKEWED FILLET WELDS

A special condition exists when members come together at an angle other than 90 degrees, and fillet welds are to be used to make the connection. Ordinary specifications for the weld leg at some joint angles could result in excessive waste of weld metal, along with difficulty in depositing the weld on the acute side of the joint.

Figure 1.37 shows skewed fillet welds and the relationships between the dihedral angle, ψ, the leg size, b, and the effective throat, t, of each weld. Formulas are given to determine the proper effective throat for each weld to deposit a minimum area, A_t, of weld metal in the joint. The leg sizes, b_1 and b_2 can be determined for the respective effective throats.

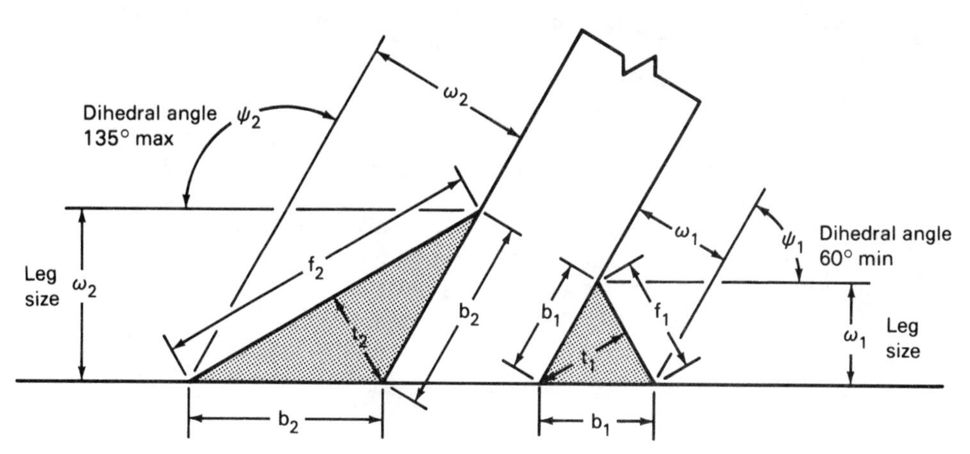

For each weld:

$$t = \frac{\omega}{\sin\left(\frac{\psi}{2}\right)} \text{ or } \omega = 2t\sin\left(\frac{\psi}{2}\right) \qquad b = \frac{t}{\cos\left(\frac{\psi}{2}\right)}$$

$$f = \frac{\omega}{\cos\left(\frac{\psi}{2}\right)} = 2t\tan\left(\frac{\psi}{2}\right) \qquad A = \frac{\omega^2}{4\sin\left(\frac{\psi}{2}\right)\cos\left(\frac{\psi}{2}\right)} = t^2\tan\left(\frac{\psi}{2}\right)$$

When $b_1 = b_2$ and $t = t_1 + t_2$:

$$t_1 = t\,\frac{\cos\left(\frac{\psi_2}{2}\right)}{\cos\left(\frac{\psi_1}{2}\right) + \cos\left(\frac{\psi_2}{2}\right)} \qquad t_2 = t\,\frac{\cos\left(\frac{\psi_1}{2}\right)}{\cos\left(\frac{\psi_1}{2}\right) + \cos\left(\frac{\psi_2}{2}\right)}$$

For minimum total weld metal:

$$t_1 = \frac{t}{1 + \tan^2\left(\frac{\psi_1}{2}\right)} \qquad t_2 = \frac{t}{1 + \tan^2\left(\frac{\psi_2}{2}\right)} \qquad A_t = \frac{t^2\tan\left(\frac{\psi_1}{2}\right)}{1 + \tan^2\left(\frac{\psi_1}{2}\right)}$$

Fig. 1.37—Formulas for analyzing fillet welds joining skewed T-joints to minimize weld metal requirements

TREATING A WELD AS A LINE

When the total length of weld in a connection is large compared to its effective throat, the weld can be assumed to be a line having a definite length and configuration rather than an area. The proper size of weld required for adequate strength can be determined using this concept. Referring to Fig. 1.38, the welded connection is considered as a single line having the same outline as the connection area. The welded connection now has length, not effective area. Instead of determining the stress on a weld, which cannot be done until the weld size is known, the problem becomes simply one of determining the force per unit length on the weld.

The property of a welded connection, when the weld is treated as a line, can be substituted in the standard design formula used for the particular type of load, as shown in Table 1.12. The force per unit length on the weld may be calculated with the appropriate modified formula.

In problems involving bending or twisting loads, the needed properties of common structural joint geometries can be calculated with the appropriate formulas in Table 1.13. Moment of inertia, I, section modulus, S, polar moment of inertia, J_w, and distance from the neutral axis or center of gravity to the extreme fibers, c, are included.

For any given connection, two dimensions are needed, width b, and depth d. Section modulus, S_w, is used for welds subjected to bending loads; polar moment of inertia, J_w, and distance, c, for twisting loads. Section moduli are given for maximum force at the top and bottom or right and left portions of the welded connections. For the unsymmetrical connections shown in Table 1.13, maximum bending force is at the bottom.

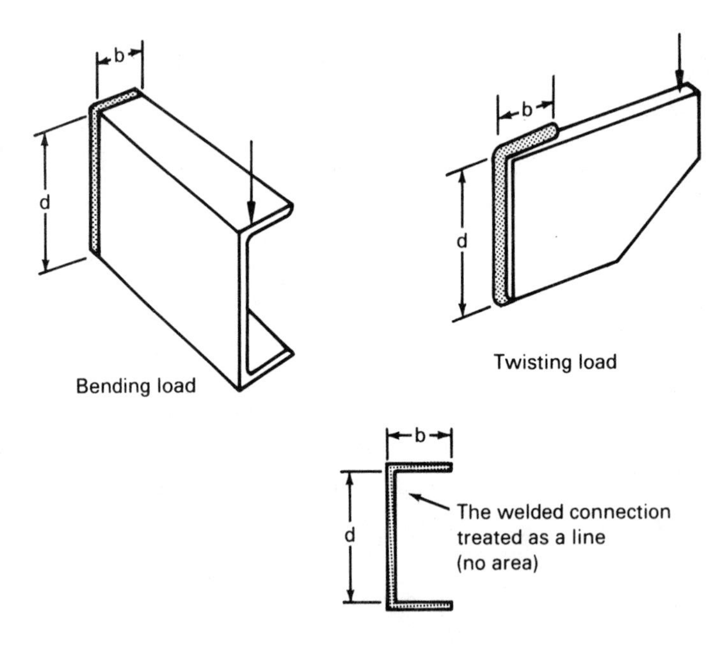

Fig. 1.38—Treating the weld as a line

Table 1.12
Formulas for calculating force per unit length on welds

Type of loading	Standard formula for unit stress	Formula for force per unit length
Tension or compression	$\sigma = \frac{P}{A}$	$f = \frac{P}{L_w}$
Vertical shear	$\tau = \frac{V}{A}$	$f = \frac{V}{L_w}$
Bending	$\sigma = \frac{M}{S}$	$f = \frac{M}{S_w}$
Torsion	$\tau = \frac{Tc}{J}$	$f = \frac{Tc}{J_w}$

σ = normal stress
τ = shear stress
f = force per unit length
P = concentrated load
V = vertical shear load
A = total area of cross section
L_w = total length of a line weld
c = distance from neutral axis to the extreme fibers of a line weld
T = torque on the weld joint
S = section modulus of an area
S_w = section modulus of a line weld
J = polar moment of inertia of an area
J_w = polar moment of inertia of a line weld
I = moment of inertia

If there is more than one force applied to the weld, they are combined vectorially. All forces that are combined must be vectored at a common location on the welded joint.

Weld size is found by dividing the resulting unit force on the weld by the allowable strength of the type of weld used.

The steps in applying this method to any welded construction are as follows:

(1) Find the position on the welded connection where the combined forces are maximum. There may be more than one combination that should be considered.

(2) Find the value of each of the forces on the welded connection at this position.

(3) Select the appropriate formula from Table 1.12 to find the unit force on the weld.

(4) Use Table 1.13 to find the appropriate properties of the welded connection treated as a line.

(5) Combine vectorially all of the unit forces acting on the weld.

(6) Determine the required weld size by dividing the total unit force by the allowable stress in the weld.

Table 1.13
Properties of welded connections treated as a line

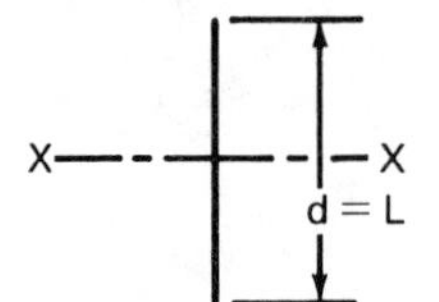

$$I_X = \frac{d^3}{12} \qquad S_X = \frac{d^2}{6}$$

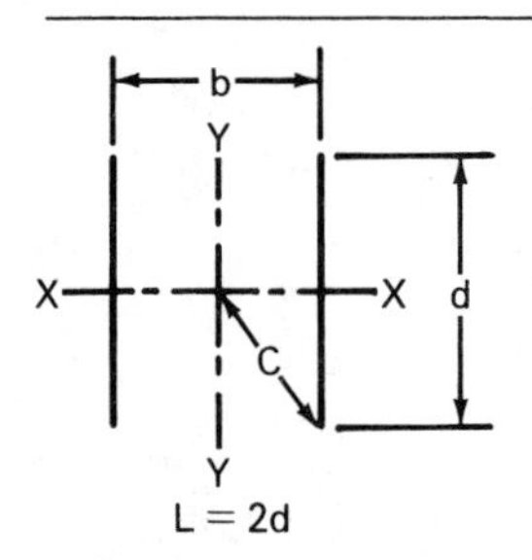

$$I_X = \frac{d^3}{6} \qquad S_X = \frac{d^3}{3} \qquad J_W = \frac{d}{6}(3b^2 + d^2)$$

$$I_Y = \frac{b^2 d}{2} \qquad S_Y = bd \qquad C = \frac{(b^2 + d^2)^{1/2}}{2}$$

L = b + d

$$I_X = \frac{d^3}{12}\left(\frac{4b + d}{b + d}\right) \qquad S_{XT} = \frac{d}{6}\left(\frac{4b + d}{b + d}\right) \qquad S_{XB} = \frac{d^2}{6}\left(\frac{4b + d}{b + d}\right)$$

$$I_Y = \frac{b^3}{12}\left(\frac{b + 4d}{b + d}\right) \qquad S_{YL} = \frac{b}{6}\left(\frac{b + 4d}{b + d}\right) \qquad S_{YR} = \frac{b^2}{6}\left(\frac{b + 4d}{b + d}\right)$$

$$J_W = \frac{b^3 + d^3}{12} + \frac{bd\,(b^2 + d^2)}{4\,(b + d)}$$

$$C_{XT} = \frac{d^2}{2\,(b + d)} \qquad C_{XB} = \frac{d}{2}\left(\frac{2\,b + d}{(b + d)}\right) \qquad C_1 = (C_{XT}{}^2 + C_{YR}{}^2)^{1/2}$$

$$C_{YL} = \frac{b^2}{2\,(b + d)} \qquad C_{YR} = \frac{b}{2}\left(\frac{b + 2d}{b + d}\right) \qquad C_2 = (C_{XB}{}^2 + C_{YL}{}^2)^{1/2}$$

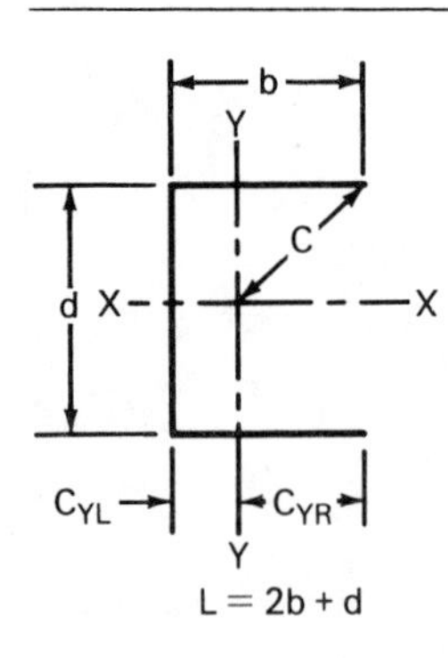

$$I_X = \frac{d^2}{12}(6b + d) \qquad S_X = \frac{d}{6}(6b + d)$$

$$I_Y = \frac{b^3}{3}\left(\frac{b + 2d}{2b + d}\right) \qquad S_{YL} = \frac{b}{3}(b + 2d)$$

$$C_{YL} = \frac{b^2}{2b + d} \qquad C_{YR} = \frac{b\,(b + d)}{2b + d} \qquad S_{YR} = \frac{b^3}{3}\left(\frac{b + 2d}{b + d}\right)$$

$$C = \left[C_{YR}{}^2 + \left(\frac{d}{2}\right)^2\right]^{1/2} \qquad J_W = \frac{b^3}{3}\left(\frac{b + 2d}{2b + d}\right) + \frac{d^2}{12}(6b + d)$$

Table 1.13 (cont.)

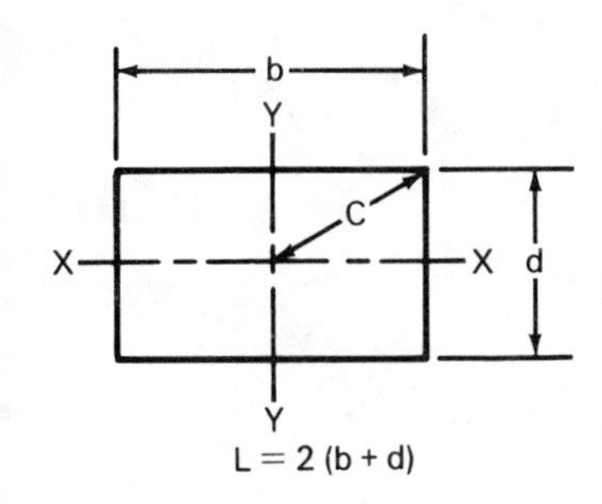

L = 2 (b + d)

$$I_X = \frac{d^2}{6}(3b + d) \qquad S_X = \frac{d}{3}(3b + d)$$

$$I_Y = \frac{b^2}{6}(b + 3d) \qquad S_Y = \frac{b}{3}(b + 3d)$$

$$J_W = \frac{(b + d)^3}{6} \qquad C = \frac{(b^2 + d^2)^{1/2}}{2}$$

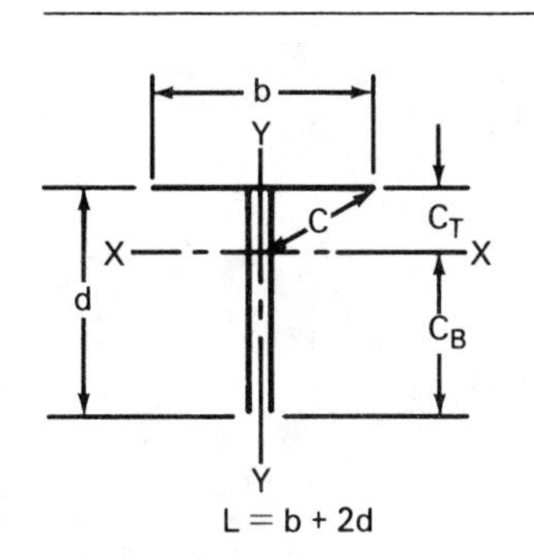

L = b + 2d

$$I_X = \frac{d^3}{3}\left(\frac{2b + d}{b + 2d}\right) \qquad S_{XT} = \frac{d}{3}(2b + d) \qquad S_{XB} = \frac{d^2}{3}\left(\frac{2b + d}{b + d}\right)$$

$$I_Y = \frac{b^3}{12} \qquad S_Y = \frac{b^2}{6} \qquad C_T = \frac{d^2}{b + 2d}$$

$$J_W = \frac{d^3}{3}\left(\frac{2b + d}{b + 2d}\right) + \frac{b^3}{12} \qquad C_b = d\left(\frac{b + d}{b + 2d}\right)$$

$$C = C_T{}^2 + \left(\frac{b}{2}\right)^2 {}^{1/2}$$

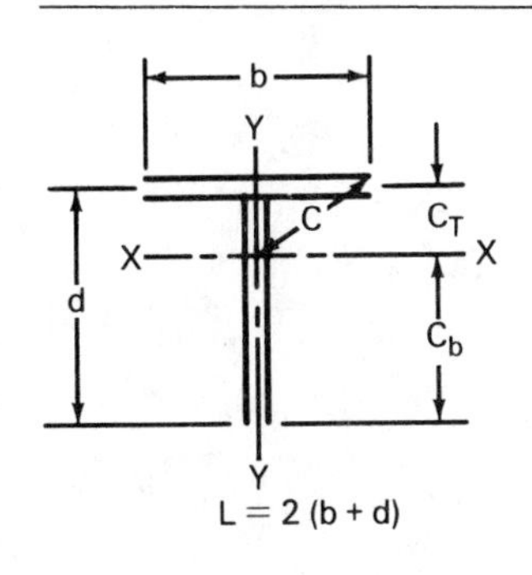

L = 2 (b + d)

$$I_X = \frac{d^3}{6}\left(\frac{4b + d}{b + d}\right) \qquad S_{XT} = \frac{d}{3}(4b + d) \qquad S_{XB} = \frac{d^2}{3}\left(\frac{4b + d}{b + d}\right)$$

$$I_Y = \frac{b^2}{6} \qquad S_Y = \frac{b}{3} \qquad C_T = \frac{d^2}{2(b + d)}$$

$$C_B = \frac{d}{2}\left(\frac{2b + d}{b + d}\right)$$

$$J_W = \frac{d^3}{6}\left(\frac{4b + d}{b + d}\right) + \frac{b^2}{6} \qquad C = \left[C_T{}^2 + \left(\frac{b}{2}\right)^2\right]^{1/2}$$

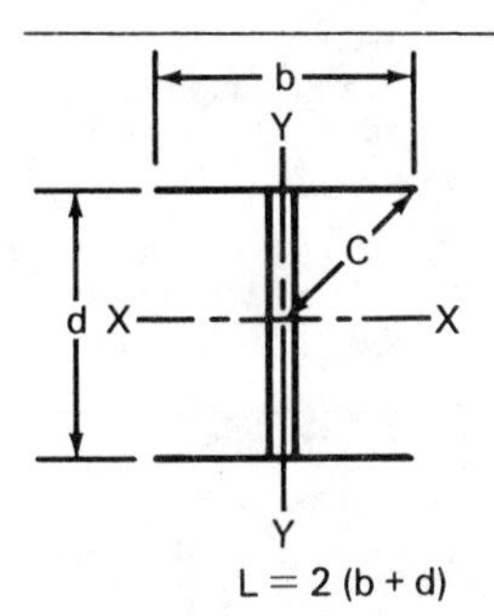

L = 2 (b + d)

$$I_X = \frac{d^2}{6}(3b + d) \qquad S_X = \frac{d}{3}(3b + d)$$

$$I_Y = \frac{b^3}{6} \qquad S_Y = \frac{b^2}{3}$$

$$J_W = \frac{d^2}{6}(3b + d) + \frac{b^3}{6} \qquad C = \frac{(b^2 + d^2)^{1/2}}{2}$$

Table 1.13 (cont.)

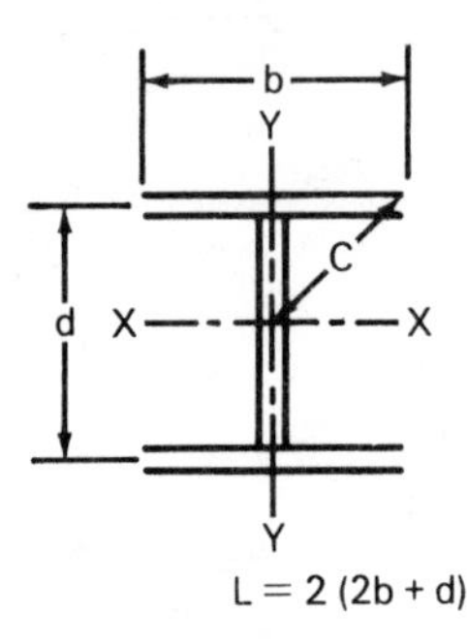

$$I_X = \frac{d^2}{6}(6b + d) \qquad S_X = \frac{d}{3}(6b + d)$$

$$I_Y = \frac{b^3}{3} \qquad S_Y = \frac{2}{3}b^2$$

$$J = \frac{d^2}{6}(6b + d) + \frac{b^3}{3} \qquad C = \frac{(b^2 + d^2)^{1/2}}{2}$$

$$L = 2\pi r$$

$$I = \pi r^3 \qquad S_W = \pi r^2 \qquad J_W = 2\pi r^3$$

The following example illustrates the steps in calculating the size of a weld considered as a line.

Assume that a bracket supporting an eccentric load of 18,000 lbs is to be fillet welded to the flange of a vertical column, as shown in Fig. 1.39.

Step 1. The point of maximum combined unit forces is at the right ends of the top and bottom horizontal welds.

Step 2. The twisting force caused by the eccentric loading is divided into horizontal (f_h) and vertical (f_v) components. The distance from the center of gravity to the point of combined stress, c_{yr}, is calculated from the formula in Table 1.13 for this shape of connection (the fourth configuration).

$$c_{yr} = \frac{b\,(b + d)}{2b + d} = \frac{5\,(5 + 10)}{(2)\,(5) + 10} = \frac{75}{20} = 3.75 \text{ in.}$$

The polar moment of inertia:

$$J_w = \frac{b^3}{3}\frac{(b + 2d)}{(2b + d)} + \frac{d^2}{12}(6\,b + d)$$

$$= \frac{5^3}{3}\frac{(5 + 20)}{(10 + 10)} + \frac{10^2}{12}(30 + 10)$$

$$= 385.4 \text{ in.}^3$$

Horizontal component of twisting:

Torque, $T = 18{,}000 \times 10 = 180{,}000$ in. · lb

$$f_h = \frac{(T)\,(d/2)}{J_w} = \frac{(180{,}000)\,(10/2)}{385.4} = 2{,}340 \text{ lb/in.}$$

Vertical component of twisting:

$$f_v = \frac{T\,c_{yr}}{J_w} = \frac{(180{,}000)\,(3.75)}{385.4} = 1{,}750 \text{ lb/in.}$$

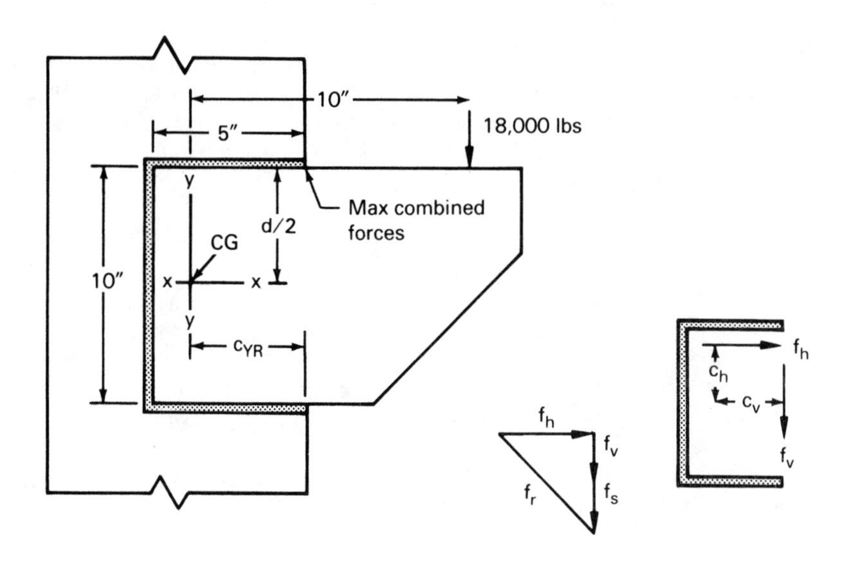

Fig. 1.39—Bracket joined to a column face with a fillet weld

Vertical shear force:

$$f_s = \frac{P}{L_w} = \frac{18{,}000}{20} = 900 \text{ lb/in.}$$

Step 3. Determine the resultant force:

$$f_v = [f_h{}^2 + (f_v + f_s)^2]^{1/2}$$

$$= [(2{,}340)^2 + (1{,}750 + 900)^2]^{1/2}$$

$$= 3{,}540 \text{ lb/in.}$$

Step 4. The allowable shearing stress on the effective area of weld metal having an ultimate tensile strength of 60,000 psi is as follows (Table 1.7):

$$\tau = 0.30\,(60{,}000)$$

$$\tau = 18{,}000 \text{ psi}$$

The effective throat:

$$(E) = \frac{f_v}{\tau} = \frac{3{,}540}{18{,}000} = 0.197 \text{ in.}$$

Assuming an equal leg fillet weld, the fillet weld size:

$$S = \frac{(E)}{0.707} = \frac{0.197}{0.707} = 0.279 \text{ in.}$$

A 5/16-in. fillet weld should be specified on the welding symbol.

Procedures for determining the allowable eccentric loads for various welded connections used in structural steel construction are given in the *Manual of Steel Construction*, published by the American Institute of Steel Construction. Appropriate data are presented in tabular form.

STRUCTURAL TUBULAR CONNECTIONS

Tubular members are being used in structures such as drill rigs, space frames, trusses, booms, and earthmoving and mining equipment.[19] They have the advantage of minimizing deflection under load because of their greater rigidity when compared to standard structural shapes. Various types of welded tubular connections, the component designations, and nomenclature are shown in Fig. 1.40.

With structural tubing, there is no need for cutting holes at intersections and, as a result, the connections can have high strength and stiffness. However, if a connection is to be made with a complete-joint-penetration groove weld, the weld usually must be made from one side only and without backing because the size or configuration, or both, will prevent access to the root side of the weld. Special skill is required for making tubular connections using complete-joint-penetration welds from one side.

With relatively small, thin-wall tubes, the end of the brace tube may be partially or fully flattened. The end of the flattened section is trimmed at the appropriate angle to abut against the main member where it is to be welded. This design should only be used with relatively low load conditions because the load is concentrated on a narrow area of the main tube member. The flattened section of the brace member must be free of cracks.

WELD JOINT DESIGN

When tubular members are fit together for welding, the end of the branch member or brace is normally contoured to the shape of the main member. The members may be joined with their axes at 80 to 110 degrees in the case of T-connections [Fig. 1.40(C)], or at some angle less than 80 degrees in Y- and K-connections [Figs. 1.40(D) and (E)]. The tubes may have a circular or box shape, and the branch member may be equal or smaller in size than the main member.

19. The welding of steel tubular structures is covered by ANSI/AWS D1.1, *Structural Welding Code — Steel*, latest edition, Miami: American Welding Society.

Consequently, the angle between the adjacent outside tube surfaces or their tangents, in a plane perpendicular to the joint (local dihedral angle), can vary around the joint from about 150 to 30 degrees. To accommodate this, the weld joint design and welding procedures must be varied around the joint to obtain a strong and sound weld.

Tubular joints are normally accessible for welding only from outside the tubes. Therefore, the joints are generally made with single groove or fillet welds. Groove welds may be designed for complete or partial joint penetration, depending upon the load conditions. To obtain adequate joint penetration, shielded metal arc, gas metal arc, and flux cored arc welding are generally used to make tubular joints in structures.

Suggested groove designs for complete joint penetration with four dihedral angle ranges are shown in Fig. 1.41. The sections of circular and box connections to which the groove designs apply are shown in Figs. 1.42(A) and (B), respectively. The specified root opening, *R*, or the width of a back-up weld, *W*, in Fig. 1.41 depends upon the welding process and the groove angle.

Suggested groove designs for partial-joint-penetration groove welds for circular and box connections are shown in Fig. 1.43. The sections of circular and box connections to which they apply are shown in Fig. 1.44. The joint design for side connections in equal box sections is shown in Fig. 1.45.

An allowance, *Z*, must be made for incomplete fusion at the root of a partial-joint-pentration weld when establishing the design effective throat, (*E*). The allowance depends upon the groove angle, position of welding, and the welding process. Recommended allowance values are given in Table 1.14 for various groove angles.

Suggested fillet weld details for T-, K-, and Y-connections in circular tubes are shown in Fig. 1.46. The recommended allowable stress on the effective throat of partial-joint-penetration groove welds and fillet welds in steel T-, K-, and Y-connections is 30 percent of the specified minimum tensile strength of the weld metal. The

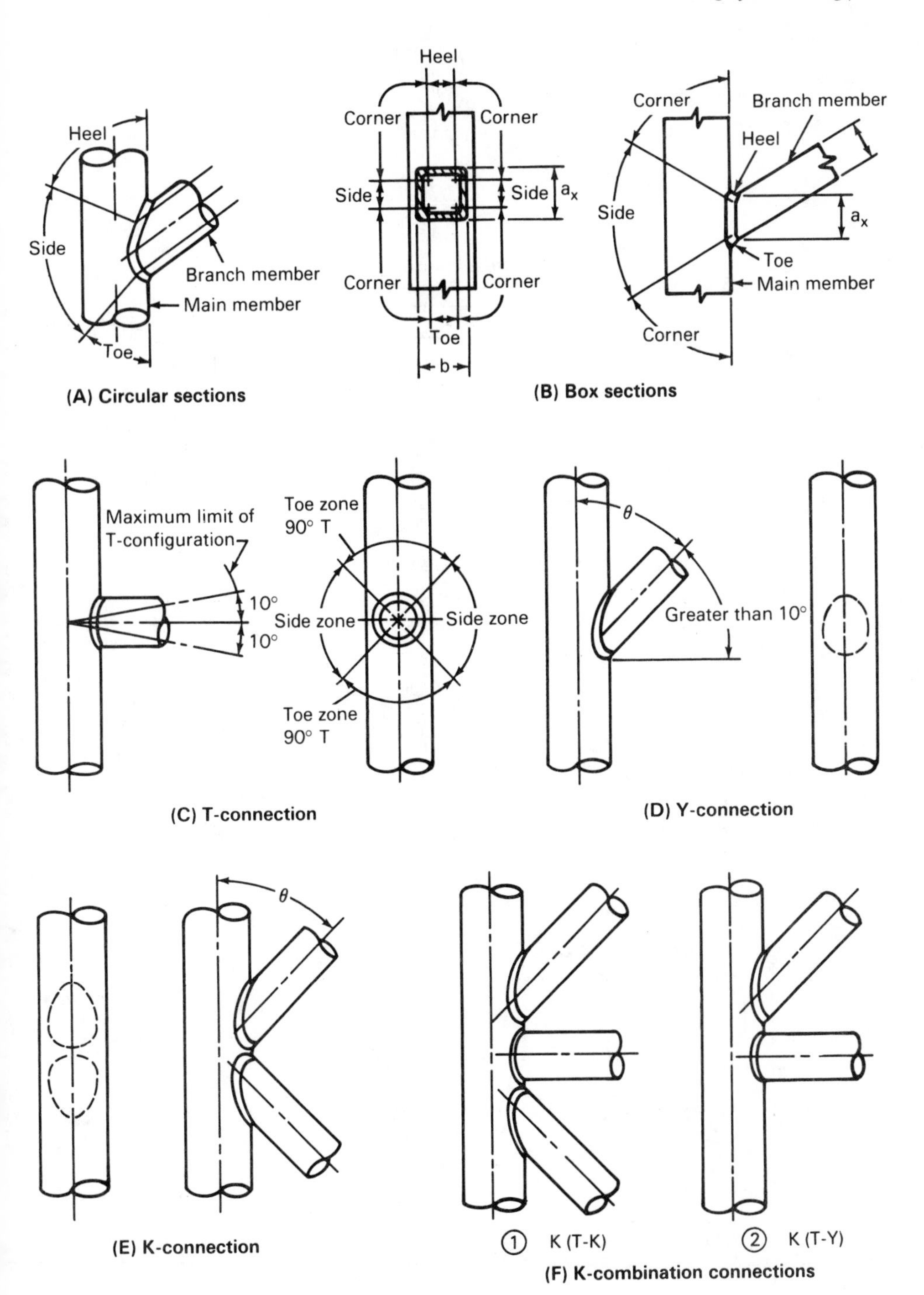

Fig. 1.40—Welded tubular connections, components, and nomenclature

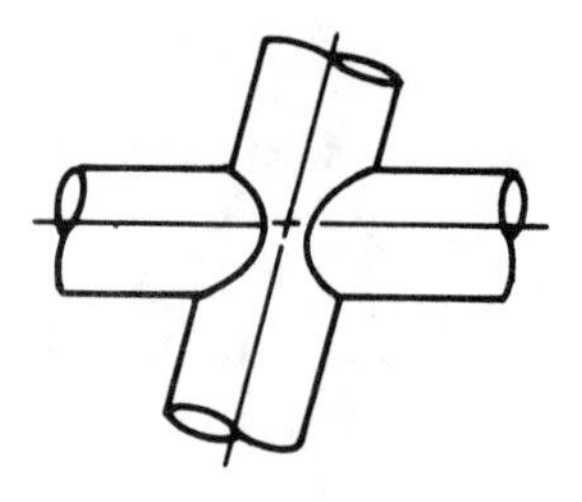

(G) Cross connection

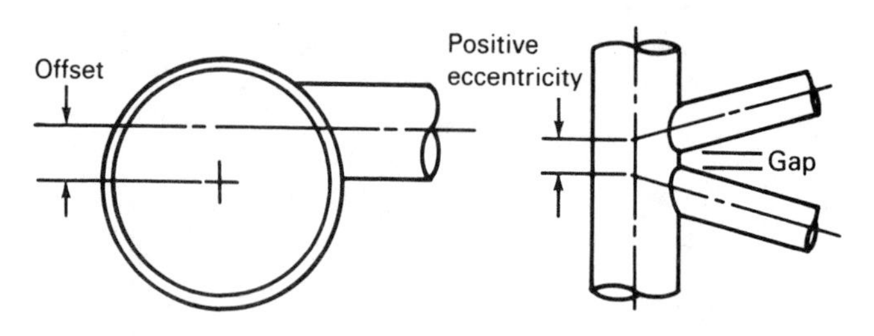

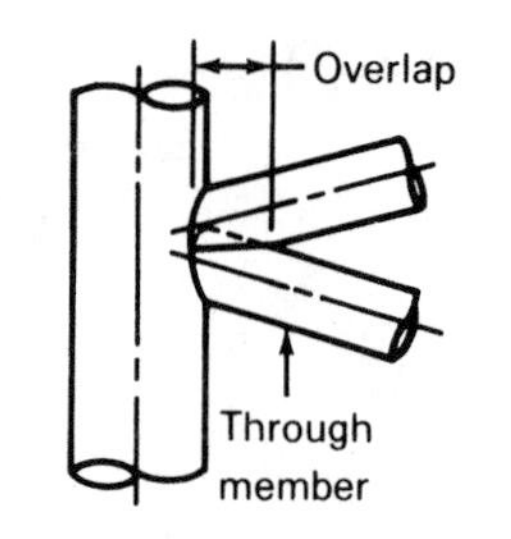

(H) Deviations from concentric connections

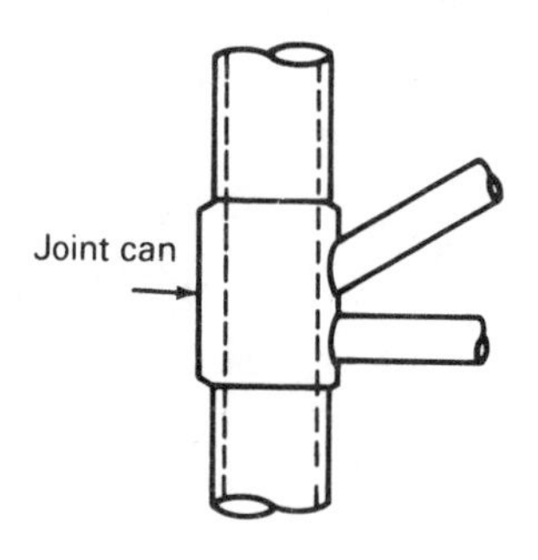

(I) Simple tubular connections

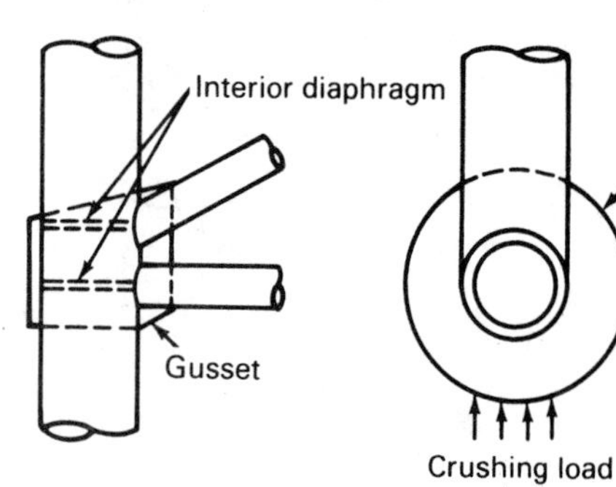

(J) Examples of complex reinforced connections

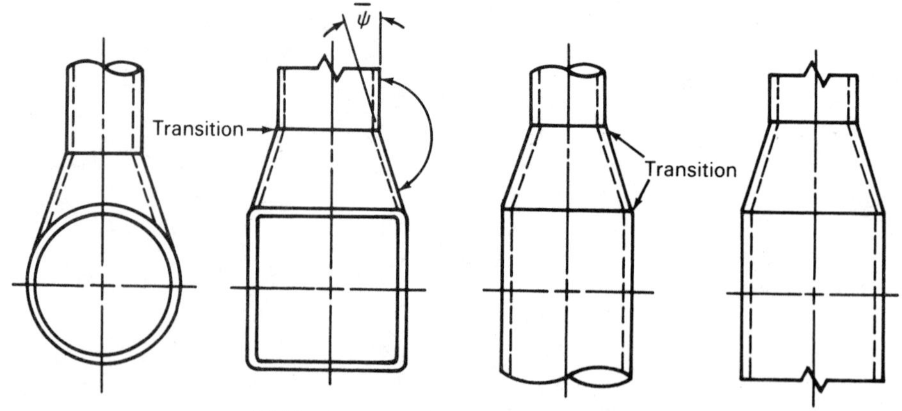

(K) Flared connections and transitions

Fig. 1.40 (cont.)—Welded tubular connections, components, and nomenclature

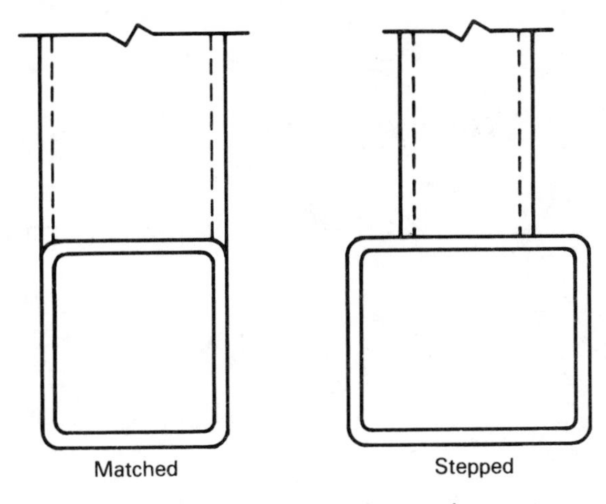

(L) Connection types for box sections

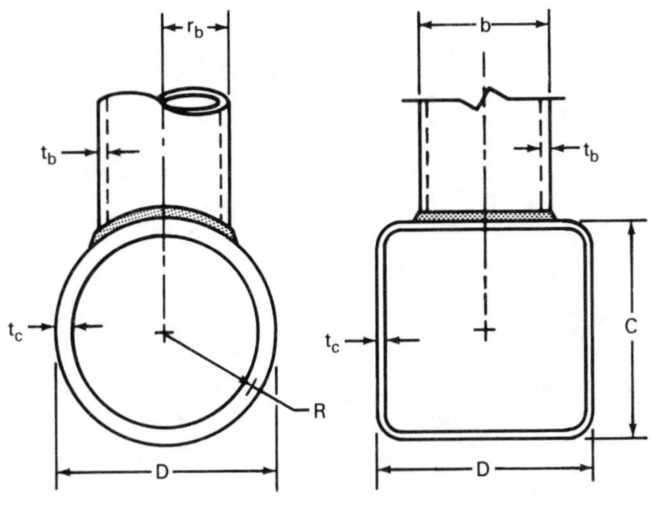

(M) Geometric parameters

Parameter	Circular sections	Box sections
β	r_b/R	b/D
η	—	a_x/D [See (B)]
γ	R/t_c	$D/2t_c$
τ	t_b/t_c	t_b/t_c
θ	Angle between member center lines	
ψ	Local dihedral angle at given point on welded joint	

Fig. 1.40 (cont.)—Welded tubular connections, components, and nomenclature

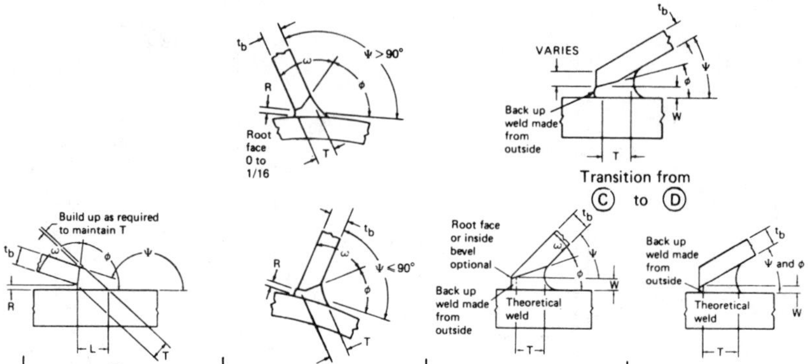

		Ⓐ Ψ = 180°–135°	Ⓑ Ψ = 150°–50°	Ⓒ Ψ = 75°–30°	Ⓓ Ψ = 37-1/2°–15°
End preparation (ω)	max	90°	90°	(a)	
	min	45°	10° or 45° for Ψ > 105°	10°	
Fitup or root opening (R), FCAW SMAW (1)	max	3/16 in.	1/4 in.	W max. (b): 1/8 in.; 3/16 in.	φ: 22-1/2°–37-1/2; 15°–20-1/2°
	min	1/16 in. No min for φ > 90°	1/16 in.		
Fitup or root opening (R), GMAW FCAW (2)	max	3/16 in.	1/4 in. for φ > 45°; 5/16 in. for φ ≤ 45°	W max. (b): 1/8 in.; 1/4 in.; 3/8 in.; 1/2 in.	φ: 30°–37-1/2; 25°–30°; 20°–25°; 15°–20°
	min	1/16 in. No min for φ > 120°	1/16 in.		
Joint included angle φ	max		60° for Ψ ≤ 105°	37-1/2° if more use Ⓑ	
	min		37-1/2° if less use Ⓒ	1/2 Ψ	
Completed weld	T	≥ t_b	≥ t for Ψ > 90°	≥ t/sin Ψ but need not exceed 1.75t	≥ $2t_b$
	L	≥ t/sin Ψ but need not exceed 1.75t	≥ t/sin Ψ for Ψ ≤ 90°	Weld may be built up to meet this	

(a) Otherwise as needed to obtain required φ.
(b) Initial passes of back up weld discounted until width of groove (W) is sufficient to assure sound welding; the necessary width of weld groove (W) provided by back up weld.

Notes:
1. These root details apply to SMAW and FCAW (self-shielded).
2. These root details apply to GMAW (short circuiting transfer) and FCAW (gas shielded).

Fig. 1.41—Joint designs for complete-joint-penetration groove welds in simple T-, K-, and Y-tubular connections

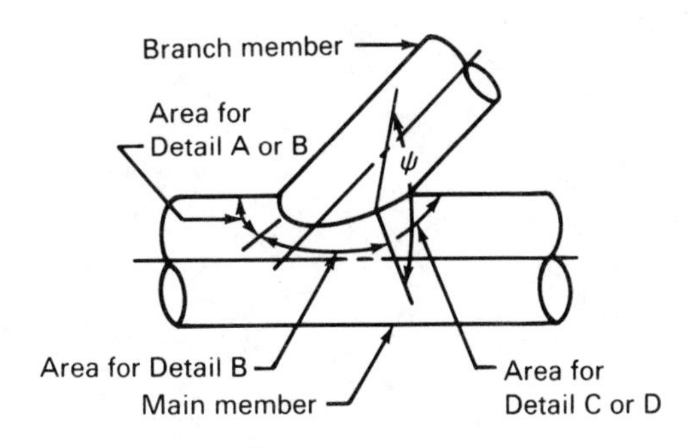

(A) Circular sections

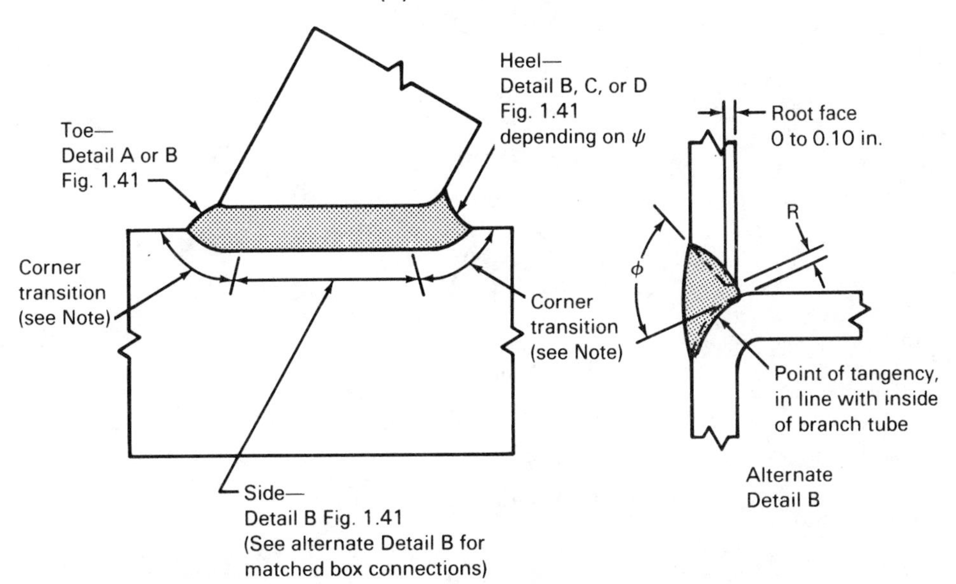

Note:
Joint preparation for welds at corner shall provide a smooth transition from one detail to another. Welding shall be carried continuously around corners, with corners fully built up and all starts and stops within flat faces.

(B) Box sections

Fig. 1.42—Locations of complete-joint-penetration groove weld designs on tubular connections

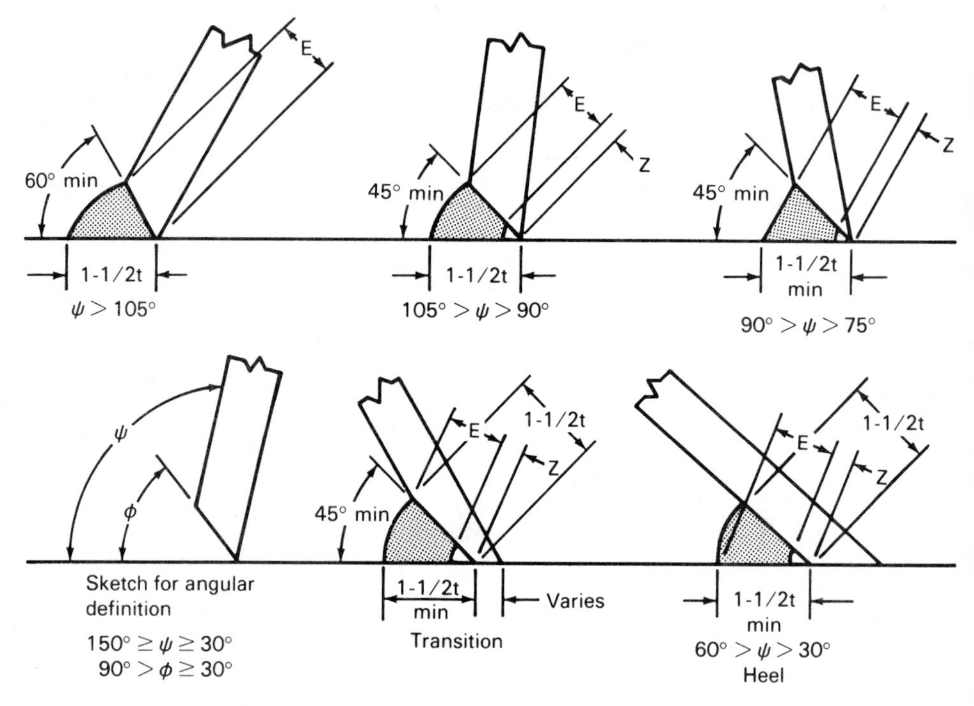

Notes:

1. t = thickness of thinner section.
2. Bevel to feather edge except in transition and heel zones.
3. Root opening: 0 to 3/16 in.
4. Effective throat = E, where E ≥ t.
5. Joint preparation for corner transitions shall provide a smooth transition from one detail to another. Welding shall be carried continuously around corners, with corners fully built up and all starts and stops within flat faces.

Fig. 1.43—Joint designs for partial-joint-penetration groove welds in simple T-, K-, and Y-tubular connections

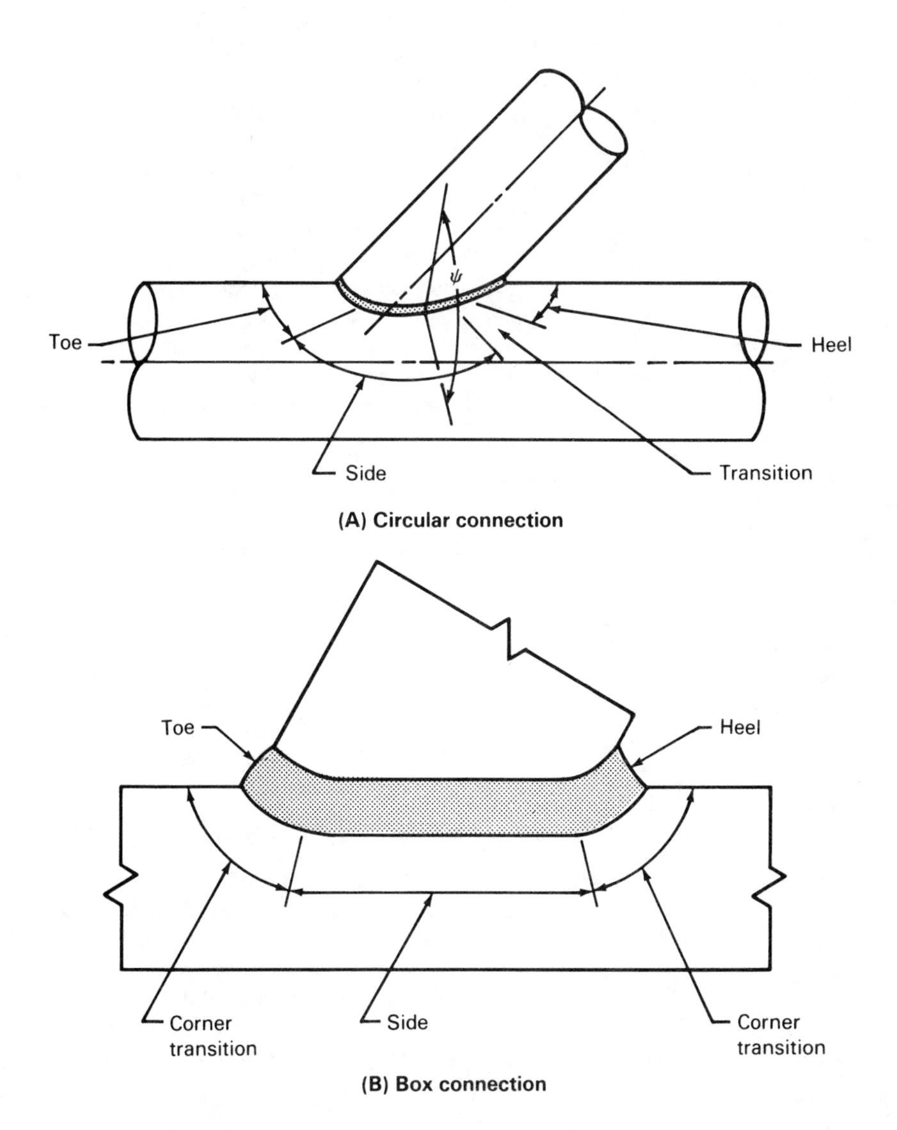

Fig. 1.44—Applicable locations of partial-joint-penetration groove weld designs

stress on the adjoining base metal should not exceed that permitted by the applicable code.

A welded tubular connection is limited in strength by four factors:

(1) Local or punching shear failure

(2) Uneven distribution of load on the welded connection

(3) General collapse

(4) Lamellar tearing

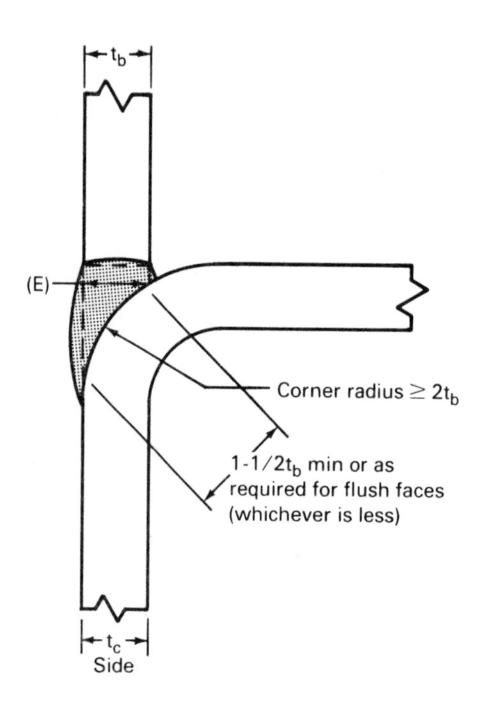

Fig. 1.45—Joint design for the sides of connections in equal box sections

LOCAL FAILURE

Where a circular or stepped-box T-, K-, or Y-connection (Fig. 1.40) is made by simply welding the branch member to the main member, local stresses at a potential failure surface through the main member wall may limit the useable strength of the welded joint. The shear stress at which failure can occur depends upon both the geometry of the section and the strength of the main member. The actual localized stress situation is more complex than simple shear; it includes shell bending and membrane stress as well. Whatever the mode of main member failure, the allowable punching shear stress is a conservative representation of the average shear stress at failure in static tests of simple welded tubular connections. The method for determining the punching shear stress in the main member is given in *AWS D1.1, Structural Welding Code—Steel*, latest edition.

The term *punching shear* tends to be confusing because punching indicates pushing in rather than pulling out. However, the forces capable of pulling a connection apart would be equal to the forces capable of pushing a connection apart, should the latter be feasible. The term *punching* relates the phenomenon to pre-existing analytical experience.

The actual punching shear stress in the main member caused by the axial force and any bending moment in the branch member must be determined and compared with the allowable punching shear stress. The effective area and length of the weld as well as its section modulus must be determined to treat the axial force and bending moment on the joint. These joint properties are factored into the stress and force calculations as described in *AWS D1.1, Structural Welding Code—Steel.*

UNEVEN DISTRIBUTION OF LOAD

Another condition that can limit the strength of a welded connection is uneven distribution of load on the weld. Under load, some bending of the main member could take place, which might cause uneven distribution of the force applied to the weld. As a result, some yielding and redistribution of stresses may have to take place for the connection to reach its design load. To provide for this, welds in T-, K-, and Y-connections [Figs. 1.40(C), (D), and (E)] must be capable, at their ultimate breaking strength, of developing the lesser of (1) the yield strength of the branch member or (2) the ultimate punching shear strength of the shear area of the main member. The locations of these are illustrated in Fig. 1.47. This particular part of the design is best handled by working with terms of unit force (pounds per linear inch).

Table 1.14
Allowance for incomplete fusion at the root of partial-joint-penetration groove welds

	Position of welding:	V or OH[a]	Position of welding:	H or F[a]
Groove angle, ϕ	Process[b]	Z[c], in.	Process[b]	Z[c], in.
$\phi \geq 60°$	SMAW	0	SMAW	0
	FCAW	0	FCAW	0
	FCAW-G	0	FCAW-G	0
	GMAW	NA[d].	GMAW	0
	GMAW-S	0	GMAW-S	0
$60° > \phi \geq 45°$	SMAW	1/8	SMAW	1/8
	FCAW	1/8	FCAW	0
	FCAW-G	1/8	FCAW-G	0
	GMAW	NA[c]	GMAW	0
	GMAW-S	1/8	GMAW-S	1/8
$45° > \phi \geq 30°$	SMAW	1/4	SMAW	1/4
	FCAW	1/4	FCAW	1/8
	FCAW-G	3/8	FCAW-G	1/4
	GMAW	NA[c]	GMAW	1/4
	GMAW-S	3/8	GMAW-S	1/4

a. Postion of welding: F = Flat; H = Horizontal; V = Vertical; OH = Overhead.
b. Processes: FCAW = Self shielded flux cored arc welding GMAW = Spray transfer or globular transfer
FCAW-G = Gas shielded flux cored arc welding GMAW-S = Short circuiting transfer
c. Refer to Fig. 1.43.
d. NA = Not applicable.

The ultimate breaking strength, Fig. 1.47(A), of fillet welds and partial-joint-penetration groove welds is computed at 2.67 times the basic allowable stress for 60 ksi and 70 ksi tensile strength weld metal and at 2.2 times for higher strength weld metals.

The unit force on the weld from the brace member at its yield strength, Fig. 1.47(B), is as follows:

$$f_1 = \sigma_y t_b$$

where

f_1 = unit force, lb/in.

σ_y = yield strength of brace, psi

t_b = thickness of brace, in.

The ultimate shear on the main member shear area, Fig. 1.47(C), at failure is as follows:

$$f_2 = 1.8 \tau_a t$$

where

f_2 = ultimate unit shear normal to the weld, lb/in.

τ_a = allowable shear stress, psi

t = thickness of the main member, in.

The unit shear force per in. on the weld, f_3, is:

$$f_3 = f_2/\sin\theta \text{ or } 1.8 \tau_a t/\sin\theta$$

where θ is the angle between the axes of the two members.

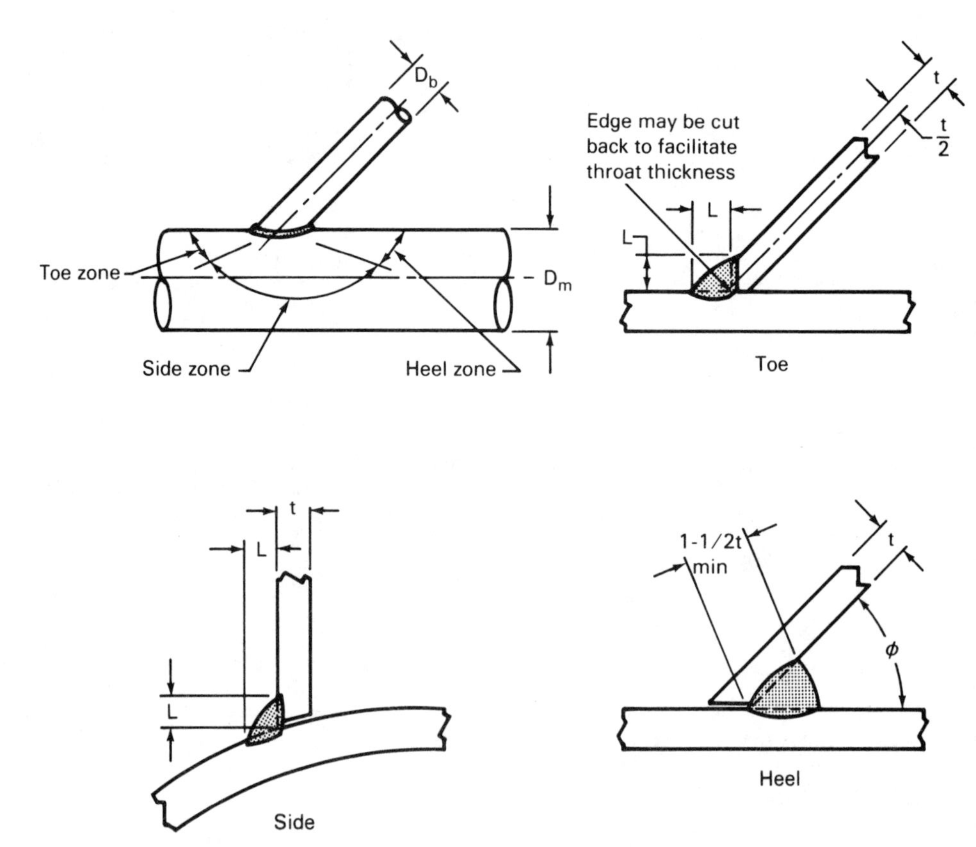

Notes:
1. t = thickness of thinner part
2. L = minimum size = t
3. Root opening 0 to 3/16 in.
4. ϕ = 15° min.

Fig. 1.46—Fillet weld details for circular T-, K-, and Y-connections

GENERAL COLLAPSE

As noted previously, the strength of the connection also depends on what might be termed *general collapse*. The strength and stability of the main member in a tubular connection should be investigated using the proper technology and in accordance with the applicable design code. If the main member has sufficient thickness required for punching shear and this thickness extends beyond the branch members for a distance of at least D/4, general collapse should not be a limiting factor.

LAMINATIONS AND LAMELLAR TEARING

In tubular connections where the branch member is welded to the outside surface of the main member, the capacity to transmit through-thickness stresses is essential to the proper functioning of the joint. Laminations (pre-existing planes of weakness) or lamellar tears (cracks parallel to the tube surface caused by high localized through-thickness thermal strains induced at restrained corner and T-joint welds) may impair this capacity.

Consideration of the problem of lamellar tearing must include design aspects and welding procedures that are consistent with the properties of the connected material. In connections where lamellar tearing might be a problem, consideration should be given in design to provide for maximum component flexibility and minimum weld shrinkage strain.

The following precautions should help to minimize the problems of lamellar tearing in highly restrained welded connections during fabrication. It is assumed that procedures producing low-hydrogen weld metal would be used in any case.

(1) On corner joints, where feasible, the bevel should be on the through-thickness member.

(2) The size of the weld groove should be kept to a minimum consistent with the design, and overwelding should be avoided.

(3) Subassemblies involving corner and T-joints should be fabricated completely prior to final assembly of connections. Final assembly should preferably be at butt joints.

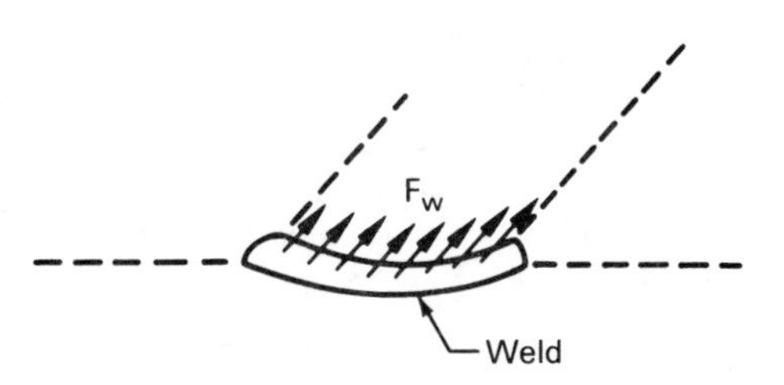

(A) Ultimate strength of welded connection

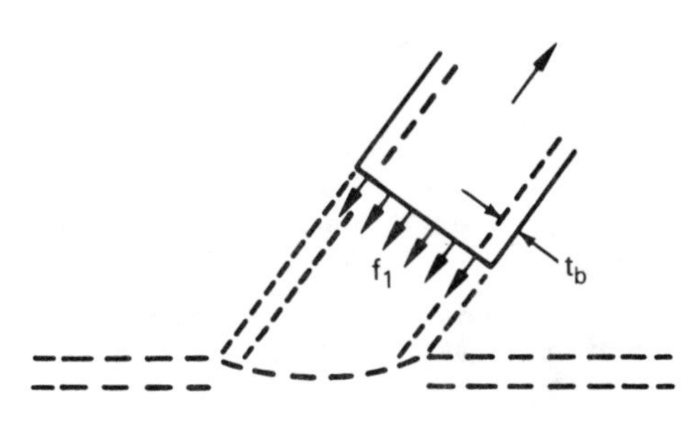

(B) Brace member at yield strength

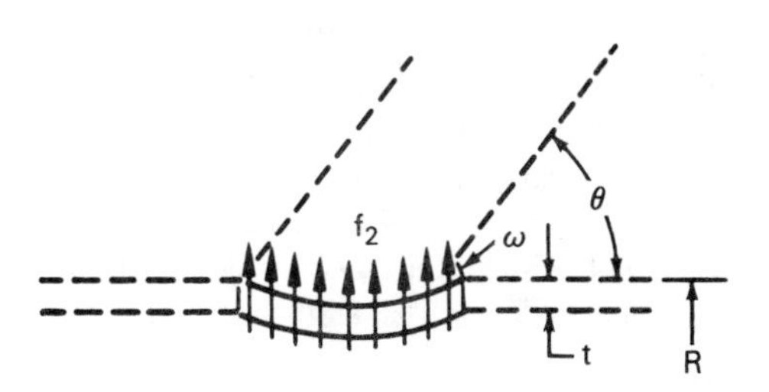

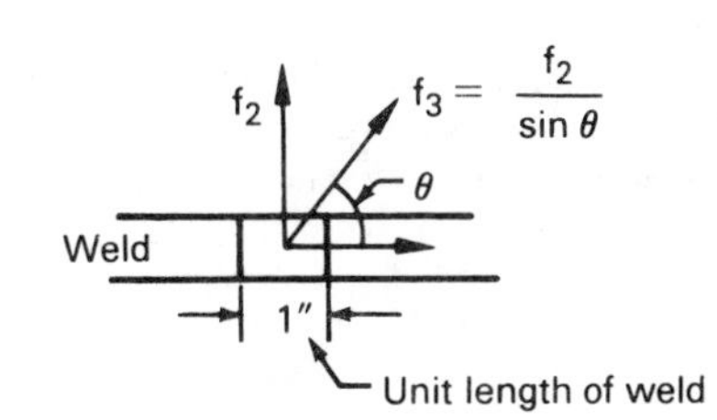

(C) Ultimate shear on the main member shear area at failure

Fig. 1.47—Loads on welded tubular connections

(4) A predetermined welding sequence should be selected to minimize overall shrinkage of the most highly restrained elements.

(5) The lowest strength weld metal available, consistent with design requirements, should be used to promote straining in the weld metal rather than in the more sensitive through-thickness direction of the base metal.

(6) Buttering with low-strength weld metal, peening, or other special weld procedures should be considered to minimize through-thickness shrinkage strains in the base metal.

(7) Material with improved through-thickness ductility should be specified for critical connections.[20]

In critical joint areas subject to through-thickness direction loading, material with pre-existing laminations and large metallic inclusions should be avoided. In addition, the following precautions should be taken:

(1) The designer should specify ultrasonic inspection, after fabrication or erection, or both, of those specific highly-restrained connections that could be subject to lamellar tearing and which are critical to the structural integrity.

(2) The designer must consider whether minor weld flaws or base metal imperfections can be left unrepaired without jeopardizing structural integrity. Gouging and repair welding will add additional cycles of weld shrinkage to the connection, and may result in the extension of existing flaws or the generation of new flaws by lamellar tearing.

(3) When lamellar tears are identified and repair is deemed advisable, rational consideration should be given to the proper repair required. A special welding procedure or a change in joint detail may be necessary.

FATIGUE

The design of welded tubular structures subject to cyclic loading is handled in the same manner as discussed previously. The specific treatment may vary with the applicable code for the structure.[21] Stress categories are assigned to various types of pipe, attachments to pipe, joint designs, and loading conditions. The total cyclic, fatigue stress range for the desired service life of a particular situation can be determined.

Fatigue behavior can be improved by one or more of the following actions:

(1) A capping layer can be added to provide a smooth contour with the base metal.

(2) The weld face may be ground transverse to the weld axis.

(3) The toe of the weld may be peened with a blunt instrument to cause local plastic deformation and to smooth the transition between the weld and base metals.

---

20. Improved quality steel does not eliminate weld shrinkage and, by itself, will not necessarily avoid lamellar tearing in highly restrained joints. Thus, it should not be specified in the absence of comprehensive design and fabrication considerations.

21. Such codes include *ANSI/AWS D1.1, Structural Welding Code—Steel,* and *API RP 2A, Recommended Practice for Planning, Designing, and Constructing Fixed Offshore Platforms,* 11th Ed., Dallas: American Petroleum Institute, 1980.

ALUMINUM STRUCTURES

DESIGNING FOR WELDING

The methods employed to design structures in aluminum are generally the same as those used with steel or other metals.[22] The methods and stress values recommended for structural aluminum design are set forth in the *Specifications for Aluminum Structures, Construction Manual Series*, published by The Aluminum Association.

Aluminum is available in many product forms and shapes, both cast and wrought. The designer can take advantage of the light weight of aluminum by utilizing available aluminum structural forms. Proper engineering design minimizes the number of joints and amount of welding without affecting product requirements. This, in turn, provides for good appearance and proper functioning of the product by limiting distortion caused by heating. To eliminate joints, the designer may use castings, extrusions, forgings, or bent or roll-formed shapes to replace complex assemblies. Special extrusions that incorporate edge preparations for welding may provide savings in manufacturing costs. Typical designs are shown in Fig. 1.48. An integral lip can be provided on the extrusion to facilitate alignment or serve as a weld backing, or both.

For economical fabrication, the designer should employ the least expensive metal-forming and metal-working processes, minimize the amount of welding required, and place welds at judicious locations of low stress. A simple example is the fabrication of an aluminum tray, Fig. 1.49. Instead of using five pieces of sheet and eight welds located at the corners, Fig. 1.49(A), such a unit could be fabricated from three pieces of sheet, one of which is formed into the bottom and two sides, Fig. 1.49(B). This reduces the amount of welding.

Further reduction in welding could be achieved by additional forming, as in Fig. 1.49(C). However, some distortion would likely take place in the two welded sides because all of the welds are in those two planes. The refinement of a design to limit only the amount of welding could lead to problems in fabrication, end use, or appearance. Therefore, the extent of welding should not be the single consideration in weldment design.

22. Welding requirements applicable to welded aluminum structures are given in *ANSI/AWS D1.2, Structural Welding Code—Aluminum*, latest edition, published by the *American Welding Society*.

WELD JOINTS

Butt, lap, T-, edge, and corner joints may be used in aluminum design. For structural applications, edge and corner joints should be avoided because they are harder to fit, weaker, and more prone to fatigue failure than the other joints. However, they are commonly used in sheet metal fabrications.

Butt Joints

Butt joints are generally easy to design, present good appearance, and perform better under cyclic loading than other types of joints. However, they require accurate alignment and usually require joint edge preparation on thicknesses above 1/4 in. to permit satisfactory root penetration. In addition, back chipping and a back weld are recommended to ensure complete fusion on thicker sections.

Sections of different thicknesses may be butted together and welded. However, it is better to bevel the thicker section before welding, as shown in Fig. 1.50, to reduce stress concentration, particularly when the joint will be exposed to cyclic loading in service.

When thin aluminum sheets are to be welded to thicker sections, it is difficult to obtain adequate depth of fusion in the thicker section without melting away the thin section. This difficulty can be avoided by extruding or machining a lip on the thicker section equal in thickness to that of the thin part. If the thicker section is an extrusion, a welding lip can be incorporated in the design as described previously. This arrangement improves the heat balance across the joint.

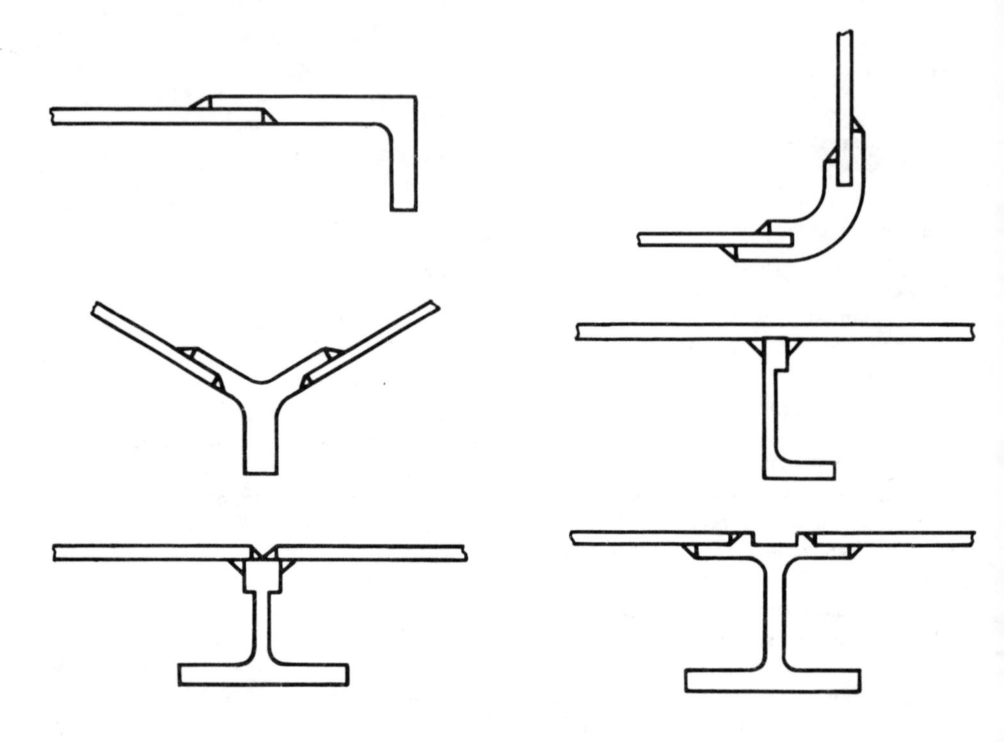

Fig. 1.48—Typical extrusion designs incorporating desired joint geometry, alignment, and reinforcement

Lap Joints

Lap joints are used more frequently on aluminum alloys than is customary with most other metals. In thicknesses up to 1/2 in., it is more economical to use single-lap joints with fillet welds on both sides rather than butt joints welded with complete joint penetration. Lap joints require no edge preparation, are easy to fit, and require less jigging than butt joints. The efficiency of lap joints ranges from 70 to 100 percent, depending on the base metal composition and temper. Preferred types of lap joints are shown in Fig. 1.51.

Lap joints do create an offset in the plane of the structure unless the members are in the same plane and strips are used on both sides of the joint. Those with an offset tend to rotate under load. Lap joints may be impractical if the joint is not accessible on both sides. In large structures, it may be more economical to weld a butt joint from one side in the flat position than to weld one side of a lap joint in the overhead position (Fig. 1.51).

T-Joints

T-joints seldom require edge preparation because they are usually connected by fillet welds. The welds should have complete fusion to or beyond the root (corner) of the joint. A single- or double-bevel groove weld in combination with fillet welds may be used with thicknesses above 3/4 in. to reduce the amount of weld metal.

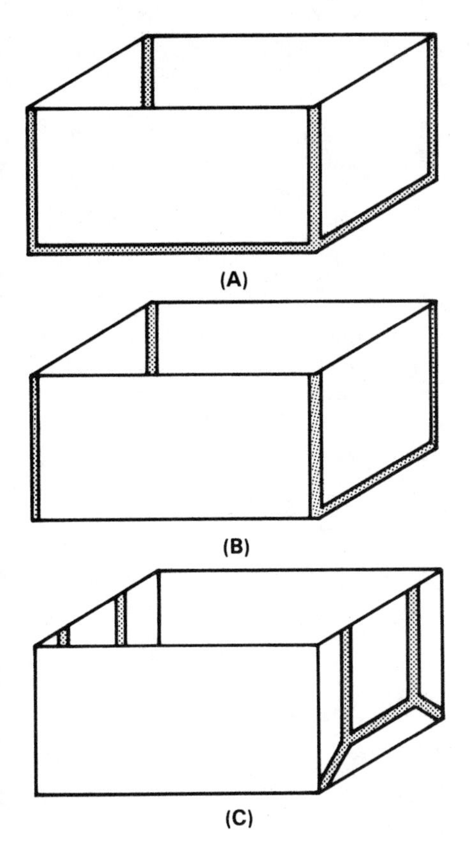

Fig. 1.49—Designs for an aluminum tray

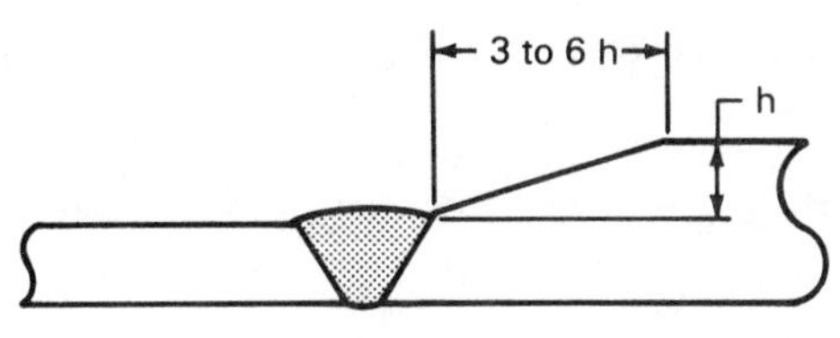

Fig. 1.50—Beveling the thicker member to reduce stress concentration at the weld

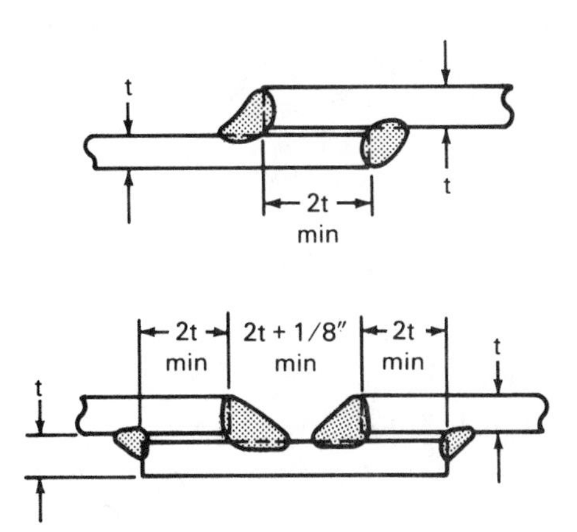

Fig. 1.51—Preferred types of lap joints

T-joints are easily fitted and normally require no back chipping. Necessary jigging is usually quite simple.

A T-joint with a single fillet weld is not recommended. Although the joint may have adequate shear and tensile strength, it is very weak when loaded with the root of the fillet weld in tension. Small continous fillet welds should be used on both sides of the joint, rather than either large intermittent fillet welds on both sides or a large fillet weld on one side. Continous fillet welding is recommended for better fatigue life and for avoiding crevice corrosion and crater cracks. Suggested allowable shear stresses in fillet welds for building and bridge structures are given in the *Specification for Aluminum Structures* published by The Aluminum Association.

JOINT DESIGN

In general, the design of welded joints for aluminum is quite consistent with the recommendations for steel joints.[23] However, smaller

23. Suggested groove weld joint designs are given in the *Welding Handbook*, Vol. 4, 7th Ed., 1982: 334 ff, and in ANSI/AWS D1.2, *Structural Welding Code—Aluminum,* latest edition, published by the American Welding Society.

root openings and larger groove angles are generally used because of the higher fluidity of aluminum under the welding arc and the larger gas nozzles on welding guns and torches. The excellent machinability of aluminum makes J- and U-groove preparations economical to reduce weld metal volume, especially on thick sections.

EFFECTS OF WELDING ON STRENGTH

Aluminum alloys are normally used in the strain-hardened or heat-treated condition, or a combination of both, to take advantage of their high strength-to-weight ratios. The effects of strain hardening or heat treatment are negated when aluminum is exposed to the elevated temperatures encountered in welding. The heat of welding softens the heat-affected zone in the base metal. The extent of softening is related to the section thickness and heat input. The soft heat-affected zone must be considered in design; its orientation with respect to the direction of stress and its proportion of the total cross section determines the allowable load on the joint.

The variation in tensile or yield strength across a welded joint in aluminum structures is illustrated in Fig. 1.52. With plate, the extent of decreased properties is considered to be a 2-in. wide band with the weld in the center. When joining sheet gages with an automatic welding process, the band will be narrower. The orientation of the band with respect to the direction of stress and its proportion of the total cross section determine its effect on the allowable load on the joint. The minimum mechanical properties for welded aluminum alloys are given in the *Specification for Aluminum Structures* published by The Aluminum Association.

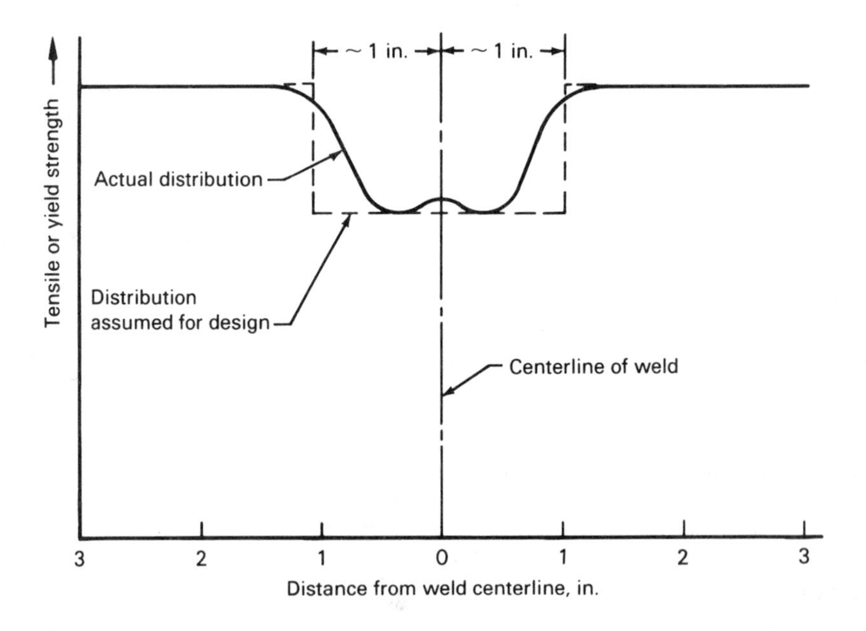

Fig. 1.52—Distribution of tensile or yield strength across a weld in aluminum

Transverse welds in columns and beams should be located at points of lateral support to minimize the effects of welding on buckling strength. The effects of longitudinal welds in structural members can be neglected if the softened zone is less than 15 percent of the total cross-sectional area. Circumferential welds in piping or tubing may reduce bending strength; longitudinal welds usually have little effect on buckling strength when the heat-affected zone is a small percentage of the total area of cross section.

With the proper choice of filler metal, a weldment of a heat-treatable aluminum alloy can be solution heat treated and aged after welding. The welded assembly will regain substantially full strength with some loss in ductility. This is the best method of providing maximum weld strength, but it is usually uneconomical or impractical because of the size of the furnace required, cost of heat treating, or the resultant distortion caused by the quenching operation.

It may be practical, at times, to weld a heat-treatable alloy in the solution-treated condition and age after welding. This can increase the strength over that in the as-welded condition while avoiding the distortion problem associated with solution heat treating.

There is no method of overcoming softening in nonheat-treatable alloys, other than further rolling or stretching of the parts after welding, and this is seldom practical.

The weakest location in a welded assembly is the annealed heat-affected zone. Tensile specimens will fail either in the weld metal or in the heat-affected zone even though a recommended filler metal was used and the weld was sound.

In general, with the use of recommended filler metal, the minimum as-welded strengths of nonheat-treatable alloys are the annealed strengths of the base metals. The minimum as-welded strengths of heat-treatable alloys have been established by testing a large number of welds and statistically analyzing the results.

STRESS DISTRIBUTION

Where welds are located in critical areas but do not cover the entire cross section, the strength of the section depends on the percentage of the cross-sectional area affected by the heat of welding. When members must be joined at locations of high stress, it is desirable that the welds be parallel to the principal member and to the main stress in that member. Transverse welds should be avoided. For example, longitudinal welds that join a web to a flange of an I-beam usually have very little effect on the strength of the member because most of the beam cross section has original base metal properties.

Frequently, welds are more highly stressed at the ends than in the central portions. To avoid using thicker sections, areas of high stresses in welds can be minimized by *sniping*. This consists of beveling the end of a member to limit stress concentration in the weld at that end. The weld, however, should wrap around the end of the member. This type of member termination is illustrated in Fig. 1.53.

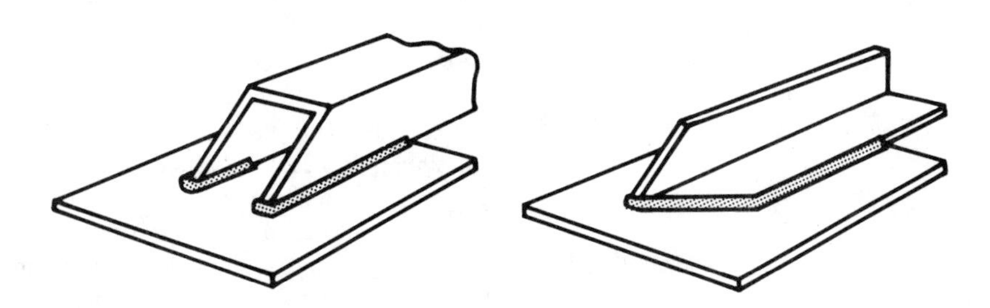

Fig. 1.53—Beveling the end of the member and welding around the end to limit stress concentration

In many weldments, it is possible to locate the welds where they will not be subjected to high stresses. It is frequently possible to make connections between a main member and accessories such as braces by welding at the neutral axis or other point of low stress. Joints of low efficiency may be reinforced, but this involves an additional cost.

SHEAR STRENGTH OF FILLET WELDS

The shear strength of fillet welds is controlled by the composition of the filler metal. Typical shear strengths of longitudinal and transverse fillet welds made with several aluminum filler metals are shown in Figs. 1.54 and 1.55, respectively. The highest strength filler metal is alloy 5556. Use of a high-strength filler metal permits smaller welds. For example, assume a longitudinal fillet weld having a strength of 4000 lb/in. is desired. If 5356 filler metal is used, a 1/4-in. fillet weld can be applied in a single pass. However, if 4043 filler metal is used, it would require a 3/8-in. fillet weld that probably would need to be deposited with three weld passes. The use of the stronger filler metal has obvious economic advantage.

The minimum practical fillet weld sizes depend on the thickness of the base metal and the welding process. The minimum sizes are about the same as those specified for steel (Table 1.10). Where minimum weld sizes must be used, a filler metal with the lowest suitable strength for the applied load should be selected to take advantage of its ductility.

By applying the appropriate safety factor to the shear strength of a filler metal, the designer can determine the allowable shear stress in a fillet weld. Appropriate factors of safety and allowable shear stresses in fillet welds for aluminum structures are given in the *Specification for Aluminum Structures* published by The Aluminum Association.

FATIGUE STRENGTH

The fatigue strength of welded aluminum structures follows the same general rules that apply to fabricated assemblies of other metals. Fatigue strength is governed by the peak stresses at points of stress concentration, rather than by nominal stresses. Eliminating stress raisers to reduce the peak stresses tends to increase the fatigue life of the assembly.

Average fatigue strengths of as-welded joints in four aluminum alloys are shown in Fig. 1.56. These are average test results for butt joints in 3/8-in. plate welded by one of the inert gas shielded metal arc welding processes. Specimens were welded on one side, back gouged, and then back welded. The stress ratio of zero means that the tensile stress went from zero to the plotted value and back to zero during each cycle. Other weldable alloys of intermediate static strengths perform in similar fatigue tests with generally proportional intermediate curves.

The fatigue strengths of the various alloys are markedly different and below 10^4 cycles the designer may prefer one alloy over another for a particular application. However, above 10^6 cycles, the differences among various alloys are very small. A solution to fatigue problems in the 10^6 cycle range and above is primarily found in a change of design rather than by a change of alloy.

The designer should utilize symmetry in the assembly for balanced loading, and should avoid sharp changes in direction, notches, and other stress raisers. The fatigue strength of a groove weld may be significantly increased by such means as removing weld reinforcement or by peening the weldment. If such procedures are not practical, the weld reinforcement should fair into the base metal gradually to avoid abrupt changes in thickness. With welding processes that produce relatively smooth weld beads, there is little or no increase in fatigue strength gained by smoothing the weld faces. The benefit of smooth weld beads can be nullified by excessive spatter during welding. Spatter marks sometimes create severe stress raisers in the base metal adjacent to the weld.

While the residual stresses from welding are not considered to affect the static strength of aluminum, they can be detrimental in regard to fatigue strength. Several methods can be employed to reduce residual welding stresses including shot peening, multiple-pin gun peening, thermal treatments, and hydrostatic pressurizing of pressure vessels beyond the yield strength. Shot peening or hammer peening is beneficial when it changes the residual stresses at the weld

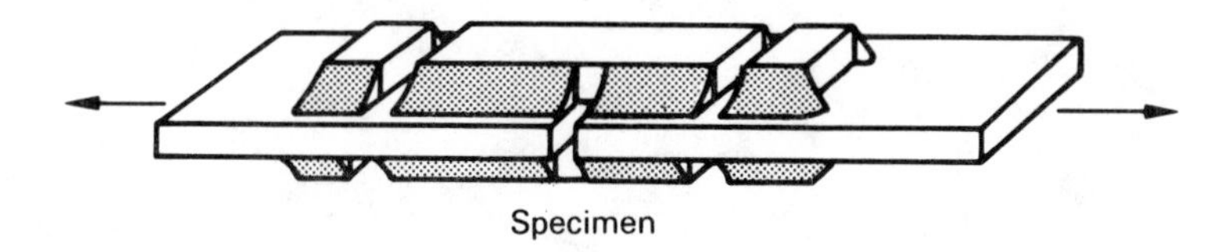

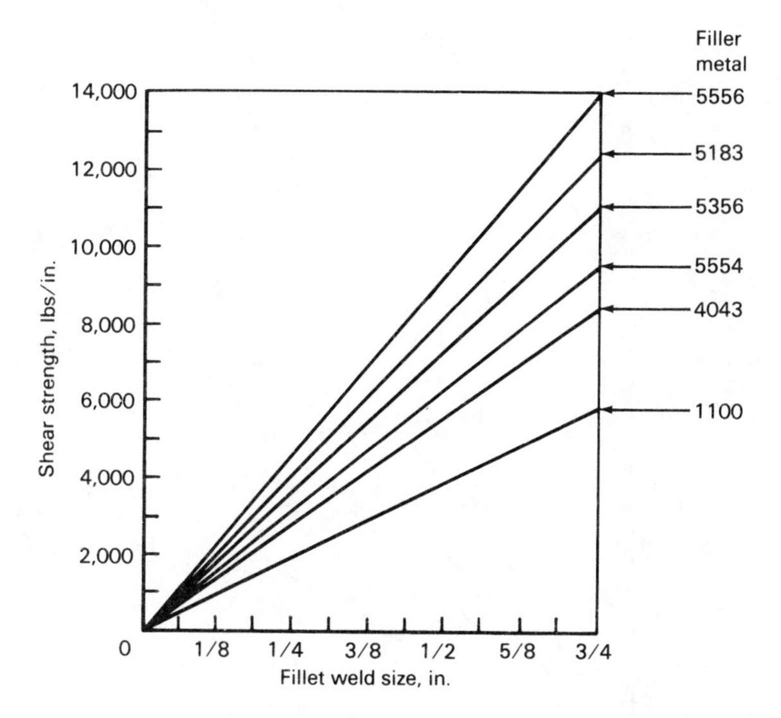

Fig. 1.54—Typical shear strengths of longitudinal fillet welds with various aluminum filler metals

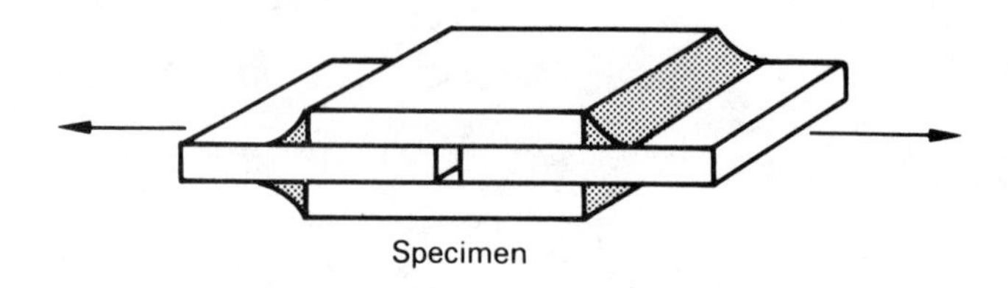

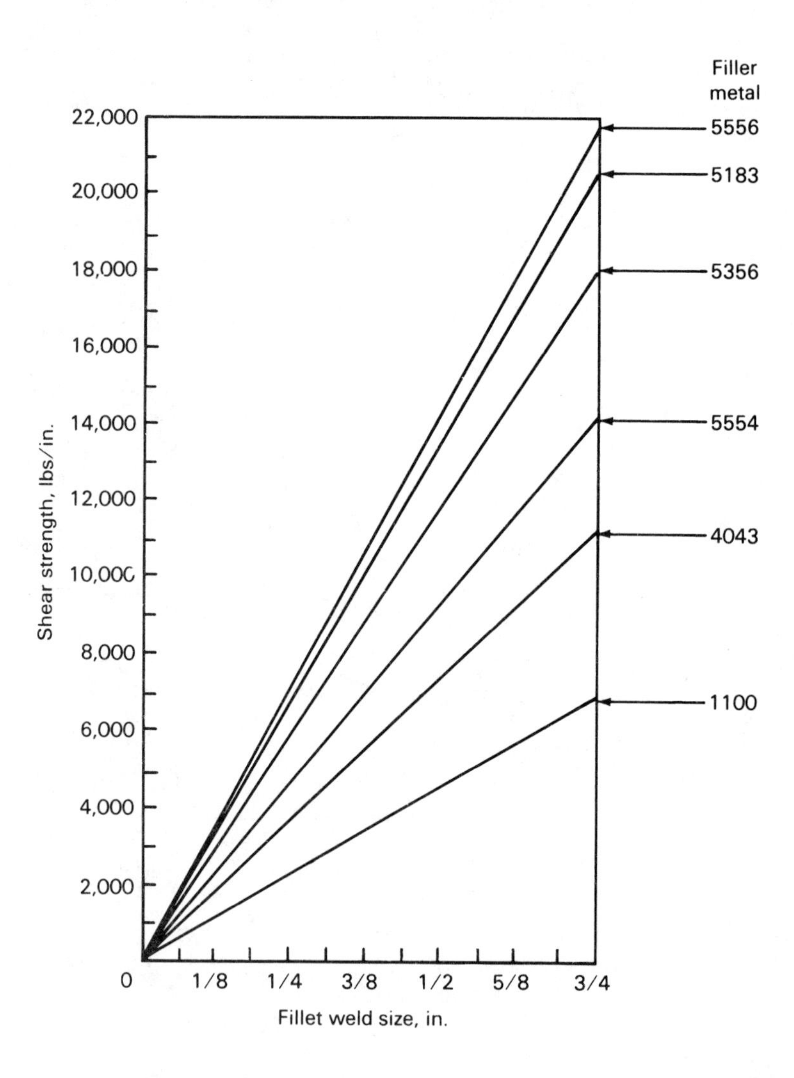

Fig. 1.55—Typical shear strengths of transverse fillet welds with various aluminum filler metals

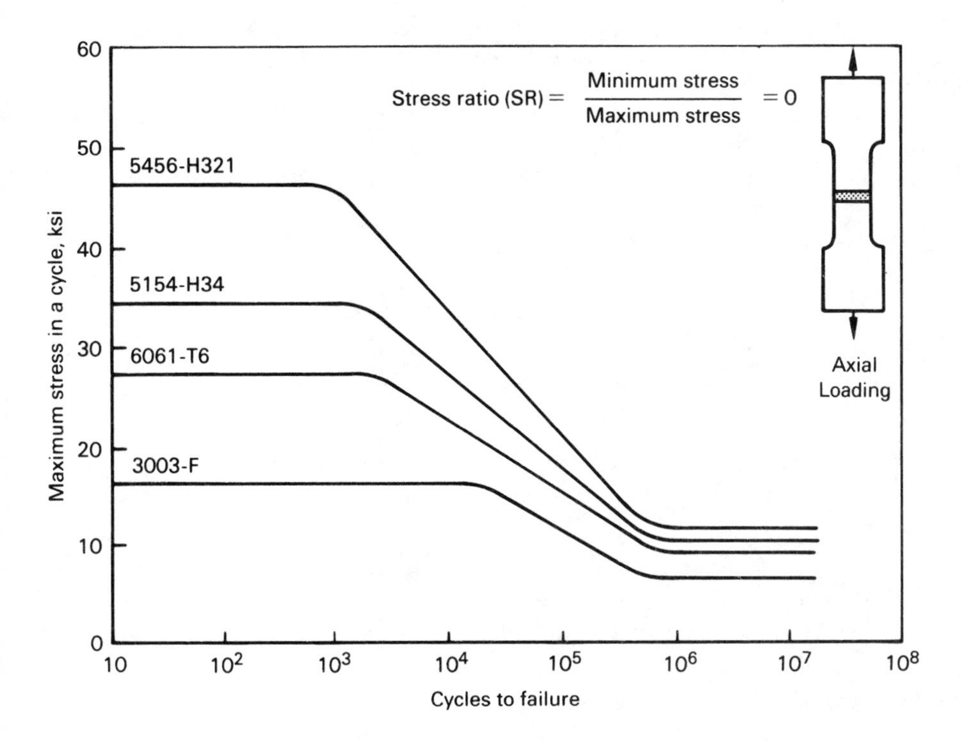

Fig. 1.56—Fatigue test results of butt joints in four aluminum alloys, 3/8 in. thick shielded gas arc welded plates

face from tension to compression for a depth of 0.005 to 0.030 inch. Shot peening may more than double the life of direct-tension fatigue test specimens with longitudinal butt joints.

Thermal treatments to relieve residual stresses are beneficial for increasing the fatigue characteristics as well as the dimensional stability of a welded aluminum assembly during subsequent machining. In the case of nonheat-treatable alloys, such as the 5000 series, thermal treatments for the various alloys can relieve up to 80 percent of the residual welding stresses with little decrease in the static strength of the base metal. Heat-treatable alloys are not so well suited to thermal treatments for relieving residual stresses because temperatures that are high enough to cause significant reductions in residual stress also tend to substantially diminish strength properties. However, a reduction in residual welding stresses of about 50 percent is possible if a decrease in strength of about 20 percent can be tolerated.

EFFECT OF SERVICE TEMPERATURE

The minimum tensile strengths of aluminum arc welds at other than room temperature are given in Table 1.15. The performance of welds in nonheat-treatable alloys follows closely those of the annealed base metals.

Most aluminum alloys lose a substantial portion of their strength at temperatures above 300° F. Certain alloys, such as Type 2219, have better elevated temperature properties, but there are definite limitations on applications. The 5000 series alloys with magnesium content of 3.5 per-

Table 1.15
Minimum tensile strengths at various temperatures of arc-welded butt joints in aluminum alloys

Base alloy designation	Filler metal	Ultimate tensile strength, ksi					
		−300° F	−200° F	−100° F	100° F	300° F[a]	500° F[a]
2219-T37[b]	2319	48.5	40.0	36.0	35.0	31.0	19.0
2219-T62[c]	2319	64.5	59.5	55.0	50.0	38.0	22.0
3003	ER1100	27.5	21.5	17.5	14.0	9.5	5.0
5052	ER5356	38.0	31.0	26.5	25.0	21.0	10.5
5083	ER5183	54.5	46.0	40.5	40.0	—	—
5086	ER5356	48.0	40.5	35.5	35.0	—	—
5454	ER5554	44.0	37.0	32.0	31.0	26.0	15.0
5456	ER5556	56.0	47.5	42.5	42.0	—	—
6061-T6[b]	ER4043	34.5	30.0	26.5	24.0	20.0	6.0
6061-T6[c]	ER4043	55.0	49.5	46.0	42.0	31.5	7.0

a. Alloys not listed at 300° F and 500° F are not recommended for use at sustained operating temperatures of over 150° F.
b. As welded.
c. Heat treated and aged after welding.

cent or higher are not recommended for use at sustained temperatures above 150° F. Alloy 5454 with its comparable filler metal ER5554 is the strongest 5000 series alloy recommended for such applications as hot chemical storage and tank trailers.

Aluminum is an ideal material for low temperature applications. Most aluminum alloys have higher ultimate and yield tensile strengths at temperatures below room temperature. The 5000 series alloys have good strength and ductility at very low temperatures together with good notch toughness. Alloys 5083 and 5456 have been used extensively in pipelines, storage tanks, and marine vessel tankage for handling cryogenic liquids and gases.

Metric Conversion Factors

$t_C = 0.56\,(t_F - 32)$
1 in. = 25.4 mm
1 in./min = 0.42 mm/s
1 in. · lbf = 0.11 N · m or 0.11 J
1 ft · lbf = 1.36 N · m or 1.36 J
1 lbf/in. = 175 N/m
1 lb/in.^{3} = 27.7 Mg/m^3
1 psi = 6.89 kPa
1 ksi = 6.89 MPa
1 hp = 746 W

SUPPLEMENTARY READING LIST

ANSI/AWS D1.1, *Structural Welding Code — Steel*, latest edition, Miami, FL: American Welding Society.

ANSI/AWS D1.2, *Structural Welding Code — Aluminum*, latest edition, Miami, FL: American Welding Society.

Blodgett, O.W., *Design of Welded Structures*, Cleveland, OH: Lincoln Arc Welding Foundation, 1966.

Cary, H.B., *Modern Welding Technology*, Englewood Cliffs, NJ: Prentice-Hall, 1979: 15-1 ff.

Economic Design of Weldments, AWRA Technical Note 8, Australian Welding Research Association, 1979 March.

Manual of Steel Construction, Chicago: American Institute of Steel Construction, latest edition.

Procedure Handbook of Arc Welding, 12th Ed., Cleveland, OH: The Lincoln Electric Co., 1973 June: 2.1-1 ff.

Rolfe, S., Fatigue and fracture control in structures, *AISC Engineering Journal*, 14 (1), 1977.

Solutions to the Design of Weldments, D810.17, Cleveland, OH: The Lincoln Electric Co., 1975 Jan.

Specification for Aluminum Structures, Wash. DC: The Aluminum Association, 1976 Apr.

Welded Structural Design, Toronto, Canada: Canadian Welding Bureau, 1968.

Welding Kaiser Aluminum, Oakland, CA: Kaiser Aluminum and Chemical Sales, Inc., 1967.

2
Symbols for Welding, Brazing, and Nondestructive Examination

Purpose *94*

Welding Symbols *94*

Brazing Symbols *113*

Nondestructive Examination Symbols .. *116*

Chapter Committee

C.D. BURNHAM, *Chairman*
General Electric Company

J.T. BISKUP
Canadian Institute of Steel Construction

W.L. GREEN
Ohio State University

F.H. GRIGG
FMC Corporation

E.A. HARWART
Consultant

J.G. ROBERTS
Southern California Drafting Services

M.W. ROTH
Hobart School of Welding Technology

J.J. STANCZAK
Steel Detailers and Designers

Welding Handbook Committee Member

E.H. DAGGETT
The Babcock and Wilcox Company

2

Symbols for Welding, Brazing, and Nondestructive Examination

PURPOSE

Standard symbols are used to indicate desired welding and brazing information on engineering drawings. They are used universally to convey the design requirements to the shop in a concise manner. For example, a symbol can be used to specify the type of weld, groove design, welding process, face and root contour, sequence of welding, length of weld, effective throat, and other information. However, there are cases where all information cannot be conveyed by a symbol alone. Supplementary notes or dimensional details, or both, are sometimes required to provide the shop with complete requirements. The designer must be sure that the requirements are fully presented on the drawing or specifications.

Nondestructive examination requirements for welded or brazed joints can also be called out with symbols. The specific inspection methods[1] to be used are indicated on the symbols. The appropriate inspection methods depend upon the quality requirements with respect to discontinuities in welded or brazed joints.

The complete system of symbols is described in *AWS A2.4, Symbols for Welding and Nondestructive Testing,* 1979 Edition, published by the American Welding Society. This publication (latest edition) should be referred to when actually selecting the appropriate symbols for describing the desired joint and the inspection requirements. In practice, most designers will use only a few of the many available symbols. The information presented here describes the fundamentals of the symbols and how to apply them.

WELDING SYMBOLS

BASIC WELD SYMBOLS

The terms *weld symbol* and *welding symbol* have different meanings. A weld symbol, Fig. 2.1, indicates the required type of weld (or braze). The welding symbol, Fig. 2.2, includes the weld symbol and supplementary information. A complete welding symbol consists of the following elements:

(1) Reference line (always shown horizontally)

(2) Arrow

(3) Basic weld symbol

(4) Dimensions and other data

(5) Supplementary symbols

1. Nondestructive testing methods, procedures, and the type of discontinuities that each method will reveal are discussed in *AWS B1.0, Guide for Nondestructive Inspection of Welds,* and *Welding Inspection,* Second Edition, 1980, published by the American Welding Society.

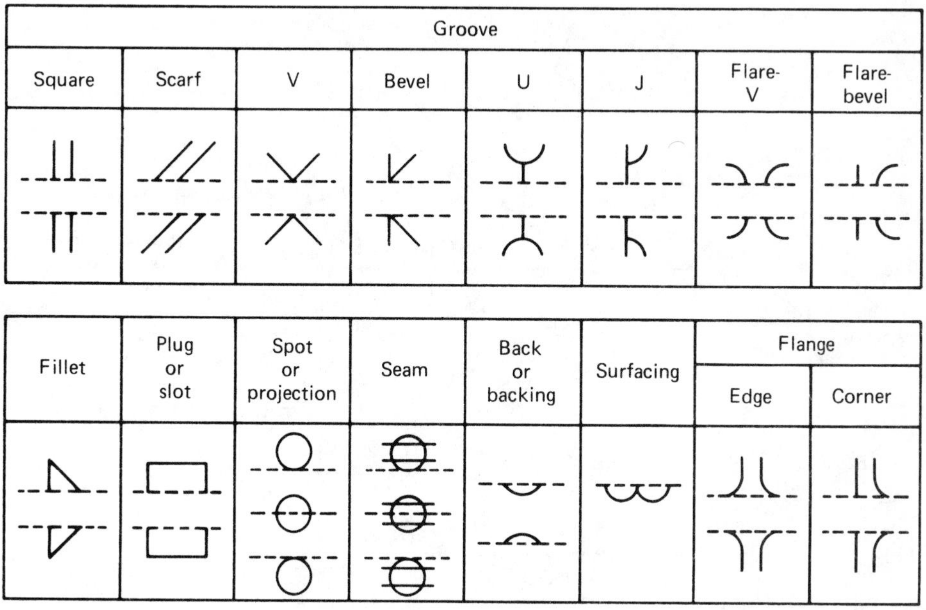

Groove							
Square	Scarf	V	Bevel	U	J	Flare-V	Flare-bevel

Fillet	Plug or slot	Spot or projection	Seam	Back or backing	Surfacing	Flange	
						Edge	Corner

Note: The dashed line is not part of the basic weld symbol, but represents a reference line (see Fig. 2.2).

Fig. 2.1—Basic weld symbols

(6) Finish symbols
(7) Tail
(8) Specification, process, or other references

All elements need not be used unless required for clarity.

LOCATION OF ELEMENTS

The elements of a welding symbol have standard locations with respect to each other (Fig. 2.2).

Location Significance of Arrow

The arrow element in a welding symbol in conjunction with the reference line determines the *arrow side* and *other side* of a weld, as shown in Fig. 2.3(A).

The symbol depicting an *arrow side weld* is always placed below the reference line, Fig. 2.3(B). The arrow side is always closest to the reader when viewed from the bottom of the drawing. The weld symbol depicting an *other side weld* is placed above the reference line, i.e., away from the reader, Fig. 2.3(C). Welds on both sides of a joint are shown by placing weld symbols on both sides of the reference line, Fig. 2.3(D).

Some weld symbols have no arrow or other side significance. However, supplementary symbols used in conjunction with these weld symbols may have such significance. For example, welding symbols for resistance spot and seam welding have no side significance, Fig 2.3(E).

References

When a specification, process, test, or other reference is needed to clarify a welding symbol, the reference is placed in a tail on the welding symbol, as shown in Fig. 2.3(F). The letters *CJP* in the tail of the arrow are used to indicate that a complete-joint-penetration weld is required. The type weld or joint preparation may be optional. The tail may be omitted when no specification, process, or other reference is required with a welding symbol.

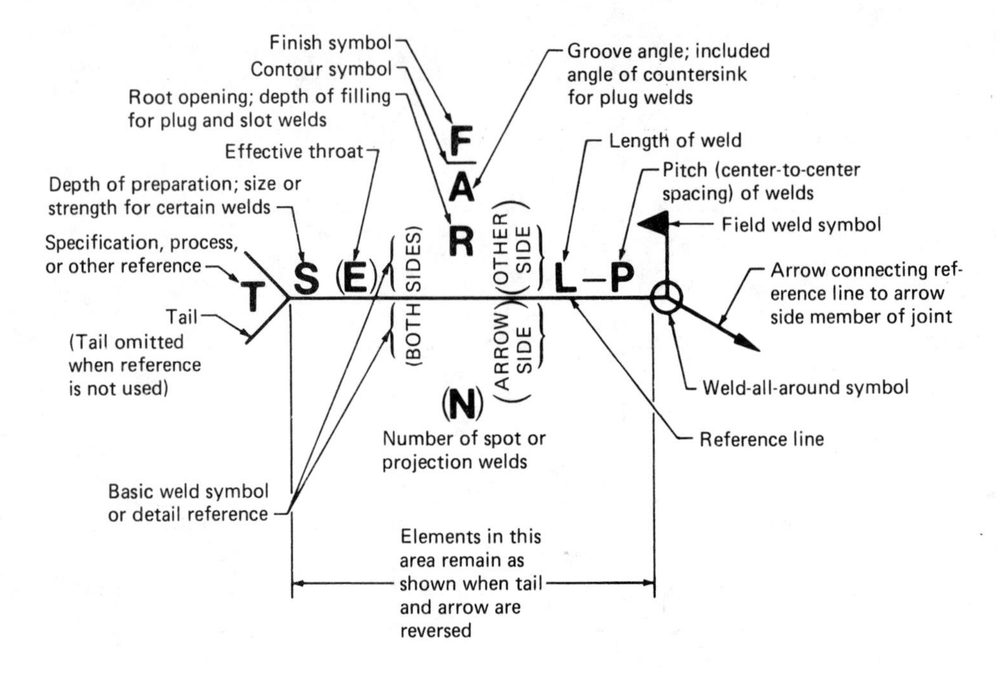

Fig. 2.2—Standard location of elements of a welding symbol

DIMENSIONS

Dimensions of a weld are shown on the same side of the reference line as the weld symbol. The size of the weld is shown to the left of the weld symbol, and the length of the weld is placed on the right. If a length is not given, the weld symbol applies to that portion of the joint between abrupt changes in the direction of welding or between specified dimension lines. If a weld symbol is shown on each side of the reference line, dimensions must be given for each weld even though both welds are identical.

Either US Customary or SI units may be used when specifying dimensions. However, only one of the two should be used for a product or project. Examples of dimensioning for typical fillet welds are shown in Fig. 2.4.

If a weld in a joint is to be intermittent, the length of the increments and the pitch (center-to-center spacing) are placed to the right of the weld symbol, Fig. 2.5.

The location on the symbol for specifying groove-weld root opening, groove angle, plug or slot weld filling depth, the number of welds required in a joint, and other dimensions are shown in Fig. 2.2. Additional information on the dimensioning of welds can be found in *AWS A2.4, Symbols for Welding and Nondestructive Testing,* latest edition.

SUPPLEMENTARY SYMBOLS

Figure 2.6 shows supplementary symbols that are used on a welding symbol. They complement the basic symbols and provide additional requirements or instructions to the welding shop.

Weld-All-Around Symbol

A weld that extends completely around a joint is indicated by the weld-all-around symbol. Figure 2.7 shows examples of its use. The weld can be in more than one plane, Fig. 2.7(C). This symbol should not be used if more than one type of weld is required to make the joint.

Field Weld Symbol

Field welds are made at the erection site, not

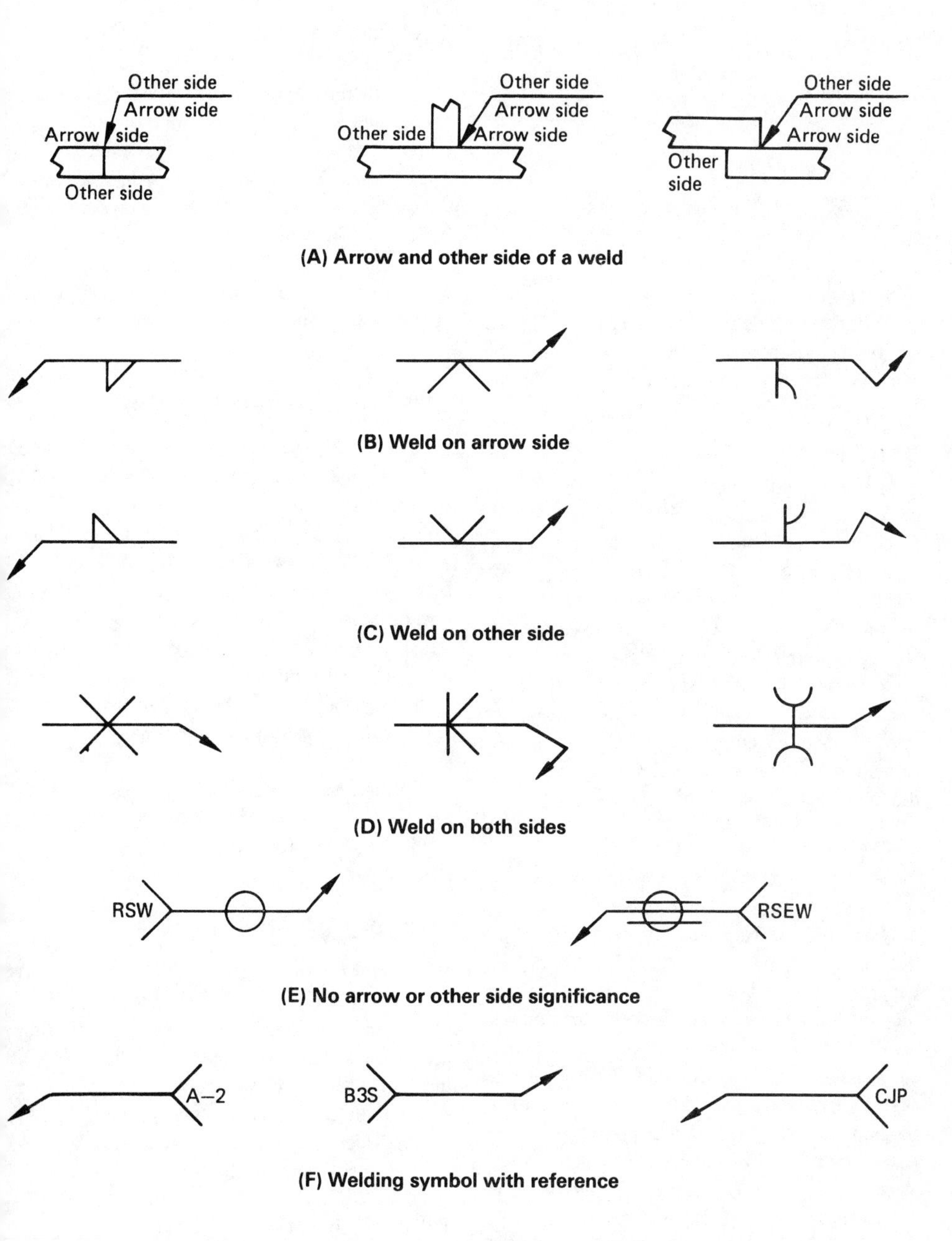

Fig. 2.3—Significance of arrow

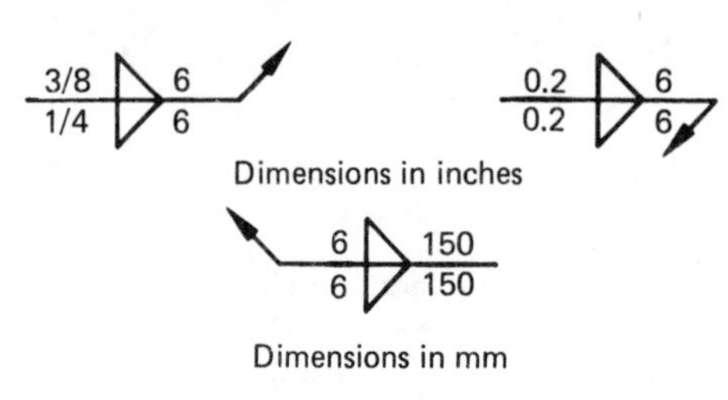

Fig. 2.4—Weld size and length

in the shop or at the place of initial construction. Each of these welds is designated by a field weld symbol (flag) that is always placed above the reference line and pointed away from the arrow toward the tail, Fig. 2.8.

Melt-Thru Symbol

The melt-thru symbol is used to show complete joint penetration (CJP) with root reinforcement on the back side of welds to be made from one side only. The reinforcement is shown by placing the melt-thru symbol on the side of the reference line opposite the weld symbol, as shown in Fig. 2.9(A). The height of root reinforcement can be specified to the left of the symbol, as shown in Fig. 2.9(B), if the amount of reinforcement is critical. Control of the reinforcement height must be practical with the specified joint design and welding process.

Backing and Spacer Symbols

A backing symbol is placed above or below the reference line to indicate that a backing ring, strip, or consumable insert is to be used in making the weld. It must be used in combination with a groove weld symbol to avoid interpretation as a plug or slot weld. A welding symbol for a typical joint with backing is shown in Fig. 2.10(A). It is a combination of a groove weld symbol on one side of the reference line and a backing symbol on the opposite side. An *R* may be placed within the backing symbol if the backing is to be removed after welding. The backing type, material, and dimensions should be specified in a note.

A welding symbol for a typical joint with a spacer strip inserted in the root of the joint is shown in Fig. 2.10(B). It is a modified groove weld symbol having a rectangle within it. The material and dimension of the spacer strip should be specified in a note.

Contour Symbol

A contour symbol is used on a welding symbol to indicate the shape of the finished weld. Welds that are to be made approximately flat, convex, or concave without subsequent finishing are represented by adding the flush, convex, or concave contour symbol to the weld symbol, Fig. 2.11(A). Welds that are to be finished by mechanical means are depicted by adding both the appropriate contour symbol and the user's standard finish symbol to the weld symbol, Fig. 2.11(B).

CONSTRUCTION OF SYMBOLS

Bevel-, J-, and flare-bevel-groove, fillet, and corner-flange weld symbols are constructed with the perpendicular leg always to the left. When only one member of a joint is to be prepared for welding, the arrow is pointed with a definite break toward that member unless the preparation is obvious. The arrow need not be broken if either member may be prepared. These features are illustrated in Fig. 2.12. Suggested size dimensions for welding symbol elements are given in *AWS A2.4, Symbols for Welding and Nondestructive Testing,* latest edition.

When a combination of welds is to be specified to make a joint, the weld symbol for each weld is placed on the welding symbol. Examples of such symbols are shown in Fig. 2.13.

MULTIPLE REFERENCE LINES

Two or more reference lines may be used with a single arrow to indicate a sequence of operations, see Fig. 2.14(A) and (B). The first operation is shown on the reference line nearest the joint. Subsequent operations are shown sequentially on other reference lines joining the arrow. Reference lines may also be used to show data supplementary to the welding symbol.

TYPES OF JOINTS

A joint is the junction of members or the edges of members that are to be joined or have been joined. Joining can be done by any number

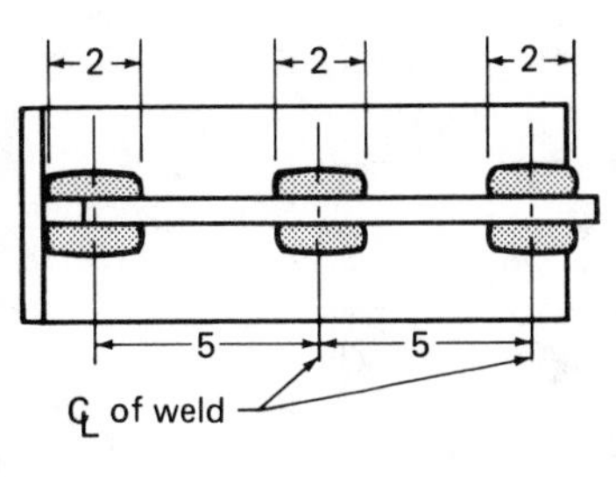

Desired welds

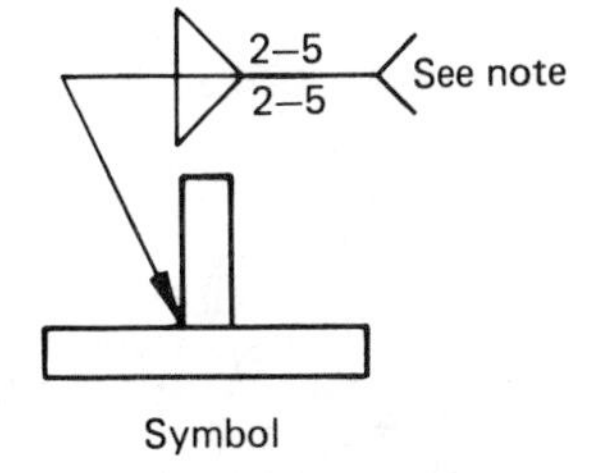

Symbol

Note: If required by actual length of the joint, the length of the increment of the welds at the end of the joint should be increased to terminate the weld at the end of the joint.

(A) Length and pitch of increments of chain intermittent welding

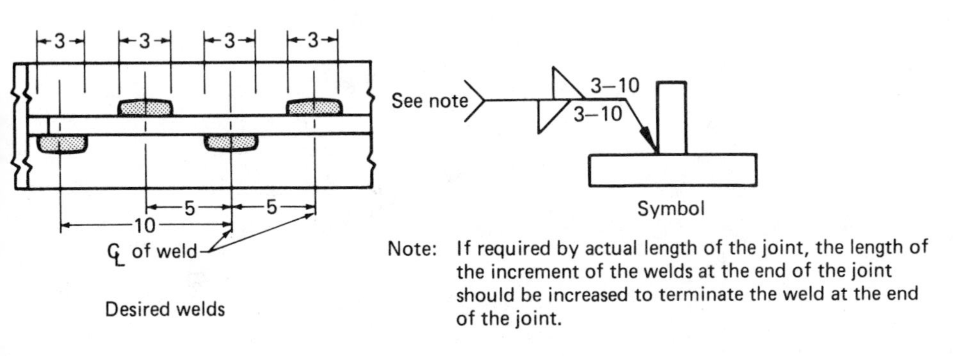

Desired welds

Symbol

Note: If required by actual length of the joint, the length of the increment of the welds at the end of the joint should be increased to terminate the weld at the end of the joint.

(B) Length and pitch of increments of staggered intermittent welding

Fig. 2.5—Dimensioning intermittent fillet welds

Examine or weld all around	Field weld	Melt-thru	Backing, spacer, or consummable insert	Contour		
				Flush or flat	Convex	Concave

Fig. 2.6—Supplementary symbols

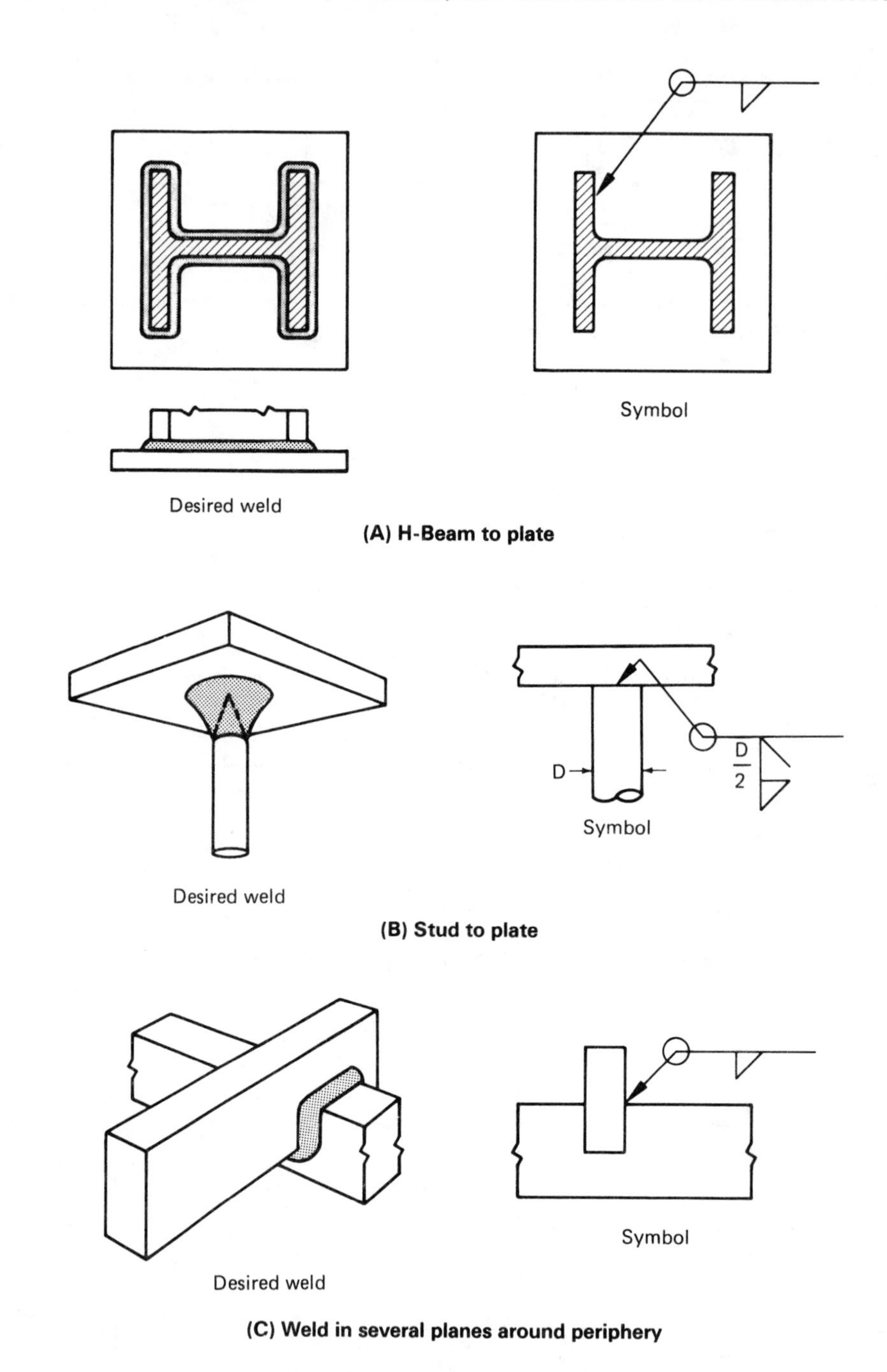

Fig. 2.7—Weld-all-around symbol

Fig. 2.8—Field weld symbol

of methods such as adhesives, bolts, screws, rivets, welds, or brazes. The five basic joints used in welding and brazing are butt, corner, lap, edge, and T-joints. These joints and the applicable welds to join them are shown in Fig. 2.15.

PROCESSES

Letter designations are used in the tail of a welding symbol to indicate the appropriate welding or brazing process. The more frequently used welding designations are listed in Table 2.1. A complete listing of designations for welding, brazing, and allied processes is given in *AWS A2.4, Symbols for Welding and Nondestructive Testing,* latest edition.

Table 2.1
Frequently used welding process designations

Letter designation	Welding process
SMAW	Shielded metal arc welding
SAW	Submerged arc welding
GMAW	Gas metal arc welding
FCAW	Flux cored arc welding
GTAW	Gas tungsten arc welding
PAW	Plasma arc welding
OFW	Oxyfuel gas welding
EBW	Electron beam welding
RSW	Resistance spot welding
RSEW	Resistance seam welding

EXAMPLES

After the joint is designed, a welding symbol can generally be used to specify the required welding. Figures 2.16 through 2.22 show examples of welded joints and the proper symbols to describe them. When the desired weld cannot be adequately described with the basic weld symbols, the weld should be detailed on the drawing. Reference is made to the detail on the reference line.

For a single-V-groove weld, both members are beveled equally to form a groove at the joint. Figure 2.16(A) shows such a weld and the appropriate welding symbol.

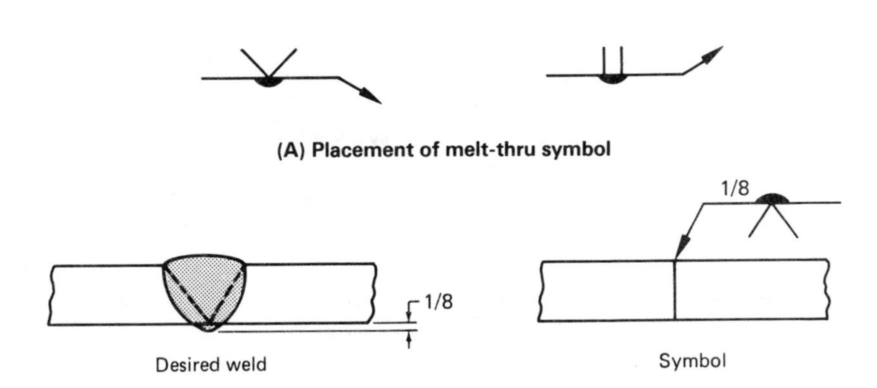

Fig. 2.9—Melt-thru symbol

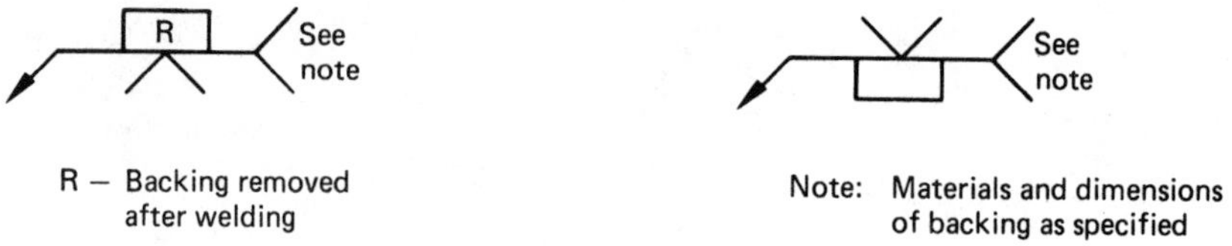

R – Backing removed after welding

Note: Materials and dimensions of backing as specified

(A) Backing symbol

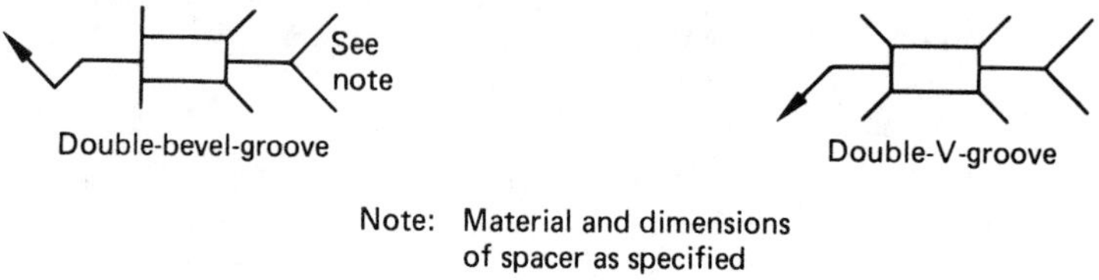

See note

Double-bevel-groove

Double-V-groove

Note: Material and dimensions of spacer as specified

(B) Spacer symbol

Fig. 2.10—Backing and spacer symbols

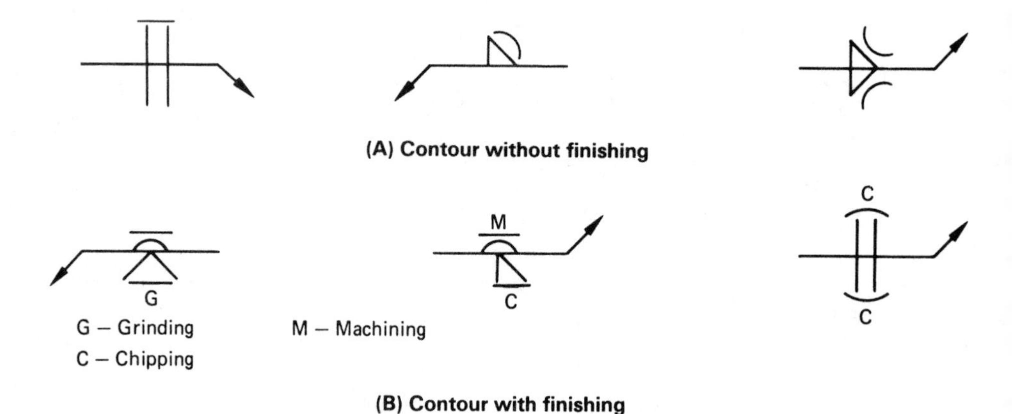

(A) Contour without finishing

G – Grinding

C – Chipping

M – Machining

(B) Contour with finishing

Fig. 2.11—Contour symbols

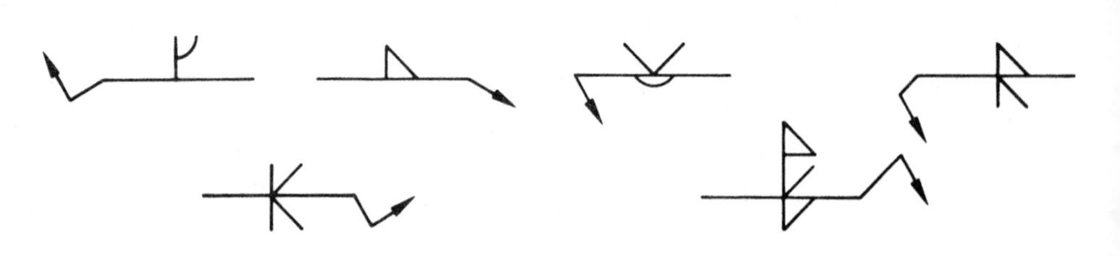

Fig. 2.12—Weld symbol construction

Fig. 2.13—Combined weld symbols

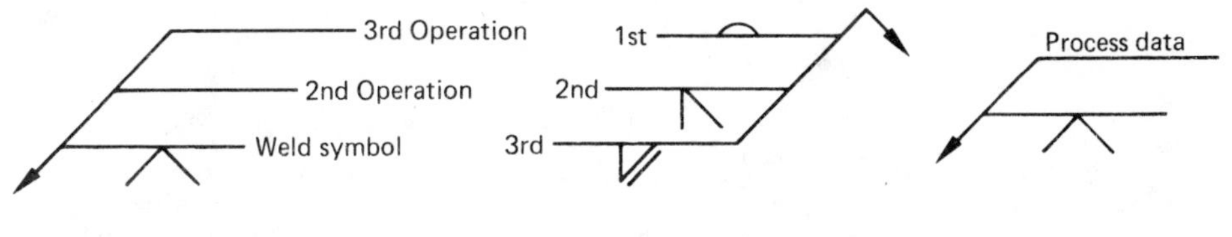

(A) Multiple reference lines

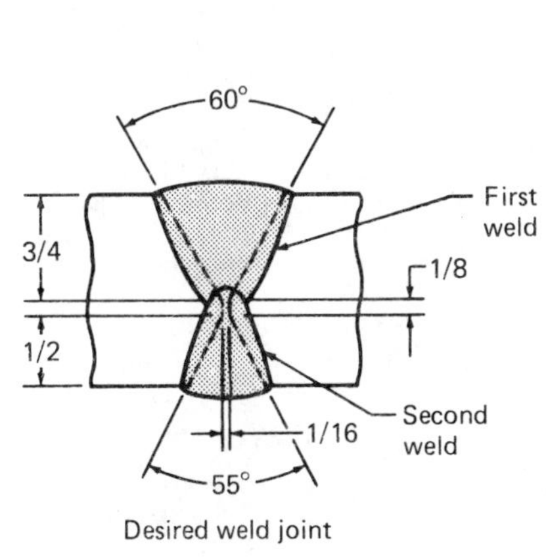

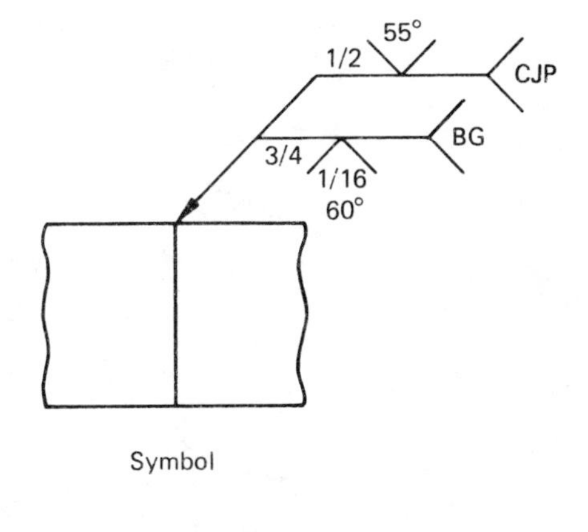

Depth of preparation (arrow side) – 3/4 in.
Depth of preparation (other side) – 1/2 in.
Root opening – 1/16 in.
Groove angle (arrow side) = 60°
Groove angle (other side) = 55°

Note: BG – Back gouge to sound metal
CJP – Complete joint penetration

(B) Application of multiple reference lines

Fig. 2.14—Multiple reference lines

If a V-shaped groove is to be prepared on both sides of the joint, the weld is a double-V-groove type. The symbol for this type of weld is shown in Fig. 2.16(B).

When a round member is placed on a flat surface and a weld is made lengthwise along one side, the weld is a single-flare-bevel-groove type. The weld and the appropriate symbol are shown in Fig. 2.17(A). If two round members are placed side by side and welded together lengthwise, the weld is a single-flare-V-groove type. The weld and symbol are shown in Fig. 2.17(B). The round shapes may be bent or rolled plates, pipes, or tubes.

Fillet Welds

Joints that can be joined by fillet welds are lap, corner, and T-types.[2] Fillet welds are also used in conjunction with groove welds as reinforcement in corner and T-joints. Examples of fillet weld symbols are shown in Fig. 2.18.

Plug and Slot Welds

Plug and slot welds are similar in design but

2. A fillet weld has an approximate triangular cross section and joins two surfaces at about 90 degrees to each other. When the surfaces are at a greater or lesser angle, the weld should be specified with appropriate explanatory details and notes.

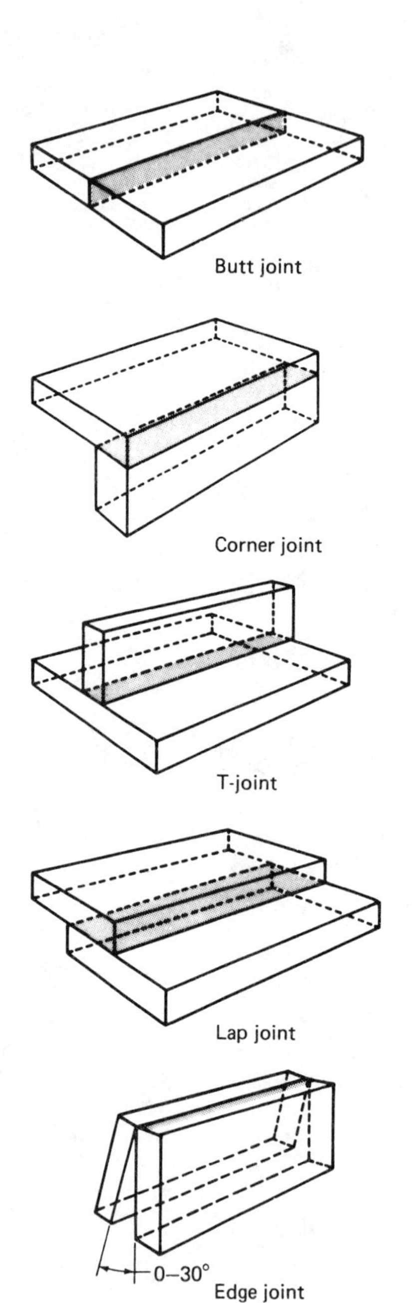

Applicable welds

Square-Groove
V-Groove
Bevel-Groove
U-Groove
J-Groove
Flare-V-Groove
Flare-Bevel-Groove
Edge-Flange
Braze

Applicable welds

Fillet
Square-Groove
V-Groove
Bevel-Groove
U-Groove
J-Groove
Flare-V-Groove
Flare-Bevel-Groove
Edge-Flange
Corner-Flange
Spot
Projection
Seam
Braze

Applicable welds

Fillet
Plug
Slot
Square-Groove
Bevel-Groove
J-Groove
Flare-Bevel-Groove
Spot
Projection
Seam
Braze

Applicable welds

Fillet
Plug
Slot
Bevel-Groove
J-Groove
Flare-Bevel-Groove
Spot
Projection
Seam
Braze

Applicable welds

Square-Groove
Bevel-Groove
V-Groove
U-Groove
J-Groove
Edge-Flange
Corner-Flange
Seam
Edge

Fig. 2.15—Basic types of joints

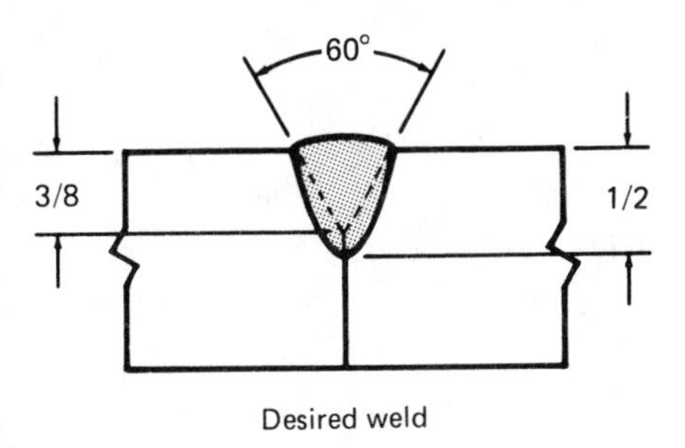

Desired weld

3/8 (1/2)
0
60°

Symbol

Depth of preparation – 3/8 in.
Effective throat (always shown in parentheses) – 1/2 in.

Root opening – 0
Groove angle – 60°

(A) Single V-groove weld from arrow side

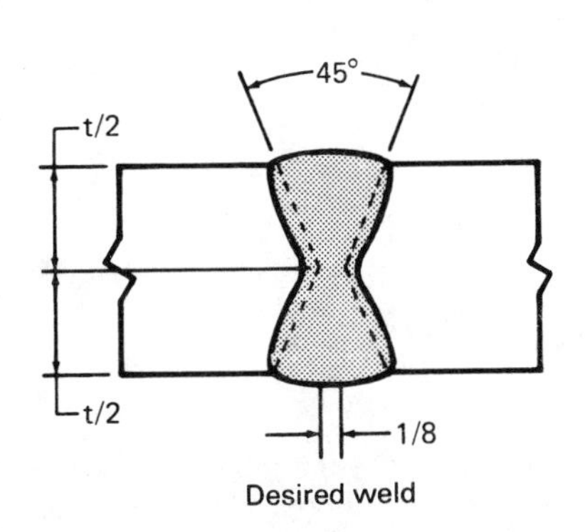

Desired weld

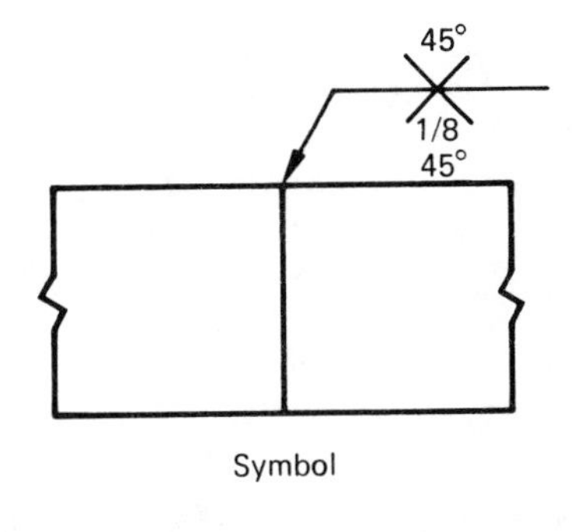

Symbol

Depth of preparation (each side) – t/2
Groove angle (each side) – 45°
Root opening – 1/8 in.

(B) Double V-groove weld

Fig. 2.16—Single- and Double-V-groove welds

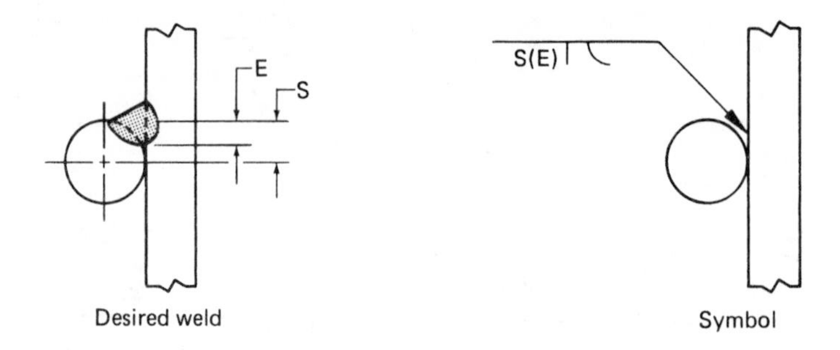

(A) Single flare-bevel-groove weld

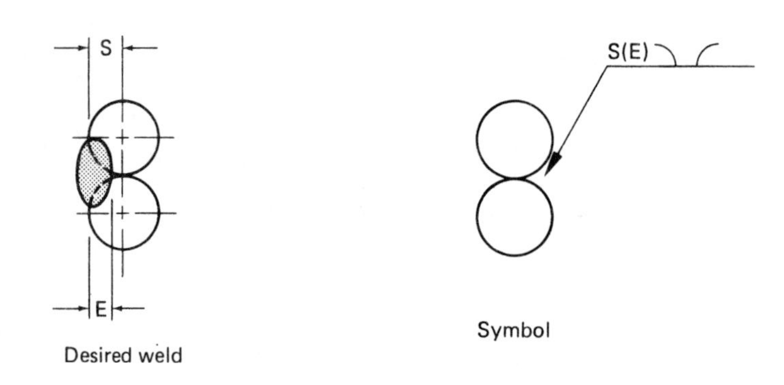

(B) Single flare-V-groove weld

Fig. 2.17—Flare-bevel and flare-V welds

different in shape. In either case, the hole or slot is made in only one member of the joint. A plug or slot weld is not to be confused with a fillet weld in a hole. Plug and slot welds require definite depths of filling. An example of a plug weld and the welding symbol are shown in Fig. 2.19(A), and of a slot weld and the symbol in Fig. 2.19(B). The dimensions, spacing, orientation, and location of slot welds cannot be given on the welding symbol. These data are shown on the drawing or a detail with a reference to them in the tail of the welding symbol.

Flange Welds

A flange weld is made on the edges of two or more members that are usually light-gage sheet metal. At least one of the members has to be flanged by bending it approximately 90 degrees. Examples of flange welds and welding symbols are shown in Fig. 2.20.

Spot Welds

A spot weld is made between or upon overlapping members. Coalescence may start and continue over the faying surfaces, or may proceed from the surface of one member. The weld cross section (plan view) is approximately circular. Fusion welding processes that have the capability of melting through one member of a joint and fusing with the second member at the faying surface may be used to make spot welds. Resistance welding equipment is also used. Examples

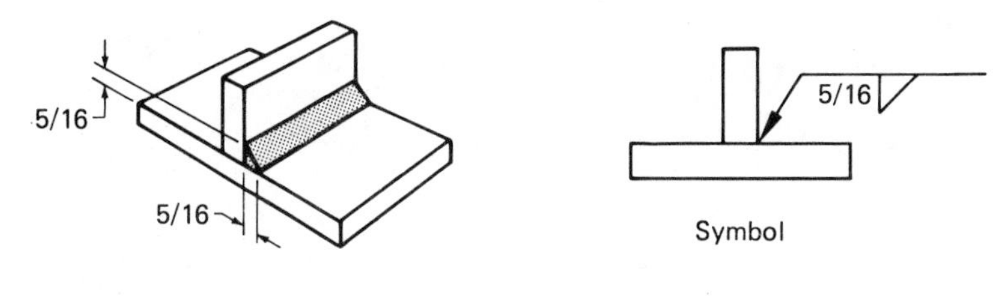

Desired weld

Size of weld – 5/16 in.

(A) Fillet weld with equal legs

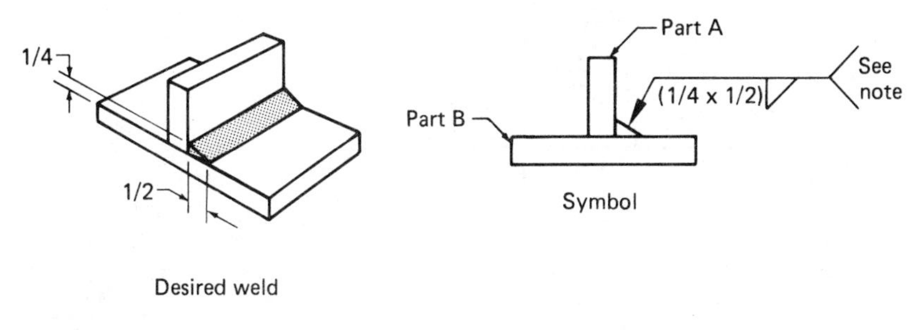

Size of vertical leg – 1/4 in.
Size of horizontal leg – 1/2 in.

Note: The 1/4-in. leg applies to Part A. Weld orientation must be shown on the drawing.

(B) Fillet weld with unequal legs

Fig. 2.18—Fillet welds

of arc and resistance spot welds are shown in Fig. 2.21, together with the proper welding symbols.

Seam Welds

A seam weld is a continuous weld made between or upon overlapping members. Coalescence may start and occur on the faying surfaces, or may proceed from the surface of one member. The continuous weld may be a single weld bead or a series of overlapping spot welds. Seam welds are made with processes and equipment that are similar to those used for spot welding. A means of moving the welding head along the seam must be provided. Examples of arc and resistance seam welds and the appropriate welding symbols are shown in Fig. 2.22.

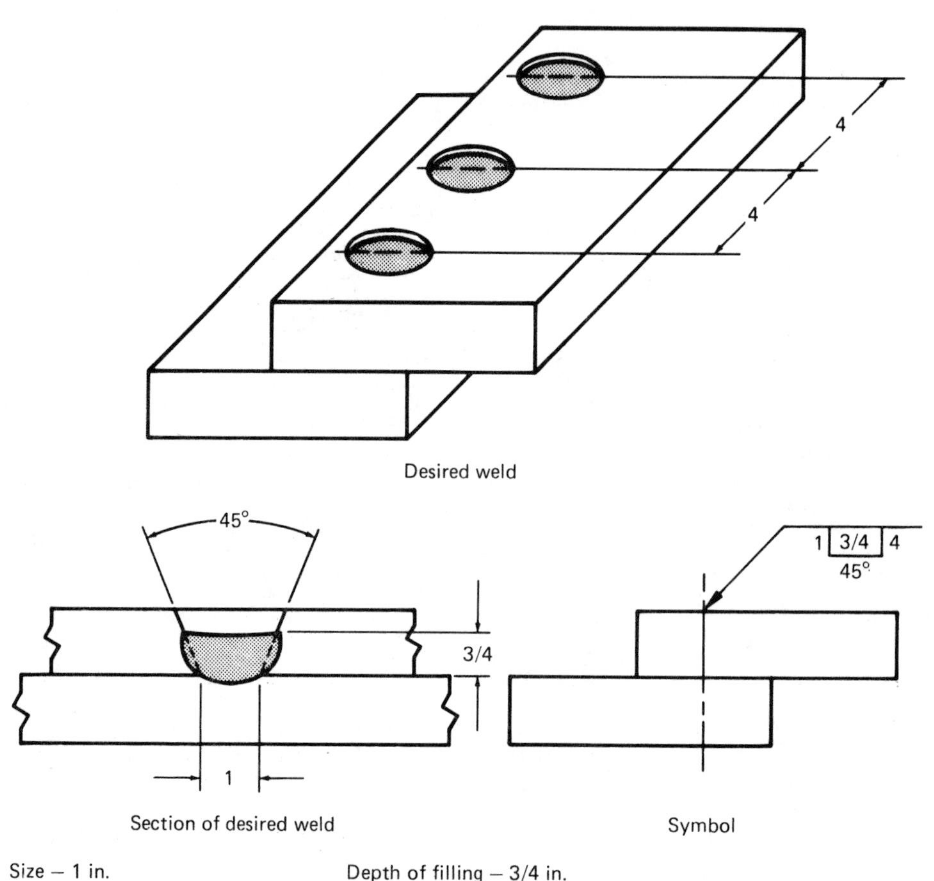

Size – 1 in.
Angle of countersink – 45°
Depth of filling – 3/4 in.
Pitch (center-to-center spacing) – 4 in.

Fig. 2.19A—Plug welds

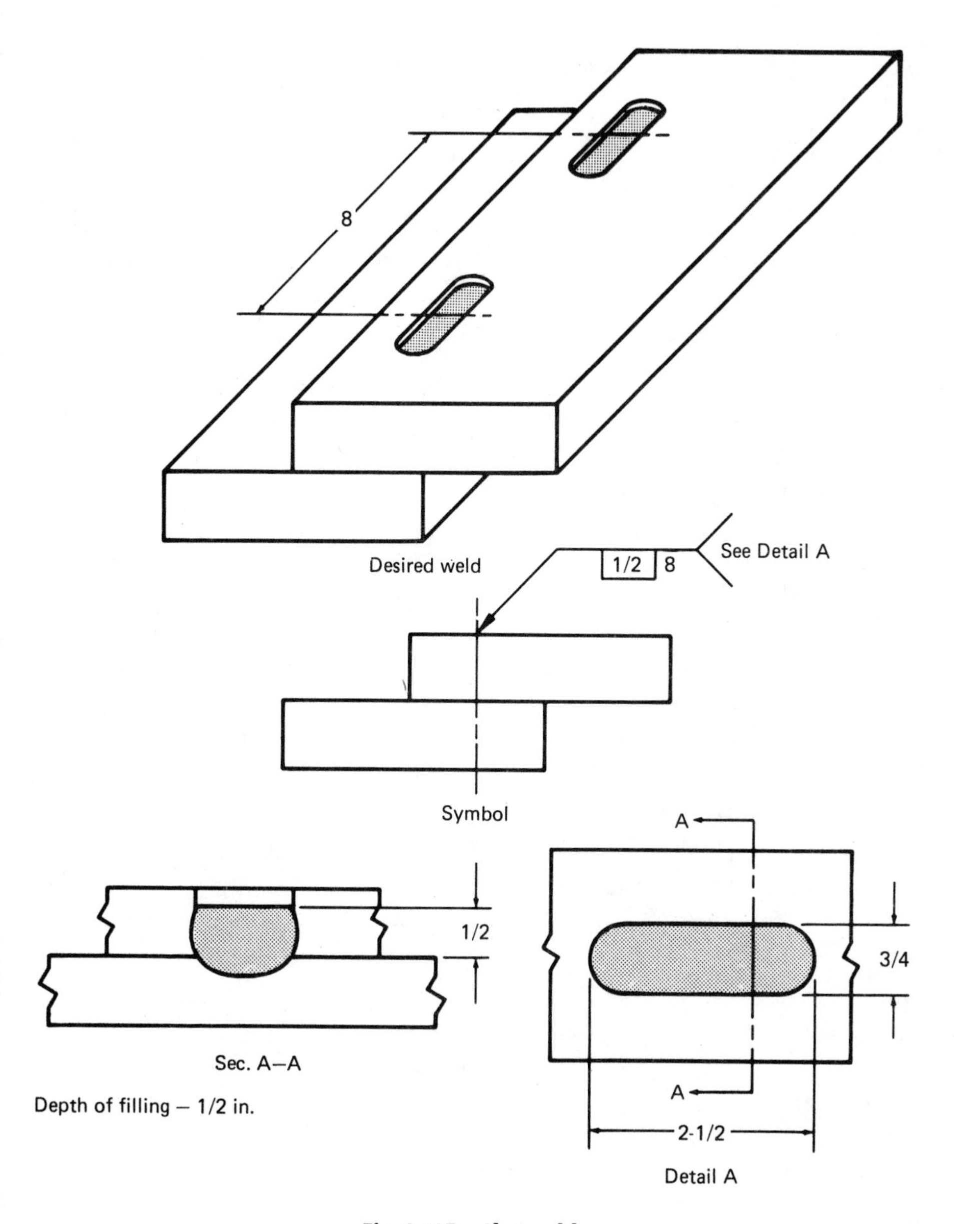

Fig. 2.19B—Slot welds

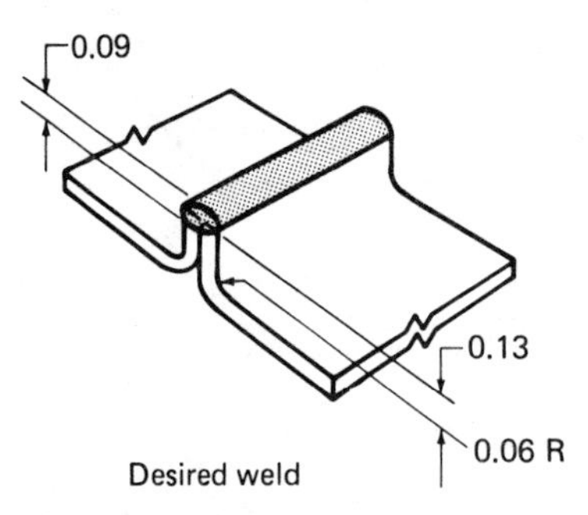

Desired weld

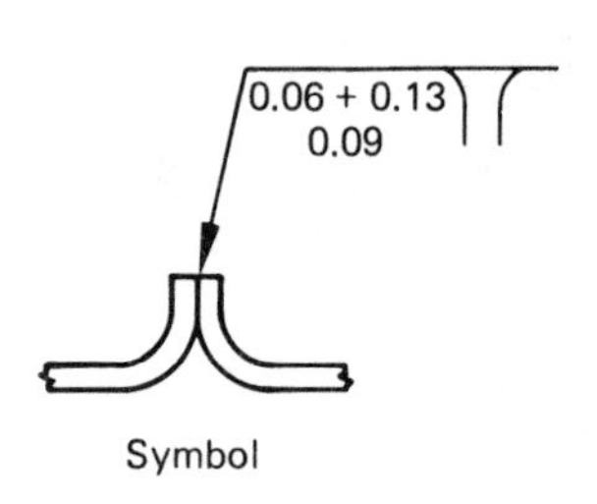

Symbol

Radius of flange – 0.06 in.
Height of flange above point of tangency – 0.13 in.
Weld thickness – 0.09 in.

(A) Edge-flange weld

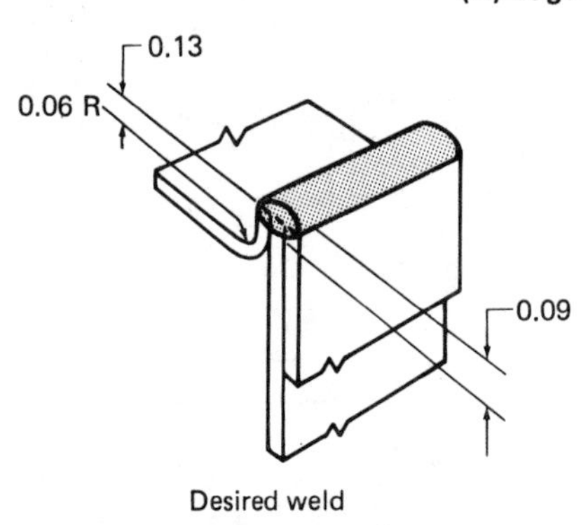

Desired weld

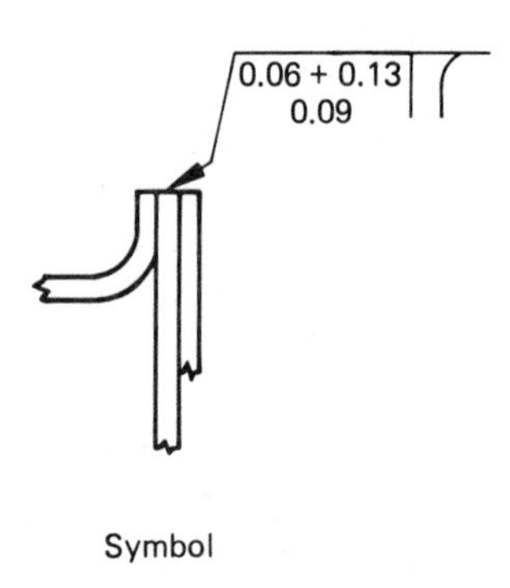

Symbol

Radius of flange – 0.06 in.
Height of flange above point of tangency – 0.13 in.
Weld thickness – 0.09 in.

(B) Corner-flange weld

Fig. 2.20—Flange welds

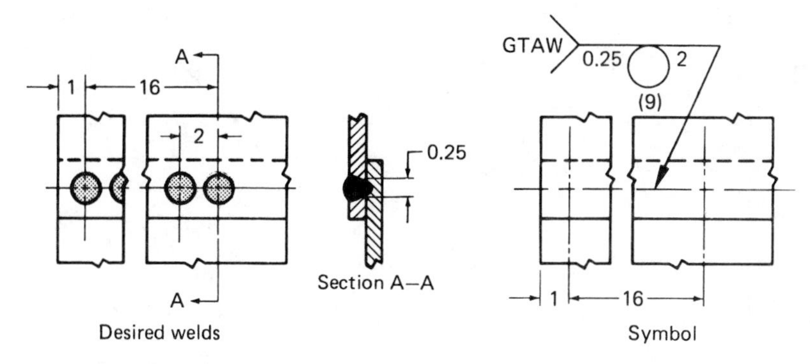

Size (at faying surface) – 0.25 in.

Number of spot welds – 9

Pitch (center-to-center spacing) – 2 in.

Note: Size can be given in pounds or newtons per spot rather than the diameter.

(A) Arc spot weld

Desired welds

Section A–A

Symbol

Size (at faying surface) – 0.25 in.

Number of spot welds – 5

Pitch (center-to-center spacing) – 1 in.

Distance from center of first spot weld to edge – 1/2 in.

Note: Size can be given in pounds or newtons per spot rather than the diameter.

(B) Resistance spot welds

Fig. 2.21—Spot welds

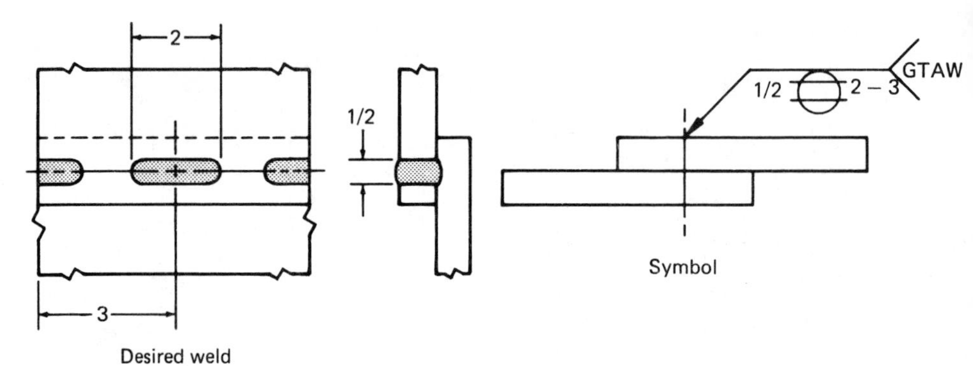

Size (at faying surface) — 1/2 in.
Length — 2 in.
Pitch (center-to-center spacing) — 3 in.

Note: Size can be given in pounds per linear in. or newtons per millimeter.

(A) Arc seam weld

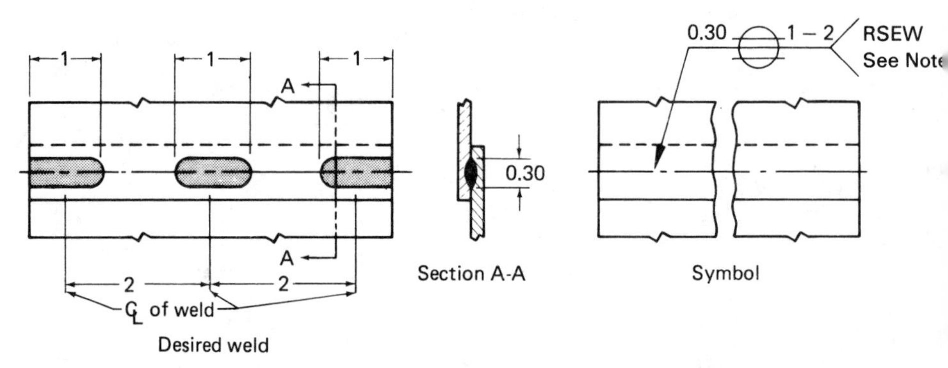

Size (at faying surface) — 0.30 in.
Length — 1 in.
Pitch (center-to-center spacing) — 2 in.

Note: If required by actual length of the joint, the length of the increment of the welds at the end of the joint should be increased to terminate the weld at the end of the joint.

(B) Resistance seam weld

Fig. 2.22—Seam welds

BRAZING SYMBOLS

BRAZED JOINTS

If no special preparation other than cleaning is required for making a brazed joint, the arrow and reference line are used with the brazing process designation indicated in the tail. See Fig. 2.23(A). Application of conventional weld symbols to brazed joints is illustrated in Figs. 2.23(B) through (H). Joint clearance can be indicated on the brazing symbol.

BRAZING PROCESSES

Brazing processes to be used in construction can be designated by letters. The commonly used brazing processes and their letter designations are given in Table 2.2.

Table 2.2
Letter designations for brazing processes

Process	Designation
Diffusion brazing	DFB
Dip brazing	DB
Furnace brazing	FB
Induction brazing	IB
Infrared brazing	IRB
Resistance brazing	RB
Torch brazing	TB

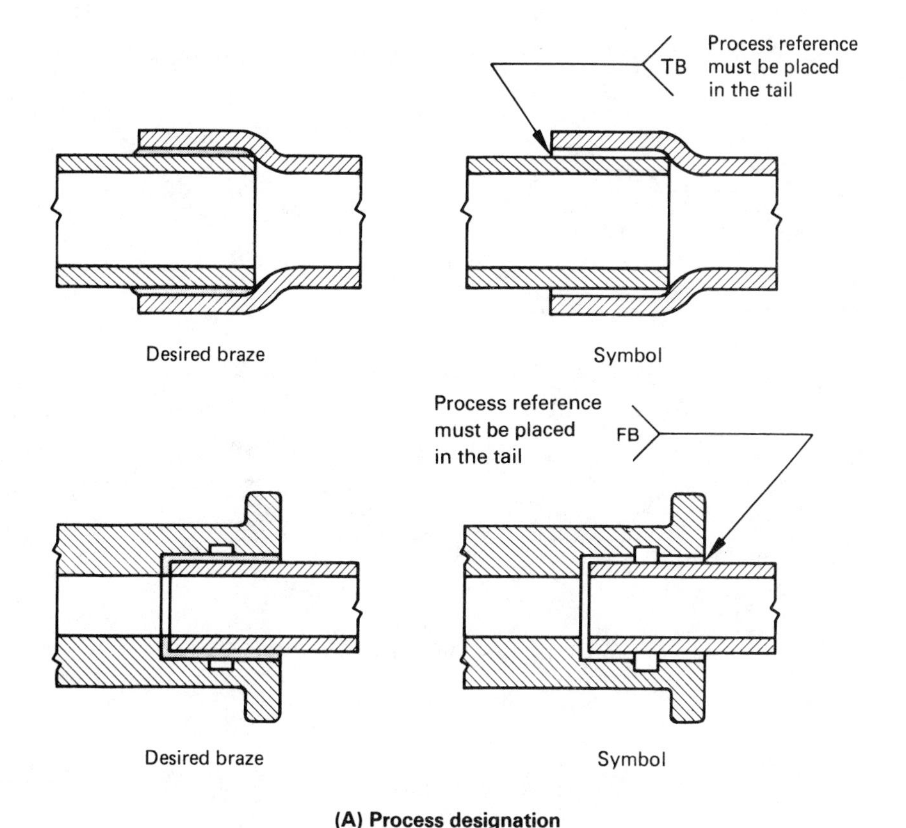

(A) Process designation

Fig. 2.23—Application of brazing symbols

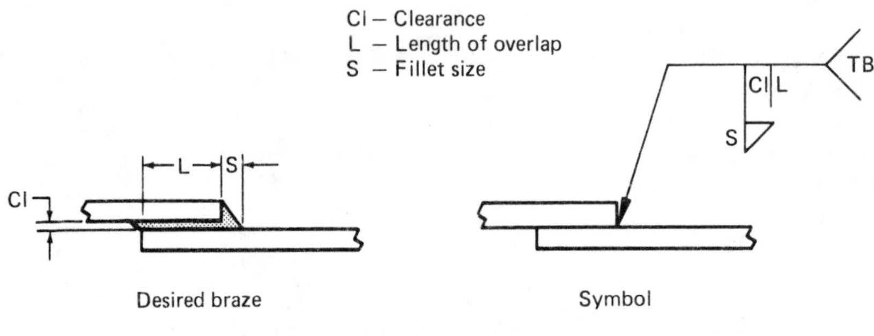

(B) Location of elements of brazing symbol

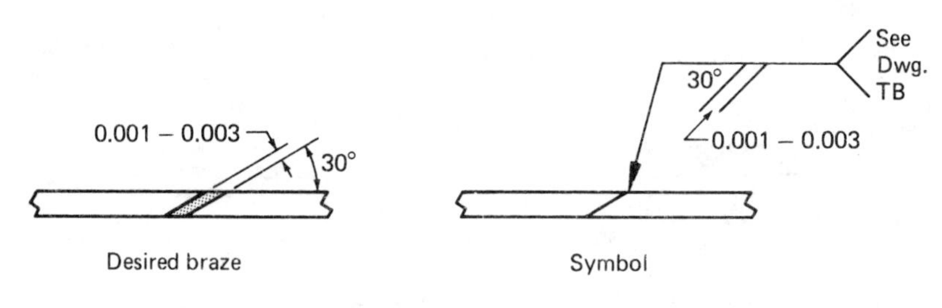

(C) Scarf joint

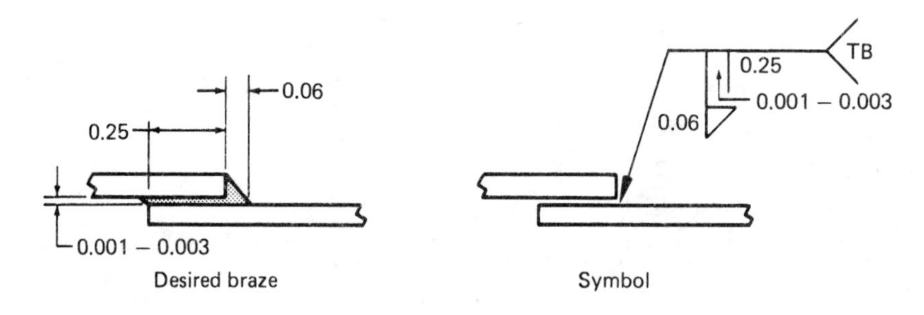

(D) Lap joint with fillet

Fig. 2.23 (cont.)—Application of brazing symbols

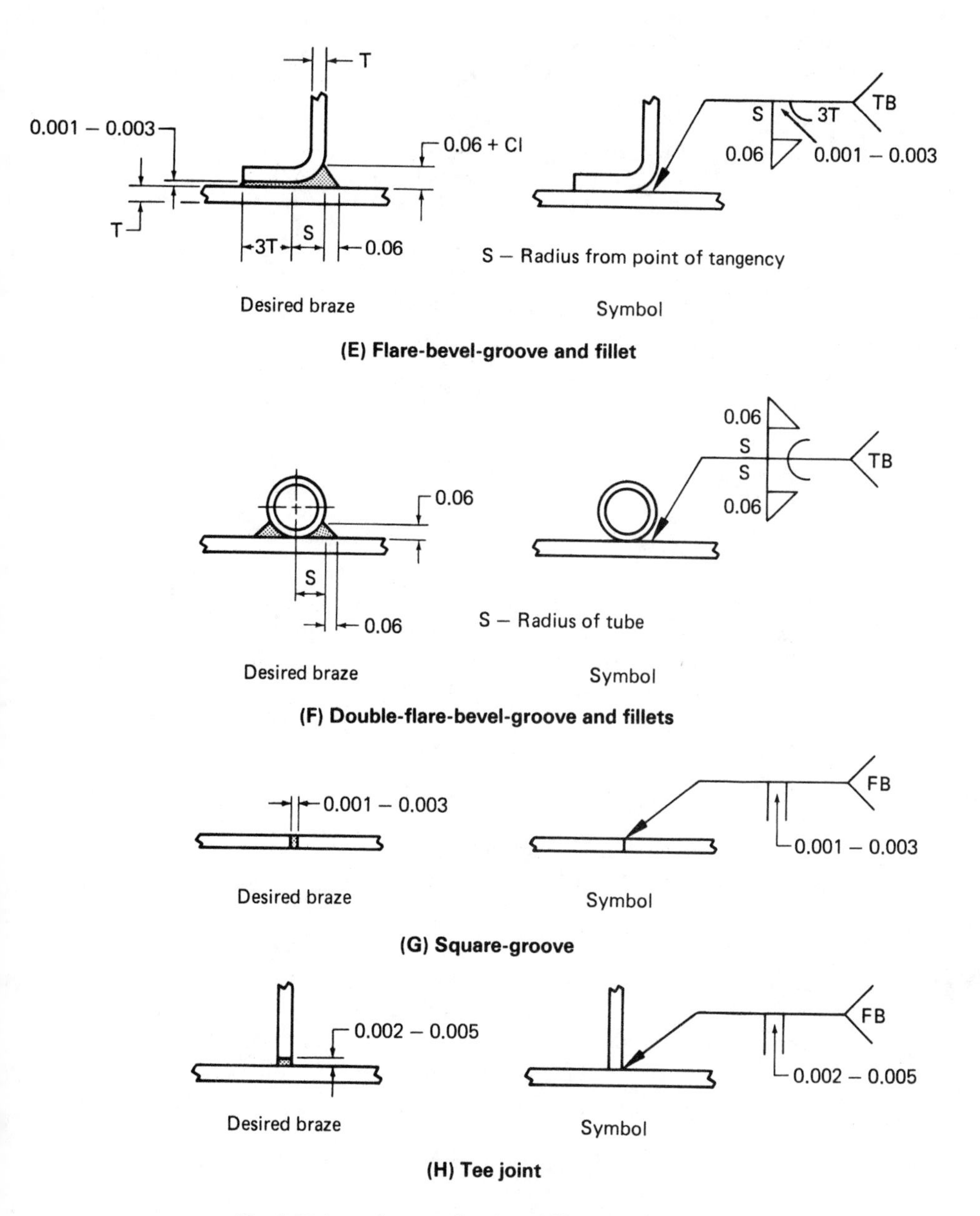

Fig. 2.23 (cont.)—Application of brazing symbols

NONDESTRUCTIVE EXAMINATION SYMBOLS

Symbols for nondestructive examinations (NDE) provide means for specifying on engineering drawings the method of examination to be used. Nondestructive examination symbols may be combined with welding symbols by using an additional reference line or by specifying the examination method in the tail of the welding symbol.

EXAMINATION SYMBOLS

The symbols for the various nondestructive examination processes are shown in Table 2.3. The elements of a nondestructive examination symbol are as follows:

(1) Reference line
(2) Arrow
(3) Basic examination symbol
(4) Examine-all-around symbol
(5) Number of examinations (N)
(6) Examine in field
(7) Tail
(8) Specifications or other references

The standard locations of these elements with respect to each other are shown in Fig. 2.24.

Table 2.3
Nondestructive examination symbols

Type of examination	Symbol	Type of examination	Symbol
Acoustic emission	AET	Penetrant	PT
Eddy current	ET	Proof	PRT
Leak	LT	Radiographic	RT
Magnetic particle	MT	Ultrasonic	UT
Neutron radiographic	NRT	Visual	VT

Significance of Arrow Location

The arrow connects the reference line to the part to be examined. The side of the part to which the arrow points is the *arrow side*. The side opposite from the arrow side is the *other side*.

Arrow Side Examination

Examinations to be made on the arrow side of a joint are indicated on the NDE symbol by placing the basic examination symbol below the reference line, i.e., toward the reader. Fig. 2.25(A) illustrates this type of symbol.

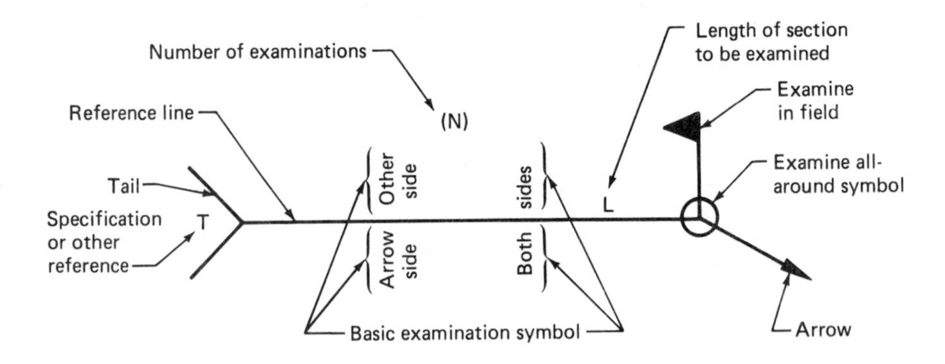

Fig. 2.24—Standard location of NDE symbol elements

Other Side Examination

Examinations to be made on the other side of the joint are indicated on the NDE symbol by placing the basic examination symbol above the reference line, i.e., away from the reader. This position is illustrated in Fig. 2.25(B).

Examinations on Both Sides

Examinations that are to be made on both sides of the joint are indicated by basic NDE symbols on both sides of the reference line. See Fig. 2.25(C).

No Side Significance

When the examination may be performed from either side or has no arrow or other side significance, the basic examination symbols are centered in the reference line. Fig. 2.25(D) shows this arrangement.

(A) Examine arrow side **(B) Examine other side**

(C) Examine both sides **(D) No side significance**

Fig. 2.25—Significance of symbol placement

Radiographic Examination

The direction of radiation may be shown in conjunction with radiographic (RT) and neutron radiographic (NRT) examination symbols. The direction of radiation may be indicated by a special symbol and line located on the drawing at the desired angle. Figure 2.26 shows the symbol together with the NDE symbol.

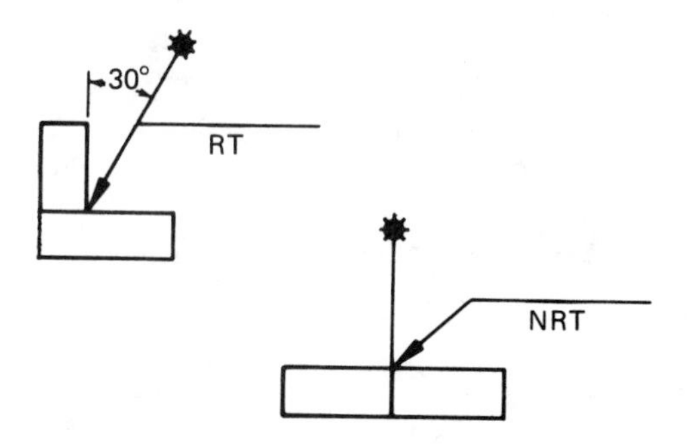

Fig. 2.26—Specifying the direction of radiographic examination

COMBINED SYMBOLS

Nondestructive examination symbols may be combined with welding symbols, Fig. 2.27(A), or with each other if a part is to be examined by two or more methods, Fig. 2.27(B).

Where an examination method with no arrow or other side significance and another method that has side significance are to be used, the NDE symbols may be combined as shown in Fig. 2.27(C).

REFERENCES

Specifications or other references need not be used on NDE symbols when they are described elsewhere. When a specification or other reference is used with an NDE symbol, the reference is placed in the tail, Fig. 2.27(D).

EXTENT OF EXAMINATION

Length

To specify the extent of examinations of welds where only the length of a section need be considered, the length is shown to the right of the NDE symbol, Fig. 2.28(A).

Exact Location. To show the exact location of a section to be examined as well as its length, appropriate dimensions are shown on the drawing, Fig. 2.28(B).

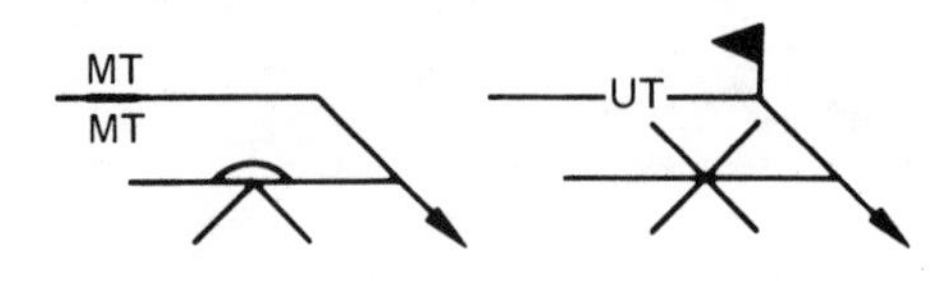

(A) Combined NDE and welding symbols

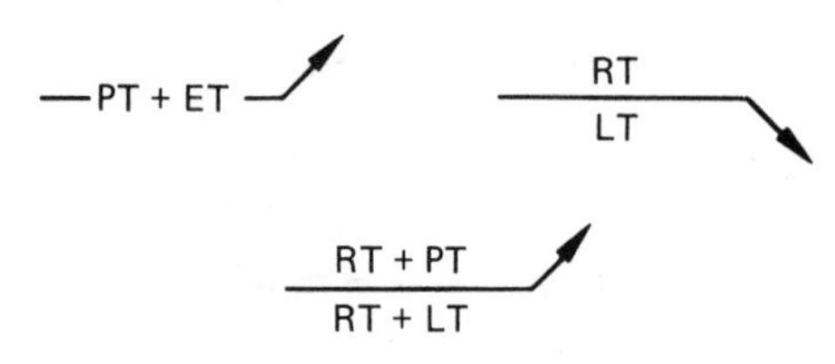

(B) Combined NDE symbols

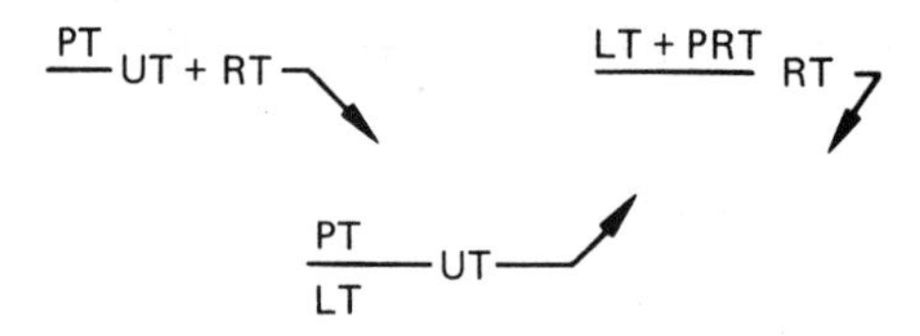

(C) Combined side and no-side significant symbols

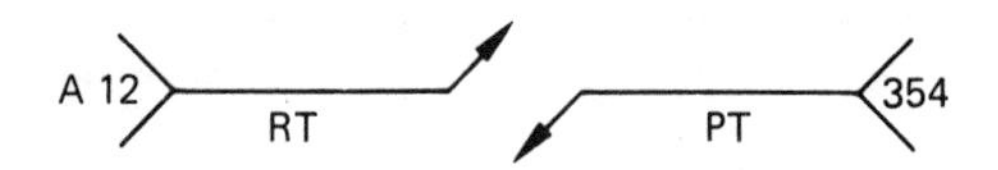

(D) NDE symbol with reference

Fig. 2.27—Combined nondestructive examination symbols

Partial Examination. When a portion of the length of a weld or part is to be examined with the locations determined by a specified procedure, the percentage of the length to be examined is indicated on the symbol, as illustrated in Fig. 2.28(C).

Full Length. The full length of a part is to be examined when no length dimension is shown on the NDE symbol.

Number of Examinations

When several examinations are to be made on a joint or part at random locations, the number of examinations is given in parentheses, as shown in Fig. 2.28(D).

All-Around Examination

Figure 2.28(E) shows the use of the examine-all-around symbol to indicate that complete examination is to be made of a continuous joint, such as a circumferential pipe joint.

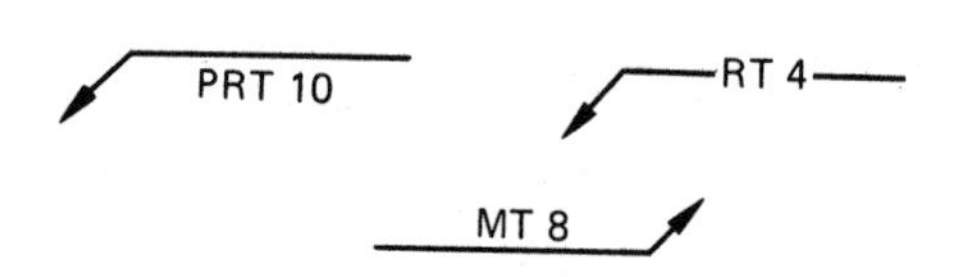

(A) Length of examination

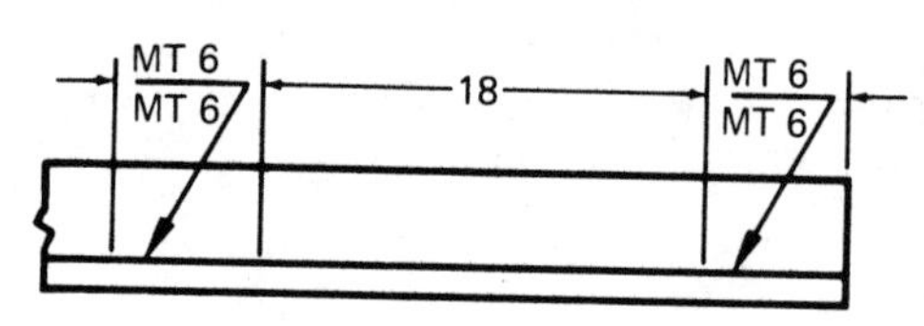

(B) Exact location of examination

(C) Partial examination

(D) Number of examinations

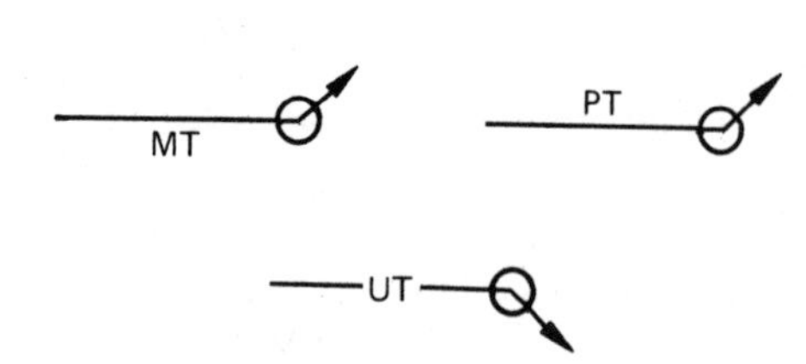

(E) All-around examination

Fig. 2.28—Extent of examination

Area

Nondestructive examination of areas of parts is indicated by one of the following methods.

Plane Areas. To indicate a plane area to be examined on a drawing, the area is enclosed by straight broken lines with a circle at each change of direction. The type of nondestructive examination to be used in the enclosed area is designated with the appropriate symbol, as shown in Fig. 2.29(A). The area may be located by coordinate dimensions.

Areas of Revolution. For nondestructive examination of areas of revolution, the area is indicated by using the examine-all-around symbol and appropriate dimensions. In Figure 2.29(B), the upper right symbol indicates that the bore of the hub is to be examined by magnetic particle inspection for a distance of three inches from the flange face. The lower symbol indicates an area of revolution is to be examined radiographically. The length of the area is shown by the dimension line.

The symbol shown in Fig. 2.29(C) indicates that a pipe or tube is to be given an internal proof examination and an external eddy current examination. The entire length is to be examined because no limiting dimensions are shown.

Acoustic Emission

Acoustic emission examination is generally applied to all or a large portion of a component, such as a pressure vessel or a pipe. The symbol shown in Fig. 2.30 indicates acoustic emission examination of the component without specific reference to locations of the sensors.

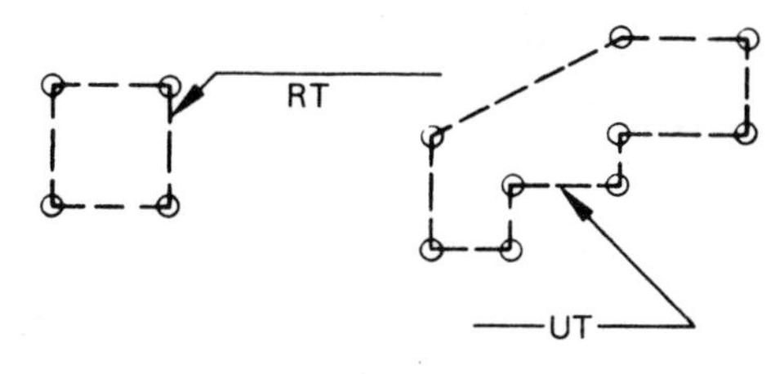

(A) Plane areas

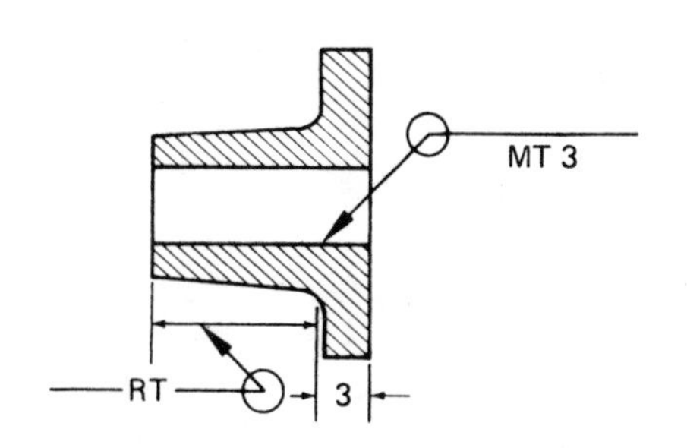

(B) Area of revolution — one side

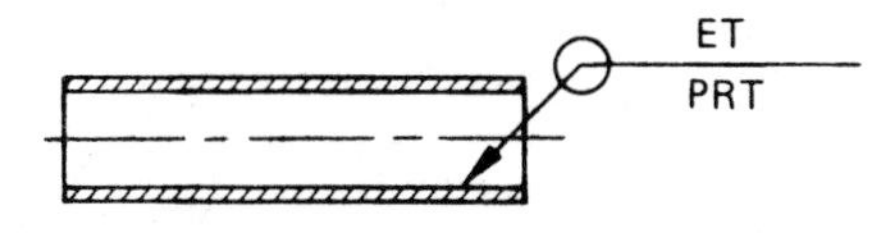

(C) Area of revolution — both sides

Fig. 2.29—Examining specified areas

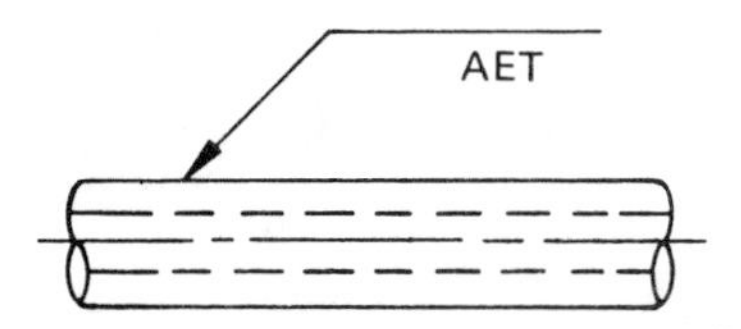

Fig. 2.30—Acoustic emission

3
Economics and Cost Estimating

The Cost Estimate *122*

Welding Costs *123*

Economics of Arc Welding *134*

Economics of Welding Automation and Robotics *135*

Economics of New Joining Processes . . . *139*

Economics of Brazing and Soldering . . . *142*

Economics of Thermal Cutting *143*

Supplementary Reading List *148*

Chapter Committee

S. CORICA, *Chairman*
Consultant

G.E. COOK
CRC Welding Symbols, Incorporated

J. KAROW
MG Industries, Incorporated

Welding Handbook Committee Member

C.W. CASE
Huntington Alloys, Incorporated

3

Economics and Cost Estimating

THE COST ESTIMATE

A cost estimate is a forecast of expenses that may be incurred in the manufacture of a product. The estimate may include only the manufacturing costs. Other elements are added to the manufacturing cost to obtain the selling price including administrative expense, handling, warehousing or storage, and profit. The additional expenses over manufacturing costs are sometimes called *general and administrative expense.*

Cost estimating is different from cost accounting in that it is a forecast or a scientific analysis of all the cost factors that enter into the production of a proposed product. Cost accounting is recording, analyzing, and interpreting the historical data of dollars and hours that were used to produce a particular product under existing conditions at a given time. Cost accounting tells management the effectiveness of current or past production while cost estimating predicts the probable costs for making new parts or products, or adopting new processes or operations.

Some purposes and uses of cost estimates are as follows:

(1) Provide information to be used in establishing the selling price for quotation, bidding, or evaluating contracts.

(2) Ascertain whether a proposed product can be made and marketed at a profit considering existing prices and future competition.

(3) Assemble data for make-or-buy decisions, that is, determine whether parts and assemblies can be made or purchased from a vendor more cheaply.

(4) Determine how much may be invested in tools and equipment to produce a product or component by one process as compared to another.

(5) Ascertain the best and most economical method, process, and materials for the manufacturing of a product.

(6) Establish a basis for a cost reduction program showing savings that are or may be made by changing methods or processes, or by the application of value analysis techniques.

(7) Predetermine standards of production performance that may be used at the start of production in the control of operating costs.

(8) Predict the effect of production volume changes on future profits by the introduction of automation, mechanization, or other improvements suitable to mass production.

In the case of new products, the estimate details should include the first formal process planning that may later become the basis for the following:

(1) Establishing personnel requirements to meet future work plans.

(2) Predicting material needs over the length of a contract.

(3) Setting the overall schedule or time table for meeting company goals.

(4) Specifying the equipment, machines, and facilities required for production of a proposed product in the time and quantities required.

ELEMENTS OF A COST ESTIMATE

All expense items to a company must be considered in the established selling price. The general and administrative (G & A) expenses as well as profit must be added to the manufacturing costs. The accounting department usually establishes G & A expenses, and management may establish the profit. Only the manufacturing cost is of prime interest to a cost estimator.

MANUFACTURING COSTS

Manufacturing cost includes direct materials; direct labor; small tools, jigs, and fixtures; and factory overhead (OH), sometimes called *burden.* In the case of weldments or brazements, the welding or brazing costs are included, as are thermal cutting costs.

Direct materials are product ingredients that become a part of the finished product, and can readily and inexpensively be traced to the product in terms of units of material per product made. They also include materials consumed in manufacturing when the costs per unit can be accurately estimated.

Direct labor is work that is actually done in producing a finished article and is readily charged to it. This may include inspection labor.

Small tools, jigs, and fixtures are the accessories used in production processes that are expendable and readily charged to the product.

Factory overhead, or burden, consists of all indirect labor, materials, and other indirect manfacturing expenses that cannot be readily allocated to a product on a per part basis. Facilities, equipment, power, air, utilities, and manufacturing services, such as maintenance, may be a part of these costs. A portion of the overhead costs may be fixed, and the remainder may vary with the production rate (number of parts per unit of time).

WELDING COSTS

The success of a business is usually measured by its profitability, based on the ability of the company to hold costs to the defined limits permitted by competitive selling prices. Welding and associated costs can be readily approximated for any job when the factors affecting those costs are known and appropriate steps are taken to determine costs. Welding costs must be accurately determined if they are to be successfully used in cost estimates for bidding, for setting rates for incentive programs, or for comparing welded construction to a competing process.

The procedures for determining brazing, soldering, and thermal cutting costs are generally similar to those for welding costs. The elements that contribute to costs with a specific process may be somewhat different but the basic concepts for determining costs are the same. Therefore, the discussion on welding costs is pertinent to the other processes.

The procedures for costing the production of welds and weldments must fit into the general accounting practices of an enterprise. Costs for welding will include the same basic elements common to other manufacturing processes and activities: labor, materials, and overhead. Each of these factors has many variables, some of which are unique or proprietary to a company. Defining these variables requires the knowledge and experience of individuals familiar with the disciplines of engineering, metallurgy, manufacturing, and quality control. There are no simple formulas that will take the place of experience and knowledge of the welding field in estimating true welding costs.

The basis for estimating welding costs historically has been an evolution of gathering data and tailoring it to fit the close economic borders of a particular operation. Gathering and evaluating data is a costly and cumbersome chore. With the proliferation of computers and application software, quick and convenient analyses of many variables of operation are readily available. Computerization and data processing have become an important part in the determination of costs. Variables generally considered to obtain the costs of welding are listed in Table 3.1, and those for brazing in Table 3.2.

Table 3.1
Typical variables for estimating welding costs

Electrode or flller wire
Type
Size
Electrode deposition efficiency
Type of joint
Type of weld
Weld size
Type of shielding
Shielding gas flow rate
Flux consumption ratio
Welding current
Arc voltage
Power source efficiency
Welding time
Operator factor
Labor rates
Overhead rate
Filler metal cost
Shielding gas cost
Flux cost
Electric power cost
Edge preparation costs
Setup costs
Weld finishing costs
Inspection costs

ELEMENTS IN WELDING COSTS

All cost systems include the same basic elements of labor, materials, and overhead. To obtain a cost for welding, the finite time required to accomplish a weld is used to determine the cost of labor, which is added to the costs of materials and overhead. Overhead costs are usually obtained by apportionment as a percentage of the labor cost. Whichever system is employed to estimate a welding cost, each of the basic elements must be analyzed, and as many of the operating variables as are known must be included for a reliable value.

The welding procedure is the starting point for estimating a cost for welding. The procedure will define the welding variables, and provide a basis for repeatability and consistency in produc-

Table 3.2
Typical variables for estimating brazing costs

Brazing process
Method of application (manual, mechanical, automatic)
Type of joint
Joint clearance
Length of overlap
Length of joint
Filler metal
Type
Form and size
Method of application
Brazing temperature range
Brazing flux or atmosphere
Assembly time
Loading time
Furnace brazing cycle
Heating time
Cooling time
Power or fuel requirements
Unloading time
Postbraze cleaning
Method
Time
In-process inspection
Labor rates
Overhead rate
Filler metal cost
Flux cost
Atmosphere cost
Electric or fuel cost
Cleaning material costs

tion. Many companies have standardized welding procedures that are used for various jobs of a similar nature. A typical welding procedure form is shown in Fig. 3.1. The welding procedure provides the basic data needed to calculate the cost of the weld.

Labor Costs

Labor costs are based on the times that it takes to perform all of the steps in making a weldment. These times can be grouped into arc time, handling time, and miscellaneous workplace time.

Arc time depends on factors controlled by the power source and associated equipment, such as electrode or filler wire feed speed, arc voltage, welding current, travel speed, type of welding power and polarity. There are many independent variables that affect the rate at which the weld is made, Welding process, joint design, weld size, electrode type and size, and welding position are a few of these variables.

Handling time includes all of the workplace functions of picking up the part, placing it in a fixture, clamping and positioning it before and during welding, and finally moving the weldment to the next location. Handling time can be calculated in an estimate with reasonable accuracy only for those operations that are repetitive. Variations in handling time of a nonrepetitive nature are best included in miscellaneous workplace time. Input from industrial engineering may be necessary when analyzing handling time.

Miscellaneous workplace time includes the many nonrepetitive, nonreccurring times that cannot easily be measured but must be costed. This will include elements such as stamping, applying antispatter compound, tack welding runoff tabs or backing strips in place, movements to reposition the body between weld passes, and any variable time increment not directly involved in making the weld.

Material Costs

Material costs cover those materials consumed in the workplace while making the weld. For those processes in which filler metal is deposited, the amount of deposited metal can be the basis of a suitable material cost. This cost will reflect deposition efficiency, which is the ratio of the weight of the weld metal deposited to the weight of the consumable used. Efficiency of deposition decreases as a result of losses such as stub ends, metal vaporization in the arc, conversion to slag of core wire components, and weld spatter. Typical deposition efficiencies for various welding electrodes and processes are given in Table 3.3.

Other material costs include those for gases, fluxes, backing rings or strips, anti-spatter compounds, etc. Gases may be for shielding for arc welding and brazing, or fuel and oxygen for oxyfuel gas welding, brazing, and soldering. Fluxes may be those used with submerged arc and electroslag welding, or those used with brazing and soldering operations.

Table 3.3
Deposition efficiency for welding processes and filler metals

Filler metal form and process	Deposition efficiency, %
Covered electrodes, SMAW	
14 in. long	55 to 65
18 in. long	60 to 70
28 in. long	65 to 75
Bare solid wire	
Submerged arc	95 to 99
Gas metal arc	90 to 95
Electroslag	95 to 99
Gas tungsten arc	99
Flux cored electrodes	
Flux cored arc	80 to 85

Overhead Costs

Overhead is the cost of the many items or operations in the factory and in the office not directly assignable to the job or weldment. These costs are apportioned pro rata among all of the work going on in the plant or a department. The main categories may include some or all of the following:

(1) Salaries of the plant executives, supervisors, inspectors, maintenance personnel, janitors, and others that cannot be directly charged to the individual job or weldment.

ARC WELDING PROCEDURE SPECIFICATION

Base metal specification ____________________
Welding process ____________________
Manual or machine welding ____________________
Position of welding ____________________
Filler metal specification ____________________
Filler metal classification ____________________
Flux ____________________
Weld metal grade* ____________________ Flow rate ____________________
Single or multiple pass ____________________
Single or multiple arc ____________________
Welding current ____________________
Polarity ____________________
Welding progression ____________________
Root treatment ____________________
Preheat and interpass temperature ____________________
Postheat treatment ____________________

*Applicable only when filler metal has no AWS classification.

WELDING PROCEDURE

Pass no.	Electrode size	Welding power		Travel speed	Joint detail
		Amperes	Volts		

This procedure may vary due to fabrication sequence, fit-up, pass size, etc., within the limitation of variables given in 4B, C, or D of AWS D1.1, (________ year) Structural Welding Code.

Procedure no. ____________________ Manufacturer or contractor ____________________

Revision no. ____________________ Authorized by ____________________

Date ____________________

Fig. 3.1—Typical welding procedure specification

(2) Fringe benefits for employees, such as premiums for life and medical insurance, and social security and pension fund contributions.

(3) Rent and depreciation of the plant and facilities.

(4) Depreciation or lease costs of plant equipment including welding machines, handling equipment, overhead cranes, and all other equipment that is not charged directly to the job or specific weldment.

(5) Maintenance cost of the buildings, grounds, etc.

(6) All taxes on the plant, real estate, equipment, and payroll.

(7) Heat, light, water, and other utilities used in the operation of the plant.

(8) Small tools, such as wrenches, chipping hammers, and electrode holders.

(9) Safety and fire equipment.

(10) Supporting departments including chemical and metallurgical laboratories and data processing.

All enterprises have some system for handling and determining overhead costs. The assignment of overhead costs is usually a function of the accounting department. Distribution of the overhead costs will vary with the system in use. A commonly used system prorates the overhead costs in accordance with the direct labor cost, which must be accurate. This system is practical and applicable to plants with labor intensive operations. In those instances where mechanization and robotics are heavily employed, overhead costs may be assigned on the amount of weld metal that is deposited, the speed of welding, or on each weldment.

METHODS FOR DETERMINING WELDING COSTS

Operator Factor

Operator factor or duty cycle in arc welding is the percentage of actual arc time during a specified length of time, which may be a minute, an hour, an 8-hour shift, or some other time span. The higher the percentage of arc time (operator factor), the greater is the amount of weld metal produced, and the higher is the efficiency of the welding operation. Other operations that a welder or machine operator has to perform, such as part cleaning, fixture loading, or tack welding, will result in a low operator factor.

A high operator factor may not indicate low cost because the welding procedure may require a larger amount of arc time than is necessary. For example, it may specify a small electrode when a larger size could be used, or a large weld groove when a smaller one would suffice.

Every effort should be made to increase the operator factor. A welder's performance determines the appearance and quality of the weld; therefore, every obstacle hampering the welder should be removed. Work should be planned and positioned to minimize physical strain and to ensure maximum comfort and safety.

Changing from the vertical or overhead position to the flat position, for example, can significantly increase welding speed. Single-pass welds on plates can be made at maximum speed when the joint is in a flat position with the weld axis slanting downhill. A 10° slant can increase welding speeds up to 50 percent.

The operator factor is higher with the use of positioners and fixtures, and with semiautomatic or automatic welding processes where the welder or machine operator does not perform other job elements, such as chipping and electrode changes. In the absence of studies or dependable data, operator factor can be estimated from Table 3.4. The actual operating factor depends on the type and size of weld, the welding position, adequacy of fixturing, location of welding, and other operating conditions. It may vary from plant to plant or even part to part if the amount of welding is significantly different. A more positive determination of the operating factor is made by time studies or by the installation of time recorders.

It may be economical to provide a welder with a helper to set up the work for welding. Every operation that a welder has to perform, apart from actual welding, reduces arc time and thus, the operator factor.

It is not unusual for a welder to spend 50 percent of the time setting up the work for welding. If a helper and an additional fixture are added, production may be doubled. The helper unloads and loads one fixture while the welder is welding at the other fixtures. With multiple-pass welds, the helper can remove oxide or slag while the welder places a pass on another weld or at another location in the same joint.

Table 3.4
Operator factor for various welding methods

Method of welding	Operator factor range, percent
Manual	5-30
Semiautomatic	10-60
Machine	40-90
Automatic	50-95

Labor Costs

Welding costs are determined by several methods. The cost of labor is usually based on an hourly rate that each worker is paid for performing the job. Sometimes, workers are paid by the number of parts produced in an hour, called the *piece rate* in an incentive system. Hourly paid employees are usually required to meet certain standards of performance, usually based on standard data of time for welding in inches per minute. Time studies are often taken at the workplace to determine a normal welding time. Other sources for the data are the welding procedure and the weld joint design. The welding procedure provides the process variables, such as travel speed and electrode type and size. The weld joint design determines the number of passes, weld size, and the amount of weld metal to be deposited. The data are modified according to the operator factors expected.

Labor costs for manual or machine welding may be expressed as the cost per unit length of weld. For single-pass welds:

$$LC = 0.2PR/(TS \times OF)$$

where:

LC = labor cost, \$/ft
PR = pay rate, \$/h
TS = travel speed, in./min
OF = operator factor, decimal value

The travel speed is obtained from the welding procedure specification. The operator factor depends upon the welding process and method of welding (see Table 3.4). The welder's pay rate is commonly the basic hourly pay plus the cost of fringe benefits, such as insurance and pay for holidays and vacation.

For multiple-pass welds:

$$LC = (PR \times WM)/(DR \times OF)$$

where:

LC = labor cost, \$/ft
PR = pay rate, \$/h
WM = deposited metal, lb/ft
DR = deposition rate, lb/h
OF = operating factor, decimal value

The weight of deposited weld metal may be calculated or determined by trial welds. Deposition rate can be obtained from standard data or by trial welds.

Standard Data

Standard data can be in many forms, but usually in table[1] or graphic form. It may be stored in a computer memory system and retrieved for use. Examples of standard data are shown in Fig. 3.2 and Table 3.5. Judicious use of standard data is essential in obtaining accurate welding costs.

Tables and graphs that provide final costs when labor rates are inserted should be used with caution because the data may apply to only a specific set of conditions. For example, the weld joint design may vary with respect to root opening or groove angle, or the weld reinforcement may differ. The deposition efficiency of the selected electrode may be different from the electrode used in the data construction, or the operator factor may vary significantly from the value assigned in the data. The data may be biased to favor one welding process over another.

1. Standard data for arc welding are given in the following publications: *Standard Data for Arc Welding* (of steel), Cambridge, England: The Welding Institute, 1975. *Procedure Handbook of Arc Welding*, 12th Ed., Cleveland, OH: The Lincoln Electric Co., 1973.

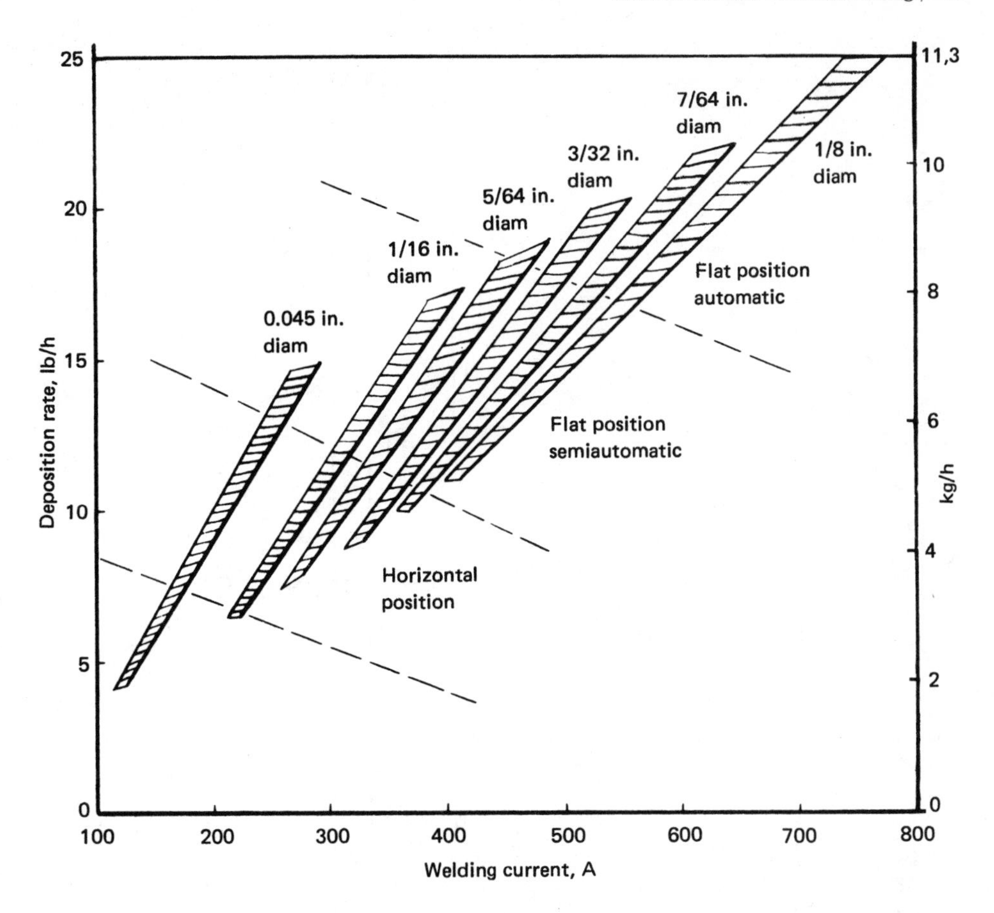

Fig. 3.2—Deposition rate vs welding current for E70T-1 mild steel electrodes with CO_2 shielding

Material Costs

Filler Metal. Standard data for filler metal costs in welding are based on the amount of deposited metal per unit length of weld or per unit for small weldments. Every weld joint design that requires the addition of filler metal has a geometric cross section. Neglecting weld reinforcement, the cross sectional area of a weld joint can be approximated by dividing it into one or more simple geometric areas including rectangles, triangles, and segments of a circle. The area of weld face reinforcement can be calculated or assumed as a percentage of the total area, for example, 10 percent for single-groove and fillet welds, and 20 percent for double-groove welds.

The weight of deposited metal per unit length can be calculated by multiplying the unit volume by the density of the metal. For example:

$$DM = 12AD$$

where:

DM = deposited metal, lb/ft
A = cross sectional area of weld, in.^{2}
D = density of the deposited metal, lb/in.^{3}

The weight of filler metal required per unit of weld length is dependent on the deposition efficiency according to the following formula:

$$FC = DM/DE$$

where:

FC = filler metal consumption, lb/ft
DM = deposited metal, lb/ft
DE = deposition efficiency (decimal value)

Table 3.5
Typical welding conditions for single electrode, machine submerged arc welding of steel plate using one pass (square groove)

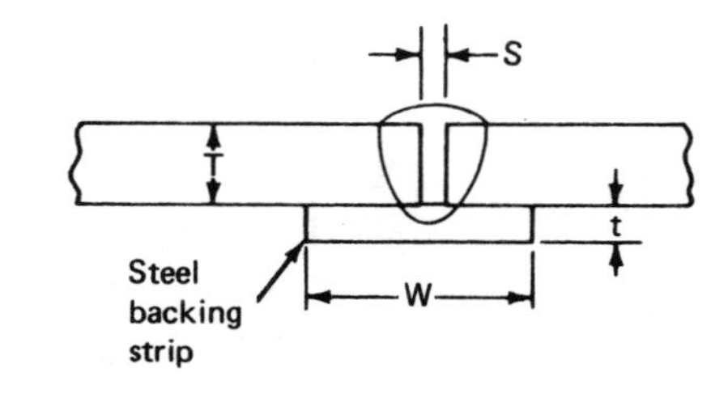

Plate thickness T		Root opening S		Current	Voltage	Travel speed		Electrode diam		Electrode consumption		Backing strip			
												t, min		W, min	
mm	in.	mm	in.	A[a]	V[a]	mm/s	in./min	mm	in.	kg/m	lb/ft	mm	in.	mm	in.
3.6	10 ga.	1.6	0-1/16	650	28	20	48	3.2	1/8	0.104	0.070	3.2	1/8	15.9	5/8
4.8	3/16	1.6	1/16	850	32	15	36	4.8	3/16	0.194	0.13	4.8	3/16	19.0	3/4
6.4	1/4	3.2	1/8	900	33	11	26	4.8	3/16	0.248	0.20	6.4	1/4	25.4	1
9.5	3/8	3.2	1/8	950	33	10	24	5.6	7/32	0.357	0.24	6.4	1/4	25.4	1
12.7	1/2	4.8	3/16	1100	34	8	18	5.6	7/32	0.685	0.46	9.5	3/8	25.4	1

a. DC, reverse polarity (electrode positive)

For greater accuracy, representative weld samples can be made, and weight measurements can be taken before and after welding. The amount of weld reinforcement must be within specification limits. In any case, the amount of deposited metal may vary in production because of variations in fit-up and weld reinforcement. Appropriate allowances should be made for these variations in the cost estimate.

To determine the cost of the filler metal consumed in making a weld, the amount of weld metal deposited must reflect the deposition efficiency or filler metal yield (see Table 3.3). Deposition efficiency accounts for the losses from stub ends, slag, vaporization in the arc, and spatter. The wide variations in deposition efficiencies are a result of many factors, some of which are inherent in the type of electrode and welding power source as well as the experience and skill of the welder. The amount of stub-end loss can affect the efficiency significantly. For example, a stub end of 2 inches reduces the amount of covered electrode deposited in a weld to less than 87 percent of the original electrode weight. The loss resulting from the covering of a covered electrode can vary from 10 to 15 percent of the original weight.

Filler metal cost can be calculated in several ways; the most common way is based on the cost per unit length of weld, as in the following formula:

$$CF = (FP \times FC)/DE$$

where:

CF = filler metal cost, \$/ft
FP = filler metal price, \$/lb
FC = filler metal consumption, lb/ft
DE = deposition efficiency, (decimal value)

The filler metal cost is the delivered cost to the plant.

A method used for calculating the amount of filler wire or electrode required with a continuous feed system, such as with gas metal arc, flux cored arc, and submerged arc welding, requires three simple steps. The first step is to determine the weight of electrode used per unit of time using the following formula:

$$EW = 60FS/WL$$

where:

EW = electrode or wire required, lb/h

FS = feed speed, in./min

WL = length of electrode or wire per unit weight, in./lb

The length of filler wire per unit weight, WL, can be calculated with the following formula:

$$WL = 1/(A \times D)$$

where:

A = wire cross sectional area, in.^{2}
D = density of the filler wire, lb/in.^{3}

The length of bare wire (filler metal) per unit weight for several AWS filler metal classifications is given in Table 3.6. Similar information for steel flux cored electrodes is given in Table 3.7.

An accurate determination of the wire feed (the melting rate of the electrode) can be simply done by allowing the wire to feed thru the gun and measuring the amount of wire fed in 5 to 10 seconds, and multiplying the results by the appropriate time ratio.

The second step is to determine the welding travel speed. It may be specified in the welding procedure or established by trial welds.

The third step is to determine the weight of electrode wire required per unit length of weld using the following formula:

$$FC = EW/(5TS)$$

where:

FC = filler metal consumption, lb/ft
EW = electrode-wire required, lb/h
TS = travel speed, in./min

The cost of the filler (electrode) wire per foot of weld is found by multiplying the filler wire consumption, FC, by the cost per pound of the filler wire.

Flux. When flux is used, its cost must be included in material costs. The cost can usually be related to the weight of deposited filler metal by a ratio, called *flux ratio*. With submerged arc welding, normally one pound of flux is used with each pound of electrode deposited, giving a flux

Table 3.6
Length per unit weight of bare electrode wire of various AWS classifications, in./lb

Diam, in.	ER-1100 (Al)	ERCuAl-AX (Cu-Al)	ERCuSi-A (Cu-Si)	ERCu (Cu)	ERCuNi (Cu-Ni)	ERAZXXA (Mg-Zn)	ERNi-1 (Ni)	ER70S-X (Steel)	ER3XX (Sst)
0.020	32400	11600	10300	9800	9950	50500	9900	11100	10950
0.025	22300	7960	7100	6750	6820	34700	6820	7680	7550
0.030	14420	5150	4600	4360	4430	22400	4400	4960	4880
0.035	10600	3780	3380	3200	3260	16500	3240	3650	3590
0.040	8120	2900	2580	2450	2490	12600	2480	2790	2750
0.045	6410	2290	2040	1940	1970	9990	1960	2210	2170
0.062	3382	1120	1070	1020	1040	5270	1030	1160	1140
0.078	2120	756	675	640	650	3300	647	730	718
0.093	1510	538	510	455	462	2350	460	519	510
0.125	825	295	263	249	253	1280	252	284	279
0.156	530	189	169	160	163	825	162	182	179
0.187	377	134	120	114	116	587	115	130	127
0.250	206	74	66	62	64	320	63	71	70

Table 3.7
Inches per pound of steel flux cored electrodes

Electrode size, in.	in./lb
0.045	2400
1/16	1250
5/64	1000
3/32	650
7/64	470
0.120	380
1/8	345
5/32	225

ratio of one. For accurate data, welding tests should be made.

For electroslag welding, the flux ratio is about 0.05 to 0.10. With oxyfuel gas welding and torch brazing, the amount of flux used can be related to the amount of filler metal consumed. This ratio is also in the range of 0.05 to 0.10. The flux cost is determined by the formula:

$$CF = FP \times FC \times FR$$

where:

CF = flux cost, \$/ft
FP = flux price, \$/lb
FC = filler metal consumption, lb/ft
FR = flux ratio

Shielding gas. The cost of shielding gas is related to welding travel speed or arc time to make a weld. The gas is used at a specific flow rate as measured by a flow meter. The gas cost per foot of weld is calculated as follows:

$$GC = (GP \times RF)/5TS$$

where:

GC = gas cost, \$/ft
GP = gas price, $\$/ft^3$
RF = rate of flow, ft^3/h
TS = travel speed, in./min

The gas cost per weld is determined as follows:

$$GC = (GP \times RF \times WT)/60$$

where:

GC = gas cost, \$/weld
GP = gas price, $\$/ft^3$
RF = rate of flow, ft^3/h
WT = arc time, min

The price of a gas is the delivered price at the welding station.

Miscellaneous Materials. Some welding processes use expendables other than filler metal, gas, and flux. These other expendables may include such items as ceramic ferrules and studs in stud welding, backing strips, and consumable inserts. These must also be included in material costs for welding.

Overhead Costs

Overhead costs are those that cannot be directly charged to a specific job. They may include management, facilities, depreciation, taxes, and sometimes small tools and safety equipment of general use.

The overhead costs must be distributed to the welding work in some manner. They are usually prorated as a percentage of direct labor costs for the job. If the overhead costs per foot of weld are needed, they can be calculated in the same manner as the labor costs per foot. The overhead rate (\$/h) is substituted for the labor rate in the appropriate calculations.

ECONOMICS OF ARC WELDING

The arc welding operations using filler metals are dependent on many economic factors that are related to the amount of filler metal consumed to make a weld. Every effort must be made to reduce the amount of filler metal required for each weld joint, consistent with quality requirements. Analyses of the areas of weld joint design, welding fabrication, welding procedures, and preparation of parts for welding can provide opportunities for minimizing costs. The following suggestions should be considered to minimize welding costs.

WELDMENT DESIGN

(1) Minimize the cross sectional areas of weld joints, consistent with quality requirements, by using narrow root openings, small groove angles and, if practical, double-groove welds instead of single-groove welds (see Fig. 3.3). Changing a V-groove angle from 90° to 60° will reduce filler metal needs by 40 percent. Required joint groove angles and root openings vary for different alloy systems and welding processes. These dimensions should be considered in the design phase.

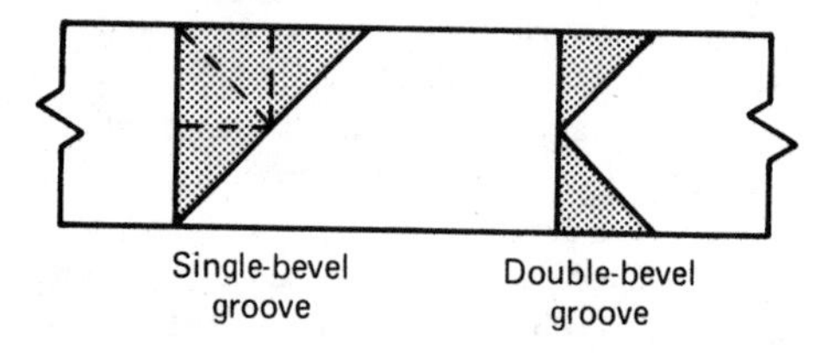

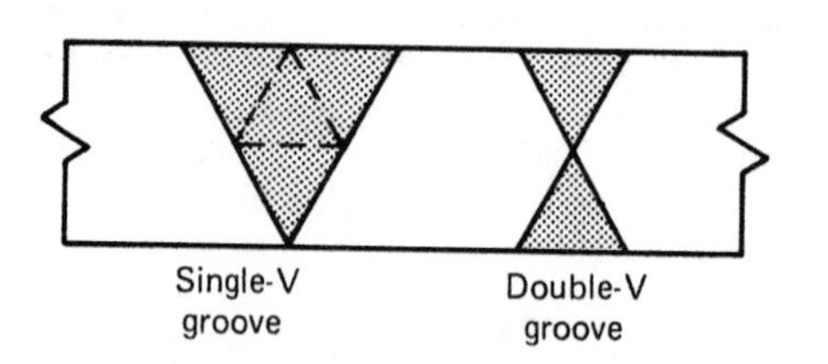

Fig. 3.3—Effect of joint design on filler metal requirements

(2) Accurately size fillet welds. Overwelding is a very common practice that probably originates from a belief that additional weld metal will add strength to a fillet weld. The effect of overwelding is clearly seen in Fig. 3.4 where a properly made weld is compared to one that has been overwelded. If a specified 1/4 in. weld is increased in size only 50 percent, over twice the needed amount of weld metal is used. Overwelding, which can increase greatly the cost of welding, may not be foreseen when the cost estimate is prepared.

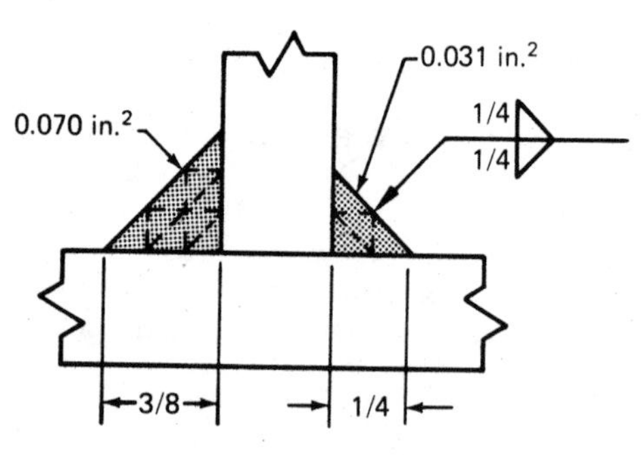

Fig. 3.4—Effect of overwelding on the weld metal requirements for a fillet weld

(3) Design for ready access to all welds. Poor conditions for making a weld increase costs and contribute to poor quality.

(4) If a choice is available, select readily weldable metals. Metals requiring complex welding procedures are more expensive to weld.

(5) Use the proper welding symbols with specific notation of weld size.

WELDING FABRICATION

(1) Provide the proper equipment for the process specified in the welding procedure, including proper ventilation.

(2) Provide the correct welding materials specified in the welding procedures.

(3) Provide accurate fit-up of parts using fixtures where possible, and inspect the fit-up before welding.

(4) Overwelding must be carefully controlled. It increases the cost of filler metal and the labor to deposit it, and may be cause for rejection of the weld on inspection.

(5) Avoid excessive reinforcement on welds because it uses unnecessary filler metal and labor.

(6) Use positioning equipment when possible. Positioning of the weldment for welding in the flat position greatly increases efficiency and reduces costs.

(7) Use power tools to remove slag and to finish the weld surfaces.

(8) Encourage the use of proper amperage and voltage settings for maximum efficiency. Excessive welding current or voltage may cause high spatter loss; low welding current gives inefficient rates of deposition.

(9) In the case of shielded metal arc welding, stub end loss adds significantly to electrode cost. Encourage welding with electrodes to a minimum practical stub end length.

WELDING PROCEDURE

(1) Provide the shop with written, qualified welding procedures for all welds.

(2) Process selection should be made on the basis of the maximum deposition rates consistent with weld soundness and availability of equipment.

(3) Filler metal selection, considering both size and costs, should be made on the basis of deposition efficiency even though the initial cost of a filler metal of high efficiency may be greater than alternates of lower efficiency.

(4) Specify the longest practical covered electrode to minimize stub losses.

(5) Purchase bare filler wire or flux cored electrode in the largest suitable packaging form available. Large packaging reduces the base price of filler metal, and requires fewer interruptions to change packages.

MANUFACTURING OPERATIONS

(1) Prepare parts as accurately as needed, particularly those parts that are bent or formed to shape.

(2) Prepare parts by shearing or blanking where possible. This may be more economical than thermal cutting.

(3) Avoid designs that require machining of edge preparations for welding.

(4) Evaluate automatic thermal cutting equipment. Hand layout and manual cutting are costly.

(5) Inspect all parts for accuracy before delivery to the welding department.

(6) Schedule delivery of materials so that all parts of an assembly arrive at the welding department as needed.

ECONOMICS OF WELDING AUTOMATION AND ROBOTICS

Where there is need to reduce costs, investment in capital equipment can provide the maximum benefit for improving a manufacturing operation. Capital equipment investment can be a simple refinement, such as replacing shielded metal arc welding with a continuous wire feed welding process (gas metal arc or flux cored arc welding) or substitution of a programmable, microprocessor-controlled robot for manual labor. The projected return on investment is the key for a valid capital expenditure.

All costs and benefits must be identified and

included in a cost analysis. Benefits may include income from increased sales or savings from a proposed change to automated equipment. Other benefits may include one or more of the following:

(1) Increased productivity

(2) Higher product quality resulting in fewer repairs and field problems

(3) Safer working conditions for the machine operator

(4) Improvement in operator factor

(5) A general environment improvement relatively free from heat, smoke, and radiation

Identified costs will cover the initial cost of the asset plus an annual cost for operating it, including material costs, gases, and other expenses. Costs must also predict the effect from inflationary rates and varying interest rates in the pay back period. Consideration should be given to the average cost of capital, or time value of money, and the realized return if the capital were invested elsewhere.

Various factors must be considered when evaluating the economics of investing in automation or robotics. Several of these factors are discussed below.

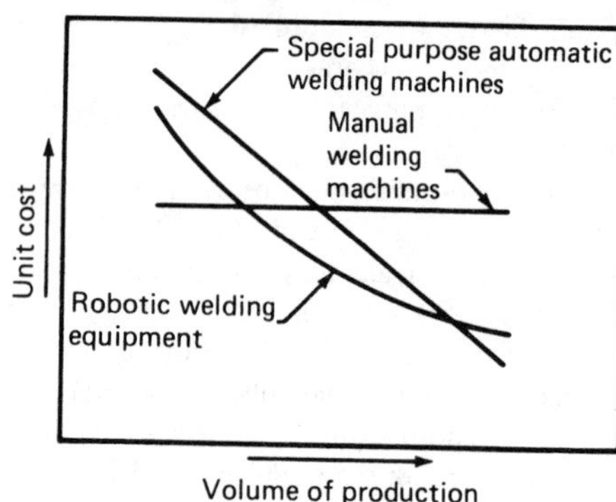

Fig. 3.5—Effect of production volume on unit cost of manual, robotic, and special purpose welding equipment

PRODUCTION VOLUME ESTIMATES

First and foremost in planning for welding automation is a firmly established production volume forecast. A volume of some minimum size will be required. There is a general relationship between production volume and the unit cost of equipment, ranging from manual arc welding power units to automated special-purpose welding machines, as shown in Fig. 3.5. In the case of long production runs, special purpose automation is frequently more efficient to use. However, for relatively short production runs, flexible, multipurpose automation can be used to advantage. An example of the latter is a microprocessor-controller robot. A properly prepared economic analysis will clearly establish the numerical data needed to make a proper choice. Thorough consideration must be given to all factors relating to the welding operation.

PART GEOMETRY

Special purpose automation may require geometric simplicity, such as straight or circular welds. For example, tubes and pipes are welded relatively fast and at low cost, in most cases, because of the simplicity of the geometry. Parts having complicated geometrical weld paths will require special purpose equipment that can be preprogrammed to track the weld path, is equipped with a seam tracking device, or is designed with permanent mechanical motion paths using cams, gears, and templates. The thickness and size of the part, and a requirement for mobility may favor special purpose equipment.

WELDING SITE AND ENVIRONMENT

The welding site and environment will greatly influence the type of welding equipment required. Open-air sites require protection against atmospheric elements including rain, snow, pollution, heat, and cold.

Economic considerations not associated with the welding operation may play a dominant role in a decision to automate. In cases involving welding in nuclear power plants, a high cost exists for each day a power plant is shut down for repairs. With such high downtime costs, the decision to automate is based on the speed of welding repair rather than the cost of repair. All other economic considerations become secondary.

ACCURACY OF PARTS

Automated welding equipment frequently requires better component accuracy than would be needed for manual welding. A welder visually senses variations in parts and makes corrections to compensate for them, a difficult task for a machine to do. Technology is available for adding sensory systems and adaptive feedback controls to a welding machine to correct for joint location and width. In most cases, component accuracy can be readily improved within specified limits by exercising better control of component manufacture. The inherent accuracy of the welding machine is an important factor. Accuracy in the components is meaningless if the welding machine accuracy is not in agreement.

MATERIAL HANDLING

The continuous character of automated welding requires associated systems for material handling that many times are integrated to work in sequence with the welding machine. Manually-loaded welding machines are often paced by the operator. In many cases of automation, ejection of parts is incorporated in the welding machine. Conveyors or large containers are used to bring parts to the machine and to remove them to the next step in manufacturing.

SAFETY

The consideration for safety requires more care than usual in operating, setting up, and maintaining an automated welding machine. Personnel usually work close to the machine during setup, programming, and maintenance. Most automatic welding machines have drive systems and rapidly moving components. Typical safety requirements of those machines are a fail-safe system and easy access to emergency stop buttons that stop the machine immediately. In normal production operations, the operator and others must be kept remote from the machine's operating envelope. This requires guards that may be interlocked with the machine to prevent careless operation of the machine if a guard is removed.

DEPOSITION EFFICIENCY

One of the benefits of automation is the efficiency achieved in filler metal deposition rates, even with processes normally employed with manual or semiautomatic welding. As indicated in Fig. 3.6, the automated welding processes typically show a 40 percent improvement over semiautomatic gas metal arc welding (GMAW), and a 250 percent improvement over shielded metal arc welding (SMAW). Often, the cost savings derived from welding process improvement may justify the investment in automated equipment.

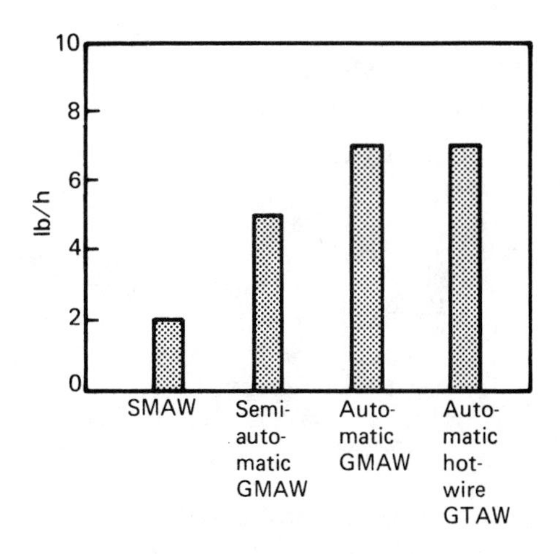

Fig. 3.6—Expected weld metal deposition rates for several arc welding processes when welding pipe in horizontal fixed (5G) position

OPERATOR FACTOR

Improvement in operator factor can be significant with automatic welding equipment. Referring to Table 3.4, the operating factor for manual welding processes ranges from 5 to 30 percent, while for automatic welding, the range is

50 to 100 percent. A signficant but lesser improvement is obtained over semiautomatic welding (35 percent average operating factor).

WELD JOINT DESIGN

Another benefit that can be realized by the use of automation is the capability for higher filler metal deposition rates which, in turn, permits the use of more efficient weld joint designs. For example, it might be possible with automation to change from a 60° V-groove to a square-groove joint design, such as that used for narrow gap welding. Such a change may permit a higher welding speed together with less deposited metal and smaller variations in weld size.

Smaller variations in weld size will result in less filler metal being used and in avoidance of overwelding. Overwelding is a negative factor in operator and material costs.

OVERALL IMPROVEMENT FACTOR

Several factors have been considered that may result in increased productivity with automated equipment. The overall improvement factor is the product of those factors including operator factor, factor for avoidance of overwelding, and welding process improvement factor.

Typically, substitution of an automated GMAW system for semiautomatic GMAW to produce fillet welds can result in improvement by a factor of about 1.4 for the welding process; the improvement for overweld reduction is about 2.25; and the operator factor improvement at 30 percent is about 2.5. Overall, automatic GMAW may be nearly 8 times more productive than semiautomatic GMAW.

COMBINED OPERATIONS

Automation permits design of equipment to utilize multiple welding guns or to do other simultaneous operations. Thus, one or more welding guns can be in operation at the same time as other operations, such as drilling, scarfing, and punching. The multifold increase in productivity can lead to rapid payback for automated equipment.

INSTALLATION COSTS

There are several resource requirements which are unique to automated equipment. Planning must include provisions for adequate floor space, foundations, special installation equipment, and utilities. It is not uncommon for the cost of installing automated equipment to be 5 percent to 10 percent of the initial cost of the machine. It is, therefore, important to plan for these costs and include them in the initial economic analyses.

ECONOMICAL EQUIPMENT AND TOOLING SELECTION

The economic responsibility of the welding engineer for welding equipment and tooling costs is to obtain the minimum cost of these items per unit of a product over the period of manufacture. To determine the most effective use of capital, management must know the percentage return on investment. The elements to be considered in these calculations are as follows:

(1) *Annual gross savings* include the savings in direct labor, overhead costs, and maintenance and repair costs expected from the proposed replacement as compared with the corresponding costs for the present installation.

(2) *Direct-labor savings* refer only to those savings in operator costs which can be utilized.

(3) *Savings in overhead costs* include savings in only those incremental overhead cost factors affected by the replacement. Such cost factors include inspection, service, power costs, and similar overhead items. Depreciation rates should not be considered in analyzing the operating cost factors.

(4) *Maintenance and repair costs* include day-to-day equipment maintenance and normal repair costs exclusive of major overhaul.

(5) *Rate of return on investment* refers to the annual interest on the unpaid balance of invested capital. On this basis, the annual capital charge against the replacement consists of two parts: (1) the annual amortization or reduction of the capital invested in the equipment and (2) the annual interest on the unamortized portion of the investment.

This concept becomes more meaningful when viewed as a mortgage taken by management on the equipment. The annual payments on this "mortgage" are made by the annual savings or profit increase. The interest portion of these equal annual payments is the rate of return to be realized from the investment in equipment, and is a measure of the earnings potential of the proposed replacement, as compared with the investment of that money in the day-to-day operation of business. It is a more fundamental measure than annual gross savings, since it is directly related to the investment and the time period involved.

The mathematical relationship among the annual gross savings, the investment, the desired percentage return on investment, and the period of capital recovery is expressed by an annuity (mortgage) formula, which has wide application to many problems involving the time-interest consideration of money.

Typical formulae used in the cost justification of new equipment and tooling are given in Table 3.8.

ECONOMICS OF NEW JOINING PROCESSES

One method of reducing manufacturing costs is the adoption of a more economical joining process when production requirements warrant investment in new equipment. Friction, high-frequency resistance, nonvacuum electron beam, laser beam, and ultrasonic welding are some processes that might be considered during the design or redesign phase to replace existing welding or brazing operations. As with other industrial processes, the successful selection and implementation of these processes requires a logical and systematic approach. They may provide an alternative to the known processes, and there will be tradeoffs to consider for a particular application. The process selection decision must be well planned if the processes represent nonconventional technology that requires a different approach than that for familiar manufacturing techniques.

The decision to use a new joining process must be made with great care. The selection for the first application is critical because future usage and acceptance in a factory may be greatly influenced by the success or failure of the first application. Many early industrial users of robots, for example, either installed the robots in inappropriate applications or overestimated the benefits that would result from using them.

In assessing the feasibility of a new process for a particular manufacturing operation, a potential user must first determine if the basic capabilities of the process are compatible with the requirements of the operation and capable of performing it. Secondly, the new process must be compared with alternative processes to determine if the process can perform the task better, at a lower cost, and with better quality than any alternatives. For assessing process capabilities, the following factors should be considered:

(1) Type of operation to be performed

(2) Nature of the material to be joined

(3) Geometry, thicknesses, and fit-up of parts

(4) Quality requirements

(5) Processing speed or parts per unit of time

(6) Economic requirements: initial investment costs, operation costs, and payback period

(7) Safety requirements: enclosures, shielding, and other safety costs

(8) Past experience with other similarly controlled equipment, such as numerical- or computer-controlled machine tools

(9) Material handling requirements with a new process and higher production rates

Once the decision is made to purchase new joining equipment, a specific type and model

Table 3.8
Typical formulae for determining economic equipment selection

A. Mathematical Analyses for Economic Equipment Selection

(1) *Calculation for Replacement of Production Machines*

Let X = number of years required for the new machine to pay for itself
C_n = cost of new machine installed and tooled
R_o = scrap value of old machine
R_n = probable scrap value of new machine at end of its useful life
K_o = present book value of old machine[a]
P_o = labor and machine cost per unit on old machine
P_n = labor and machine cost per unit on new machine
N_n = estimated annual output on new machine
I = annual allowance on investment, percent
D = annual allowance for depreciation and obsolescence, percent
T = annual allowance for taxes, percent
M = annual allowance for upkeep, percent

Then:

$$X = \frac{(C_n - R_n) + (K_o - R_o)}{N_n(P_0 - P_n) - C_n(I + D + T + M)}$$

a. The present book value of the old machine is determined by the straight-line equation

$$K_o = C_o - Y_o \frac{C_o - R_e}{Y_e}$$

Where:
C_o = cost of old machine installed and tooled
Y_o = present age of old machine
Y_e = original estimated old machine life
R_e = original estimated scrap value of old machine (this may or may not equal present value)

(2) *Which of Two Suitable Machines to Select*

Let C = first cost first machine, installed
c = first cost second machine, installed
N = number of pieces produced per year by first machine
n = number of pieces produced per year by second machine
L = labor cost per year on first machine
l = labor cost per year on second machine
B = percentage labor burden per year on first machine
b = percentage labor burden per year on second machine
I = annual allowance on investment, percent
D = annual allowance for depreciation and obsolescence on first machine, percent
d = annual allowance for depreciation and obsolescence on second machine, percent
T = annual allowance for taxes and insurance, percent
M = annual allowance for upkeep on first machine, percent
m = annual allowance for upkeep on second machine, percent
E = annual cost of power for first machine
e = annual cost of power for second machine
X = cost per unit of production of first machine
x = cost per unit of production of second machine
U = saving in floor space per year of first machine over second machine, dollars

Table 3.8 (continued)

Then:

$$X = \frac{E + L + LB + C(I + D + T + M) - U^b}{N}$$

$$x = \frac{e + l + lb + c(I + d + T + m)}{n}$$

(3) *Number of Years Required for Complete Amortization of Investment out of Savings*

$$\text{Number of years} = \frac{1}{\dfrac{S - C(I + D + T + M)}{C} + I + D}$$

b. Should the use of the first machine result in a loss rather than a savings in floor-space cost per year, the sign before U must be changed to plus (+).

B. Mathematical Analyses for Economic Fixture Selection

These formulas are particularly applicable when fixturing for small fixed-production rates.

Let N = number of pieces manufactured per year
C = first cost of fixture
I = annual allowance for interest on investment, percent
M = annual allowance for repairs, percent
T = annual allowance for taxes, percent
D = annual allowance for depreciation, percent
S = yearly cost of setup
a = saving in labor cost per unit
t = percentage of overhead applied on labor saved
H = number of years required for amortization of investment out of earnings

(1) *Number of Pieces Required to Pay for Fixture*

$$N = \frac{C(I + T + D + M) + S}{a(1 + t)}$$

(2) *Economic Investment in Fixtures for Given Production*

$$C = \frac{Na(1 + t) - S}{I + T + D + M}$$

(3) *Number of Years Required for a Fixture to Pay for Itself*

$$H = \frac{C}{Na(1 + t) - C(I + T + M) - S}$$

must be selected. This requires contacting suppliers, attending trade shows, requesting demonstrations and trial welds for evaluation, and studying comparative manufacturer's literature. Another option is to visit shops with similar equipment in operation, relying on their experience for advice.

Upon purchase of the equipment, planning should provide for installation and implementation. Sometimes, many company personnel are not prepared to accept a new process as a standard manufacturing tool. Therefore, education must be included in the planning and provided in the implementation stage. This will include seminars, sending personnel to the manufacturer's plant for training, and involving both management and factory personnel in the implementation process.

ECONOMICS OF BRAZING AND SOLDERING

Brazing and soldering processes generally require the use of filler metal. Many filler metals for these processes are costly because they contain noble or rare elements to provide desired joint properties. Every effort must be made to minimize the amount of filler metal for each joint, consistent with the quality level required. The amount of filler metal required is dependent on a number of factors. Analyses of joint designs, fabrication techniques, procedures, and operations in preparation and finishing of parts can provide information leading to reduction of costs.

JOINT DESIGN

Many variables must be considered in the design of brazed and soldered joints. From the mechanical standpoint, the joints must be capable of carrying the design loads. Rules applying to concentrated loads, stress concentration, static loading, and dynamic loading are the same as with other machined or fabricated parts.

The design of a brazed or soldered joint does have specific requirements that must be met. Some of the important factors are as follows:

(1) *Compositions*. The compositions of the base and filler metals must be compatible. The properties of the filler metal in the joint must be considered when designing for a given service condition, as well as the cost of the filler metal.

(2) *Type and Design of Joint*. There are two basic types of joints used in brazing and soldering operations, the lap joint and the butt joint. Excessive overlap in a lap joint is generally a cause for excessive costs. Butt joints are less expensive in the use of filler metal but may lack strength, and may require costly preparation and fixturing.

(3) *Service Requirements*. Mechanical performance, electrical conductivity, pressure tightness, corrosion resistance, and service temperature are some factors to be considered in service. Economics are affected by these requirements, such as the case of brazing an electrical joint where a highly conductive but costly filler metal is needed.

(4) *Joint Clearance or Fit-up*. Improper joint clearance or fit-up can affect costs by requiring excessive amounts of filler metal or repair of defective joints.

PRECLEANING AND SURFACE PREPARATION

Clean, oxide-free surfaces are imperative to ensure sound joints of uniform quality. The costs of precleaning and surface preparation generally include chemical and labor costs. Selection of a cleaning method depends on the nature of the contaminant, the base metal, the required surface condition, and the joint design.

ASSEMBLY AND FIXTURING

The cost of assembly tooling and fixturing plus the added processing times for fixturing of assemblies make it advantageous to design parts that are easily assembled and self-fixtured. Some methods employed to avoid fixtures are resistance welding; arc tack welding; interlocking tabs and slots; and mechanical methods, such as staking, expanding, flaring, spinning, swaging, knurling, and dimpling.

TECHNIQUES OF BRAZING AND SOLDERING

Many variables affect the cost of making a sound joint with brazing and soldering processes (see Table 3.2). Of fundamental importance is the technique selected, based on the following factors:

(1) Selection of base metal
(2) Joint design
(3) Surface preparation
(4) Fixturing
(5) Selection of flux or atmosphere
(6) Selection of filler metal
(7) Method of heating

These factors, singly or in combination, have a direct bearing on the costs of the operation. A correctly applied technique based on sound engineering will result in a quality product at minimum cost. The desired goal is to heat the joint or assembly to brazing temperature as uniformly and as quickly as possible, while avoiding localized overheating. Mechanized equipment must be of suitable construction to provide proper control of temperature, time, and atmosphere. In many cases, consideration must be given to thermal expansion of the base metal to preserve correct joint clearance while the assembly is raised to the brazing temperature.

INSPECTION AND QUALITY CONTROL

Inspection methods have different costs for a particular application. Two basic cost factors that should be considered in the selection of a nondestructive testing method are the cost and availability of the equipment, and the cost of performing the inspection. Visual inspection is usually the least expensive, but it is limited to the detection of filler metal discontinuities at the edges of the joint. In general, the costs of radiographic and ultrasonic testing are greater than the less complicated visual and liquid penetrant methods of inspection.

The selection of the proper method of inspection is complex. The method must be capable of meeting the inspection standards established in the purchase specifications, usually referring to a specific code or standard. It is suggested that the expertise of a qualified nondestructive testing engineer or technician be utilized in the planning stages.

ECONOMICS OF THERMAL CUTTING

Oxyfuel gas and plasma arc cutting are commonly used for shape cutting of parts from sheet and plate.[2] Oxyfuel gas cutting (OFC) is generally limited to carbon and low alloy steels. Special process modifications are required to cut high alloy steels. Plasma arc cutting (PAC) can be used to cut any metal. Most applications are for carbon steel, aluminum, and stainless steel. Both processes can be used for stack cutting, plate beveling, shape cutting, and piercing.

Shape cutting machines for PAC and OFC are similar in design. Generally, plasma arc shape cutting machines can operate at higher travel speeds than similar OFC machines.

Carbon steel plate can be cut faster with plasma arc cutting than with OFC processes in thicknesses below 3 in. if the appropriate equipment is used. For thicknesses under 1 in., PAC speeds can be up to five times those for OFC. Over 1-1/2 in. thickness, the selection of PAC or OFC will depend on other factors such as equipment costs, load factor, and applications for cutting thinner steel plates and nonferrous metals.

The economic advantages of PAC, as compared to OFC, are more likely to be apparent where long, continuous cuts are made on a large number of pieces. This type of cutting might be used in shipbuilding, tank fabrication, bridge construction, and steel service centers. The comparative economy of short cut lengths, which require frequent starts, will depend on the number of torches that can be used practically in multiple piece production.

Some limits on the number of PAC torches that can operate simultaneously are (1) the power demand on the plant utility lines, (2) the limited visibility imposed on the operator by lens shade requirements for high current PAC, and (3) the

2. For information on oxyfuel gas and plasma arc cutting, refer to the *Welding Handbook*, Vol. 2, 7th Ed., 459-516.

potential damage to the equipment and material from delayed detection of a torch malfunction. Therefore, for simultaneously cutting several small shapes from a carbon or stainless steel plate, the reliability of a PAC installation must be compared to a similar OFC installation.

The qualifying variables required to determine thermal cutting costs include:

(1) Available operating equipment
 (a) Type of control, numerical or photoelectric
 (b) Machine speed range and performance capabilities
(2) Available cutting process
 (a) Oxyfuel gas
 (b) Plasma arc
(3) Number of cutting heads available for each process
(4) Cutting head minimum spacing and maximum travel
(5) Maximum equipment availability
(6) Available operating and material handling personnel
(7) Available part programming or template preparation equipment
(8) Available material
 (a) Grade
 (b) Chemistry
 (c) Length
 (d) Width
 (e) Thickness
 (f) Number of plates
(9) Material handling capabilities
 (a) Crane
 (b) Conveyor
 (c) Number of cutting tables
(10) Parts to be cut
 (a) Quantity
 (b) Length and width for separation
 (c) Area per part
 (d) Number of inside cutouts per part
 (e) Total linear inches of cut per part
(11) Fuel gas type for OFC
(12) Power available for PAC
(13) Cost per foot of cut, or per part, for:
 (a) Fuel gas
 (b) Oxygen
 (c) Nitrogen
 (d) Argon and hydrogen
 (e) Electricty
 (f) Oxyfuel gas torch consumables
 (g) Plasma arc consumables
 (h) Direct labor — machine operator
 (i) Direct labor — material handler
 (j) Direct labor — programmer or template maker
 (k) Incremental overhead — operator and equipment
 (l) Incremental overhead — material handler
 (m) Incremental overhead — programmer or template maker
 (n) Equipment maintenance
 (o) Miscellaneous expenses — torches, cutting table slats, slag disposal, etc.
 (p) Material
(14) Torch on-time duty cycle
(15) Material utilization
(16) Material handling efficiency
(17) Scrap part allowance

The variables associated with cost estimating will produce an infinite number of possible combinations. The net effect of the variables can be divided into three categories:

(1) Consumables required for a quantity of parts
(2) Programming or template time required per part regardless of quantity
(3) Operating costs per hour

Consumables include all gases, expendable torch tip parts, and electric power. Programming or template preparation cost, once established, may be prorated over the total parts required. Once accounted for, this one time cost can be eliminated on subsequent production runs of the same parts.

Per-hour operating costs will dramatically affect the per-part cost. These costs include incremental overhead, direct labor, prorated maintenance and miscellaneous expenses, and equipment power usage.

If choices can be made in favor of PAC over OFC or of multiple torch operation over single torch operation, or both, they will show the most dramatic effect on the per-part cost. In the case of steel using OFC, the alloy content will affect the cutting speed, which may have a significant effect on the per-part cost.

To illustrate the cost allocation for OFC and PAC, the following model will be used:

Material
- Grade — AISI 1020 hot-rolled steel
- Thickness — 1 in.
- Plate Size — 84 in. x 240 in.

Piece Part
- Quantity required — 50 each
- Width for torch spacing — 12.5 in.
- Length for repeat rows — 24.5 in.
- Number of inside cutouts per part — 1
- Combined linear inches of cut per part — 80 in.
- Time required to make the tape program (on a per-piece basis) — 0.5 h

Equipment
- CNC controlled gantry
- Number of OFC torches available — 6
- Number of PAC torches available — 6
- Fuel gas used — natural gas
- PAC torch — water injection type with nitrogen plasma, 600 A maximum

Personnel Available
- Machine operator — 1
- Material handler — 1
- Programmer — 1

Consumable Costs
- Average oxyfuel gas tip — $11.50
- Average plasma arc electrode — $22.00
- Average plasma arc nozzle — $22.00
- Natural gas — $0.64/100 ft^3
- Oxygen — $1.00/100 ft^3
- Nitrogen — $1.00/100 ft^3
- Electric power — $0.078/kwh

Personnel Costs
- Direct labor operator — $9.35/h
- Direct labor material handler — $8.15/h
- Direct labor programmer — $9.75/h
- Incremental overhead, operator and machine — $25.00/h
- Incremental overhead, material handler — $8.00/hr

General Expenses
- Machine maintenance based on 1 shift operation at 50 percent torch "on" time — $3.00/h
- Miscellaneous expenses including torches, cutting table slats, etc. — $2.00/h

Based on typical cutting speeds, expected service lives of cutting accessories, gas consumption, material handling capabilities, and power requirements, the cost distribution effects of the various possible cutting torch combinations are shown in Fig. 3.7.

The consumables required per-foot-of-cut remain constant whether one or more torches are used to complete the required cut. The use of multiple torches reduces the time to complete the cutting requirement and, therefore, reduces the per-hour operating costs.

The process selection, OFC or PAC, will affect the per-hour operating cost due to the difference in cutting speeds. The consumables required also change with the process, giving different per-foot-of-cut bases. Both processes, however, have a constant *consumables required per foot of cut* regardless of the number of torches used.

The remaining variable is the cost associated with the preparation of the NC part program for a specific part or group of parts. This is a one-time cost, and will be distributed over the quantity of parts required. If the part geometry is unique and only one piece is required then the entire cost of preparing the NC part program will be reflected in the cost of that piece. Therefore, one-of-a-kind piece parts may be considerably more expensive to process than a quantity of identical parts.

The common denominator for per-hour operating cost is *combined cutting speed.* Increasing the number of cutting heads used simultaneously or selecting the faster cutting process, or both, will reduce the per-piece part cost. This will not result in a linear cost reduction, however, due to material handling considerations.

There will be a combined cutting speed where one person will be unable to remove cut piece parts as fast as the machine can cut them. At this point, the machine torch "on" time must not be increased unless an assistant is provided to help load plate and unload cut parts. It is reasonable to expect a higher torch "on" time with both the machine operator and the material handler working together. As the combined cutting speed is increased, eventually a point is again reached where parts cannot be unloaded and plates loaded

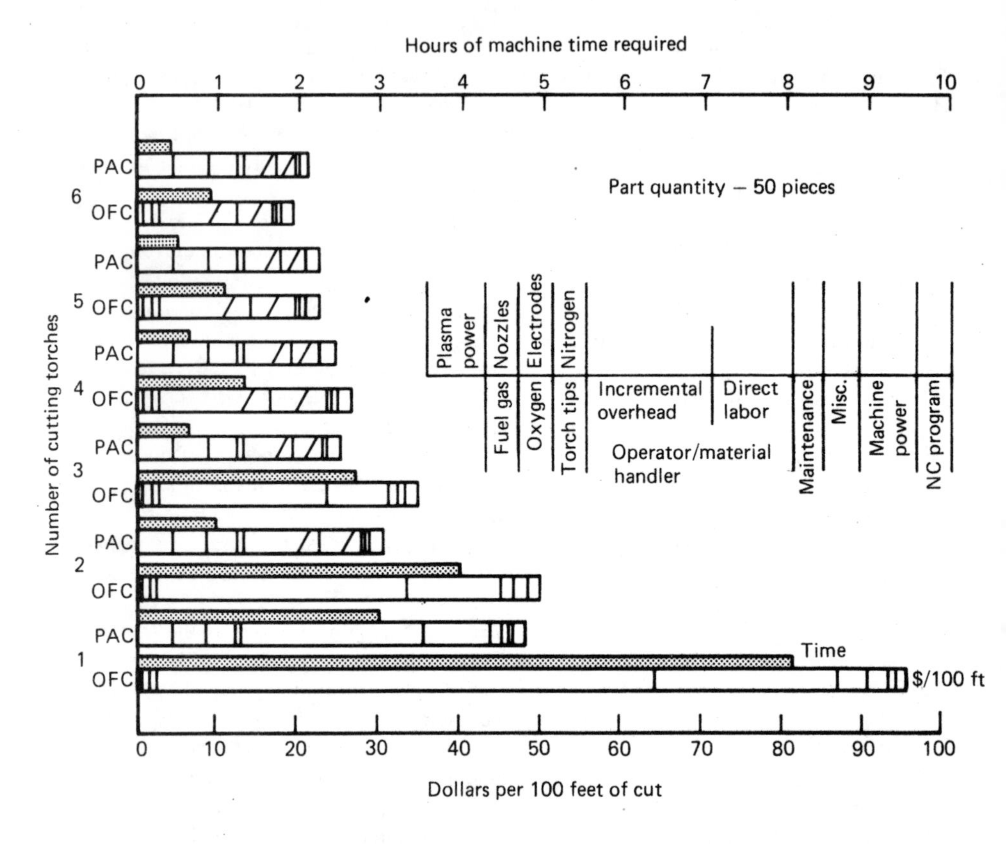

Fig. 3.7—Cost and time comparisons for thermal cutting 1-inch AISI 1020 steel plate

onto the cutting tables fast enough to keep up with the cutting machine. At this point, it is unlikely that the machine torch "on" time will be increased.

Adding more people to the group will have little effect unless additional cutting tables with material handling equipment are avilable. The material handling bottleneck must be addressed to improve throughput capabilities as the combined cutting speed crosses this threshold.

Providing that a material handling bottleneck does not exist, a higher torch "on" time will generally offset the direct and indirect labor costs for the material handling assistant.

As the per-hour operating cost is reduced by increased combined cutting speed, the consumable costs become more apparent. The costs for consumables will normally be higher for PAC than for OFC. As the material thickness increases, the power or current required for PAC increases, and the life expectancy of the consumable hardware will decrease. If the material can be cut using either OFC or PAC, the consumables cost per-part quantity for PAC will normally be two to five times as much as for OFC. Because of this, the magnitude of the reduction in per-hour operating costs resulting from increasing the number of cutting heads will be less apparent for PAC than for OFC. This becomes more noticeable as the arc current and the material thickness increase.

The cost per 100 feet of cut based on the model graph (Fig. 3.7) gives an advantage to four OFC torches at $27 per 100 ft. as opposed to two PAC torches at $31 per 100 ft. However, the scheduled machine time for the production run favors PAC at 1 hour over OFC at 1.3 hours. If machine capacity presents a problem, the decision likely would be to sacrifice the per-part cost in favor of the increased throughput capability.

A change in one or more variables may affect the economic conditions. An example is illustrated in Fig. 3.8 for cutting 0.5-in. thick AISI 1020 steel plate. The only change from the previous model is the reduced plate thickness. Steel of 0.5 in. thickness can be cut with PAC at a much higher speed than 1-in. thick plate (105 vs 50 in./min), while the difference in OFC speed may be only 22 vs 18 in./min. Other advantages of the thinner plate include reduced arc current and increased nozzle and electrode lives. A possible disadvantage lies in an increased commitment to material handling. A bottleneck may occur as soon as two plasma arc torches are used to simultaneously cut parts.

The cost per 100 feet of cut gives a cost advantage to six OFC torches at $15.32 over two PAC torches at $16.00. However, the scheduled machine time for the production run favors plasma arc at 40 minutes over oxyfuel at 44 minutes.

Any general statements will only apply to a specific set of conditions. A detailed analysis of specific models will be required to determine accurately the economics and estimated costs of thermal cutting. The proliferation of computers and application software offers quick and convenient analysis of operating variables.

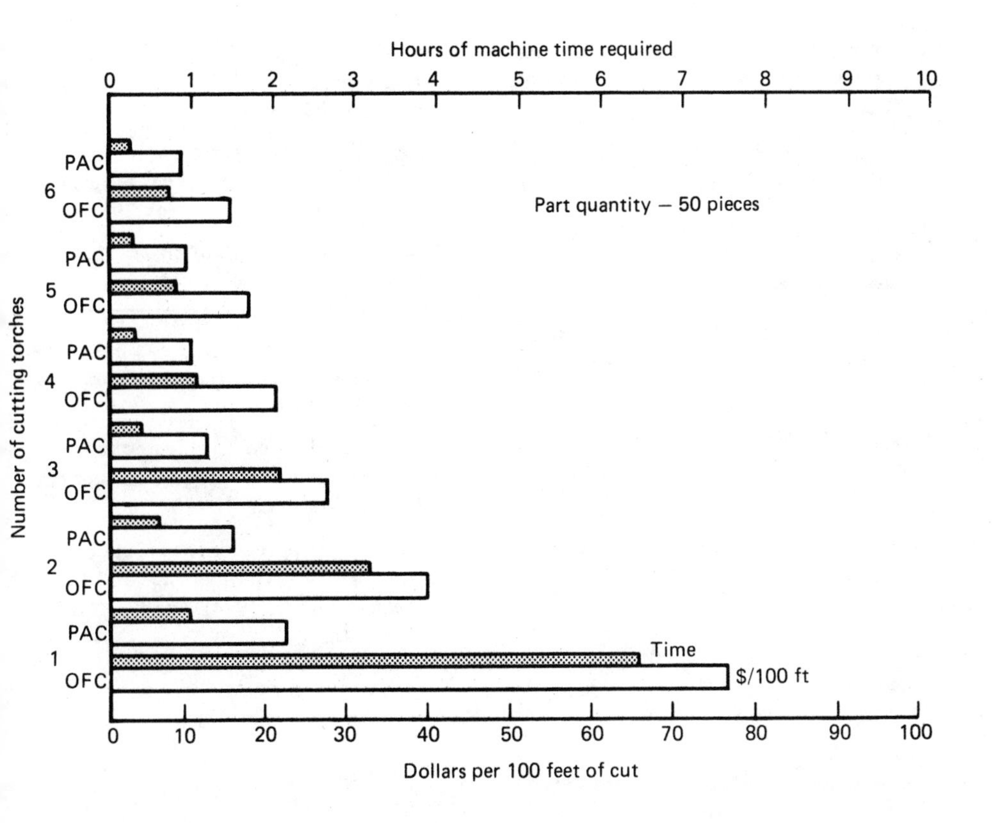

Fig. 3.8—Cost and time comparisons for thermal cutting 0.5-in. AISI 1020 steel plate

Metric Conversion Factors

1 in. = 25.4 mm	1 lb/ft = 1.5 kg/m
1 $in.^2$ = 645 mm^2	1 lb/h = 0.454 kg/h
1 in./min = 0.42 mm/s	1 lb/$in.^3$ = 2.77 x 10^4 kg/m^3
1 in./lb = 55.9 mm/kg	$1/ft = $3.28/m
1 ft/lb = 0.67 m/kg	$1/$ft^3$ = $35.31/$m^3$
1 ft^3/h = 0.472 L/min	$1/lb = $2.21/kg

SUPPLEMENTARY READING LIST

An easy way to compute cost of weld metal, *Welding Design and Fabrication,* July 1982: 54-56.

Cary, H., *Modern Welding Technology*, Englewood Cliffs, NJ: Prentice-Hall, 1979; 541-59.

Hines, W.G. Jr., Selecting the most economical welding process, *Metal Progress*, 102: Nov 1972: 42-44.

Lesnewich, A., The real cost of depositing a pound of weld metal, *Metal Progress*, 121: Apr. 1982: 52-55

Necastro, N.P., How much time to weld that assembly, *Welding Design and Fabrication*, Feb. 1983: 50-53.

Oswald, P., *Cost Estimating for Engineers and Managers,* Englewood Cliffs, NJ: Prentice-Hall, 1974.

Pandjiris, A.K., Cooper, N.C., and Davis, W.J., Know costs - then weld, *Welding Journal,* 47 (7): July 1968: 561-68.

Pavone, V.J., Methods for economic justification of an arc welding robot installation, *Welding Journal*, 62 (11): Nov. 1983: 40-46.

Procedure Handbook of Arc Welding, Cleveland, OH: The Lincoln Electric Co, 1973: Section 12.

Rudolph, H.M., The hidden costs of thermal cutting, *Welding Design and Fabrication,* Aug. 1979: 76-80.

Smith, D., and Layden, L., Figure the way to cutting economy, *Welding Design and Fabrication*, Mar. 1978: 106-107.

Sullivan, M.J., Application considerations for selecting industrial robots for arc welding, *Welding Journal*, 59 (4): Apr 1980: 28-31.

4
Fixtures and Positioners

Fixtures . . . . *150* *Positioners* *159*

Supplementary Reading List *175*

Chapter Committee

D.R. SPISIAK, *Chairman*
Aronson Machine Company

P.H. GALTON
Big Three Industries, Inc.

E.B. GEMPLER
United Aircraft Products, Inc.

W.H. KEARNS
American Welding Society

L.M. LAYDEN
Airco Welding Products

S.J. LEVY
Teledyne Readco

Welding Handbook Comittee Member

E.H. DAGGETT
Babcock and Wilcox Company

4
Fixtures and Positioners

FIXTURES

GENERAL DESCRIPTION

A *fixture* is a device for holding parts in their correct positions during assembly. It is sometimes referred to as a *jig*. For welding and brazing, a fixture generally locates the parts to provide the specified joint geometry for subsequent joining.

When making a small number of assemblies, temporary fixturing may be used, provided specified dimensions can be maintained. Such fixturing may consist of bars, clamps, spacers, etc. Parts may be positioned with rules, squares, and other measuring devices. For quantity production, design and construction of accurate and durable fixtures is justified for economical assembly or precise part location, or both.

DESIGN

There are only a few standard commercial welding fixtures. Examples are fixtures for welding linear seams in sheet and tubing, as shown in Fig. 4.1, and clamping fixtures to align pipe sections while welding the circumferential joints, Fig. 4.2.

When designing a special fixture, commercial tool and die components and clamps should be used wherever practical. For like assemblies, a fixture may sometimes be designed to accommodate two or more sizes by repositioning certain locators on the fixture.

Desirable features of fixtures for welding or brazing include, but are not limited to the following:

(1) All joints in the assembly must be convenient and accessible while the parts are fixtured.

(2) Fixtures should be strong and sufficiently rigid for accurate location of parts, proper alignment of joints, and resistance to distortion.

(3) A fixture should be easily manipulated into the optimum position for the joining operation; balancing of a loaded fixture may be advisable.

(4) For economic reasons, the fixture need not provide greater accuracy of part position than required by specifications.

(5) The fixture design should be as simple and inexpensive as practical, while performing its functions properly.

(6) Tack welding or staking and welding or brazing may be done in a single fixture. Where appropriate, the assembly may be removed from the fixture for final welding or brazing.

(7) Fixtures should permit expansion and contraction of parts without binding, in at least one plane, to minimize stressing of the assembly during the joining operation.

(8) Clamps, cams, rams, and other moving components should be quick-acting. Screws and moving parts should be protected from flux and spatter.

(9) Fixture design must permit easy and rapid unloading of the completed assembly.

(10) A fixture should be mounted on a positioner, spindle, wheels, or another device to place the assembly in the optimum position for welding or brazing.

(11) Fixtures must be designed so that the operators are protected from pressure and pinch points that may cause injury. Controls for air, hydraulic, and electrically operated fixture com-

ponents must be inoperable while the fixture is being loaded and unloaded.

When designing a fixture, the designer needs certain information concerning the assembly including the following:

(1) Manufacturing drawings of the component parts and assembly, and accessibility to actual parts

(2) Engineering specifications for the assembly

(3) Joining (welding or brazing) procedure specifications

(4) Manufacturing operations documentation

(5) Production requirements and schedule

From these documents, the designer can determine the requirements for the fixturing including the following:

(1) Type of fixturing

(2) Tolerance and distortion requirements of the assembly

(3) Joining process to be used

(4) Accessibility to the joint during joining

(5) Limitations on fixture materials

(6) Service life

(7) Number of fixtures needed to meet production requirements

APPLICATIONS

Permanent fixtures are normally used for production when their costs can be justified by savings in operating costs. Fixtures may be needed when close dimensional tolerances must be maintained during assembly even with low production. Fixture costs must be planned and included in the project cost estimate. Self-fixturing joint designs should be used where possible to minimize fixturing requirements. Even with such joint designs, fixturing may be needed to maintain alignment of parts.

Fig. 4.1—Typical clamping arrangement for automatic welding of butt joints in sheet

Fig. 4.2—External clamping fixture to align pipe joints for welding

WELDING FIXTURES

Design Considerations

Welding fixtures may be used with manual, machine, or automatic welding. Arc welding (except electrogas welding) is best done in the flat position to take advantage of high deposition rates. Therefore, a fixture should place the joint in the flat position for welding wherever possible. If the joints are located in several planes, provisions should be made to reposition the fixture so that most or all of the joints can be welded in the flat position. Joints must be readily accessible for welding from one or both sides, as required.

With machine or automatic welding, the work may move and the welding torch or gun remain stationary or vice versa. With circumferential joints, the work is normally rotated horizontally under a fixed arc welding head. An exception is welding bosses or pins to large, horizontal surfaces. The vertical boss or pin is placed on the stationary surface and the welding torch rotates around the boss or pin to produce a fillet weld in the horizontal position. Long linear joints are normally welded with the work stationary while the welding head moves along the joint.

The welding procedures for an assembly must be established before commencing fixture design. The welding gun or torch design must also be known to avoid interference between it and the fixture during welding.

In some applications, provisions are needed in the welding fixture to control root face contour. A nonconsumable backing is commonly used to do this. The backing may be made of either ceramic blocks, a bed of flux, or a copper bar with the desired groove in the surface next to the joint. A water-cooled copper bar may be appropriate for high production, but not for quench-hardenable steels. Inert gas may be directed to the root of the joint from passages in a copper bar backing bar to minimize oxidation. In any case, the welding arc must not impinge on the nonconsumable backing.

A clamping arrangement is generally required to hold the parts in proper alignment and against a backing, particularly with sheet gages (see Fig. 4.1). The edges to be joined are clamped against a backing by two rows of nonmagnetic hold-down fingers.

If the welding procedure calls for tack welding, clamps may be needed only to hold alignment during this operation. With some designs, the assembly can be tack welded in the fixture and then removed for welding provided distortion can be held to acceptable limits.

Arc Blow

Current flowing in a welding fixture may produce magnetic fields near the welding arc. These fields may cause the arc to deflect from axial alignment with the electrode. This behavior is known as *arc blow*. It results in erratic arc action and inconsistent joint penetration. During the fixture design phase, precautions should be taken to avoid potential causes of arc blow. Some points to be considered are as follows:

(1) Fixture materials located within 1 to 2 in. of the joint must be nonmagnetic.

(2) Steel in other locations on fixtures should be low carbon or austenitic stainless type to avoid residual magnetism of the fixture.

(3) The amount of steel in the fixture on each side of but near to the joint should be approximately equal.

(4) Space for run-off tabs at the ends of the seam, when required, should be provided in the fixture.

(5) The work lead connection should be at the starting end of linear joints.

Examples of Welding Fixtures

Welding fixture designs may range from simple, manually operated types to large, complicated air or hydraulic operated types. A simple fixture for aligning nozzles in a cylindrical tank is shown in Fig. 4.3(A), (B), and (C). In View A, the tank is located axially in the fixture by a pin entering one nozzle port and then the tank is clamped in position. The tank and clamp are rotated 90 degrees against a stop, as shown in View B. The nozzle, held in a chuck, is moved from above into the port for welding, as shown in View C. The gas metal arc welding gun is mechanically rotated about the nozzle to make a horizontal fillet weld between the parts.

A large internal welding fixture for aligning circumferential joints in 260-in. diameter cases is shown in Fig. 4.4. The segmented alignment shoes are pressed against the inside diameter of the joint by hydraulic cylinders.

BRAZING FIXTURES

General Considerations

Assembly of parts for brazing depends on the brazing process to be used, the materials being joined, and the configuration of individual details in a brazed component. Components to be joined must be assembled in a fixed position relative to each other and maintained in position throughout the brazing cycle. Surfaces to be joined must be properly spaced to provide the desired joint clearance at room temperature and at brazing temperature. The method of assembly and fixturing for brazing can be critical.

The design of parts that are easily assembled and self-fixturing during brazing is economical because the costs of assembly fixtures plus processing time for loading and unloading brazing fixtures are avoided. Sheet metal structures often are held together for brazing by intermittent spot welds, rivets, or thin metal straps. Positioning of parts by welding methods requires the use of a flux or an inert atmosphere at welded areas to prevent surface oxidation that would inhibit brazing filler metal flow. Aluminum sheet structures are sometimes held together for brazing with interlocking tabs and slots.

Assembly Methods

Cylindrical parts, tubing, and solid members can be assembled for brazing by a number of methods. The method should, if possible, provide a uniform joint clearance that can be maintained throughout the brazing operation. Some of the methods employed for assembling cylindrical parts are staking, expanding or flaring, spinning or swaging, knurling, and dimpling.

More complex structures usually require elaborate assembly methods. One requirement is that a structure being brazed must have acceptable braze joint clearance at brazing temperature consistent with the brazing filler metal, the base metal, and the brazing process used. Joint clearances can be maintained using shims of wire, ribbon, or screen that are compatible with the base metals.

Examples of methods used to position parts for brazing without the use of fixtures are shown in Fig. 4.5. In some cases, assemblies with self-fixtured joints may require the use of fixtures because of the complexity of the parts, location of the brazed joints, or the distribution of the mass. Fixturing may consist of clamps, springs, support blocks, channels, rings, etc., to keep parts in proper attitude for brazing. Mild steel, stainless steel, machinable ceramics, and graphite are often used for fixturing. Brazing of complex shapes often needs fixtures to locate part details and to maintain surface contours.

Fixture Materials

Selecting the fixture design best suited for a particular brazing application requires consideration of fixture material. The choice must be based on (1) the brazing method to be used, (2) the base metals in the assembly, and (3) the required brazing temperature and atmosphere. When brazing in a vacuum, the only limitations on fixture materials are that they be stable at brazing temperature, possess compatible thermal expansion properties, and do not outgas and contaminate the brazing atmosphere.

Whenever metal fixtures are used for brazing, fixture surfaces contacting the part being brazed should be adequately protected to prevent

(A) Tank located by pin in nozzle port

Fig. 4.3—Fixture for welding a nozzle in a tank

(B) Tank rotated 90 degrees

(C) Nozzle lowered into port for welding

Fig. 4.3 (Cont.)—Fixture for welding a nozzle in a tank

Fig. 4.4—Large internal hydraulic fixture for circumferential welds in 260-in. diameter cases

brazing the part to the fixture. Brazing fixture materials must not interact with the brazing atmosphere to produce undesirable reactions detrimental to filler metal flow or to the mechanical properties of the base metal.

In most cases, unheated fixtures are made of carbon steel, stainless steel, nonferrous alloys, and nonmetals. These fixtures are used for positioning the components for tack welding. They need not be massive or heavy, but should be sturdy enough to position components properly as required by the brazement design.

Fixtures for short production runs at moderate temperatures are commonly made of low carbon steel. Carbon steel is economical but it has low strength at brazing temperatures. For long production runs, stainless steel and heat resistant alloys are preferred for fixture construction.

Fixtures used in furnace brazing at high temperatures must have good stability at temperature and the ability to cool rapidly. Most metals are not stable enough to maintain tolerances during a high temperature brazing cycle. Therefore, ceramics or graphite is used for hot fixturing. Ceramics, due to their high processing cost, are limited to small fixtures and for spacer blocks to maintain joint clearance during brazing of small components.

Graphite has good thermal conductivity, good dimensional stability, but low abrasion resistance. It diffuses into some metals at brazing temperatures, and reacts with water vapor in a hydrogen atmosphere to produce contaminants.

Silicon carbide is also used for braze fixturing, but it has lower thermal conductivity than graphite and the same detrimental characteristics. Molybdenum and tungsten may be used, but they are high in cost and must not be heated in an oxidizing atmosphere.

Induction brazing often requires that the fixture be located in proximity to the inductor. In such cases, the choice of materials for the fixture is limited to those that remain relatively unaffected by the magnetic field. Nonmetals, such as graphite and ceramics, are commonly used. The use of metals readily heated by induced currents should be avoided.

In chemical bath dip brazing, fixtures must be free of moisture to avoid explosion. Fixture materials must not absorb moisture when exposed to the atmosphere. The buoyancy effect of the molten salt must be considered, and the fixture material must not react with the salt bath.

The high thermal expansions of aluminum and magnesium require that fixtures for brazements of these metals be designed to provide for expansion and contraction of the assembly during heating and cooling. Springs are sometimes used to permit such movement. High temperature nickel alloy springs are preferred because they have good characteristics in the 1000°-1200°F brazing temperature range, and good corrosion resistance.

Fixture Design

Applications requiring brazing fixtures vary widely. Therefore, it is difficult to be specific about fixture designs. However, some basic rules to follow when designing brazing fixtures are as follows:

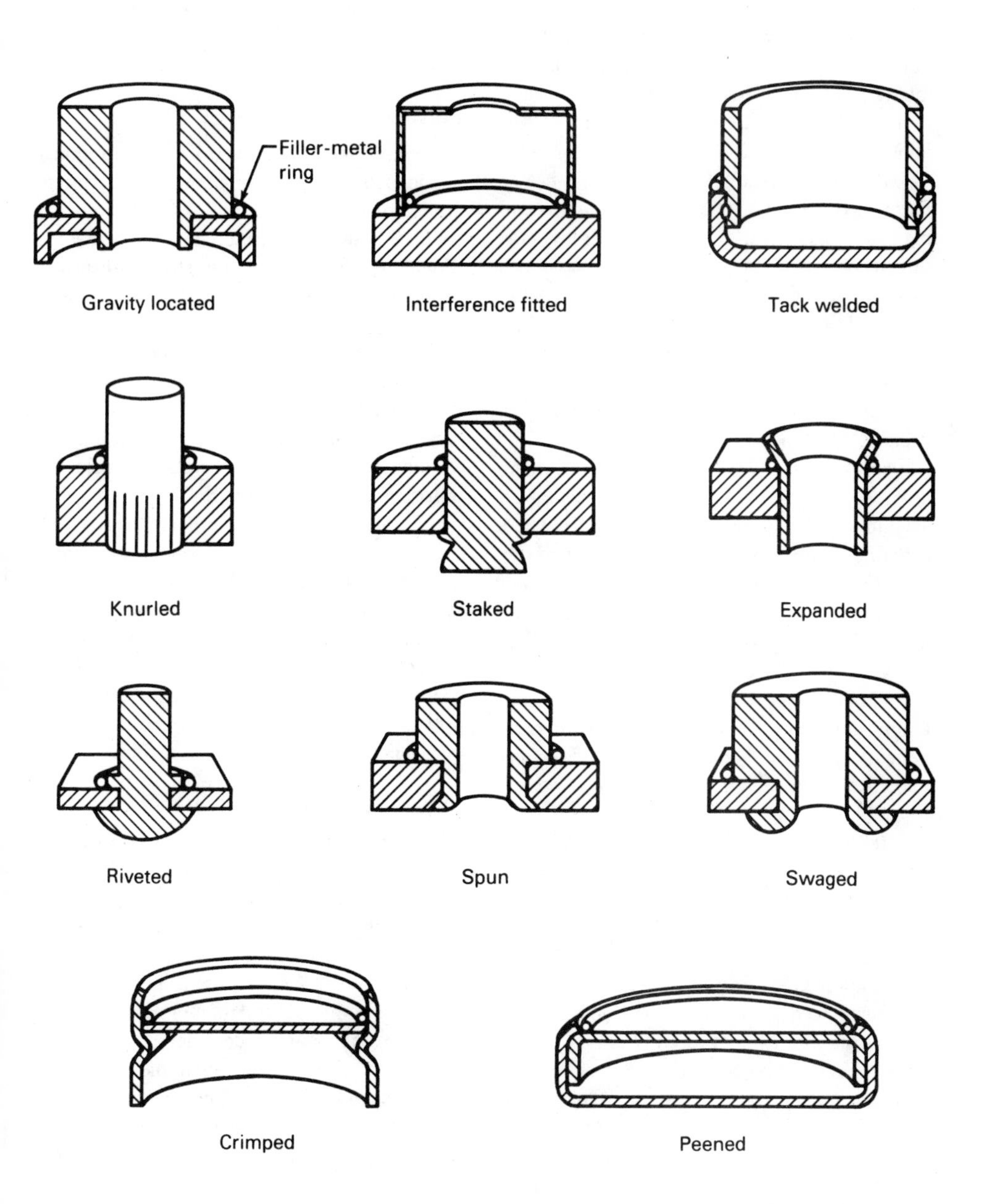

Fig. 4.5—Typical self-fixturing methods for brazed assemblies

(1) Keep fixture design simple. Use thin sections consistent with rigidity and durability requirements. Fixtures for torch brazing must not interfere with the torch flame, the brazer's view, or the application of the brazing filler metal to the joint.

(2) Avoid the use of bolts or screws in heated fixtures. These fasteners tend to relax upon heating and often pressure weld in place. If used, the threads should have loose fit, and be coated with a magnesia-alcohol mixture to prevent sticking.

(3) Design springs or clamps to withstand brazing temperatures when used in heated areas.

(4) Fabricate furnace brazing fixtures with components of uniform thicknesses for even heating. Avoid heavy sections.

(5) Avoid the use of dissimilar metals where differences in thermal expansion could affect assembly dimensions. An exception is where expansion differences are used advantageously.

(6) Subject a fixture to the brazing environment and temperature before it is used for brazing hardware to ensure stability and to relieve stresses.

(7) Where practical, use nickel alloys containing aluminum and titanium for fixture components because they develop protective oxides upon heating in air. These oxides inhibit wetting by molten brazing filler metals.

(8) Do not specify fixture materials that might react at elevated temperatures with the assembly being brazed while they are in intimate contact. For example, nickel alloys are generally unsatisfactory fixture materials for furnace brazing titanium, or vice versa, because a nickel-titanium eutectic forms at about 1730° F.

Furnace brazing of assemblies that require close tolerances, such as contoured shapes, requires very stable fixturing. Metal fixturing normally lacks the stability required for repeated heating and cooling. Therefore, ceramic fixturing is often used for holding close tolerances. Castable ceramics, such as fused silica, are excellent. They possess superior dimensional stability, and can be cast to close tolerances without machining. Their primary drawbacks are poor thermal conductivity and porosity that tends to absorb moisture. Usually a baking cycle to remove moisture is required before each use. Alumina is also used for braze fixturing, but it is diffficult to fabricate to close tolerances, hard to machine, and has poor thermal shock properties.

Furnace Brazing Fixtures

Components of fixtures should be as thin as possible for rapid heating, consistent with required rigidity and durability. The number of contacts between the fixture and the brazement should be a minimum, consistent with proper support. Point or line contact is preferable to area contact. Even with a minimum number of contacts between a fixture and the components to be brazed, it is sometimes difficult to prevent filler metal from flowing and wetting nearby contacts. If a contact area between a fixture and a component must be large, a contact material that resists wetting must be used.

In general, fixtures external to the assembly should expand faster than the assembly and internal fixtures more slowly. For applications where tight clamping is required, the reverse is true. To equalize the load from shrinkage on the brazement during cooling, systems of levers, cams, and weights can be used. Wedges and weights often provide good joint follow-up when the brazing filler metal melts and flows in the joints.

Induction Brazing Fixtures

Most assemblies that are brazed by induction require some fixturing, even though components may be held securely together mechanically before being positioned in the inductor. Fixtures can range from a simple locating pin to hold the assembly in the center of the inductor to an elaborate clamping and holding arrangement. Every assembly must be considered separately in relation to its fixturing requirements. Fixture design is influenced not only by the size and shape of the assembly, but also by its production rate. Functions of a fixture may include supporting the assembly, positioning the assembly in the inductor, and holding the assembly securely together during the brazing cycle until the filler metal has solidified.

Selection of materials for fixtures is especially important. If all components of a fixture are sufficiently remote from the inductor (several inches away) to be unaffected by its magnetic field, materials used for fixtures for other brazing processes are suitable. If all or part of a fixture is within about 2 in. of the inductor or the leads to the inductor, the fixture must be made of a nonmetallic material and preferably, one that is heat-resistant. Several such materials are commercially available, including heat-resistant glass, ceramics, quartz, and plaster. When use of a metal near the inductor or the leads is unavoidable, aluminum, copper, and brass are the preferred metals. Water-cooled fixtures may be necessary.

Dip Brazing Fixtures

Assemblies with self-fixtured joints minimize fixturing requirements during dip brazing. A fixture should not touch any joint being brazed to avoid the risk of brazing the assembly to the fixture.

Fixturing of assemblies can create problems including distortion, difficulty in maintaining dimensional tolerances, and the necessity of heating the added mass of the fixture. However, these difficulties can be overcome by proper fixture design including use of springs, fixture construction of a material having thermal expansion characteristics compatible with those of the base metal, and attention to maintenance of tolerances during brazing.

The fixture materials must be compatible with the molten salt and brazing flux, and have reasonably long lives. The fixture design must provide for draining of molten salt when the fixture is removed from the bath.

POSITIONERS

GENERAL DESCRIPTION

A *positioner* is a mechanical device that supports and moves a weldment to the desired position for welding and other operations. In some cases, a positioner may move a weldment as welding progresses along a joint. A welding fixture may be mounted on a positioner to place the fixture and weldment in the most advantageous positions for loading, welding, and unloading.

Some assemblies may be fixtured on the floor and the joints tack welded to hold the assembly together. Then, the weldment is removed from the fixture and mounted on a positioner for welding the joints in the best positions for economical production.

A weldment on a positioner may be repositioned during welding or upon completion for cleaning, machining a specified weld contour, nondestructive inspection, and weld repairs.

ECONOMIC CONSIDERATIONS

There are both positive and negative considerations for the use of positioners for welding. They are primarily governed by welding and handling costs.

Deposition Rates

The highest deposition rates in arc welding can be obtained when welding is done in the flat position because gravity keeps the molten metal in the joint. The next best position is the horizontal. The overall result of positioning of weldments should be lower welding costs.

Welder Skill

It takes less skill to arc weld in the flat position than in other welding positions because it is easier for the welder to control the molten weld pool. Therefore, welding labor costs should be lower when the assembly can be easily manipulated for welding joints in the flat position.

Operator Factor and Set-Up Costs

Operator factor is the ratio of arc time to total time that a welder applies to a weldment. When the welder must manually reposition a weldment, wait for a crane operator to move it, or weld in positions other than flat, the operator factor will be lower and welding costs higher than when a positioner a used. Operator factor should be higher when a weldment can be rapidly positioned for welding. However, labor costs for safely loading and unloading a heavy weldment on a positioner, or for repositioning a weldment with a crane or other lift must also be considered.

For relatively short, small welds, it may be more economical to weld the joints in fixed positions than to reposition the weldment for ease of welding. The welding costs may be somewhat higher but the overall labor costs may be lower because of savings in handling costs. The cost of positioning equipment and handling must be offset by labor savings.

Weld Quality

In general, a qualified welder should be capable of producing welds with fewer defects in the flat position than in other welding positions. The result is fewer repairs. Also, a joint can be filled with fewer passes by taking advantage of the high deposition rates in the flat position. Welds made with a minimum of passes generally have lower welding stresses and associated distortion. However, heat input limitations must be observed with some alloy steels.

TYPES OF POSITIONING

Positioning can be done with one, two, or three different motions. One motion is rotation about one axis. This is normally accomplished with turning rolls or headstock and tailstock arrangements, both of which rotate the assembly about a horizontal axis.

Two motion positioning is a combination of rotation and tilting. It is normally accomplished with a positioner that has a tilting table as well as rotation. Three motion positioning is accomplished by adding vertical movement with an elevating device in the machine base, thus providing rotation, tilt, and elevation.

TURNING ROLLS

Turning rolls are used in sets, as shown in Fig. 4.6. Each set consists of one powered roll and one or more idler rolls, each roll having two or more wheels. Turning roll design is quite simple; a set of rolls normally consists of a fabricated steel frame, wheels, drive train, drive motor, and controls. Simplicity of design offers low initial cost as well as low maintenance and repair costs. Standard models can be used for many applications.

Turning rolls can be manufactured to specific requirements, such as wheel construction and surface composition, weldment weight and diameter, special motions, and unitized frames.

Usage of turning rolls is normally limited to rotation of a cylindrically shaped weldment about its horizontal axis. Noncylindrical assemblies can also be rotated on turning rolls using special round fixturing to hold the assembly. The fixtures rest on the turning rolls.

Turning rolls can position a seam for manual, semiautomatic, or machine welding in the flat position. With machine welding, circumferential welds may be rotated under a fixed welding head. Longitudinal welds can be made with a welding head mounted on a traveling carriage or manipulator.

HEADSTOCK - TAILSTOCK POSITIONERS

General Description

A *headstock* is a single-axis positioning device providing complete rotation of a vertical table about the horizontal axis. This device provides easy access to all sides of a large weldment for welding in the flat position and for other industrial operations.

A headstock is sometimes used in conjunction with a tailstock, as shown in Fig. 4.7. A tailstock is usually of the same configuration as the headstock, but it is not powered. A simple trunnion, an outboard roller support, or any other free-wheeling support structure best suited to the configuration of the weldment may be used in place of a tailstock.

In all instances, precise installation and alignment of a headstock-tailstock positioner is essential so that the axes of rotation are aligned. Otherwise, the equipment or the weldment may be damaged.

Generally, the concept and application of headstock-tailstock positioners are much the same as lathes used in machine shop operations.

Fig. 4.6—Turning rolls

However, in welding fabrication, the applications can be more versatile with the use of special fixtures and the addition of horizontal or vertical movement, or both, to the bases of the headstock and tailstock.

A headstock may be used independently for rotating weldments about the horizontal axis provided the overhanging load capacity is not exceeded with this type of usage. An example is the welding of elbows or flanges to short pipe sections.

Applications

Headstock-tailstock positioners are well suited for handling large assemblies for welding. Typical assemblies are structural girders, truck and machinery frames, truck and railroad tanks and bodies, armored tank hulls, earthmoving

Fig. 4.7—Headstock-tailstock positioner

and farm equipment components, pipe fabrication, and large transformer tanks. A headstock-tailstock positioner is sometimes used in conjunction with one or more welding head manipulators or other machine welding equipment. Appropriate fixtures and tooling are needed with a headstock-tailstock positioner.

Sound engineering principles and common sense must be applied in the joining of an assembly held in a headstock-tailstock positioner. Most long and narrow weldments supported at the ends tend to sag under their own weight or distort slightly as a result of fabrication. Rigid mounting of a weldment between the two positioner tables is to be avoided when the headstock and tailstock are installed in fixed locations. Distortion from welding may damage the equipment when the weldment is rotated.

A flexible connection must be provided at each end of a clamping and holding fixture designed to accommodate a weldment between the headstock and tailstock tables. A universal joint or pivot trunnion in the fixture is recommended to compensate for tolerances in the weldment components.

Consideration must be given to the inertia of rotation of a large weldment when selecting the proper capacity of the headstock. The larger the polar moment of inertia, the greater is the required torque to start and stop rotation.[1] When one or more more components of a weldment are at considerable distances from the rotational axis, the flywheel effect created by this condition can have serious effects on the headstock drive and gear train. Under such circumstances, rotation speed should be low to avoid excessive starting, stopping, and jogging torque, and possible damage to the drive train.

Where usages involve weldments with a significant portion of the mass located at some distance from the rotation centerline, it is advisable to select a headstock capacity greater than that normally needed for symmetrical weldments of the same mass.

Features and Accessories

To provide versatility to headstock-tailstock positioners, many features and accessories can be added. The following are the more common variations used with this equipment:

(1) Variable speed drive

(2) Self-centering chucks

(3) Automatic indexing controls

(4) Powered elevation

(5) Power and idler travel carriages with track and locking devices

(6) Through-hole tables

(7) Powered tailstock movement to accommodate varying weldment lengths and to facilitate clamping

(8) Digital tachometers

TURNTABLE POSITIONERS

Description

A *turntable* is a single-axis positioning device that provides rotation of a horizontal table or platform about a vertical axis, thus allowing rotation of a workpiece about that axis. They may range in size from a small bench unit to large floor models. See Fig. 4.8.

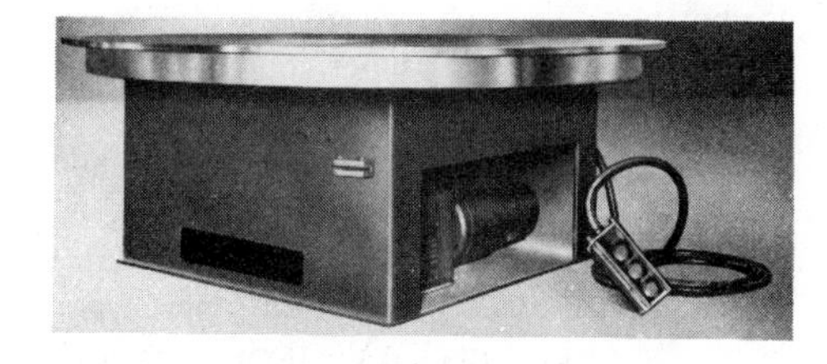

Fig. 4.8—A typical turntable positioner

Applications

Turntable positioners are available in capacities from 50 lb to 500 tons and larger. They are used extensively for machine welding, scarfing and cutting, cladding, grinding, polishing, assembly, and nondestructive testing. Turntables can be built in a "lazy susan" concept to index several assemblies on a common table. Several smaller turntables are mounted on the main table together with an indexing mechanism.

Features and Accessories

A turntable positioner can be equipped with a number of optional features and accessories. The following are some popular adaptations:

(1) Constant or variable speed drive

(2) Manual rotation of the table

(3) Indexing mechanism

(4) Tilting base

(5) Travel carriage and track

(6) Through hole in the center of the table

(7) Digital tachometer

TILTING-ROTATING POSITIONERS

General Description

Where versatility in positioning is required for shop operations, a tilting-rotating positioner is recommended. In most cases, the positioner table can be tilted through an angle of 135 degrees including the horizontal and vertical positions. The turntable mechanism is mounted on an assembly that is pivoted in the main frame

---

1. The unbalanced torque required to accelerate or decelerate a rotating body is related to the polar moment of inertia of the body and the angular acceleration by the following equation:

$$T_o = J_M \alpha$$

where

T_o = unbalanced torque

J_M = polar moment of inertia of the mass about the axis of rotation

α = acceleration or deceleration, radians /s/s

of the machine to provide a tilt axis. The tilt angle is limited by the design of the frame. Two common configurations of gear-driven tilting positioners differ by the limitation imposed by the tilt mechanism.

A *flat-135°* positioner can tilt the table from the horizontal through the vertical to 45 degrees past the vertical. A typical positioner of this type is shown in Fig. 4.9.

To provide clearance for the table and weldment when tilted past the vertical position, the base may be mounted on legs which enable the positioner to be raised to an appropriate height for the load to clear the floor.

The other common configuration for gear-driven tilting positioners is the *45°-90°* design. This type can tilt the table from 90 degrees forward through the horizontal to 45 degrees backward. A typical 45°-90° positioner is shown in Fig. 4.10. This design is most common in large sizes, and is available in capacities from about 40,000 to several million pounds.

The basic capacity rating of a positioner is based on (1) the total load (combined weight of the workpiece and the fixturing), and (2) the location of the center of gravity (CG)[2] of the combined load. Two dimensions are used to locate the CG of the combined load: (1) the distance from the positioner table surface to the CG of the combined load, and (2) the distance from the axis of rotation to the CG of the combined load, referred to as eccentricity (ECC). Thus, it is common practice to rate a positioner for a given combined load at a given CG and ECC. The load that can be placed on a positioner will be less than the rated basic capacity if the CG of the load is further than 12 in. from either axis.

There is a broad overlap in basic capacity of the two common types of positioners. The selection depends upon the application.

A positioner can have two functions. One function is to position or place the weldment where the welds can be made in the flat position and are easily accessible for welding. The other function is to provide work travel for welding with a stationary welding head.

If work travel is intended, the drive system for rotation must have variable speed. Commonly, travel about the rotation axis is used to make circumferential welds. The tilt axis is less frequently used for work travel, and is usually equipped with a constant speed drive.

Drop-Center Tilting Positioners

For special work where it is advantageous to use the tilt drive for work travel, a tilting positioner of special design can be used. An example of using work travel about the tilt axis is the fabrication of hemispherical pressure vessel heads from tapered sections. The welding fixture is mounted on the positioner table so that the center of a fixtured head coincides with the positioner tilt axis. Rotation around the tilt axis must be provided for welding the meridian seams.

The drop-center variation of the 45°-90° positioner is particularly useful for this application. In a conventionally designed positioner, the surface of the table is not on the tilt axis but is offset by a distance called *inherent overhang*. This distance may vary from a few inches to 18 inches depending on positioner size.

For rotation about the tilt axis, the center of the hemispherical head must coincide with that axis. This often poses problems in mounting such a weldment on a standard tilting positioner because of its inherent overhang.

To avoid this difficulty, a drop-center positioner is used. It is usuallyof the 45°-90° configuration, but the trunnion assembly is modified so that the tilt axis lies in or below the plane of the table face. Then, a hemispheric head can be mounted so the center is on the tilt axis. The arrangement of a hemisphere on a typical drop-center positioner is shown in Fig. 4.11.

The maximum size of weldments that can be handled on a drop center positioner is limited by the distance between the vertical side members of the base. Also, this type of positioner requires more floor space than a conventional positioner with the same table size because the tilt journals must be outside the table rather than beneath it.

---

2. The center of gravity of a body is that point where the body would be perfectly balanced in all positions if it were suspended at that location.

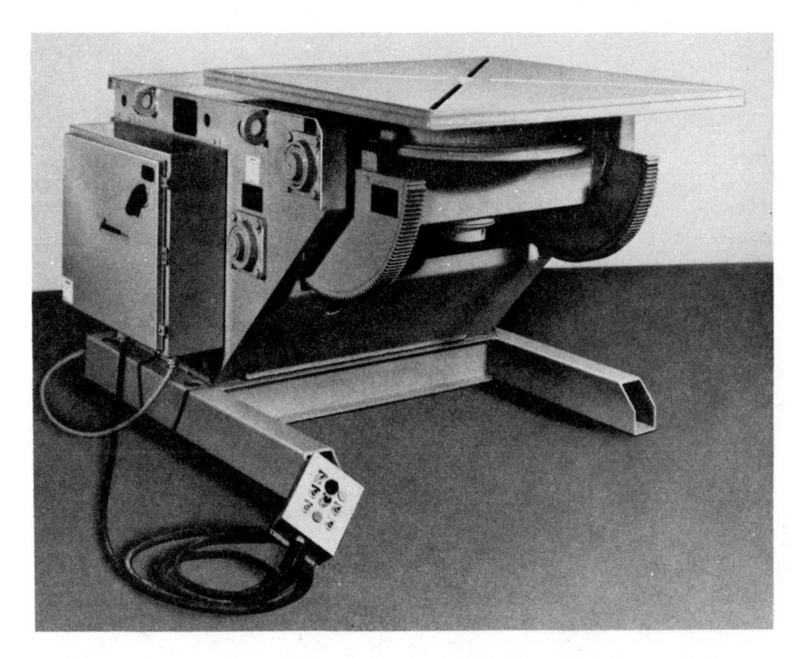

Fig. 4.9—A typical flat-135° tilting positioner

Fig. 4.10—Typical 45°-90° tilting positioner

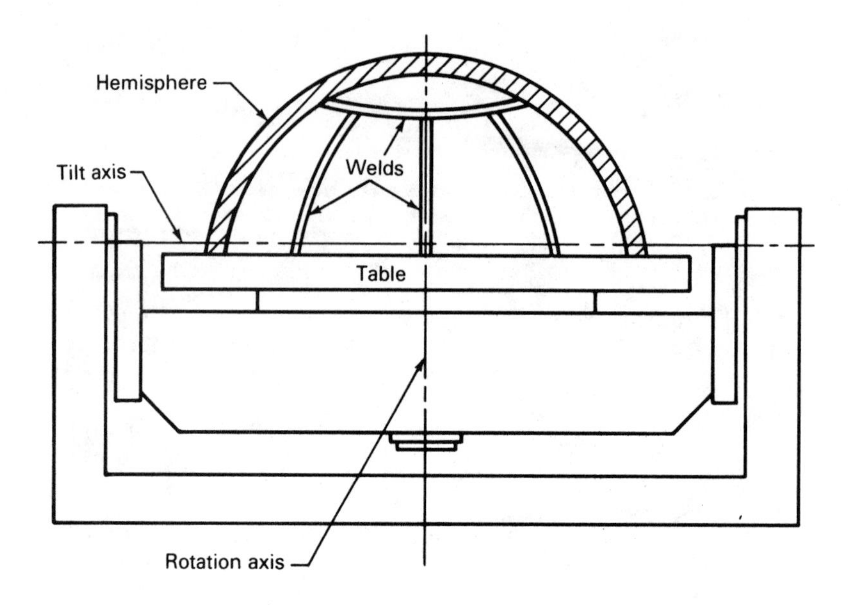

Fig. 4.11—Arrangement of a hemispherical head on a drop-center tilting positioner

Powered Elevation Positioners

When a positioner is intended for general purpose application in a welding shop, a flat-135° positioner with powered elevation generally is recommended because it is more versatile than a 45°-90° positioner. The ability to raise the chassis of a flat-135° positioner to permit full tilting of large weldments is particularly useful. A standard positioner must be raised and lowered with a shop crane while the positioner is unloaded. This limitation is avoided when a powered elevation positioner is used for the application. The positioner, loaded or unloaded, can be raised and lowered under power with integral rack and pinion or jack screw drives. A typical powered elevation positioner is shown in Fig. 4.12. Powered elevation positioners are the most versatile of the gear driven types, but they are the most costly. They are available with capacities up to about 60 tons.

Figure 4.13 shows how a powered elevation positioner can be used in the fabrication of a flanged cylinder. In View A, the axis of the spool is horizontal and the girth welds are made. The operator then tilts the table up 45 degrees and makes the first fillet weld in the optimum position, shown in View B. In View C, the chassis is elevated and the table tilted down 45 degrees to make the second fillet weld.

Special Positioners

There are a number of variations of the basic tilt-plus-rotation positioners that are useful for specific applications. A *sky hook* positioner, shown in Fig. 4.14, is basically a turntable mounted on the end of a long arm. The arm is attached to a headstock that provides tilting capability. This configuration provides unlimited rotation about both the tilt and the rotation axes.

A positioner can be part of an automatic weld station to provide movement of the weldment during welding. Positioners can be integrated with welding robots to provide additional programmable axes for the robot system.

SAFETY

The use of mechanical positioners rather than manual labor to position weldments during welding should reduce the likelihood of injury during the operation. For safety, a weldment

Fig. 4.12—A typical powered elevation positioner with the weldment in a 135 degree position

must be firmly mounted on a positioner so that it will not move or fall off during welding and allied operations.

The weldment or the fixture may be bolted or welded to the positioner table. In either case, the fastening means must be strong enough to hold the work secure under any condition of tilt or rotation. The reaction loads at the fastening locations must be calculated or estimated to ensure that the load is secured by fasteners of proper design. Methods for attaching a weldment or fixture to a positioner table are discussed later.

Positioners are designed to be stable when loaded within their rated capacity, but may be unstable if overloaded. Tilting positioners may be unstable during tilting because of the inertia of the load. Injury to personnel or damage to equipment could be serious if an overloaded positioner suddenly tipped during operation. For this reason, positioners must be fastened to the floor or other suitable foundation for safe operation. The instructions provided by the manufacturer concerning foundation design and fastening methods must be carefully followed to assure a safe installation.

Overloading of positioners, which is frequently practiced by users, cannot be condoned. It is possible for a user to mount an overload on a positioner table, tilt it to the 90 degree position, and be unable to return the table to the flat position because of insufficient power. Shop management must make every effort to ensure that the equipment is operated within its rated capacity to assure safe operation and long life. It is imperative to accurately determine the load, center of gravity, and eccentricity to avoid overload. The safety features needed in positioning equipment include thermal- or current-limiting overload protection of the drive motor, low voltage operator controls, a load capacity chart, and emergency stop controls.

The area surrounding the work station must be clear to avoid interference between the weldment and other objects in the area. Sufficient height must be provided for weldments to clear the floor. Manual or powered elevation in positioner bases is recommended for this purpose.

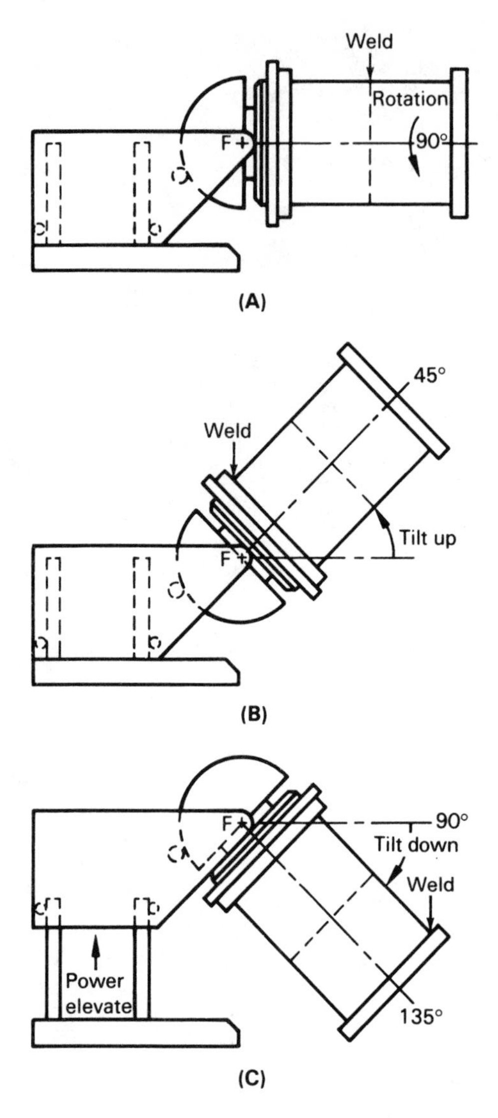

Fig. 4.13—Welding a flanged cylinder with a powered elevation positioner

When using powered weldment clamping fixtures, a safety electrical interlock should prevent rotation (or elevation where applicable) until the clamping mechanism is fully locked to the weldment components. The use of proximity and limit switches is also advisable, especially in the positioning of oversized loads to avoid overtravel or collision of the weldment with the floor or other obstructions.

TECHNICAL CONSIDERATIONS

Center of Gravity

Correct use of a positioner depends on knowing the weights of the weldment and fixture, if used, and the location of the center of gravity (CG) of the total load. The farther the center of gravity of the load is from the axis of rotation (ECC), the greater is the torque required to rotate the load. Also, the greater the distance from the table surface to the CG, the larger is the torque required to tilt the load.

Consider a welding positioner that has the worktable tilted to the vertical position. Figure 4.15(A) shows the worktable with no load. In Fig. 4.15(B), a round, uniform weldment has been mounted on the worktable concentric with the axis of rotation. The center of gravity coincides with the axis of rotation. The torque required to maintain rotation of the weldment is only that needed to overcome friction in the bearings. However, torque is required to accelerate or decelerate the weldment.

In Fig. 4.15(C), an irregular weldment is mounted with its center of gravity offset from the rotation axis. In this case, torque is required to rotate the weldment, and is equal to the weight multiplied by the offset distance (moment arm). If a 5000-lb weldment has its center of gravity located 10 in. from the rotation axis, it will require a minimum torque of 50,000 lbf · in. to rotate it. The positioner must have enough drive capacity to provide this torque.

The same consideration would apply to a vessel being rotated by turning rolls. If the center of gravity of the vessel is not on the rotation axis, the turning rolls must provide enough traction to transmit the required torque for rotation.

Location of the center of gravity of the load is also important in the tilting action of welding positioners. When the loaded table is in the vertical position, the load exerts a torque around the tilt axis as shown in Fig. 4.16. The torque (T) is equal to the weight (W) of the load multiplied by the distance from its center of gravity to the tilt axis of the positioner. The tilt axis of most con-

Fig. 4.14—A sky hook positioner

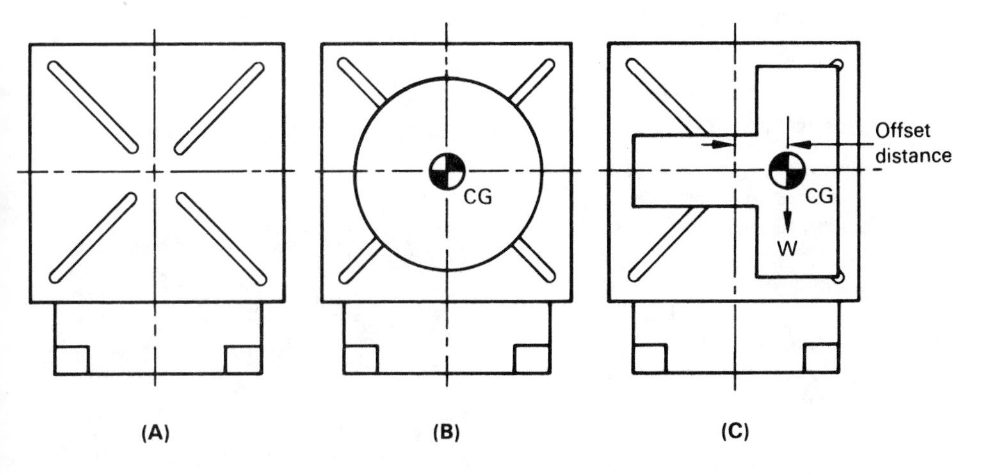

Fig. 4.15—Effect of location of the load center of gravity on rotational torque

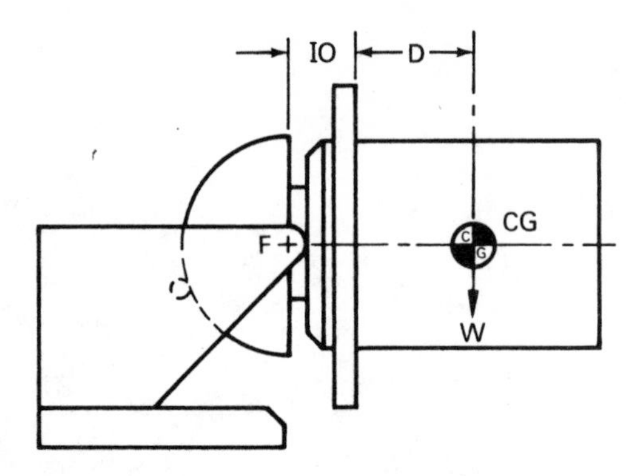

Fig. 4.16—Effect of location of load center of gravity on tilt torque

ventional positioners is located behind the table face for a length called the *inherent overhang* (IO). This length must be added to the distance (D) between the center of gravity of the load and the table face when calculating the tilt torque with the following equation:

$$T = W(IO + D)$$

where

T = tilt torque, lbf · in.

W = weight of the load, lb

IO = inherent overhang, in.

D = distance from the table to center of gravity of the load, in.

Manufacturers of positioners usually attach a rating table to their positioners that specifies the torque ratings in tilt and rotation, the inherent overhang, and the load capacity at various distances between the center of gravity of the load and the table. An example of such a table is shown in Fig. 4.17.

For turning rolls, the tractive effort required to be supplied by the rolls can be calculated with the following formula:

$$TE = WD/R$$

where

TE = tractive effort, lb

W = weight of the load, lb

D = the distance from the axis of rotation of the load to the center of gravity, in.

R = radius of the load or holding fixture, in.

The tractive effort rating of the rolls should be about twice the calculated value to allow for drag and misalignment.

To select a proper positioner for a given weldment, the location of the center of gravity has to be known to determine the required torque. This can often be obtained from the engineering office, but if not, it can be calculated or be determined experimentally from a sample weldment.

The center of gravity of a symmetrical weldment of uniform density is at the geometric

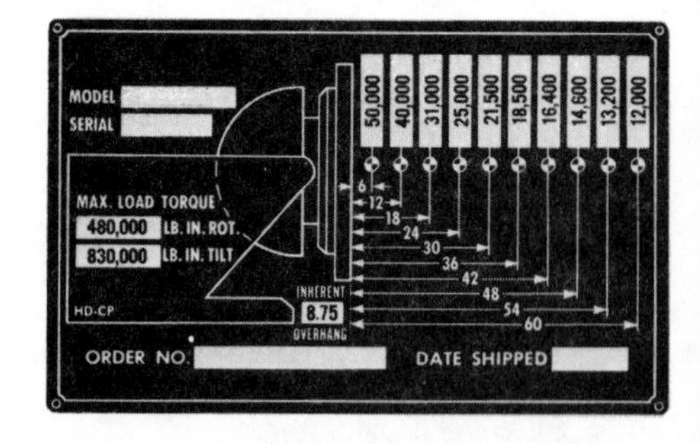

Fig. 4.17—Typical nameplate for a tilting positioner

center. For example, the center of gravity of a cylinder is located on the center axis at the midpoint of the length.

Calculating the Center of Gravity. The center of gravity with respect to mutually perpendicular *X, Y,* and *Z* axes of a nonsymmetrical weldment can be determined by dividing it into simple geometric shapes. The center of gravity of each shape is calculated by formulas found in standard mechanical engineering or machinery handbooks. The location of the center of gravity of the total weldment with respect to selected *X, Y,* and *Z* axes is determined by a transfer formula of the form:

$$y = \frac{Aa_1 + Bb_1 + Cc_1 + \ldots..}{A+B+C+\ldots..}$$

where

y = distance from the X axis to the common center of gravity

A = weight of shape A

B = weight of shape B

C = weight of shape C

a_1 = distance from X axis to center of gravity of A

b_1 = distance from X axis to center of gravity of B

c_1 = distance from X axis to center of gravity of C

Additional calculations as required are made to determine the location of the common center of gravity from the *Y* and *Z* axes. The shop must be able to physically locate the positions of the selected *X, Y,* and *Z* axes with respect to the weldment.

Finding the Center of Gravity by Experimentation. There are two practical methods of finding the center of gravity without the need for calculations in the shop. The first method can be used to find the center of gravity of small box-shaped weldments. The weldment is balanced on a rod under one flat surface, as shown in Fig. 4.18(A). A vertical line is made on the weldment above the rod. The weldment is rotated to an adjacent flat surface and the procedure repeated, Fig. 4.18(B). The center of gravity lies on a line passing through the intersection of the two lines. The location of the center of gravity in a third direction can be found by repeating the procedure with a flat surface perpendicular to the other two.

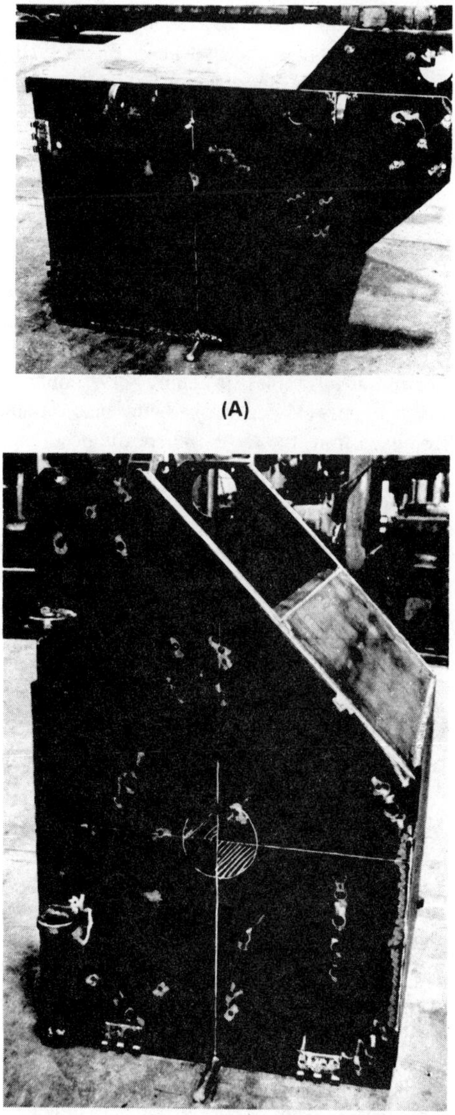

(A)

(B)

Fig. 4.18—Method of finding the center of gravity by balancing on a rod

In the second method of experimentally finding the center of gravity, a hoist can be used. The center of gravity of the weldment will settle under a single lifting point. When a plumb bob is suspended from the lift hook, a chalk mark along the line will pass through the center of gravity, as shown in Fig. 4.19(A). A second lift from a different point on the weldment will establish a second line that pinpoints the center of gravity at the intersection of the two lines, Fig. 4.19(B). (The marks from the balancing bar method were left on the weldment to show that both methods indicate the center of gravity at the same location.) A scale also might be used during lifting to obtain the weight of the weldment.

(A)

(B)

Fig. 4.19—Method of locating the center of gravity with a hoist

Attachment of the Weldment

With any type of positioning equipment involving a rotating or tilting table, the weldment must be fastened to the table. In the case of horizontal turntables, fastening serves only to prevent the weldment from being accidentally dislodged from the table. Where the positioner table will be used in positions other than horizontal, the fastening means must firmly hold the weldment to the table in any work position.

When the positioner table is horizontal, as in Fig. 4.20(A), the only force tending to move the weldment from the table is the centrifugal force of rotation. Attachment requirements are minimal at low rotation speeds. As the table is tilted, the weldment tends to slide off the table because a component of the weight acts parallel to the table, Fig. 4.20(B). The attachments must have sufficient strength in shear to prevent sliding of the weldment. The shearing force increases as the table tilt angle increases, and is equal to the weight of the weldment at 90° tilt, Fig. 4.20(C).

During tilting, a point is reached where the weldment would rotate about lower fasteners if not restrained by upper fasteners. The restraint results in tensile force on the upper fasteners [F_1 and F_2 in Figs. 4.20(C) and (D)]. The amount of tensile force depends on the weight and geometry of the weldment.

The moment of the weldment about the lower fasteners P is the product of the weight W and the horizontal distance x between the center of gravity and the lower fasteners. As the tilt

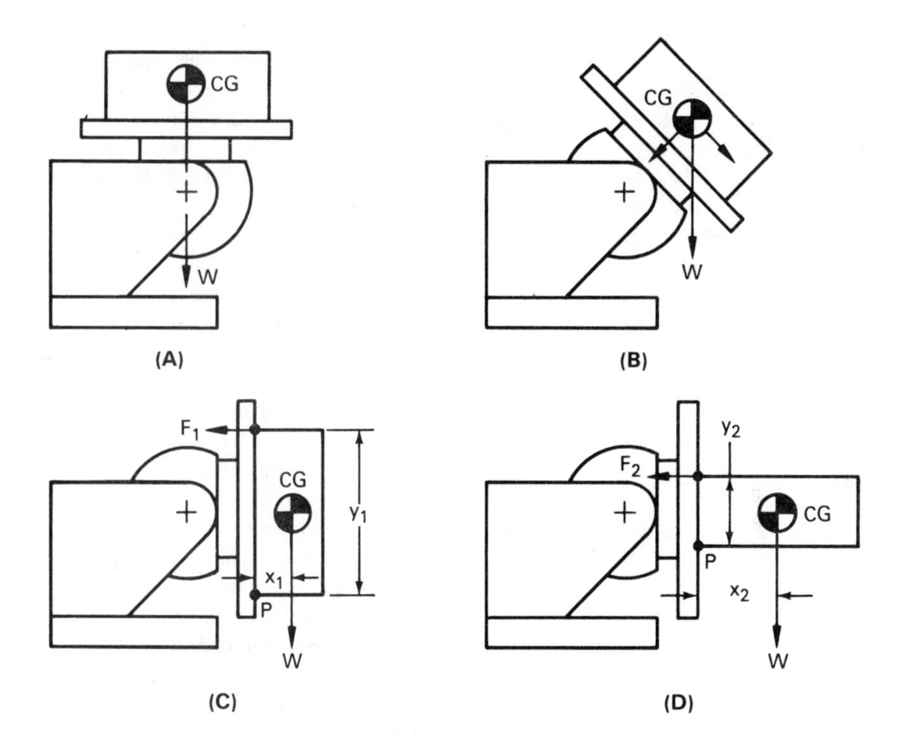

Fig. 4.20—Effects of tilting and weldment geometry on positioner fastener requirements

angle increases, the horizontal distance increases. The weight moment is balanced by the moment of the tensile force on the upper fasteners.

Assume that the weldments in Figs. 4.20(C) and (D) have equal weights, *W*, but the one in (D) is longer and narrower than the one in (C). The force, *F*, in both cases, can be determined by the following equation:

$$F = Wx/y$$

where

x = horizontal distance between P and CG

y = vertical distance between P and F

In the two cases, x_2 is greater than x_1, and y_2 is less than y_1. Therefore, F_2 is greater than F_1. Accordingly, the attachments for the weldment in Fig. 4.20(D) must be stronger than those in Fig. 4.20(C). In both cases, all attachments must have sufficient strength to withstand both the shearing force and the tensile force.

There are two common methods employed to hold a weldment to a positioner table: bolting and welding. Most tables are provided with slots for mounting bolts and some manufacturers supply T-nuts which match the slots. The largest bolts that will fit the slots should always be used to avoid failure of the attachment. As a general rule, a weldment with similar base dimensions and height can be safely mounted using four bolts of a diameter that is slightly less than the slot width in the table. If a weldment is unusually tall with respect to its base dimensions, the force, *F*, on the mounting bolts should be calculated as described previously. The distance *y* should be the minimum bolt spacing to be used. The calculation assumes that all the load is on a single bolt, and is therefore conservative. Generally, neither the table nor the workpiece are flat, which makes this assumption valid.

The mounting bolts, properly tightened, will normally provide adequate shear resistance to prevent the weldment from sliding off the table. If there is any doubt that the load is adequately contained against sliding, steel stop blocks should be placed against the base of the weldment and tack welded to the table.

A weldment can be fastened to the table with temporary welds. The required size and length of each tack weld to carry the applied force, *F*, or shear load, or both should be determined as described in Chapter 1 of this Volume. When the job is completed, the tack welds are cut loose after the table is returned to the flat position or after the weldment is adequately supported by shoring or a hoist.

If welding directly to the positioner table is objectionable, a waster plate can be securely bolted to the table and the weldment welded to the plate. The waster plate can be replaced, when necessary.

Work Lead Connection

Welding positioners that support a weldment on rotating members should provide some means of carrying the welding current from the weldment to a point on the chassis where the work lead is connected. It would be impractical to continually relocate the work lead on the weldment as the table is rotated.

Two common methods for conducting welding current from the table to the base are sliding brushes and bearings. Either method can be satisfactory if properly designed, constructed, and maintained.

Sliding brushes are usually made of copper-graphite matrix, graphite, or copper. Several units are employed in parallel to provide adequate current capacity. The brushes are spring-loaded to bear against a machined surface on a rotating element. For proper operation, the brushes and rotating surface must be clean, the brushes must have the proper length, and the spring load must be adequate. Brushes should not be lubricated.

Conducting bearings must be properly designed and adequately preloaded. If the bearing preload is lost by mechanical damage, improper adjustment, or overheating of the spindle, the bearings can be damaged by internal arcing between bearing surfaces. In normal usage, this system requires little maintenance.

The user's work lead connections must be adequate in size and properly attached to the positioning equipment in accordance with the manufacturer's instructions.

Poor connections in the welding circuit will cause explicable fluctuations in arc length, varying joint penetration, or changes in bead shape. If the positioner rotating connection is suspected, it can be checked by measuring the voltage drop under actual arc welding conditions. The voltage drop should be low (0.1 to 0.25V), but more important, it should be constant. The cause of excessive voltage drop or fluctuation across the connection must be determined and corrected.

Metric Conversion Factors

1 lb = 0.454 kg
1° = 0.0175 radians
1 in. = 25.4 mm
1 lbf · in. = 0.113 N · m
1 ton = 907 kg

SUPPLEMENTARY READING LIST

Boyes, W.E., ed., *Jigs and Fixtures,* Dearborn, MI: Society of Manufacturing Engineers, 1979.

Cary, H.B., *Modern Arc Welding*, Englewood Cliffs, NJ: Prentice-Hall, 1979: 317-22.

Fixtures, positioners, and manipulation, *Canadian Welder and Fabricator*, 66(10): 14-18; 1975 Oct.

Gallup, E., Cutting costs with jigs, fixtures, and positioners, *Welding Engineer*, 59(3): 22-24; 1974 June.

Nissley, L., Adapting positioners for robotic welding, *Welding Design and Fabrication*, 56(8): 45-47; 1983 Aug.

Payne, S., The engineering of arc welding fixtures, *Manufacturing Engineering Management*, 64(1): 47-53; 1970 Jan.

Procedure Handbook of Arc Welding, 12th Ed., The Lincoln Electric Company, 1973: 4.4-1-4.4-12.

Still, J.R., Positioning for welding, *Metal Construction*, 9(6): 246-47; 1977 June.

5
Automation and Control

Introduction *178*

Fundamentals of Welding Automation *178*

Extent of Automation *186*

Planning for Automation *187*

Resistance Welding Automation *193*

Arc Welding Automation *200*

Brazing Automation *208*

Problems of Automation *210*

Supplementary Reading List *213*

Chapter Committee

R.C. REEVE, *Chairman*
Reeve Consultants, Incorporated

K.L. JOHNSON
FMP Consultant Services

R.W. RICHARDSON
Ohio State University

W. RODEN
General Dynamics Corporation

R.L. SZABO
Szabo and Associates

Welding Handbook Comittee Member

J.R. CONDRA
E.I. DuPont de Nemours and Company

5
Automation and Control

INTRODUCTION

MEANING OF AUTOMATION

The term *automation* in manufacturing implies that some or all of the functions or steps in an operation are performed automatically by mechanical or electronic devices. Some operations may be automated to a degree but with certain functions or steps performed manually (partial automation). Other operations may be fully automatic with all functions and steps performed by the equipment in proper sequence without adjustment by an operator (total automation). Loading and unloading of automatic equipment may be done manually or mechanically.

Automation can be applied to many welding processes, brazing, soldering, thermal cutting, and auxiliary operations. Automatic equipment may be designed to accommodate a specific assembly or family of similar assemblies with minor adjustments (fixed automation). On the other hand, the equipment may be flexible in that the sequence of functions or steps can be readily programmed to perform the same operation on different components or assemblies (flexible automation).

WHEN TO USE AUTOMATION

Regardless of the degree of sophistication, the objectives of automation are to reduce manufacturing costs by increasing productivity, and to improve product quality and reliability by reducing or eliminating human errors. Other possible benefits include reductions in overall floor space, maintenance, and inventory. More balanced and controlled production scheduling often results from a consistently uniform high quality output.

Automation can be a success or failure depending on the application. A successful application requires careful planning, economic justification, and the full cooperation and support of management, product designers, manufacturing engineers, labor, and maintenance. Major factors that should be carefully analyzed to determine if automation is feasible include the product, plant and equipment, and costs.[1]

FUNDAMENTALS OF WELDING AUTOMATION

Automation involves more than equipment or computer controls, it is a way of organizing, planning, and monitoring a production process. The procedures, systems, and production controls for the manual welding that is to be replaced are usually insufficient for automatic welding. Most manual welding methods rely heavily upon the knowledge, skill, and judgement of the welder.

---

1. The economics of welding automation and robotics are discussed in Chapter 3.

To successfully produce sound welds automatically, the equipment must be able while making a weld to perform in much the same manner as a welder or welding operator. Automatic welding equipment is not intelligent, nor can it make complex judgements about welds. Nevertheless, methods and procedures for automatic welding must anticipate and deal with judgemental factors. Many of the topics that should be considered when acquiring automatic welding equipment and in developing automatic welding procedures, methods, and production controls are discussed here.

MACHINE WELDING

In machine welding, the operation is done by the welding equipment, not a welder, under the constant observation and control of a welding operator. The equipment may or may not perform the loading and unloading of the work. The work may be stationary while a welding head is moved mechanically along the weld joint, or the work may be moved under a stationary welding head.

Machine welding can increase welding productivity by reducing operator fatique, and by increasing the consistency and quality of the welds. The equipment should control the following variables when arc welding:

(1) Initiation and maintenance of the welding arc

(2) Feeding of the filler metal into the joint

(3) Travel speed along the joint

In machine welding, the welding operator must be near to the point of welding to closely observe the welding operation. The operator interacts continuously with the equipment to assure proper placement and quality of the weld metal. Machine welding should provide the welding operator with sufficient time to monitor and control the guidance aspects of the operation as well as the welding process variables. Weld quality can be enhanced as a result of proper control of process variables. Additionally, a smaller number of weld starts and stops with machine welding reduces the likelihood of certain types of weld discontinuities.

Machine welding devices generally fit into one of the following categories:

(1) Machine carriages

(2) Welding head manipulators

(3) Side-beam carriages

(4) Specialized welding machines

(5) Positioners, turntables, and turning rolls.[2]

These travel devices provide means for moving an automatic welding head relative to the part being welded or vice versa. Machine welding can be used with most fusion welding processes.

Machine Carriages

A welding machine carriage provides relatively inexpensive means for providing arc motion. A typical carriage rides on a linear track or curved track of the same contour as the joint to be welded, as shown in Fig. 5.1. Some carriages are specifically designed to ride on the surface of material being welded in the flat position, and some use the material for joint guidance. The travel speed of the carriage is a welding variable, and uniformity is important to weld quality. Tractors are used primarily for welding in the flat or horizontal position.

Fig. 5.1—Welding machine carriage with a submerged arc welding head

2. Positioners, turntables, and turning rolls are discussed in Chapter 4.

Welding carriages may be employed for welding in the horizontal, vertical, or overhead positions. Others may be designed to follow irregular joint contours. They employ a special track upon which the welding carriage is mounted. Since the welding operator must continually interact with the controls, the carriage controls and the welding controls are typically placed within easy reach of the operator. This type of welding equipment is used for cylindrical and spherical tanks, aircraft and missile assemblies, etc. The rigidity with which the welding carriage is held to the track and the uniformity of travel are critical to good performance.

Welding Head Manipulators

A welding head manipulator typically consists of a vertical mast and a horizontal boom that carries an automatic welding head. A large welding head manipulator carrying a twin submerged arc welding unit is shown in Fig. 5.2. Manipulators usually have power for moving the boom up and down the mast, and in most units the mast will swivel on the base. In some cases,

Fig. 5.2—Welding head manipulator carrying a twin submerged arc welding unit

the welding head may move by power along the boom, while in others the boom itself may move horizontally on the mast assembly.

It is essential during operation that the boom or welding head move smoothly at speeds that are compatible with the welding process being used. The carriage itself must also move smoothly and at constant speeds if the manipulator is designed to move along tracks on the shop floor. In selecting and specifying a welding manipulator, it is important to determine the actual weight to be carried at the end of the boom. It is essential that the manipulator be rigid and deflection minimized during the welding operation.

Welding manipulators are among the most versatile types of machine welding equipment available. They are used to position the welding head for longitudinal, transverse, and circular welds. Heavy duty manipulators are specifically designed to support the weight of the operator as well as that of the welding equipment. All of the welding and manipulating controls are placed at the operator station (Fig. 5.2).

Side-Beam Carriage

A side-beam carriage consists of a welding carriage mounted on a horizontal beam, as shown in Fig. 5.3. It provides powered linear travel for the welding head.

Fig. 5.3—Side-beam carriage with gas metal arc welding head

The powered welding carriage supports the welding head, filler wire feeder, and sometimes, controls for the operator. Typically, the welding head on the carriage is adjustable for vertical height and transverse position.

During operation, the welding process is monitored by the welding operator. Welding position is manually adjusted to follow joints that are not in good alignment. Travel speed of the side-beam carriage is adjusted by the operator to accommodate different welding procedures and variations in part fit-up.

Special Welding Machines

Specialized welding machines are available with custom workholding devices, torch travel mechanism, and other specific features. Applications include equipment for making circumferential and longitudinal welds on tanks and cylinders, welding piping and tubing, fabricating flanged beams, welding studs or bosses to plates, and for performing special maintenance work, such as the rebuilding of track pads for crawler type tractors.

One common type of special welding machine rotates a welding head around the axis of a relatively small diameter part, such as the spud on a tank or a boss on a plate, as shown in Fig. 5.4. The welding gun makes a fillet weld joining the boss to the plate. These machines are quickly adjustable for different sizes of spuds or bosses.

AUTOMATIC WELDING

Automatic welding is done with equipment that performs an entire welding operation without adjustment of the controls by a welding operator. An automatic welding system may or may not perform the loading and unloading of the components. A typical automatic arc welding machine that welds a bracket to a tube is shown in Fig. 5.5.

The equipment and techniques currently used in automatic welding are not capable of controlling the operation completely. A welding operator is required to oversee the process. An important aspect of automatic welding is that the operator need not continuously monitor the operation. Compared to machine welding, this tends to increase productivity, improve quality, and reduce operator fatigue.

A welding operator, although not directly controlling the welding process, requires several skills to operate and interact with automatic equipment. The operator is responsible for the proper operation of a complex electromechanical system, and must recognize variations from normal operation. Deductive skills may be required if the welding operator cannot directly view the actual welding process.

Fig. 5.4—Joining bosses to a plate part by automatic gas metal arc welding

Fig. 5.5—Automatic arc welding machine for joining a bracket to a tube

All of the elements required for machine welding are utilized in an automatic welding system. Continuous filler wire feeding, where used, arc movement, and workpiece positioning technology are basic to automatic welding. Automatic welding requires welding fixtures to hold the component parts in position with respect to each other.[3] The welding fixtures are usually designed to hold one specific assembly and, by minor variation, may allow for a family of similar assemblies. Because of the inherent high cost of design and manufacture, the use of welding fixtures is economical only in high volume production where large numbers of identical parts are produced on a continuous basis.

Automatic welding also requires that the parts to be joined be prepared in a consistent and uniform manner. Parts must be uniform if the resultant welds are to be uniform. Therefore, part preparation for automatic welding is usually more expensive than for manual or machine welding.

Automatic welding requires a complex welding cycle controller that must provide for control of the welding operation as well as the material handling equipment and component fixturing. The controller must precisely time a complex sequence of operations.

The successful application of automatic welding offers the following advantages:

(1) Consistent weld quality

(2) Reduced variable welding costs

(3) Predictable welding production rates

(4) Integration with other automatic operations on the weldment

(5) Increased productivity as a result of increased welding speeds and filler metal deposition rates

Automatic welding has limitations, including the following:

(1) Capital investment requirements are much higher than for manual or machine welding equipment.

(2) Dedicated fixturing is required for accurate part location and orientation.

---

3. Welding fixtures are discussed in Chapter 4.

(3) Elaborate arc movement and control devices with predetermined welding sequence are required.

(4) Production requirements must be large enough to justify the costs of equipment and installation, the training of programmers, and the maintenance of the equipment.

The welding machine controller is the primary element of an automatic welding system. Many automatic welding controllers are networks of electrical and electromechanical hardware including relays, switches, contacts, timers, rheostats, potentiometers, pushbuttons, and lights. Recent designs make use of microprocessors and minicomputers for automatic welding control. A microprocessor panel is shown in Fig. 5.6. Software is used in place of the former complex electromechanical hardware.

FLEXIBLE AUTOMATED WELDING

In flexible automated welding, a numerical control or computer control program replaces complex fixturing and sequencing devices of fixed automated welding. The welding program can be easily changed to accommodate small lot production of several designs that require the same welding process, while providing productivity and quality equivalent to automatic welding. Small volume production of similar parts can be produced automatically in a short time. A flexible system can be easily reprogrammed for a new part.

An industrial robot is the most flexible of the automated systems used in manufacturing operations. A robot is essentially a mechanical device that can be programmed to perform some task of manipulation or locomotion under auto-

Fig. 5.6—Microprocessor panel for a welding machine

matic control. A welding gun or torch can be mounted on a "wrist" at the end of the robot arm.

Robots may be designed with various motion systems; two common systems, articulated (jointed) and rectilinear, are shown in Fig. 5.7. The choice of robot design depends upon the applications for which a robot is to be used.

An automated work station is composed of various pieces of equipment, totally integrated by a single control. A typical station incorporates a robot arm, a robotic work positioner, a welding process package, and required fixturing to hold the parts for welding.

A welding robot can be programmed to follow a predetermined path duplicating a seam in an assembly to be welded. With some equipment, this is done by moving the welding torch along the weld joint. Signals are sent back to the robot controller that it can memorize the taught motion. After the job is completed, certain robots provide for copying of the welding program for future production of that weldment. In addition to controlling the welding torch path, some robots can also set and change welding variables.

If distortion is a significant factor or the piece parts are not accurately prepared or located, the weld joint may be outside of the programmed location. To compensate for mislocation, the flexible or robotic welding system may require sensory feedback to correct the program. Sensors can provide information to the robot controller to make real-time adaptive corrections to the programmed path. Currently, a limited number of sensors and adaptive controls are available for use with certain welding processes. Most of these are adaptive for joint tracking only.

Flexible automation for welding can eliminate the need for complicated customized fixtures, and it is finding increased acceptance in low volume production. The inherent flexibility of robotic systems can accommodate several different types of weldments at the same work station. Furthermore, flexible welding automation can be interfaced with other automatic manufacturing operations.

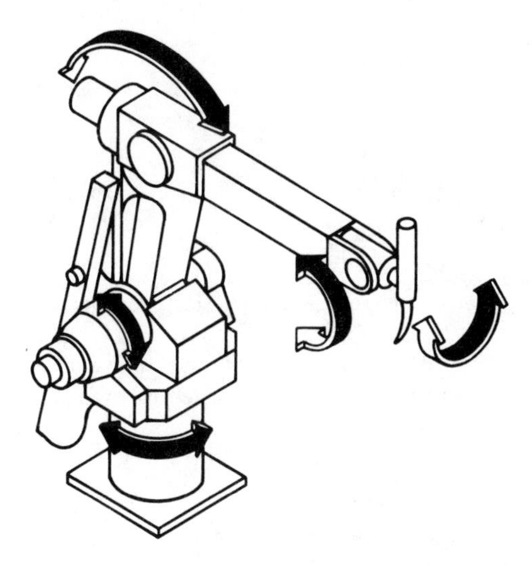

(A) Articulated robot

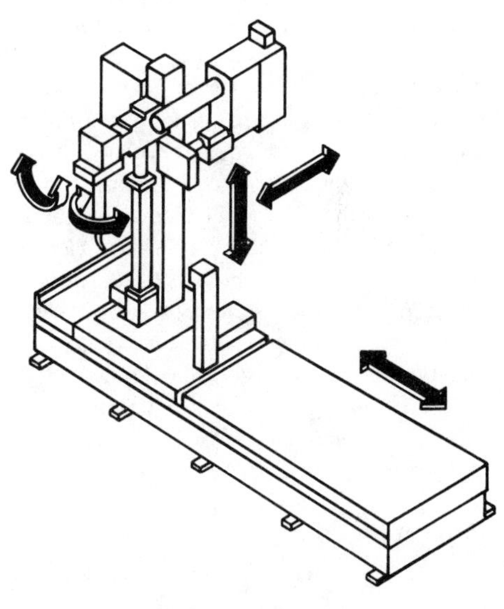

(B) Rectilinear robot

Fig. 5.7—Robotic motion systems

EXTENT OF AUTOMATION

The extent to which automation should be employed is governed by four factors:

(1) Product quality

(2) Production levels

(3) Manpower

(4) Investment

The importance of these factors will change with the application and with the type of industry. The decisions on how much and when to automate can best be evaluated by examining each factor.

PRODUCT QUALITY

Process Control

A weld is usually considered a critical element of a fabricated component if failure of the weld would result in serious injury or death, or extensive damage to and loss of service of a welded structure. For such welds, thicker sections, additional consumables, welding process development, proper part preparation, adequate supervision, or extensive inspection and testing may be specified to assure the required product quality.

Automated welding can significantly improve product quality while reducing costs. By regulating and controlling process variables, the possibility of human error is reduced. Automation also allows the welding operator to devote attention to other factors that can improve weld quality. Automation of critical welds is common in the aerospace, nuclear, and construction industries. Welding automation is also being applied in many other industries.

Product Improvement

As market pressures dictate the need for higher quality, and often at lower cost, the long term viability of a manufacturer can hinge upon welding automation. Often, reductions in repair and scrap made possible through automation can financially justify conversion to automation. A reduction in the repair and scrap costs from about 10 percent to about 1 percent is not uncommon. The financial benefits of quality products and customer satisfaction are more difficult to measure. At best, they can translate into some increase in market share. At worst, they can still mean the survival of the business in the midst of increased competition.

PRODUCTION LEVELS

Production Volume and Changeover Time

A major factor in measuring utilization of an automated welding system is the percentage of time the system is kept productive. Where high production levels are required for a single or a limited number of parts, fixed automation is appropriate. Flexible automation should be considered if frequent model or engineering changes are anticipated. Where production volumes are limited, flexible automation should be considered to produce small batches of components in economical lot sizes.

Answers to the following questions are useful when analyzing the potential cost savings of an automatic welding installation:

(1) How many welders are currently working or projected to be working on the assemblies of interest?

(2) How many shifts are involved?

(3) What is the extent of welding and what sizes of welds are needed on each assembly?

(4) How many assemblies are required per year?

(5) What is the frequency of changeovers (based on production batch size, model changes, engineering changes, etc.)?

Floor Space

Plant floor space is a part of product costs. Frequently, manual welding fabrication requires a relatively large amount of floor space because of the number of workers required for the operations. A reduction in floor space for a welding operation or an increase in output in the same floor space may be an important consideration. Therefore, automation of welding operations should be considered where floor space is limited.

Inventory

Flexible welding automation can generate considerable savings through increased rate of internal inventory turn-over. Savings can be realized through reductions in the amount of work in process and the inventory of finished components. These reductions can lead to additional savings in floor space. Flexible automation allows scheduling of smaller lot sizes as a result of reduced tooling and set up costs.

Testing and Weld Development Costs

When automation is being evaluated, a welding development and testing program should be undertaken to ensure that optimum welding performance is obtained from the proposed investment in capital equipment. The investment in development and testing programs should be proportional to the expected capital outlay.

MANPOWER

Human and Environmental Factors

Manual welders are frequently required to perform a skilled trade in a hostile environment: noisy, dirty, and hot. Welding fumes and gases are a potential health hazard as is the possibility of being burned. Arc welding automation permits the welding operator to work outside of the hazardous environment. Additionally, welding automation can relieve the operator of repetitive, monotonous, and tiring tasks. Automation directly improves welding productivity.

Availability of Manpower

Welding shops are often limited in skilled manpower. Skilled welders are normally in demand and receive relatively high labor rates. Operation of automated welding equipment by semiskilled operators can lead to labor cost savings.

Education and Training

The introduction of automation and technology to the manufacturing operation requires education and training of personnel. Introduction of machine welding usually does not materially change the skills required in the shop. However, the technology of automatic and flexible welding systems may require planned education and training programs. These programs should cover the required skills to set up, operate, and maintain the equipment and a basic understanding of the technology being introduced. The availability of trained personnel or personnel capable of being trained materially affects the cost of automation.

INVESTMENT

A cost-benefit analysis should be made of potential welding automation projects. The discounted cash flow technique is generally accepted in the metal working industry. This technique uses a proforma annual cash flow for the expected life of the system. Typically, a ten-year life is more than adequate to assess the financial impact of a welding automation system.

The estimated savings and costs that will result from the project should be clearly identified. Welding usually affects manufacturing costs both upstream and downstream from the actual welding step. Financial justification requires a review of the financial condition of the company, including availability and cost of capital.

PLANNING FOR AUTOMATION

Planning is a necessary prerequisite to successful application of automation. Such planning includes the identification and in-depth analysis of potentially successful applications as well as planning for the implementation of the selected automated welding systems. Companies often form automation teams to manage and perform the required planning and implementation tasks.

The team should study all pertinent factors concerning planning, procurement, installation, and startup of the automation project.

PRODUCT DESIGN

Designing for Automation

Most weldments are not designed for efficient automatic welding. Considerable savings can be realized by reviewing a weldment design to optimize it for automation. The costs and benefits of redesign should be included in the financial analysis of a project.

Tolerances and Part Fit-Up

Part tolerances and joint fit-up are important in automatic welding. The dimensional accuracy of the parts, so long as they are within functional tolerances, is not nearly as important as the consistency. If the weld joint geometry is consistent, automatic welding can be done without the need for adaptive feedback controls. The manufacturing processes used in the preparation of component parts must be reviewed for consistency. Costs of improving component consistency must be included in the financial analysis.

Processing and Scheduling

The procedures and scheduling in most shops were initially developed for manual welding. Conversion to automation requires new procedures and scheduling to maximize productivity. A detailed review of subassembly designs and component scheduling is in order. Often, an entire family of assemblies can be fabricated with one flexible automatic welding system.

Fixturing

When assemblies are redesigned or procedures are changed for automation, new welding fixtures are often required.[4] Flexible automation can significantly reduce the required investment in new fixturing.

There is a substantial difference between the fixturing required for fixed automatic welding and that required for flexible automatic welding. Fixed automatic welding requires customized fixtures that provide exact weld joint locations. In flexible automation, the software program can be designed to avoid complex customized fixturing. Often all that is required for flexible automatic welding is a tack welding fixture.

PROCESS DESIGN

Welding Process Selection and Welding Procedures

Automation of a welding operation often requires a change in the welding process. The selection of an appropriate welding process and associated welding consumables is an important step. Procedures for automatic welding should be qualified according to the applicable welding codes or standard, such as *AWS D1.1, Structural Welding Code—Steel.*

Selection of Welding Equipment

Automatic welding systems are normally required to operate at high duty cycles to justify the investment. High duty cycles require the use of rugged heavy duty welding equipment designed for continuous operation. Water cooling should be specified where applicable to avoid overheating of the welding equipment.

Safety

The welding operator should be prevented from entering the work envelope of an automatic welding system. However, if this is not possible, safety interlocks are necessary to prevent injury when the equipment is operational. The applicable safe practices in Chapter 10 for the welding process and equipment must be followed when installing and operating automatic equipment.

FEASIBILITY TESTING

Welding tests are normally conducted to prove the feasibility of automatic welding. The test parts may be actual production components or simulations. When designing weld test specimens, many factors, such as part weight, moments of inertia, and surface conditions, may be important to the welding process. Some of the more common test objectives are as follows:

(1) Evaluation of weld quality and consistency

4. Fixturing for welding is discussed in Chapter 4.

(2) Determination of welding conditions and production cycle times (includes set-up time, indexing times, welding time, unloading, etc.)

(3) Establishment of working relationships with equipment manufacturers

(4) Identification of operator skills and safety requirements

(5) Evaluation of the effects of distortion, fit-up tolerances, heat buildup, weld spatter, etc.

(6) Provision of welded specimens for testing and evaluation

Feasibility tests can be conducted by the user, the equipment manufacturer, or a subcontractor. The tests should be conducted in two or more phases. Initial testing should be performed by the user to determine the optimum balance between welding conditions, weld quality, and component fit-up tolerances. The results should be provided to the equipment manufacturer for subsequent testing of the automatic equipment. After installation in the user's plant, testing of the equipment to verify production capability may be warranted.

FACILITIES

Utilities

Electrical power, air, and water requirements for automatic welding equipment are usually within the ranges of typical machinery used in a fabrication shop. One exception may be resistance welding equipment that can require high electrical power demands.

Where cooling water is required, a recirculating cooling system is recommended to conserve water. A compressed air supply may require filters, driers, or oilers.

Voltage stability in the power lines should be measured with a recording voltmeter or oscillograph. If voltage fluctuations exceed the range recommended by the equipment manufacturer, heavier service lines or a constant voltage transformer can be installed.

Location

The following factors should be considered in evaluating the location of an automatic welding system:

(1) Safety — clearances, arc radiation exposure, fume control, spatter

(2) Work area conditions — lighting, temperature, humidity, cleanliness, vibrations

The effects of lighting, noise, and smoke on the operator should be considered. Many automatic welding operations require some observation by the operator or other attendant workers. Therefore, working conditions in the area should be conducive to stable, reliable performance.

Safety requirements, such as a clear area around the machine, can normally be determined from the manufacturer's specifications. Collision hazards include possible contacts with material handling equipment, fixtures, hand tools, etc., and should receive special consideration.

Specifications

User requirements are generally written for custom designed machines, and the specifications are made a part of the purchase (or lease) order. Manufacturer specifications are used to define standard equipment. Specifications serve the following purposes:

(1) Define the performance expectations of the purchaser and the supplier.

(2) Establish minimum quantitative requirements for defining the equipment.

(3) Provide uniform criteria to allow comparison of several vendor proposals against the purchaser's needs.

(4) Provide support groups (maintenance, programming, etc.) with an information base in preparing for delivery of the equipment.

Environment

Strict environmental requirements are not normally specified for automatic welding equipment. However, extended exposure to a hostile environment can reduce equipment life. When installing automatic welding equipment, it is best to provide as much protection as possible. Unlike conventional welding machines, automatic equipment often contains vulnerable electronic and mechanical devices. A clean environment will help assure good performance and long machine life. Potential environmental problems with automatic welding equipment are listed in Table 5.1.

Table 5.1
Potential environmental problems with automatic welding equipment

Environmental factor	Potential problem areas
Temperature	Electronic and mechanical accuracies
Humidity	
Common moisture	Electronic components and rust
Salt air	Electronic components and severe corrosion
Vibration	Electronics and servo stability
Dust and dirt	
Abrasive	Gears, bearings, etc.
Nonabrasive	Air-cooled components
Magnetic	Collection on solid state devices
Lighting, noise, fumes, and smoke	Operator efficiency

The environment for automatic welding operations should meet the minimum requirements specified by the equipment manufacturer. A harsh environment can rapidly degrade the performance of the machine, create costly maintenance problems, and shorten the overall useful life of the system. Considering the cost of automatic welding equipment, investment in providing a suitable environment is often justified.

PROCUREMENT SCHEDULING

Scheduling the acquisition and installation of an automatic welding machine can be straightforward for machines of standard design. These machines include those that are inventoried and have been tested with actual production parts by the manufacturer. On the other hand, custom designed systems that automate new welding processes or involve unique and difficult applications may require a lengthy debugging period that can cause scheduling difficulties. A suggested checklist for procurement scheduling is given in Table 5.2.

Table 5.2
Procurement scheduling checklist

1. Identify potential welding applications
2. Estimate potential costs savings
3. Research literature for similar application case histories
4. Identify potential equipment suppliers
5. Perform welding tests to determine feasibility, and identify limitations
6. Obtain approvals from engineering, manufacturing, quality control, and other departments, as required
7. Prepare equipment and fixturing specifications
8. Obtain budgetary proposals from potential equipment suppliers
9. Appropriate procurement funds
10. Select an equipment supplier
11. Monitor design, construction, and test of the equipment by the supplier
12. Install and checkout the equipment
13. Perform pilot production welding tests
14. Turn the equipment over to production

TRAINING AND EDUCATION

Skills Required

The basic job skills for automatic welding are a good knowledge of the welding process and training in the programming of automatic equipment.

Knowledge of the Welding Process. Fixed automatic or flexible automatic welding systems may not require a trained welder. However, a skilled welding operator may be required by the governing welding code or specification. Welding skills are an asset for applications where the operator performs a decision making role. Bead placement in multiple-pass arc welded joints and in-process adjustments of welding variables are examples of duties that may require basic welding skills.

Knowledge of Programming. The type and level of skill required to program automatic welding machines varies widely. Many robots are programmed on-line in a "teach-learn" mode by jogging through the desired motions and entering the welding conditions (speed, current, voltage). No numerical programming skills and only a nominal amount of training are required.

Other types of automatic welding machines, in particular those integrated into a computer or CNC control system, may require numerical control (NC) programming skills. Programmers experienced in APT or similar computer languages may be required.

Maintenance. Maintenance technicians must be skilled in electronics, mechanical apparatus, and often computer controls.

Training Required

Training is usually required to prepare operators for automatic welding, and it should be provided by the equipment manufacturer. Other production factors, such as part preparation, processing procedures, welding procedures, quality control, and remedial measures may also require special training.

The following factors should be considered in estimating the scope of training required:

(1) Number of controls, gauges, meters, and displays presented to the operators

(2) Degree of difficulty in criticality of the setup

(3) Number of steps in the operating procedure

(4) Degree of operator vigilance required

(5) Responsibility for evaluation of results (weld quality, distortion, etc.)

(6) Corrective actions that need to be taken or reported

(7) Required quality level

(8) Potential product liability

(9) Value of parts and completed weldments

The operator of a simple automatic welding machine, which has little more than ON and OFF switches and no variations in procedures, may not require formal training but only a few minutes of instruction by a foreman. An elaborate computer-controlled electron beam welding machine, such as the one shown in Fig. 5.8, may require formal training for the operator, the responsible engineer, and the maintenance personnel. On-the-job training can provide the

Fig. 5.8—Computer-controlled, high-vacuum electron beam welding machine

operator with a functional knowledge of the machine. However, such training is usually more costly than classroom training.

PRODUCTION CONSIDERATIONS

Space Allocation

Automatic welding equipment will rarely fit into the same floor space as that occupied by a single manual welding operation, but it usually replaces several welding stations. To obtain good production flow, floor space requirements for the entire operation should be studied in context. Often, additional space savings may be realized by automating other operations as well.

Material Handling

Established methods for material and parts handling may not be suitable for automatic welding operations. For example, an overhead crane that is suitable for transferring parts to work tables for manual welding may be unsatisfactory for moving parts to an automatic welding machine.

Automated material handling should be considered for improved efficiency, orderly parts flow, and reduced risk of parts damage. Robotic material handling offers versatility, but may exceed the capacity of a simple automated welding machine. Custom designed material handling systems are usually justifiable on high volume, dedicated production lines.

Flexibility and Expansion

The ability to change or add to the capability of an automatic welding system may be needed by some users. Examples of changes and additions that may be considered are as follows:

(1) A change of welding process

(2) An increase in welding power

(3) A change in the program storage medium, for example, from paper tape to floppy disk

(4) Addition to the system of new operating features or enhancements

(5) Addition of other controlled equipment, such as positioners, material handling devices, and markers

(6) Retrofit of new or different adaptive control sensors (TV cameras, tracking systems, etc.)

Complications can arise when making changes or additions to automatic welding stations. Hardware, software, and electronic components of different models, or equipment from different manufacturers are usually not compatible. (Compatibility will be better achieved when standards are established for interconnections and communications between components.)

Shop Acceptance

Improved productivity can be realized from a new automatic welding machine if the users have contributed to the selection and planning for the system. Shop personnel should be advised in advance that an automatic welding machine is being considered for a particular operation. Welding supervisors, welders, maintenance personnel, and welding inspectors can play an important role in making a transition from manual to automatic welding proceed smoothly. They can often formulate requirements for the machine.

Maintenance

Automatic welding machines are often complex electromechanical systems that include precision mechanisms and components. The equipment can be extremely reliable, when it is treated properly.

A preventive maintenance program is essential for long-term reliability without degradation of performance. Preventive maintenance activities may include the following:

(1) Calibration of speeds and other welding variables (voltage, current, welding wire speed, etc.)

(2) Routine replacement of finite life components (e.g., guides, hoses, belts, seals, contact tips, etc.)

(3) Cleaning of critical surfaces (bearings, optics, and similar items)

(4) Oil changes and lubrication of bearings, seals, screws, etc.

(5) Replacement of filters

(6) Monitoring of temperature indicators, flow meters, line voltage, fluid levels, etc.

(7) Inspection of structural components and mechanisms for cracks, loose bolts, worn bearings, evidence of impact or abrasion, etc.

(8) Discussion of operation with the operator

A good maintenance program should include:

(1) A file of all available documentation (wiring diagrams, manuals, etc.) located near to the machine

(2) A step-by-step troubleshooting procedure with clearly marked test points for instrumentation

(3) A generous supply of spare parts, such as circuit boards, fuses, and valves, located close to the machine

(4) A log of performed maintenance with details of problems encountered, the symptoms, and the actions taken to overcome them

RESISTANCE WELDING AUTOMATION

Automation of resistance welding operations may range from an indexing fixture that positions assemblies between the electrodes of a standard resistance welding machine to dedicated special machines that perform all welding operations on a particular assembly.[5] An example of the latter type is shown in Fig. 5.9.

When flexibility is required in an automatic resistance welding operation to handle various assemblies, a robotic installation is a good choice. An example is the robotic spot welding line for automobile bodies shown in Fig. 5.10. Such an installation can be programmed to spot weld similar car bodies for various models.

MATERIAL HANDLING

When applying automation to a resistance spot welding application, it may be practical to move either the assembly or the welding gun to each weld location. Obviously, weight has to be considered in this decision. A typical resistance spot weld gun weighs about 50 to 100 lb, and the welding cable another 50 lb. A typical portable resistance spot welding gun, cable, and power source are shown in Fig. 5.11.

Moving the Assembly

Joining of relatively thick sheets of low resistance alloys, such as aluminum, may require large welding current conductors. Large, heavy welding cables may be required to carry such high currents and reach from the power source to the welding gun. Such conductors may exceed the load carrying capability of a robot. One solution is to move the assembly into position and weld with a stationary or indexing spot welding machine. Alternatively, the assembly may be indexed while single spot welds are made with a stationary welding machine.

Assemblies that weigh less than 150 lb may be moved with an industrial robot under certain conditions. Supporting tooling may be required to prevent damage to the assembly during movement.

Moving the Welding Gun

In high production spot welding, the time between welds must be a minimum. At high welding rates, moving a welding gun of low weight may be the better approach for welding relatively large, bulky assemblies.

PRODUCTION REQUIREMENTS

Quantity of Production

In high volume applications, such as automotive or appliance manufacturing, each welding gun in an automatic installation may make the same spot welds repeatedly. In low volume applications, such as aerospace work, a welding robot can be used to make a number of different spot welds before repeating an operation. A flexible, reprogrammable industrial robot can be used for both types of applications.

---

5. Resistance welding processes and equipment are discussed in the *Welding Handbook*, Vol. 3, 7th Ed.

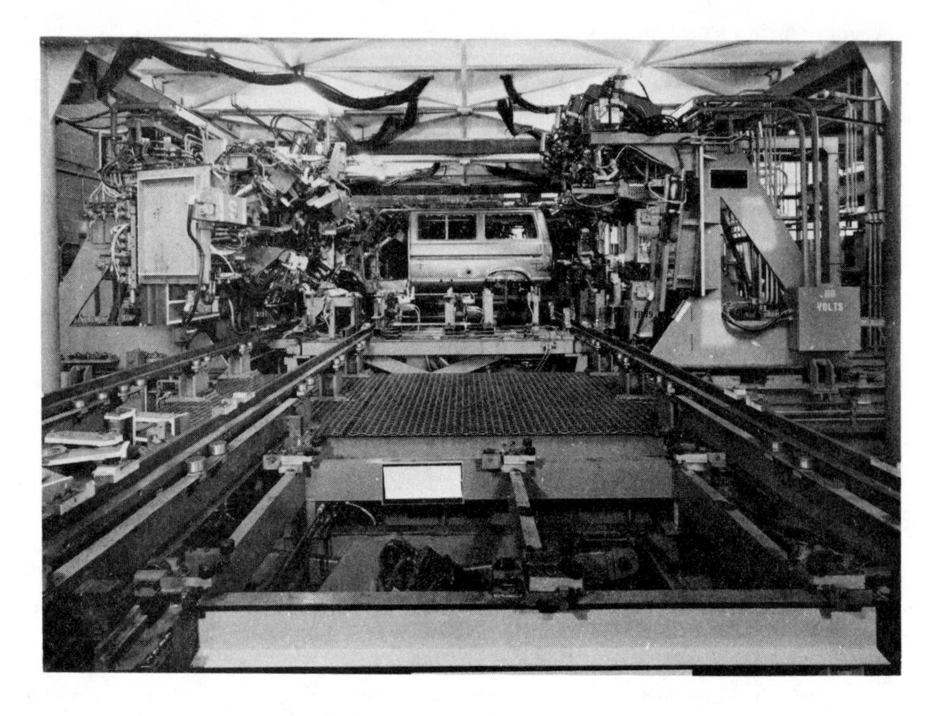

Fig. 5.9—A dedicated automatic resistance welding line

Fig. 5.10—Automotive robotic resistance welding line

Production Rates

Typically, production rates of 15 to 30 spot welds per minute can be achieved when the welds are in close proximity. As distances between spot welds increase, production rates will depend on the maximum practical travel speed for the welding gun or the assembly, whichever is moved.

Production rate computations should include the following factors:

(1) Transfer time between weld locations
(2) Time for the gun or work to stabilize in position
(3) Welding cycle time, comprising:
  (a) Squeeze time (typically 0 to 10 cycles)
  (b) Weld time (typically 5 to 20 cycles)
  (c) Hold time (0 to 10 cycles)
(4) Fixturing cycle time
(5) Material handling time

Welding cycle time can be optimized using special programming, such as upsloping the welding current during the latter portion of *squeeze* time. Likewise, welding current downslope can be applied during a portion of *hold* time. *Hold* time can be minimized by initiating the release of the gun before the time is completed, to take advantage of gun retraction delay. A properly designed welding cycle can significantly reduce the total welding cycle time for the assembly.

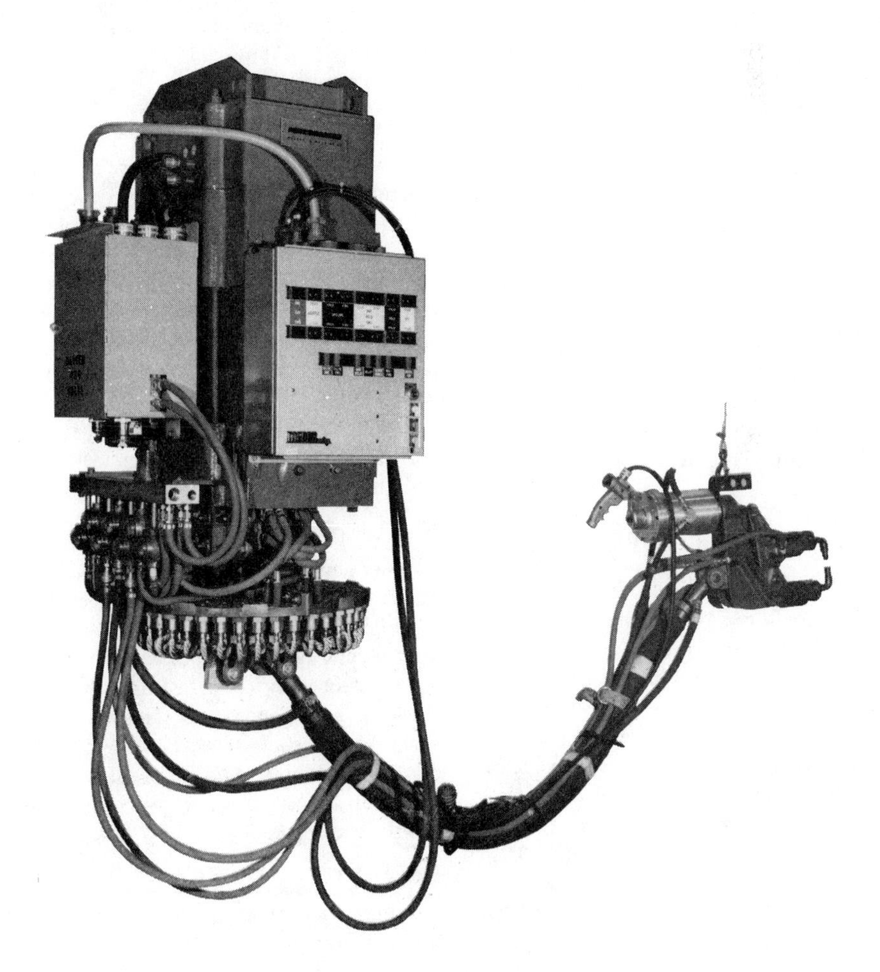

Fig. 5.11—Typical portable resistance spot welding machine

On a typical high production application, a single welding gun may make 50 spot welds on an assembly. Increasing the welding cycle time by so little as 10 cycles (60 Hz) to improve weld quality will increase the total cycle time for 50 welds by 8 seconds. Therefore, quality and production rate can be a trade-off.

Quality Control

Quality control is an important benefit of automation. However, weld quality is dependent upon process control. An automatic spot welding machine will repeat bad welds as well as good ones.

An automatic spot welding machine can maintain good spot weld quality with feedback controls and appropriate electrode maintenance. For example, a robot is normally programmed to always position the plane of the welding gun perpendicular to the weld seam. In manual spot welding, an operator can easily misalign the welding gun as a result of fatigue, boredom, or carelessness. This condition may result in improper welding pressure and weld metal expulsion from the seam.

When different thicknesses of metal are to be spot welded with the same welding gun, an automatic spot welding machine can be programmed to select a different welding schedule for each thickness. Weld quality is improved when the proper welding schedule can be used for every joint thickness.

WELDING ROBOTS

Types of Robots

There are four general geometric classifications of industrial robots. *Rectilinear (cartesian coordinate) robots* have linear axes, usually three in number, which move a wrist in space. Their working zone is box shaped. *Cylindrical coordinate robots* have one circular axis and two linear axes. Their working zone is a cylinder. *Spherical coordinate robots* employ two circular axes and one linear axis to move the robot wrist. Their working zone is spherical. The fourth classification, *anthropomorphic (jointed arm) robots* utilize rotary joints and motions similar to a human arm to move the robot wrist. The working zone has an irregular shape.

All four robot geometries perform the same basic function: the movement of the robot wrist to a location in space. Each geometry has advantages and limitations under certain conditions. (Antropomorphic and rectilinear robots are favored designs for arc welding.)

Load Capacity

Most robots for spot welding are designed to manipulate welding guns weighing between 75 and 150 lb. Manual spot welding guns are often counterbalanced using an overhead support that can limit positioning of the gun. Robots of sufficient load capacity can move spot welding guns without counterbalances. Counterbalances may be needed to support heavy welding cables unless integral spot welding gun-transformer units of low weight are used.

Accuracy and Repeatability

A robot can repeatedly move the welding gun to each weld location and position it perpendicular to the weld seam. It can also replay programmed welding schedules. A manual welding operator is less likely to perform as well because of the weight of the gun and monotony of the task. However, an operator can detect and repair poor spot welds; current spot welding robots cannot.

Robot accuracy and repeatability vary with the manufacturer. Some robot models require periodic recalibration or reprogramming.

Number of Axes

Spot welding robots should have six or more axes of motion and be capable of approaching points in the work envelope from any angle. This permits the robot to be flexible in positioning a welding gun to weld an assembly. In the case of an automobile, a door, a window, and a roof can all be spot welded at one work station. Some movements that are awkward for an operator, such as positioning the welding gun upside down, are easily performed by a robot.

Reliability

Industrial robots are currently capable of 2000 to 4000 hours of operation between failures with duty cycles of 98 percent or better. Robots are capable of operating continuously, so long as proper maintenance procedures are followed. On continuous lines with multiple robots, interruption of production can be minimized by (1) installing backup units in the line, (2) distributing the work of an inoperative robot to other nearby robots, or (3) quick replacement of the inoperable robot.

Maintenance

In addition to the recommended maintenance of the robot, maintenance of the resistance welding equipment is necessary. Typically, the following items of an automatic spot welding machine require periodic maintenance:

(1) Weekly or monthly

(a) Tighten or replace electrical shunts

(b) Replaced frayed or worn welding cables

(c) Repair or replace leaking air or hydraulic pressure lines

(d) Replace leaking or damaged air or hydraulic system regulators

(e) Replace welding transformers that show signs of deterioration

(2) Hourly or after a specific number of welds, regrind, repair, or replace the spot welding electrodes

The frequency of electrode maintenance can be reduced by automatic compensation systems, such as steppers, that periodically increase the welding current to compensate for electrode wear. Welding current upslope can minimize weld metal expulsion and reduce electrode wear.

Safety

The operator of a robot can easily avoid close proximity to jagged edges of parts, moving conveyors, weld metal expulsion, and other welding hazards. However, the movement of the robot arm creates a dangerous environment. The workers in the area must be completely prevented from entering the working envelope of the robot. Protective fences, power interlocks, and detection devices must be installed to assure worker safety.

RESISTANCE WELDING EQUIPMENT

Automatic welding imposes specific demands on resistance spot welding equipment. Often, equipment must be specially designed and welding procedures developed to meet automatic welding requirements.

Welding Electrodes

Electrode life is influenced by electrode force, electrode material, welding schedule, welding current, and the type of material being welded. Upslope of welding current can greatly reduce electrode wear. Progressively increasing the current after a fixed number of welds can compensate for mushrooming of the electrodes, thereby extending electrode life. Electrode life is also extended by proper positioning of the electrodes on the seam.

Spot Welding Guns

Portable spot welding guns are normally designed to fit the assembly. Two basic designs are scissor and C-types.

Pneumatic guns are usually preferred because of the rapid follow-up and uniform electrode force characteristics. Hydraulic spot welding guns are normally used where space is limited or where high electrode forces are required. Unfortunately, hydraulic guns do not have the consistent electrode follow-up characteristics of pneumatic guns.

Integral transformer-gun units with up to 480 V primary can be safely used for automatic welding. Power consumption is significantly reduced with these units because of the close coupling between the transformer and the welding gun.

CONTROLS

Multiple Schedule Controls

Microcomputer controls are available to program a number of different welding schedules. An automatic welding machine can be interfaced to such a control, and programmed to select the programmed schedule for each spot weld or group of welds. Applications, such as automobile bodies and home appliances which involve welding various gauges of sheet metal in various combinations, can benefit from multiple schedule controls, such as those shown in Fig. 5.12.

Supervisory Controls

Off-Flange Detectors. Automatic spot weld machines can be equipped to detect faulty or deteriorating process conditions. When teamed with optional sensory devices, such as off-flange detectors, automatic spot welding machines can monitor process conditions.

An off-flange detector measures the voltage between the electrodes or the power factor. If the voltage or the power factor is below a set threshold value, the system signals the absence of material between the electrodes. An off-flange detector can be used to sense that a particular spot weld was not made.

Steppers. Steppers can be used to automatically increase the welding heat after a set number of spot welds to compensate for electrode mushrooming, to signal the need to replace electrodes, or to indicate the need for preventive maintenance functions on the machine. The devices can also be used to measure production rates.

Current Monitoring. A current monitoring device can indicate contaminated welding electrodes by detecting a predetermined change in welding current.

Energy Monitoring. Weld energy monitoring is a variation of current monitoring. The monitor is used to indicate trends or detect deterioration in equipment performance.

Resistance Monitoring. Resistance monitoring devices measure the change in electrical resistance between the electrodes as an indication of spot weld formation. When a resistance change does not occur, it can be inferred that a spot weld was not formed, and some adjustment of the welding conditions is warranted. If an improper resistance change occurs early in the welding cycle, the welding current can be terminated to avoid making an unacceptable spot weld.

Interface Between Components

The conventional interface between an automatic machine control and the spot welding equipment is through input-output (I/O) channels. These channels enable the machine control to send and receive discrete signals. For example, an output signal can be sent from the machine to a welding control unit to initiate a weld. An input signal from the control unit to the machine can

Fig. 5.12—Master control panel for computer-controlled automatic welding of automobile bodies

signal the machine to continue the operation sequence. A typical input/output communication link is as follows:

(1) The machine control sends an output signal to the welding control unit to initiate the electrode force and the welding sequence.

(2) The machine control sends an output signal to select the programmed weld schedule.

(3) The machine control receives an input signal from the welding control unit signifying that the control initiated a sequence.

(4) The machine control receives an input signal from the welding control unit at the completion of a weld.

(5) The machine control receives an input signal from a supervisory control to turn on an alarm or interrupt the machine operation.

(6) The machine control sends an output signal to the welding control unit to confirm that it has received the weld completion signal and is moving to the next weld location.

(7) The machine control sends an output signal to a supervisory control to signal when the welding electrodes need to be redressed.

High level communication between automatic machine controls and welding controls is possible through standard communication links. These communication links permit long distance communications between central controls, welding controls, machine controls, and data acquisition centers for the following purposes:

(1) In-process maintenance, such as a detection of a malfunctioning SCR or a missed weld

(2) Continuous in-process monitoring of current, power, or electrical resistance for quality control

(3) Monitoring of equipment status and condition

(4) Coordination of equipment activities

(5) Production output control of monitoring

(6) In-process modification of programs or schedules

ACCESSORY EQUIPMENT

Monitors

Weld monitors can provide nondestructive quality assurance information. If weld quality data are difficult to obtain through random or periodic destructive testing, weld monitors may be an alternative. Data can be recorded for review at a later date or as a permanent record.

Monitors can reduce the amount of destructive testing required. Only welds of questionable quality need be examined. Monitor data and periodic destructive testing can provide an effective quality control program.

Monitoring equipment can perform some or all of the following functions:

(1) Sense position for detecting the presence of a part.

(2) Measure welding current or power to detect long-term changes to indicate the need for cable and gun maintenance.

(3) Measure welding voltage to detect the absence or presence of a part.

(4) Measure resistance or impedance to detect the formation of a weld.

(5) Sense electrode position to detect movement of electrodes signifying weld formation.

(6) Measure ultrasonic reflection from a weld to detect weld size for quality control.

(7) Perform acoustic emission detection and analysis of sound emitted from a weld as it forms to determine weld size or the presence of flaws.

Adaptive Process Control

A welding process is said to be adaptively controlled when changes in weld integrity are measured and corrected in real time. The sensors listed previously for monitors can be used to detect certain properties of spot welds. These inputs can be analyzed and acted upon by an adaptive control program to optimize or correct the welding schedule variables. Users of adaptive controls are cautioned to thoroughly test and evaluate the controls before abandoning established quality control programs.

PROGRAMMING

Assembly Program Development

The first step in the development of a program for an assembly is to establish welding schedules for each proposed material and thickness combination. A robot can be "taught" the sequence of welding positions with the plane of the welding gun perpendicular to the seam. The sequence of weld positioning should be chosen to minimize the repositioning and index time for the

the robot. Finally, a travel speed between weld locations is selected. It should not be too high to avoid oscillation or overshooting of the electrodes. Appropriate welding schedules are selected and programmed at each welding position.

Touch-up Programs

Touch-up programs are the editing changes that are made in welding programs to improve or correct automatic operation. Typical reasons for editing include the following:

(1) Part dimensional changes

(2) Optimization of welding variables to improve weld quality

(3) Optimization of gun movements and speeds to increase production rate

(4) Changes in fixturing location

(5) Correction for movement drift or calibration changes

ARC WELDING AUTOMATION

TYPES OF AUTOMATION

Various arc welding processes are used for automatic welding operations. They include gas metal arc, flux cored arc, gas tungsten arc, plasma arc, and submerged arc welding. In most cases, automatic welding is best confined to the flat or horizontal positions for easy control of the molten weld pool. Nevertheless, automatic welding can be done in other positions using special arc control programming. An example of an automatic arc welding machine that is programmed to weld continuously around a pipe point is shown in Fig. 5.13.

As with resistance welding, either the work or the welding gun or torch may be moved during automatic arc welding, depending on the shape and mass of the weldment and the path of the joint to be welded. It is easier to make circumferential welds in piping, tubing, and tanks if the work can be rotated under a stationary head for welding in the flat position. Long seams in plate, pipe sections, tanks, and structural members are easier to weld by moving the welding gun or torch along the seam. Another example of an automatic arc welding machine where the welding gun is moved while the work remains stationary is shown in Fig. 5.14.

The automatic welding machines shown in Figs. 5.4, 5.5, 5.13, and 5.14 represent fixed automation because they are designed to accommodate families of similar assemblies. Where flexible automation is desirable for arc welding several different assemblies, industrial robots are commonly used.

ARC WELDING ROBOTS

Robotic arc welding is applicable to high, medium, and low volume manufacturing operations under certain conditions. It can be applied to automation of medium and low volume production quantities where the total volume warrants the investment. Even job shop quantities can be welded where the investments in the equipment and the programming time can be justified.

Robot Features

Robots require special features and capabilities to successfully perform arc welding operations. Arc welding robots are generally high precision machines containing electric servomotor drives and special interfaces with the arc welding equipment.

An anthropomorphic (jointed arm) robot is favored for arc welding small parts where the travel distances between welds are large because the arm of this type of robot is capable of quick motion. This robot design also is preferred for nonmoveable assemblies that require the robot to reach around or inside of a part to position the robot wrist.

Rectilinear robots are favored for most other arc welding applications for safety reasons. They are particularly suited for applications where a welding operator is required to be in close proximity to the welding arc. Rectilinear robots move slower and in a more predictable path than anthropomorphic robots.

Number of Axes

Arc welding robots usually have five or six axes (see Fig. 5.7), and some may be equipped with seven axes. A complete robotic work station may contain as many as eleven axes of coordinated motion.

Three axes of motion are required to position an object in space. The motion may be circular or linear, but at least two of the axes must be perpendicular to each other. Three axes are also required to orient an object at a point in space. These orientation axes are incorporated in the *wrist* of an arc welding robot. A robot wrist generally provides mutually perpendicular circular motions.

Arc welding processes that use a consumable electrode, such as gas metal arc and flux cored arc welding, have a built-in degree of freedom because the welding gun can weld in any direc-

Fig. 5.13—Automatic arc welding machine for making pipe joints in the horizontal fixed position

tion. Hence, a robot with five axes is capable of positioning the welding gun with these processes.

Robots can be equipped with additional axes to extend their working volume. These axes will be redundant degrees of freedom, and may or may not be usable while the robot is in motion.

Arc welding often involves manipulation of both the assembly and the welding torch. The axes needed to manipulate the assembly are equivalent to those needed to manipulate the welding gun, and additional axes will expand the flexibility of the system. If there are more than six unique degrees of freedom, a point in space can be approached from multiple positions. This flexibility is important in arc welding because gravity affects the behavior of the molten weld pool.

Accuracy and Repeatability

The precision with which a robot can approach an abstract point in space is called *accuracy*. Accuracy is required in robots where the control programs are developed numerically. It is a measure of the ability of the robot to reproduce a program that was generated by a computer or a digitizer. Robots which are "taught" the programs do not have to be so accurate because they depend upon memorization and replay.

Repeatability is a measure of the ability of a robot to repeatedly approach a point in space. Arc welding robots must have repeatability, but accuracy is not mandatory. The more repeatable a robot is, the larger are the allowable component tolerances. The total allowable variation in arc welding is the sum of the robot repeatability, the component parts variations, and the part positioning equipment repeatability.

Process Control

To be an efficient arc welding machine, an industrial robot must be capable of controlling the arc welding process. As a minimum, the robot must be able to turn shielding gas on and off, initiate and terminate the welding sequence, and select the programmed welding conditions. Some robots control the welding process by selecting preset values. Other, more sophisticated robots are capable of directly controlling filler wire feeders and power supplies, and establishing the process conditions as a part of the robot program.

Robots that can select only preprogrammed welding conditions place more responsibility upon the welding equipment manufacturer, the user, and the operator. Such robots increase the costs of the welding equipment and the cost of interfacing it to the robot. Also, the welding

Fig. 5.14—An automatic arc welding machine for fillet welding stiffeners to panels

process variables must be programmed separately from the robot program. Robots with built-in process control capabilities usually offer more flexibility and integrated process control programming.

ARC WELDING EQUIPMENT

In general, welding equipment for automatic arc welding is designed differently from that used for manual arc welding. Automatic arc welding is normally done at high duty cycles, and the welding equipment must be able to operate under those conditions. In addition, the equipment components must have the necessary features and controls to interface with the main control system.

Arc Welding Power Sources

Automatic arc welding machines may require power sources different from those used for manual welding.[6] An automatic welding machine is usually designed to electronically communicate with the power source to control the welding power program for optimum performance.

The main control system can be designed to program the welding power settings as a part of the positioning and sequence program to insure that the appropriate welding power is used for each weld. The automatic welding program becomes the control of quality for the weld.

Welding power sources for automatic welding must be accurate and have a means for calibration. Manual welders often adjust the welding machine controls to establish welding conditions that are suitable under changing conditions. On the other hand, an automatic welding machine merely replays an established program, and the elements of that program must be accurately reproduced every time that it is used.

Welding Wire Feeders

An automatic arc welding machine is easily designed to directly control a motorized welding wire feed system. This allows flexibility in establishing various welding wire feed rates to suit specific requirements for an assembly. An automatic machine additionally can insure that the wire feed rate is properly set for each weld through "memorization" of the setting in the program. Special interfaces are required to allow the welding machine to control the wire feed rate. Typically, an analog output is provided from the welding machine control to interface to the wire feed system. If the welding machine is not equipped to receive a signal of actual wire feed rate, the wire feeder itself must have speed regulation. A means to calibrate the wire feeder is also required. Wire feed systems for automatic welding should have a closed-loop feedback control for wire speed.

An automatic arc welding machine normally will use a significant amount of filler wire when the welding rates and duty cycles are high. The wire consumption is typically 3 to 4 times that of manual welding.

Drawing compound and debris on the surface of the welding wire tend to collect and fill conduit liners and wire guides, necessitating easy access and frequent cleaning. Wire guides and liners must be designed to minimize buildup of dirt.

Arc Welding Guns and Torches

The high duty cycle of automatic arc welding (usually between 75% and 95%) requires a welding gun or torch that can accommodate the heat generated by the welding process. Water-cooled guns and torches are recommended for this application. Water cooling systems require maintenance, and suitable flow controls must be provided.

Flexible welding torch mounts having positive locating and emergency stop capabilities are preferred for robotic welding. A welding robot does not have the ability to avoid collisions with objects in the programmed travel path. During robot programming, the welding gun or torch may collide with the work. Collisions can cause damage to the robot and the welding gun or torch.

Flexible mounts with emergency stop capabilities minimize the possibility of damage. Typically, the mounts are springloaded with two or more tooling pins to assure proper positioning. The emergency stop switch is activated when a collision knocks the gun or torch out of position.

---

6. Arc welding power sources are discussed in the *Welding Handbook,* Vol. 2, 7th Ed., 1978.

Arc Voltage Control

The gas tungsten arc welding process requires real time adjustment of the electrode-to-work distance to control arc voltage. Automatic gas tungsten arc welding machines should be equipped with an adaptive control to adjust arc voltage. These controls usually sense the actual arc voltage and reposition the welding torch as required to maintain a preset voltage.

With gas tungsten arc welding, the arc is usually initiated with superimposed high frequency voltage or high voltage pulses. Application of the high voltage must be incorporated in the welding program, and welding travel must not start until the arc has been initiated.

MATERIAL HANDLING

For an automatic welding system to efficiently perform a task, the component parts must be reliably brought to the work station. Material handling systems are often used to move component parts into position. Material handling may be as simple as hand loading of multiple welding fixtures that are successively moved into the welding station, or as complicated as a fully automated automatic loading system.

Part Transfer

A common material handling procedure is to transfer the component parts into the working envelope of an automatic welding machine. This task can be performed by turntables, indexing devices, or conveying systems. Such a procedure is used for small assemblies where parts are relatively light in weight.

Part transfer usually requires two or more fixturing stations, increasing fixturing costs. The number of transfer stations required is determined by the relationship between the loading time, the transfer time, and the welding time.

Robot Transfer

Large assemblies and assemblies that require significant arc time can be welded by moving or rotating a robotic arc welding machine into the welding area. This procedure is usually faster than moving the assembly, but it requires a relatively large working area.

A welding robot can be transferred between multiple work stations. This allows production flexibility, and reduces inventory and material handling costs. Work stations can be left in place when not in use while the robot is utilized at other locations.

Some robots can access multiple welding stations located in a semicircle. Rectilinear robots can move to welding stations that are organized in a straight line.

Positioners

A positioner can be used to move an assembly under an automatic arc welding head during the welding operation or to reposition an assembly, as required, for robotic arc welding. The assembly can be moved so that the welding can be done in a favorable position, usually the flat position.

There are two types of positioners for automatic arc welding. One type indexes an assembly to a programmed welding position. The other type is incorporated into the welding system to provide an additional motion axis. A positioner can provide continuous motion of the assembly while the machine is welding, to improve cycle time.

CONTROL INTERFACES

An automatic arc welding system requires control interfaces for component equipment. Two types of interfaces are usually provided: contact closures, and analog interfaces.

Welding Process Equipment

Welding power sources and welding wire feeders are controlled with both electrical contact closure and analog interfaces. Contact closures are used to turn equipment on and off. Analog interfaces are used to set output levels.

Fixtures and Positioners

Fixtures for automatic welding are often automated with hydraulic or pneumatic clamping devices. The welding machine often controls the operation of the clamps. Clamps can be opened to permit gun or torch access to the weld joint. Most clamping systems and positioner movements are activated by electrical contact closures.

Other Accessories

Accessories, such as torch cleaning equipment, water cooling systems, and gas delivery equipment, are also controlled by electrical contact closures. Flow and pressure detectors are installed on gas and water lines. Signals from these detectors are fed back to the automatic welding machine control through contact closures.

PROGRAMMING

An automatic arc welding system must be programmed to perform the welding operation. Programming is the establishment of a detailed sequence of steps that the machine must follow to successfully weld the assembly to specifications.

Each assembly to be welded requires an investment in programming. Programming costs vary widely depending upon the welding system being used, the experience of the programmers, the complexity of the assembly being welded, and the characteristics of the welding process. Investment in programming must be taken into account when determining the economics of automatic welding. Once an investment is made for a specific weldment, the program can usually be stored for future use.

Welding program development involves a number of steps. The first step is to calibrate the automatic welding system. Calibration insures that future use of the program will operate from a known set point.

The second step is to establish the location of the assembly with respect to the welding machine. Often, simple fixturing is sufficient.

The third step is to establish the path to be followed by the welding gun or torch as welding progresses. Some robots can be "taught" the path while other automatic welding systems have to be programmed off-line.

The fourth step is to develop the welding conditions to be used. They must then be coordinated with the work motion program.

The fifth step is to refine the program by checking and verifying performance. Often, a program requires editing to obtain the desired weld joint.

ACCESSORIES FOR AUTOMATION

Welding Gun Cleaners

Periodic cleaning of gas metal arc and flux cored arc welding guns is required for proper operation. The high duty cycle of an automatic arc welding operation may require automated gun cleaning. Systems are available that spray an antispatter agent into the nozzle of the gun. Additionally, tools that ream the nozzle to remove accumulated spatter are available. The cleaning system can be activated at required intervals by the welding control system.

Handling of Fumes and Gases

The high duty cycle of an automatic welding system can result in the generation of large volumes of fumes and gases. An appropriate exhaust system is highly recommended to safely remove fumes and gases from the welding area for protection of personnel.

Seam Tracking Systems

One problem with automatic arc welding operations is proper positioning of the welding gun or torch with respect to the welding groove to produce welds of consistent geometry and quality in every case. Dimensional tolerances of component parts, variations in edge preparation and fit-up, and other dimensional variables can affect the exact position and uniformity of the weld joints from one assembly to the next. Consequently, some control of the welding gun or torch position is desirable as welding proceeds along a joint.

There are several systems available for guiding a welding gun or torch along a joint. The simplest one is a mechanical follower system which utilizes spring-loaded probes or other devices to physically center the torch in the joint and follow vertical and horizontal part contours. These systems are, of course, limited to weld joints with features of sufficient height or width to support the mechanical followers.

An improvement over these units are electromechanical devices utilizing lightweight electronic probes. Such probes operate motorized slides that adjust the torch position to follow the joint. These devices can follow much smaller joint features and operate at higher speeds than pure mechanical systems. However, they are limited in their ability to trace multiple pass welds and square-groove welds. Also, they are adversely affected by welding heat. The welding head of the automatic welding machine shown previously in Fig. 5.14 is equipped with electro-mechanical seam following and video monitoring, as shown in Figs. 5.15(A) and (B).

There are seam tracking systems that utilize arc sensing. The simplest form of these systems is an arc voltage control used with the gas tungsten arc welding process. This control maintains consistent torch position above the work by voltage feedback directly from the arc.

More sophisticated versions of the arc seam tracking systems employ a mechanism to oscillate the arc and interpret the variations in arc characteristics to sense the location of the joint. This information can then be fed back to an adjustment slide system or directly into a machine control to adjust the welding path. Thse systems, which depend on arc oscillation, may or may not be desirable with a particular welding process, and they can be limited in travel speed by the oscillation requirements. Also, arc characteristics

Fig. 5.15(A)—An automatic welding machine equipped with video monitoring cameras and electro-mechanical seam tracking devices

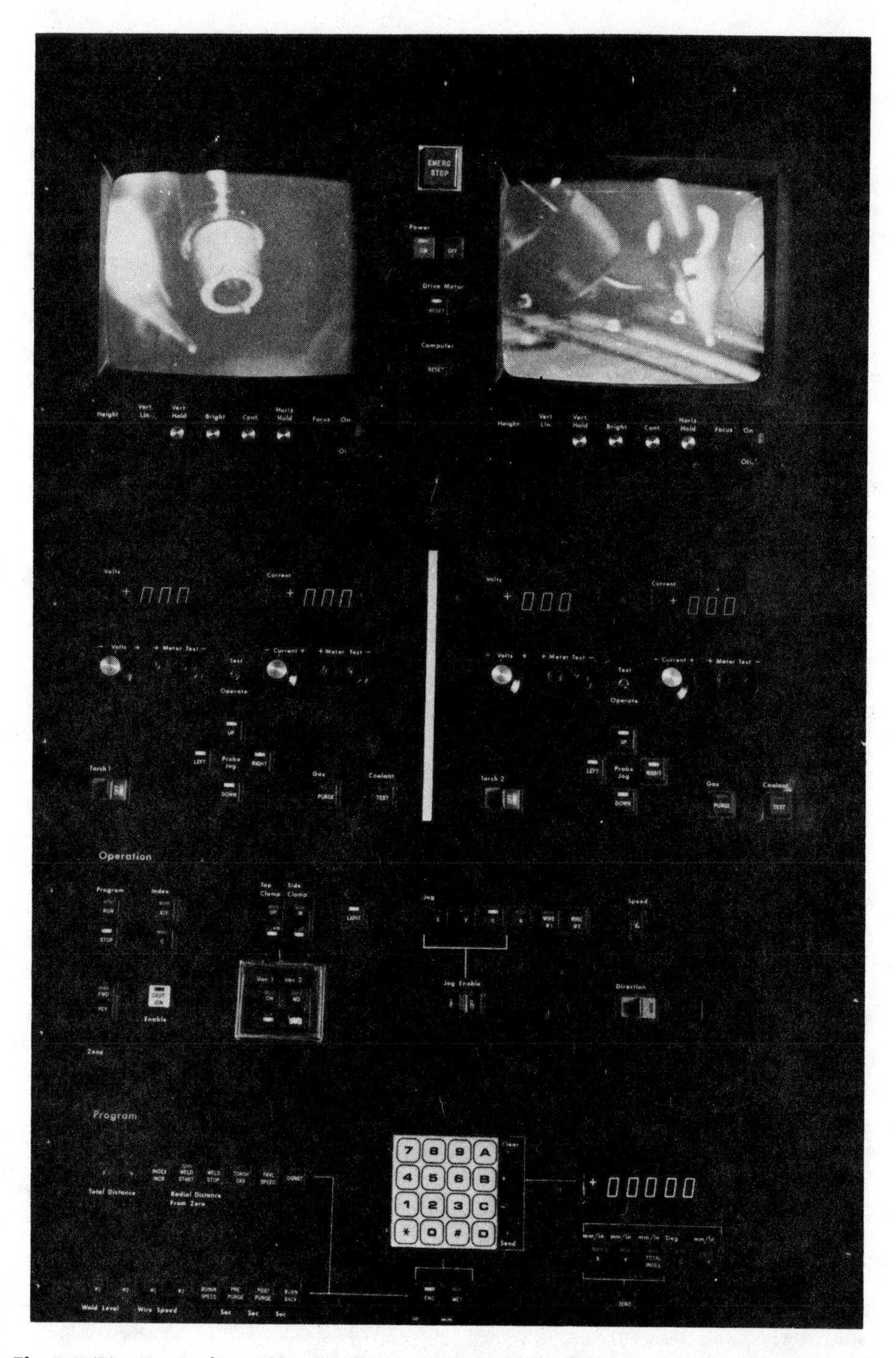

Fig. 5.15(B)—Control panel for the automatic welding machine showing the video screens for viewing the welding operation

can vary from process to process, and some tuning may be required to make an arc seam tracking system work in a given situation. The manufacturers should be consulted for details.

The most sophisticated seam tracking systems are optical types. These types utilize video cameras or other scanning devices to develop a two- or three-dimensional image of the weld joint. This image can be analyzed by a computer system that is capable of determining such information as the joint centerline, depth, width, and possibly weld volume. These systems can be used for adaptive control of welding variables as well as seam tracking.

There are several types of optical tracking systems presently available. One is a two-pass system in which a camera first is moved along the nominal weld path with the arc off. The system performs an analysis and correction. Then, a second pass is made for actual welding. This system cannot correct for any distortion that occurs during welding.

A more sophisticated version is a single-pass or real-time system. Presently, these units look slightly ahead of the actual operating arc and feed back data for correction of torch path and welding variables. These systems have some limitations, such as sharp corners, and they can be influenced by smoke and arc heat.

Welding Wire Supply

Production rates of automatic arc welding machines put heavy demands on welding wire supply systems. Welding wire spools of 60 lb or less used for semiautomatic arc welding are not satisfactory for automatic arc welding. Spools and drums of 250 lb or more are better choices.

Powered payoff packs that support the spool and reel the wire from the spool are desirable. These prevent slippage of the welding wire in the wire feed system, and resultant defects that can occur in the welds.

BRAZING AUTOMATION

The important variables involved in a brazing process include brazing temperature, time at temperature, filler metal, and brazing atmosphere. Other variables are joint fit-up, amount of filler metal, rate and mode of heating, etc.

Many brazed joints are made automatically using welding torches. Other processes that may be automated include furnace brazing (e.g., vacuum and atmosphere), resistance brazing, induction brazing, dip brazing, and infrared brazing.[7] Generally, the amount of heat supplied to the joint is automated by controlling temperature and the time at temperature. Brazing filler metal can be preplaced at the joints during assembly of components or automatically fed into the joints at brazing temperature.

Further automation of a process can be achieved by inclusion of automatic fluxing, in-line inspection and cleaning (flux removal), simultaneous brazing of multiple joints in an assembly, and continuous brazing operations.

Generally, the more automated a process becomes, the more rigorous must be its economic justification. Frequently, the increased cost of automation is justified by increased productivity. However, in the case of brazing, a further justification may well be found in the energy savings associated with the joint heating requirements.

FUNDAMENTALS

Brazing temperature is generally dictated by the base metal characteristic and the type of filler metal. Generally speaking, filler metals with higher melting temperatures correlate with higher joint strengths. However, the actual brazing temperature must be related to the melting

7. Brazing is covered in the *Welding Handbook,* Vol. 2, 7th Ed., 1980 and also the *Brazing Manual,* 3rd Ed., 1976, published by the American Welding Society.

temperature and heat treatment response of the base metal.

It is standard practice to minimize the time at brazing temperature. Brazing times of less than a minute are common. Minimal times at brazing temperatures are desirable to avoid excessive filler metal flow, base metal erosion, and oxidation.

A major consideration for filler metal type is the cost. Obviously, compatibility of the base metal and filler metal must be considered. Compatibility factors include such things as corrosion potentials, brittle compound formation, and base metal wettability.

EQUIPMENT

While manual torch brazing represents the simplest brazing technique, there is some economic justification for this process. First, the braze joint is visible to the operator who can adjust the process based on observation. Consequently, the operator represents an in-line inspection control. Second, this process directs heat only to the joint area versus heating of an entire assembly when furnace brazing techniques are used. When energy costs represent a large percentage of the cost of a braze joint, this is an important consideration.

The major drawback to torch brazing is, of course, the fact that it is labor intensive and consequently represents low productivity. A continuous belt furnace can alleviate this problem; however, the joint is not visible during the brazing operation and energy efficiency is low because the entire assembly is heated.

Automatic brazing machines can provide the attributes of both torch and furnace brazing, that is, energy economy and continuous operation. Typically, heat is directed to the joint area by one or more torches. Similar effects can be obtained by induction heating. A typical machine has provisions for assembly and fixturing, automatic fluxing, preheating (if needed), brazing, air or water quenching, part removal, and inspection.

Temperature Control

Process control and consistency that can be accomplished with automatic equipment are of primary importance. Temperature control is critical to successful brazing. When joining components of different masses, all components must attain brazing temperature at the same time. Overheating of a thinner component can cause molten filler metal to flow onto that component before it wets the thicker component. This can reduce the amount of filler metal available to fill the joint.

Automatic brazing equipment should provide a correct and predetermined heating pattern. This can be accomplished effectively with either induction or torch heating. Induction heating works best on ferrous metal parts, especially those of symmetrical shape. Induction heating coils must be designed for efficient localized heating of the joint area. Fixturing and removal of the brazed part from the heating coil must also be considered during coil design.

Torch heating involves the use of a preset heating pattern by appropriately designed torches located at one or more machine stations. Torches of various sizes, utilizing different fuel gases mixed with air or oxygen, may be used. Gas-air torches, with gentle wrap-around flames, can be used on many applications. When a pinpoint localized flame is required or when hotter and faster heating is desired, oxygen may be used in place of or in combination with air. The use of oxygen represents an additional operating cost, however, and should be considered only when its cost can be justified by increased production rates.

Heating Time

Time at temperature is an important control feature of automatic brazing equipment. Heating time is a function of the desired production rate and can be determined in advance. Once the heating time of an assembly is determined, the number of heat stations and dwell time at each can be allocated among the total number of work stations. Digital timing equipment can provide exact control of the dwell or heating time.

To maximize production, assemblies should be heated only to the temperature at which filler metal melts and flows into the joint. Holding the assembly at temperature beyond this point is not recommended in most cases because it can promote excess oxidation and other problems, such as volatilization of filler metal constituents and base metal erosion.

Filler Metal Application

Control of filler metal amounts and consequent joint quality is also possible with automatic brazing equipment. Brazing filler metal application can be effectively controlled with the use of preplaced shims or wire rings, or with timed automatic feed of wire to a joint from a filler wire feeder. Volume per joint can be controlled to provide consistency or repeatability that is not attainable with manual rod feeding.

Flux Application

Control of the amount and coverage of brazing flux can be provided by automatic equipment. Dispensers are available that can automate this operation. Dispensable fluxes are available for use in these units, and flow can be adjusted to handle a variety of applications.

ADVANTAGES OF BRAZING AUTOMATION

The main advantages of automatic brazing are as follows:

(1) High production rates
(2) High productivity per worker
(3) Filler metal savings
(4) Consistency of results
(5) Energy savings
(6) Adaptability and flexibility

Production requirements essentially determine the level or amount of automation required. Low production requires only basic equipment, such as a motorized conveyor transporting assemblies through a furnace. High production requirements may involve a fully automated system utilizing parts feeding and automatic unloading of completed assemblies. This type of equipment is able to produce in excess of 1,000 parts per hour.

Automatic brazing equipment can dramatically increase the productivity per worker. As an example, one unskilled brazing operator running an automatic brazing machine is able, in many cases, to produce as much as 4 or 5 highly skilled manual brazers. This increase in productivity alone is often enough to justify the purchase of automatic brazing equipment.

Another feature of automatic brazing equipment that reflects a major cost advantage and economic justification for purchase is savings in filler metal. Instead of feeding extra filler metal for assurance, as can be the case with manual brazing, automatic equipment can provide an exact and correct amount each and every time.

The advantage of consistent quality production is one that should not be overlooked. Rework or repairs of brazements are expensive, and can require extra operations. The best way to prevent these problems is to produce a good brazement the first time. This is expected with automatic equipment.

Automatic brazing equipment should be energy efficient. Localizing heat input only to the joint area is good energy conservation. Utilizing low cost fuels mixed with compressed or shop air is effective energy efficiency.

PROBLEMS OF AUTOMATION

RISK FACTORS

Factory Integration

The introduction of an automatic welding or brazing machine can create many problems in a manufacturing shop. For example, there can be incompatibilities with other shop operations, such as machining, cutting, or forming. Material handling for automatic operations can present problems. Tasks that are often overlooked in preparation for welding automation include the following:

Part Programming. The procedure for generating and storing machine programs is usually new to a welding shop. Unlike CNC machine tools and other automatic machines where programs are written off-line by a trained programmer, some welding robots are programmed on-line by teaching or by direct keyboard entry of a computer language.

Program Proofing. The program for automatic welding or brazing equipment is usually evaluated using actual production parts. Simulated production parts cannot duplicate the effects of distortion, heat input, and clamping, and such parts should only be used in the initial stages of process development. Decisions regarding the use of simulations and the time to switch to production parts can be difficult. An inexperienced user should solicit help from the equipment manufacturer or other experts.

Documentation. A system of numbering, filing, revising, and maintaining operating programs is required for programmable automatic welding and brazing machines. Systems for numerically controlled machine tools may not be suitable for the following reasons:

(1) Welding and brazing are operations joining two or more parts, whereas machining is done on a single part.

(2) Welding programs, which are usually small in terms of data storage, are often formulated into a single tape for fast, convenient access. Machine tool programs are usually generated individually for each operation.

(3) Programs may be edited by personnel who are not familiar with welding processes.

(4) Several different programs may be required to compensate for joint fit-up or part variations with welding and brazing operations. Machine tools usually need only one program for each part or operation.

(5) The costs of mishandling program documentation can be large. Expensive materials can be ruined using an old or incorrect program.

System Integration

The introduction of automatic welding or brazing may have widespread impact on company operations. Several aspects should be evaluated.

Design. Special design considerations are required for automatic welding and brazing. Joint designs and placements should be selected to optimize repeatability of joint location and fit-up. Component strength with automatic welding or brazing may be different from that with manual welding or brazing. Appropriate tests should be made to determine actual mechanical properties using the proposed automatic joining equipment. Corrosion resistance may also be affected. Savings in materials and consumables may be possible through design changes that recognize the increased consistency and reliability of automatic welding or brazing.

Manufacturing Planning. Component sequence may need to be changed for automatic welding or brazing. A few potential changes are:

(1) Periodic destructive or nondestructive testing and inspection may be required for quality control.

(2) Additional quality control tests, such as joint fit-up inspection, may be required.

(3) Multiple welding operations may be combined at a single work station.

(4) Welding or brazing conditions, machine control programs, and other formulated information may need to accompany components or be attached to planning sheets.

(5) New process specifications and acceptance criteria may be required.

Fixturing and Positioning. The fixturing and positioning of parts for automatic welding or brazing can be different from any previous experience of the average fabrication shop. Fixtures and positioners that are not furnished by the machine manufacturer with automatic welding or brazing equipment should be planned and designed by personnel experienced with the specific joining process.

Manufacturing. Automatic welding and brazing operations should be integrated into other factory operations. The following items should be considered:

(1) Operator job classification
(2) Operator skills and training
(3) Production backup capability
(4) Component part repeatability
(5) Repair procedures
(6) Intershift communications

(7) Safety procedure

(8) Consumables and spare parts

(9) Test parts and testing procedures

(10) Maintenance capability

(11) Material handling

(12) Calibration of equipment and procedures

(13) Production support personnel

Inspection and Quality Control. Automatic welding often complicates inspection. Fillet welds are preferred for automatic welding, but are difficult to inspect. Multiple welding operations performed at one station often restrict access for inspection, and in-process inspection may not be feasible from operational and economical stand points. New quality control procedures are usually required.

Reliability of Equipment

Generally, the reliability of a system is inversely proportional to its complexity, and automatic welding and brazing machines are complex systems. Nevertheless, the inherent precision and durability of equipment components usually exceeds the duty requirements because welding and brazing operations impose light dynamic machine loads and accuracy requirements compared to machining. The reliability problems in welding machines are usually related to environmental conditions of heat, spatter, fume, and collision. Electronic components require consideration of the electromagnetic and electric power environment.

CONSISTENCY AND PREDICTABILITY

Several of the many variables that can affect the quality of welded or brazed joints include the following:

(1) Material variations (thickness, geometry, composition, surface finish, etc.)

(2) Welding or brazing conditions (voltage, current, gas flow rates, etc.)

(3) Fixturing (part location, chill bars, materials, etc.)

(4) Procedures (alignment, preheat, heating rate, etc.)

(5) Consumables (filler metals, gases, etc.)

(6) Machine design and operation (bearings, guides, gear trains, etc.)

VENDOR ASSISTANCE

Reliable suppliers of automatic welding and brazing systems can assist the users in several ways.

Evaluation and Consulting Services

In general, it is prudent to consult with other knowledgeable persons when possible. Equipment manufacturers and engineering consultants are often more capable of determining the feasibility of a proposed automatic welding or brazing system than the user. They may be able to cite examples of similar installations and arrange for plant visits, point out potential problems, and indicate the need for special equipment.

Testing and Prototype Production

Welding or brazing tests can be performed on simple shapes, such as flat plate or pipe, when evaluating automation feasibility. Prototype production parts should be welded or brazed during equipment acceptance tests to establish production readiness. These tests can reveal unexpected problems, which are best solved before production commitments are made.

Training and Education

An appropriate school or training program should be located for training of the equipment operators. A manufacturer of automatic welding or brazing equipment can be a source of training programs. Other sources are technical societies, vocational schools, and colleges or universities.

Installation Services

Installation is normally done by or with the guidance of the equipment manufacturer. The installation of foundations, utilities, and interfaces with other equipment is usually performed by the user. The installation project should be thoroughly planned in advance of the delivery of the equipment. Engineering drawings and installation requirements of the equipment should be obtained from the manufacturer as early as possible.

Debugging the system can be a complex process and consume more time than expected. The equipment manufacturer should provide the needed test equipment and instrumentation to accomplish the job. The user can help expedite the task by providing required assistance to the manufacturer's start-up personnel.

SUPPLEMENTARY READING LIST

Computer Control and Monitoring

Bollinger, J.G., and Ramsey, P.W., Computer controlled self programming welding machine, *Welding Journal*, 58(5): 15-21; 1979 May.

Homes, J.G., and Resnick, B.J., Human combines with robot to increase welding versatility, *Welding and Metal Fabrication*, 48(1): 13-14, 17-18, 20; 1980 Jan./Feb.

Kuhne, A.H., Frassek, B., and Starke, G., Components for the automated GMAW process, *Welding Journal*, 63(1): 31-34; 1984 Jan.

Paxton, C.F., Solving resistance (spot) welding problems with mini- and micro-computers, *Welding Journal*, 58(8): 27-32; 1979 Aug.

Scott, J.J., and Brandt, H., Adaptive feed-forward digital control of GTA welding, *Welding Journal*, 61(3): 36-44; 1982 Mar.

Automatic Systems

Henschel, C., and Evans, D., Automatic brazing systems, *Welding Journal*, 61(10): 29-32; 1982 Oct.

Jones, S.B., Automan '83 - state of the art and beyond, *Metal Construction*, 15(5): 283-285; 1983 May.

Miller, L., Development of the "apprentice" arc welding robot, *Metal Construction*, 12(11): 615-619; 1980 Nov.

Nozaki, T., et al., Robot sees, decides and acts, *Welding and Metal Fabrication*, 47(9): 647, 649-651, 653-655, 657-658; 1979 Nov.

Soroka, D.P., and Sigman, R.D., Robotic arc welding: what makes a system?, *Welding Journal*, 61(9): 15-21; 1982 Sept.

Sullivan, M.J., Application considerations for selecting industrial robots for arc welding, *Welding Journal*, 59(4): 28-31; 1980 Apr.

Wolke, R.C., Integration of a robotic welding system with existing manufacturing processes, *Welding Journal*, 61(9): 23-28; 1982 Sept.

Seam Tracking

Bollinger, J.G., and Harrison, M.L., Automated welding using spatial seam tracking, *Welding Journal*, 50(11): 787-792; 1971 Nov.

Goldberg, F., and Karlen, R., Seam tracking and height sensing — inductive systems for arc welding and thermal cutting, *Metal Construction*, 12(12): 668-671; 1980 Dec.

Lacoe, D., and Seibert, L., 3-D vision-guided robotic welding system aids railroad repair shop, *Welding Journal*, 63(3): 53-56; 1984 Mar.

Pan, J.L., et al., Development of a two-directional seam tracking system with laser sensor, *Welding Journal*, 62(2): 28-31; 1983 Feb.

Pandjuris, A.K., and Weinfurt, E.J., Tending the arc, *Welding Journal*, 51(9): 633-637; 1972 Sept.

Richardson, R.W., et al., Coaxial arc weld pool viewing for process monitoring and control, *Welding Journal*, 63(3): 43-50; 1984 Mar.

6
Codes and Standards

Definitions *216*

Sources *216*

Applications *217*

American Association of State Highway and Transportation Officials *219*

American Bureau of Shipping *219*

American Institute of Steel Construction *221*

American National Standards Institute *221*

American Petroleum Institute *222*

American Railway Engineering Association *223*

American Society of Mechanical Engineers *223*

American Society for Testing and Materials *225*

American Waterworks Association *226*

American Welding Society *226*

Association of American Railroads *231*

Canadian Standards Association *231*

Compressed Gas Association *232*

Federal Government *232*

International Organization for Standardization *237*

National Board of Boiler and Pressure Vessel Inspectors *237*

National Fire Protection Association . . . *238*

Pipe Fabrication Institute *238*

Society of Automotive Engineers *239*

Manufacturers Associations *244*

Chapter Committee

G.N. FISCHER, *Chairman*
Fischer Engineering Company

H.W. EBERT
Exxon Research and Engineering

W.L. BALLIS
Colombia Gas Distribution Company

P.D. FLENNER
Consumer Power Company

J.W. McGREW
Babcock and Wilcox Company

Welding Handbook Committee Member

M. TOMSIC
Hobart Brothers Company

6
Codes and Standards

DEFINITIONS

The purpose of this chapter is to familiarize fabricators and purchasers of welded products with the basic documents that govern or guide welding activities. These documents serve to assure that (1) only safe and reliable welded products will be produced, and (2) those persons associated with welding operations will not be exposed to undue danger or other conditions that would be harmful to their health. Publications relating only to the manufacture of welding materials or equipment are not covered in this chapter. However, those publications may be referenced in the basic documents, and their relationship to safety and reliability should not be underestimated.

The American Welding Society uses the general term *standards* to refer to documents that govern and guide welding activities. Standards describe the technical requirements for a material, process, product, system, or service. They also indicate, as appropriate, the procedures, methods, equipment, or tests used to determine that the requirements have been met.

Standards include codes, specifications, guides, methods, and recommended practices. These documents have many similarities, and the terms are often used interchangeably, but sometimes incorrectly.

Codes and specifications are similar types of standards that use the verbs *shall* and *will* to indicate the mandatory use of certain materials or actions, or both. Codes differ from specifications in that their use is mandated with the force of law by one or more governmental jurisdictions. The use of specifications becomes mandatory only when they are referenced by codes or contractural documents.

Guides and recommended practices are standards that are offered primarily as aids to the user. They use verbs such as *should* and *may* because their use is usually optional. However, if these documents are referenced by codes or contractural agreements, their use may become mandatory. If the codes or agreements contain non-mandatory sections or appendices, the use of referenced guides or recommended practices is at the user's discretion.

The user of a standard should become acquainted with its scope and intended use, both of which are usually included within the *Scope* or *Introduction*. It is equally important, but often more difficult, to recognize subjects that are not covered by the document. These omissions may require additional technical consideration. A document may cover the details of the product form without considering special conditions under which it will be used. Examples of special conditions would be corrosive atmospheres, elevated temperatures, and dynamic rather than static loading.

Standards vary in their method of achieving compliance. Some have specific requirements that do not allow for alternative actions. Others permit alternative actions or procedures as long as they result in properties that meet specified criteria. These criteria are often given as minimum requirements; for example, the ultimate tensile strength of a welded specimen must meet or exceed the minimum tensile strength specified for the base material.

SOURCES

Private and governmental organizations develop, issue, and update codes and standards that apply to their particular areas of interest.

Table 6.1 lists those organizations of concern to the welding industry and their current addresses. The interests of many of these groups overlap with regard to welding, and some agreements have been made to reduce duplication of effort. Many codes and standards that are concerned with welding, brazing, and allied processes are prepared by the American Welding Society (AWS) because these subjects are of primary interest to the members. Codes and standards that apply to a particular product are usually prepared by the group that has overall responsibility. For example, those for railroad freight cars are published by the Association of American Railroads (AAR). However, freight cars are basically structures, and the applicable AAR specification refers to *AWS D1.1, Structural Welding Code—Steel* for the qualification of welders and welding operators.

Each organization that prepares codes or standards has committees or task groups perform this function. Members of these committees or groups are specialists in their field. They prepare drafts of codes and standards that are reviewed and approved by a larger group. The review group is selected to include persons with diverse ranges of interests including, for example, producers, users, and government representatives. To avoid control or undue influence by one interest group, agreement must be achieved by a high percentage of all members.

The federal government develops or adopts codes and standards for items and services that are in the public rather than the private domain. The mechanisms for developing federal or military documents are similar to those of private organizations. Standard-writing committees usually exist within a federal department or agency that has responsibility for a particular item or service.

The American National Standards Institute (ANSI) is a private organization responsible for coordinating national standards for use within the United States. ANSI does not actually prepare standards. Instead, it forms national interest review groups to determine whether proposed standards are in the public interest. Each group is composed of persons from various organizations concerned with the scope and provisions of a particular document. If there is a consensus regarding the general value of a particular standard, then it may be adopted as an American National Standard. Adoption of a standard by ANSI does not, of itself, give it mandatory status. However, if the standard is cited by a governmental rule or regulation, it may then be backed by force of law.

Other industrial countries also develop and issue codes and standards on the subject of welding. The following are examples of other national standards designations and the bodies responsible for them

- BS — British Standard issued by the British Standards Association
- CSA— Canadian Standard issued by the Canadian Standards Association
- DIN— West German Standard issued by the Deutsches Institute fuer Normung
- JIS — Japanese Industrial Standard issued by the Japanese Standards Association
- NF — French Norms issued by the Association Francaise de Normalisation

Of these, the Canadian Standards Association is discussed in a following section.

There is also an International Organization for Standardization (ISO). Its goal is the establishment of uniform standards for use in international trade. This organization is discussed at more length in a following section.

APPLICATIONS

The minimum requirements of a particular code or standard may not satisfy the special needs of every user. Therefore, a user may find it necessary to invoke additional requirements to obtain desired quality.

There are various mechanisms by which most codes and standards may be revised. These are used when a code or standard is found to be be in error, unreasonably restrictive, or not applicable with respect to new technological developments. Some codes and standards are updated on a regular basis, while others are revised as needed. The revisions may be in the form of addenda, or they may be incorporated in superseding documents.

Table 6.1
Sources of codes and standards of interest to the welding industry

American Association of State Highway and Transportation Officials (AASHTO)
444 N. Capital St., N.W.
Washington, DC 20001
(202) 624-5800

American Bureau of Shipping (ABS)
65 Broadway
New York, NY 10006
(212) 440-0300

American Institute of Steel Construction (AISC)
400 N. Michigan Ave.
Chicago, IL 60611
(312) 670-2400

American National Standards Institute (ANSI)
1430 Broadway
New York, NY 10018
(212) 354-3300

American Petroleum Institute (API)
1220 L Street, N.W.
Washington, DC 20005
(202) 457-7000

American Railway Engineering Association (AREA)
2000 L Street, N.W.
Washington, DC 20036
(202) 835-9336

American Society of Mechanical Engineers (ASME)
345 East 47th Street
New York, NY 10017
(212) 705-7722

American Society for Testing and Materials (ASTM)
1916 Race Street
Philadelphia, PA 19103
(215) 299-5400

American Water Works Association (AWWA)
6666 W. Quincy Ave.
Denver, CO 80235
(303) 794-7711

American Welding Society (AWS)
550 N.W. LeJeune Road
Miami, FL 33126
(305) 443-9353

Association of American Railroads (AAR)
1920 L Street, N.W.
Washington, DC 20036
(202) 835-9100

Canadian Standards Association (CSA)
178 Rexdale Blvd.
Rexdale, Ontario
Canada M9W 1R3
(416) 744-4000

Compressed Gas Association (CGA)
1235 Jeff Davis Hwy.
Arlington, VA 22202
(703) 979-0900

International Organization for Standardization (ISO)
(See American National Standards Institute)

National Board of Boiler and Pressure Vessel Inspectors (NBBPVI)
1055 Crupper Ave.
Columbus, OH 43229
(614) 888-8320

National Fire Protection Association (NFPA)
Batterymarch Park
Quincy, MA 02269
(617) 328-9290

Naval Publications and Forms Center[1]
5801 Taber Ave.
Philadelphia, PA 19120
(215) 697-2000

Pipe Fabrication Institute (PFI)
1326 Freeport Rd.
Pittsburgh, PA 15238
(412) 782-1624

Society of Automotive Engineers (SAE)
400 Commonwealth Dr.
Warrendale, PA 15096
(412) 776-4841

Superintendent of Documents[2]
U.S. Government Printing Office
Washington, DC 20402
(202) 783-3238

Uniform Boiler and Pressure Vessel Laws Society (UBPVLS)
2838 Long Beach Rd.
Oceanside, NY 11572
(516) 536-5485

1. Military Specifications
2. Federal Specifications

If the user has a question about a particular code or standard regarding either an interpretation or a possible error, he should contact the responsible organization.

When the use of a code or standard is mandatory, whether as a result of a government regulation or a legal contract, it is essential to know the particular edition of the document to be used. It is unfortunate, but not uncommon, to find that an outdated edition of a referenced document has been specified, and must be followed to be in compliance. If there is a question concerning which edition or revision of a document is to be used, it should be resolved before commencement of work.

Organizations responsible for preparing codes and standards that relate to welding are discussed in the following sections. The publications are listed without reference to date of publication, latest revision, or amendment. New publications relating to welding may be issued, and current ones may be withdrawn or revised. The responsible organization should be contacted for current information on its codes and standards.

Some organizations cover many product categories while others may cover only one. Table 6.2 lists the organizations and the product categories covered by their documents. The National Fire Protection Association is not listed in the table because its standards are concerned with safe practices and equipment rather than with products. The American Welding Society and the American Petroleum Institute also publish standards concerned with welding safety.

AMERICAN ASSOCIATION OF STATE HIGHWAY AND TRANSPORTATION OFFICIALS

The member agencies of this association, known as AASHTO, are the U.S. Department of Transportation and the Departments of Transportation and Highways of the fifty states, Washington, D.C., and Puerto Rico. The AASHTO specifications are prepared by committees of individuals from the member agencies. These documents are the minimum rules to be followed by all member agencies or others in the design and construction of steel highway bridges.

Standard Specifications for Highway Bridges

This AASHTO specification covers the design and construction requirements for all types of highway bridges. It refers to the welding fabrication requirements in the *AASHTO Standard Specifications for Welding of Structural Steel Highway Bridges* and the *AWS D1.1, Structural Welding Code—Steel.*

Standard Specifications for Welding of Structural Steel Highway Bridges

This AASHTO specification provides modifications to the *AWS D1.1, Structural Welding Code—Steel* deemed necessary for use by member agencies. These are referenced to the applicable sections of the AWS Code.

Guide Specifications for Fracture Critical Non-Redundant Steel Bridge Members

Fracture critical, non-redundant members or components of a bridge are tension members or components, the failure of which would likely result in collapse of the structure. The document assigns the responsibility for specifying those bridge members or components, if any, that fall into the fracture critical category. It requires that such members or components be fabricated to the required workmanship standards only by organizations having the proper personnel, experience, procedures, knowledge, and equipment. For example, all welding inspectors and nondestructive testing personnel must have demonstrated competency for assuring quality in compliance with the specifications. The document also contains requirements additional to those in the *Standard Specifications for Welding of Structural Steel Highway Bridges*.

AMERICAN BUREAU OF SHIPPING

The function of the American Bureau of Shipping (ABS) is to control the quality of ship construction. Each year, ABS reissues the *Rules*

Table 6.2
Products covered by codes and standards of various organizations

Product	AAR	AASHTO	ABS	AISC	API	AREA	ASME NBPVI UBPVLS	ASTM	AWS	AWWA	FED	PFI	SAE
Base metals			X		X		X	X			X		X
Bridges		X		X		X			X		X		
Buildings				X					X				
Construction equipment									X		X		X
Cranes									X				
Filler metals			X				X		X		X		X
Machine tools									X				
Military equipment											X		
Power generation equipment			X				X				X		
Piping			X		X		X		X	X	X	X	
Presses									X				
Pressure vessels			X		X		X						
Railway equipment	X					X			X				
Sheet metal fabrication									X				
Ships			X						X		X		
Storage tanks					X				X	X			
Structures, general				X					X				
Vehicles									X		X		X

for Building and Classing Steel Vessels. These rules are applicable to ships that are intended to have American registration.

To obtain American registration and insurance, a ship must be classified (approved) by an ABS surveyor (inspector). The survey begins with a review of the proposed design. Reviews are also made during and after construction to verify that construction complies with the ABS rules. The process is completed with the assignment and registration of a class (numerical identification) for the ship.

One section of the ABS Rules addresses welding, and is divided into the following parts:

Part 1 — Hull Construction

Part 2 — Boilers, Unfired Pressure Vessels, Piping, and Engineering Structures

Part 3 — Weld Tests

The section addresses such topics as weld design, welding procedures, qualification testing, preparation for welding, production welding, workmanship, and inspection.

ABS also publishes a list of welding consumables, entitled *Approved Welding Electrodes, Wire-Flux, and Wire-Gas Combinations.* These consumables are produced by various manufacturers around the world. They are tested under ABS supervision and approved for use under the ABS rules.

AMERICAN INSTITUTE OF STEEL CONSTRUCTION

The American Institute of Steel Construction (AISC) is a non-profit trade organization for the fabricated structural steel industry in the United States. The Institute's objectives are to improve and advance the use of fabricated structural steel through research and engineering studies and to develop the most efficient and economical design of structures. The organization also conducts programs to improve and control product quality.

Manual of Steel Construction

This manual covers such topics as the dimensions and properties of rolled structural steel shapes; beam, girder, and column design; and welded connection design. Other topics include the use of welding symbols and prequalified weld joint designs. Codes and specifications that are published by AISC and other organizations and pertain to certain aspects of structural steel design and fabrication, but are not covered by other parts of the manual, are also included in the manual.

Specification for the Design, Fabrication, and Erection of Structural Steel for Buildings

This document specifies, in detail, all principal steps required for the construction of structural steel buildings. It references the AWS filler metal specifications, and specifies the particular filler metal classification to be used with a welding process for each type of structural steel. Requirements for the types and details of fillet, plug, and slot welds are also included. The specification refers to *AWS D1.1, Structural Welding Code—Steel* for welding procedure and welder performance qualifications.

Quality Criteria and Inspection Standards

This document covers such subjects as preparation of materials, fitting and fastening, dimensional tolerances, welding, surface preparation, and painting. It discusses the practical implementation of some of the requirements of other AISC specifications. Typical problems that may be encountered in steel construction and recommended solutions are presented. The welding section provides interpretations regarding AISC requirements for prequalification of welding procedures, preheating, control of distortion, and tack welding.

AMERICAN NATIONAL STANDARDS INSTITUTE

As mentioned previously, the American National Standards Institute (ANSI) is the coordinating organization for the U.S. voluntary standards system; it does not develop standards directly. The Institute provides means for determining the need for standards, and ensures that organizations competent to fill these needs undertake the development work. Its approval procedures for American National Standards ensure that all concerned national interests (a consensus) have an opportunity to participate in the development of a standard or to comment on its provisions prior to publication. ANSI is the U.S. member of nontreaty international standards organizations, such as the International Organization for Standardization (ISO), and the International Electrotechnical Commission (IEC).

American National Standards, which now number approximately 10,000 documents, encompass virtually every field and every discipline. They deal with dimensions, ratings, terminology and symbols, test methods, and performance and safety specifications for materials, equipment, components, and products in some two dozen fields. These fields include construction; electrical and electronics; heating, air conditioning, and refrigeration; information systems; medical devices; mechanical; nuclear; physical distribution; piping and processing; photography and motion pictures; textiles; and welding.

The American National Standards provide a common language that can be used confidently by industry, suppliers, customers, business, the public, government, and labor. Each of these interests has either participated in the development of the standards or has been given the opportunity to comment on their provisions. However, these standards are developed and used voluntarily. They become mandatory only when they are adopted or referenced by a governmental body.

The ANSI federation consists of companies, large and small, and trade, technical, professional, labor, and consumer organizations that total over 1000 members. The standards are primarily developed by the member organizations.

AMERICAN PETROLEUM INSTITUTE

The American Petroleum Institute (API) publishes documents in all areas related to petroleum production. Those documents that include welding requirements are discussed below.

Pipelines and Refinery Equipment

API Std 1104, Standard for Welding Pipelines and Related Facilities. This standard applies to arc and oxyfuel gas welding of piping used in the compression, pumping, and transmission of crude petroleum, petroleum products, and fuel gases, and also to the distribution systems when applicable. It presents methods for the production of acceptable welds by qualified welders using approved welding procedures, materials, and equipment. It also presents methods for the production of suitable radiographs by qualified technicians using approved procedures and equipment, to ensure proper analysis of weld quality. Standards of acceptability and repair of weld defects are also included.

The legal authority for the use of API Std 1104 comes from *Title 49, Part 195, Transportation of Liquids by Pipeline,* of the United States Code of Federal Regulations.

RP 1107, Recommended Pipeline Maintenance Welding Practices. The primary purpose of this recommended practice is safety. It contains prohibitions of practices that are known to be unsafe, and also warnings about practices for which caution is necessary. It also includes methods for the inspection of repair welds, and for installing appurtenances on loaded piping systems being used for the transmission of natural gas, crude petroleum, and petroleum products.

The legal authority for RP 1107 comes from reference to it in *ASME B31.4, Liquified Petroleum Transportation Piping Systems* (described under the American Society of Mechanical Engineers). The latter publication, like *API Std 1104, Standard for Welding Pipelines and Related Facilities,* is also referenced by *Title 49, Part 195, Transportation of Liquids by Pipeline,* of the United States Code of Federal Regulations.

Publ 942, Recommended Practice for Welded, Plain Carbon Steel Refinery Equipment for Environmental Cracking Service. This publication proposes actions for protection against hydrogen stress cracking of welds in plain carbon steel that are exposed, under stress, to certain aqueous-phase acidic environments, such as moist hydrogen sulfide.

The Guide for Inspection of Refinery Equipment. This publication consists of 20 chapters and an appendix, all of which may be purchased separately from API. The appendix, entitled *Inspection of Welding,* is the only part that applies specifically to welding. Its objective is to guide the user in determining whether welded joints are of acceptable quality and comply with both the requirements of the contract or job specifications and the prescribed welding procedure specifications.

Storage Tanks for Refinery Service

Std 620, Recommended Rules for Design and Construction of Large, Welded, Low-Pressure Storage Tanks. These rules cover the design and construction of large, field-welded tanks that are used for storage of petroleum intermediates and finished products under pressure of 15 psig and less.

Std 650, Welded Steel Tanks for Oil Storage. This standard covers the material, design, fabrication, erection, and testing requirements for vertical, cylindrical, welded steel storage tanks that are above ground and not subject to internal pressure.

RP 510, Inspection, Rating, and Repair of Pressure Vessels in Petroleum Refinery Service. This recommended practice covers the inspection, repair, evaluation for continued use, and methods for computing the maximum allowable working pressure of existing pressure vessels. The vessels include those constructed in accordance with Section VIII of the *ASME Boiler and Pressure Vessel Code,* or other pressure vessel codes.

Safety and Fire Protection

Publ 2009, Safe Practices in Gas and Electric Cutting and Welding in Refineries, Gasoline Plants, Cycling Plants, and Petrochemical Plants. This publication outlines precautions for protecting persons from injury and property from damage by fire that might result during the operation of oxyfuel gas and electric cutting and welding equipment in and around petroleum operations.

PSD 2200, Repairs to Crude Oil, Liquified Petroleum Gas, and Products Pipelines. This petroleum safety data sheet is a guide to safe practices for the repair of pipelines for crude oil, liquified petroleum gas, and petroleum products.

PSD 2201, Welding or Hot Tapping on Equipment Containing Flammables. This petroleum safety data sheet lists procedures for welding, as well as for making hot taps (connections while in operation), on pipelines, vessels, or tanks containing flammables. This data sheet and PSD 2200 are also requirements of *ASME B31.4, Liquified Petroleum Transportation Piping Systems.*

AMERICAN RAILWAY ENGINEERING ASSOCIATION

The American Railway Engineering Association (AREA) publishes the *Manual for Railway Engineering.* This manual contains specifications, rules, plans, and instructions that constitute the recommended practices of railway engineering. Two chapters specifically cover steel construction. One of these covers the design, fabrication, and erection of buildings for railway purposes. The other addresses the same topics for railway bridges and miscellaneous steel structures.

AMERICAN SOCIETY OF MECHANICAL ENGINEERS

Two standing committees of the American Society of Mechanical Engineers (ASME) are actively involved in the formulation, revision, and interpretation of codes and standards covering products that may be fabricated by welding. These committees are responsible for preparing the *ASME Boiler and Pressure Vessel Code* and the *Code for Pressure Piping,* which are American National Standards.

Boiler and Pressure Vessel Code

The *ASME Boiler and Pressure Vessel Code (BPVC)* contains eleven sections. Sections I, III, IV, VIII, and X cover the design, construction, and inspection of boilers and pressure vessels. Sections VI, VII, and XI cover the care and operation of boilers or nuclear power plant components. The remaining Sections II, V, and IX cover material specifications, nondestructive examination, and welding and brazing qualifications, respectively.

Section I, Power Boilers covers power, electric, and miniature boilers; high temperature boilers used in stationary service; and power boilers used in locomotive, portable, and traction service. *Section III, Nuclear Power Plant Components* addresses the various components required by the nuclear power industry. *Section IV, Heating Boilers* applies to steam heating and hot water supply boilers that are directly fired by oil, gas, electricity, or coal. *Section VIII, Pressure Vessels* covers unfired pressure vessels.

Section II, Material Specifications contains the specifications for acceptable ferrous and nonferrous base metals, and for acceptable welding and brazing filler metals and fluxes. Many of these specifications are identical to and have the same numerical designation as ASTM and AWS specifications for base metals and welding consumables, respectively. *Section V, Nondestructive Examination* covers methods and standards for nondestructive examination of boilers and pressure vessels.

Section IX, Welding and Brazing Qualifications covers the qualification of (1) welders, welding operators, brazers, and brazing operators, and (2) the welding and brazing procedures that are to be employed for welding or brazing of boilers or pressure vessels. This section of the Code is often cited by other codes and standards, and by regulatory bodies, as the welding and brazing qualification standard for other types of welded or brazed products.

The *ASME Boiler and Pressure Vessel Code* is referenced in the safety regulations of most States and major cities of the USA, and also the Provinces of Canada. A number of Federal agencies include the Code as part of their respective regulations.

The Uniform Boiler and Pressure Vessel Laws Society (UBPVLS) has, as its objective, uniformity of laws, rules, and regulations that affect boiler and pressure vessel fabricators, inspection agencies, and users. The Society believes that such laws, rules, and regulations should follow nationally accepted codes and standards. It recommends the *ASME Boiler and Pressure Vessel Code* as the standard for construction and the *Inspection Code* of the National Boiler and Pressure Vessel Inspectors (discussed in a following section) as the standard for inspection and repair.

The *ASME Boiler and Pressure Vessel Code* is unique in that it requires third party inspection independent of the fabricator and the user. The National Board of Boiler and Pressure Vessel Inspectors (NBBPVI) commissions inspectors by examination. These inspectors are employed either by authorized inspection agencies (usually insurance companies) or by jurisdictional authorities.

Prior to building a boiler or pressure vessel, a company must have a quality control system and a manual that describes it. The system must be acceptable to the authorized inspection agency and either the jurisdictional authority or the NBBPVI. Based on the recommendations to ASME, a code symbol stamp and Certificate of Authorization may be issued by ASME to the fabricator. The authorized inspection agency is also involved in monitoring the fabrication and field erection of boilers and pressure vessels. An authorized inspector must be satisfied that all applicable provisions of the *ASME Boiler and Pressure Vessel Code* have been followed before allowing the fabricator to apply its code symbol stamp to the unit.

Code for Pressure Piping

The *ASME B31, Code for Pressure Piping* presently consists of six sections. Each section prescribes the minimum requirements for the design, materials, fabrication, erection, testing, and inspection of a particular type of piping system.

B31.1, Power Piping covers power and auxiliary service systems for electric generation stations; industrial and institutional plants; central and district heating plants; and district heating systems.

B31.2, Fuel Gas Piping covers piping systems for fuel gases such as natural gas, manufactured gas, liquefied petroleum gas (LPG) – air mixtures above the upper combustible limits,

LPG in the gaseous phase, or mixtures of these gases. The covered piping systems, both in and between buildings, extend from the outlet of the consumer's meter set assembly (or point of delivery) to and including the first pressure-containing valve upstream of the gas utilization device.

B31.3, Chemical Plant and Petroleum Refinery Piping covers all piping within the property limits of facilities engaged in processing or handling of chemical, petroleum, or related products. Examples are chemical plants, petroleum refineries, loading terminals, natural gas processing plants (including liquefied natural gas facilities), bulk plants, compounding plants, and tank farms. This section applies to piping systems that handle all fluids, including fluidized solids, and to all types of service including raw, intermediate, and finished chemicals; oil and other petroleum products; gas; steam; air; water; and refrigerants, except as specifically excluded.

Piping for air and other gases, which is not now within the scope of existing sections of this code, may be designed, fabricated, inspected, and tested in accordance with the requirements of this section of the Code. The piping must be in plants, buildings, and similar facilities that are not otherwise within the scope of this section.

B31.4, Liquid Petroleum Transportation Piping Systems covers piping for transporting liquid petroleum products between producers' lease facilities, tank farms, natural gas processing plants, refineries, stations, terminals, and other delivery and receiving points. Examples of such products are crude oil, condensate, natural gasoline, natural gas liquids, and liquefied petroleum gas.

B31.5, Refrigeration Piping applies to refrigerant and brine piping for use at temperatures as low as -320° F, whether erected on the premises or factory assembled. It does not include (1) self-contained or unit refrigeration systems subject to the requirements of Underwriters' Laboratories or any other nationally recognized testing laboratory, (2) water piping, or (3) piping designed for external or internal pressure not exceeding 15 psig, regardless of size. Other sections of the Code may provide requirements for refigeration piping in their respective scopes.

B31.8, Gas Transmission and Distribution Piping Systems addresses gas compressor stations, gas metering and regulating stations, gas mains, and service lines up to the outlet of the customer's meter set assembly. Gas storage lines and gas storage equipment of the closed-pipe type that is either fabricated or forged from pipe or fabricated from pipe and fittings are also included.

Boiler external piping is defined by Section I of the *ASME Boiler and Pressure Vessel Code.* This piping requires a quality control system and third party inspection similar to those required for boiler fabrication. Otherwise, the materials, design, fabrication, installation, and testing for boiler external piping must meet the requirements of *B31.1, Power Piping.* A fabricator is not required to provide a quality control system and third party inspection for the other piping systems described previously.

All sections of the *Code for Pressure Piping* require qualification of the welding procedures and performance of welders and welding operators to be used in construction. Some sections require these qualifications to be performed in accordance with Section IX of the *ASME Boiler and Pressure Vessel Code,* while in others it is optional. The use of *API Std 1104, Standard for Welding Pipelines and Related Facilities* or *AWS D10.9, Specification for Qualification of Welding Procedures and Welders for Piping and Tubing* is permitted in some sections as an alternative to Section IX. Each section of the Code should be consulted for the applicable qualification documents.

AMERICAN SOCIETY FOR TESTING AND MATERIALS

The American Society for Testing and Materials (ASTM) develops and publishes specifications for use in the production and testing of materials. The committees that develop the specifications are comprised of producers and users as well as others who have an interest in the subject materials. The specifications cover virtually all materials used in industry and commerce with the exception of welding consumables, which are covered by AWS specifications.

ASTM publishes an *Annual Book of ASTM*

Standards that incorporates new and revised standards. It is currently composed of 15 sections comprising 65 volumes and an index. Specifications for the metal products, test methods, and analytical procedures of interest to the welding industry are found in the first three sections, comprising 17 volumes. Section 1 covers iron and steel products; Section 2, nonferrous metal products; and Section 3, metal test methods and analytical procedures. Copies of single specifications are also available from ASTM.

Prefix letters, which are part of each specification's alpha-numeric designation, provide a general idea of the specification content. They include *A* for ferrous metals, *B* for nonferrous metals, and *E* for miscellaneous subjects including examination and testing. When ASME adopts an ASTM specification for certain applications, either in its entirety or in a revised form, it adds an *S* in front of the ASTM letter prefix.

Many ASTM specifications include supplementary requirements that must be specified by the purchaser if they are desired. These may include vacuum treatment, additional tension tests, impact tests, or ultrasonic examination.

The producer of a material or product is responsible for compliance with all mandatory and specified supplementary requirements of the appropriate ASTM specification. The user of the material is responsible for verifying that the producer has complied with all requirements.

Some codes permit the user to perform the tests required by an ASTM or other specification to verify that a material meets requirements. If the results of the tests conform to the requirements of the designated specification, the material can be used for the application.

Some products covered by ASTM specifications are fabricated by welding. The largest group is steel pipe and tubing. Some types of pipe are produced from strip by rolling and arc welding the longitudinal seam. The welding procedures generally must be qualified to the requirements of the *ASME Boiler and Pressure Vessel Code,* or another code.

Other types of pipe and tubing are produced with resistance welded seams. There are generally no specific welding requirements in the applicable ASTM specification. The finished pipe and tubing must pass specific tests that should result in failure at the welded seam if the welding operation is out of control.

Two ASTM specifications cover joints in piping systems. These are *ASTM A422, Standard Specification for Butt Welds in Still Tubes for Refinery Service,* and *ASTM F722, Standard Specification for Welded Joints for Shipboard Piping Systems.*

AMERICAN WATER WORKS ASSOCIATION

The American Water Works Association (AWWA) currently has two standards that pertain to the welding of water storage and transmission systems. One of these standards was developed jointly with and adopted by the American Welding Society.

C206, Standard for Field Welding of Steel Water Pipe Joints

This standard covers field welding of steel water pipe. It includes the welding of circumferential pipe joints as well as other welding required in the fabrication and installation of specials and accessories. The maximum wall thickness of pipe covered by this standard is 1.25 inch.

D100 (AWS D5.2), Standard for Welded Steel Elevated Tanks, Standpipes, and Reservoirs for Water Storage

This standard covers the fabrication of water storage tanks. An elevated tank is one supported on a tower. A standpipe is a flat-bottomed cylindrical tank having a shell height greater than its diameter. A reservoir is a flat-bottomed cylindrical tank having a shell height equal to or smaller than its diameter. In addition to welding details, this standard specifies the responsibilities of the purchaser and the contractor for such items as the foundation plans, the foundation itself, water for pressure testing, and a suitable right-of-way from the nearest public road for on-site erection.

AMERICAN WELDING SOCIETY

The American Welding Society (AWS) publishes numerous documents covering the use and quality control of welding. These documents include codes, standards, specifications, recommended practices, and guides. The general subject areas covered are:

(1) Definitions and symbols
(2) Filler metals
(3) Qualification and testing
(4) Welding processes
(5) Welding applications
(6) Safety

Definitions and Symbols

AWS A2.4, Symbols for Welding and Nondestructive Testing. This publication describes the standard symbols used to convey welding, brazing, and nondestructive testing requirements on drawings. Symbols in this publication are intended to facilitate communications between designers and fabrication personnel. Typical information that can be conveyed with welding symbols includes type of weld, joint geometry, weld size or effective throat, extent of welding, and contour and surface finish of the weld.

AWS A3.0, Welding Terms and Definitions. This publication lists and defines the preferred terms that should be used in oral and written communications conveying welding, brazing, soldering, thermal spraying, and thermal cutting information. Nonstandard terms are also included; these are defined by reference to the preferred terms.

Filler Metals

The AWS filler metal specifications cover most types of consumables used with the various welding and brazing processes. The specifications include both mandatory and nonmandatory provisions. The mandatory provisions cover such subjects as chemical or mechanical properties, or both, manufacture, testing, and packaging. The nonmandatory provisions, included in an appendix, are provided as a source of information for the user on the classification, description, and intended use of the filler metals covered.

Following is a current listing of AWS filler metal specifications.

Aluminum and Aluminum Alloy Bare Welding Rods and Electrodes, AWS A5.10

Aluminum and Aluminum Alloy Covered Arc Welding Electrodes, AWS A5.3

Brazing Filler Metal (Ag, Al, Au, Co, Cu, Mg, and Ni alloys), AWS A5.8

Cast Iron, Welding Rods and Covered Electrodes for Welding, AWS A5.15

Consumable Inserts, AWS A5.30

Copper and Copper Alloy Bare Welding Rods and Electrodes, AWS A5.7

Copper and Copper Alloy Covered Electrodes, AWS A5.6

Copper and Copper Alloy Gas Welding Rods, AWS A5.27

Corrosion-Resisting Chromium and Chromium-Nickel Steel Bare and Composite Metal Cored and Stranded Arc Welding Electrodes and Welding Rods, AWS A5.9

Corrosion-Resisting Chromium and Chromium-Nickel Steel Covered Electrodes, AWS A5.4

Flux Cored Corrosion-Resisting Chromium and Chromium-Nickel Steel Electrodes, AWS A5.22

Iron and Steel Oxyfuel Gas Welding Rods, AWS A5.2

Magnesium Alloy Welding Rods and Bare Electrodes, AWS A5.19

Nickel and Nickel Alloy Bare Welding Rods and Electrodes, AWS A5.14

Nickel and Nickel Alloy Covered Welding Electrodes, AWS A5.11

Steel, Carbon, Covered Arc Welding Electrodes, AWS A5.1

Steel, Carbon, Electrodes for Flux Cored Arc Welding, AWS A5.20

Steel, Carbon, Electrodes and Fluxes for Submerged Arc Welding, AWS A5.17

Steel, Carbon, Filler Metals for Gas Shielded Arc Welding, AWS A5.18

Steels, Carbon and High Strength Low Alloy, Consumables Used for Electrogas Welding of, AWS A5.25

Steels, Carbon and High Strength Low Alloy, Consumables Used for Electroslag Welding of, AWS A5.25

Steel, Low Alloy, Covered Arc Welding Electrodes, AWS A5.5

Steel, Low Alloy, Electrodes for Flux Cored

Arc Welding, AWS A5.29
Steel, Low Alloy, Electrodes and Fluxes for Submerged Arc Welding, AWS A5.23
Steel, Low Alloy, Filler Metals for Gas Shielded Arc Welding, AWS A5.28
Surfacing Welding Rods and Electrodes, Composite, AWS A5.21
Surfacing Welding Rods and Electrodes, Solid, AWS A5.13
Titanium and Titanium Alloy Bare Welding Rods and Electrodes, AWS A5.16
Tungsten Arc Welding Electrodes, AWS A5.12
Zirconium and Zirconium Alloy Bare Welding Rods and Electrodes, AWS A5.24

Most AWS filler metal specifications have been approved by ANSI as American National Standards and adopted by ASME. When ASME adopts an AWS filler metal specification, either in its entirety or with revisions, it adds the letters *SF* to the AWS alphanumeric designation. Thus, ASME SFA5.4 specification would be similar, if not identical, to the AWS A5.4 specification.

AWS also publishes two documents to aid users with the purchase of filler metals. *AWS A5.01, Filler Metal Procurement Guidelines* provides methods for identification of filler metal components, classification of lots of filler metals, and specification of the testing schedule in procurement documents.

The *AWS A5.0, Filler Metal Comparison Charts* assist in determining the manufacturers that supply filler metals in accordance with the various AWS specifications and the brand names. Conversely, the AWS specification, classification, and manufacturer of a filler metal can be determined from the brand name.

Qualification and Testing

AWS B2.1, Standard for Welding Procedure and Performance Qualification. This standard provides requirements for qualification of welding procedures, welders, and welding operators. It may be referenced in a product code, specification, or contract documents. If a contract document is not specific, certain additional requirements must also be specified, as listed in this standard. Applicable base metals are carbon and alloy steels, cast irons, aluminum, copper, nickel, and titanium alloys.

AWS B2.2, Standard for Brazing Procedure and Performance Qualification. The requirements for qualification of brazing procedures, brazers, and brazing operators for furnace, machine, and automatic brazing are covered by this publication. It is to be used when required by other documents, such as codes, specifications, or contracts. Those documents must specify certain requirements applicable to the production brazement. Applicable base metals are carbon and alloy steels, cast iron, aluminum, copper, nickel, titanium, zirconium, magnesium, and cobalt alloys.

AWS D10.9, Specification for Qualification of Welding Procedures and Welders for Piping and Tubing. This standard applies specifically to qualifications for tubular products. It covers circumferential groove and fillet welds, but excludes welded longitudinal seams involved in pipe and tube manufacture. An organization may make this specification the governing document for qualifying welding procedures and welders by referencing it in the contract and by specifying one of the two levels of acceptance requirements. One level applies to systems that require a high degree of weld quality. Examples are lines in nuclear, chemical, cryogenic, gas, or steam systems. The other level applies to systems requiring an average degree of weld quality, such as low-pressure heating, air-conditioning, sanitary water, and some gas or chemical systems.

AWS C2.16, Guide for Thermal Spray Operator and Equipment Qualification. The guide provides for the qualification of operators and equipment for applying thermal sprayed coatings. It recommends procedural guidelines for qualification testing. The criteria used to judge acceptability are determined by the certifying agent alone or together with the purchaser.

AWS B4.0, Standard Methods for Mechanical Testing of Welds. This document describes the basic mechanical tests used for evaluation of welded joints, weldability, and hot cracking. The tests applicable to welded butt joints are tension, Charpy impact, drop-weight, dynamic-tear, and bend types. Tests for fillet welds are limited to break and shear designs.

For welding materials and procedure qualifications, the most commonly used tests are round-tension; reduced-section tension; face-, root-, and side-bend; and Charpy impact. Fillet weld tests are employed to determine proper welding techniques and conditions, and the shear strength of welded joints for design purposes.

AWS C3.2, Standard Method for Evaluating the Strength of Brazed Joints in Shear. This standard describes a test method used to obtain reliable brazed joint shear strengths. For comparison purposes, specimen preparation, brazing practices, and testing procedures must be consistent. Production brazed joint strength may not be the same as test joint strength if the brazing practices are different. With furnace brazing, for example, the actual part temperature or time at temperature, or both, during production may vary from those used to determine joint strength.

Processes

AWS publishes recommended practices and guides for arc and oxyfuel gas welding and cutting; brazing; resistance welding; and thermal spraying. The processes and applicable documents are listed below.

Arc and Gas Welding and Cutting

Air Carbon-Arc Gouging and Cutting, Recommended Practices for, AWS C5.3

Electrogas Welding, Recommended Practices for, AWS C5.7

Gas Metal Arc Welding, Recommended Practices for, AWS C5.6

Gas Tungsten Arc Welding, Recommended Practices for, AWS C5.5

Oxyfuel Gas Cutting, Operator's Manual for, AWS C4.2

Plasma Arc Cutting, Recommended Practices for, AWS C5.2

Plasma Arc Welding, Recommended Practices for, AWS C5.1

Stud Welding, Recommended Practices for, AWS C5.4

Brazing

Design, Manufacture, and Inspection of Critical Brazed Components, Recommended Practices for, AWS C3.3

Resistance Welding

Resistance Welding, Recommended Practices for, AWS C1.1

Resistance Welding Coated Low Carbon Steels, Recommended Practices for, AWS C1.3

Thermal Spraying

Electric Arc Spraying, Recommended Practices for, AWS C2.17

Flame Spraying of Ceramics, AWS C2.13

Fused Thermal Sprayed Deposits, Recommended Practices for, AWS C2.15

Metallizing with Aluminum and Zinc for-Protection of Iron and Steel, Recommended Practices for, AWS C2.2

Welding Applications

AWS publishes standards that cover various applications of welding. The subjects and appropriate documents are listed below.

Automotive

Automotive Portable Gun Resistance Spot Welding, Recommended Practices for, AWS D8.5

Automotive Welding Design, Recommended Practices for, AWS D8.4

Machinery and Equipment

Earthmoving and Construction Equipment Specification for Welding, AWS D14.3

Industrial and Mill Cranes, Specification for Welding, AWS D14.1

Machinery and Equipment, Classification and Application of Welded Joints for, AWS D14.4

Metal Cutting Machine Tool Weldments, Specification for, AWS D14.2

Presses and Press Components, Specification for Welding of, AWS D14.5

Rotating Elements of Equipment, Specification for, AWS D14.6

Marine

Aluminum Hull Welding, Guide for, AWS D3.7

Steel Hull Welding, Guide for, AWS D3.5

Piping and Tubing

Aluminum and Aluminum Alloy Pipe, Recommended Practices for Gas Shielded Arc Welding of, AWS D10.7

Austenitic Chromium Nickel Stainless Steel Piping and Tubing, Recommended Practices for Welding, AWS D10.4

Chromium-Molybdenum Steel Piping and Tubing, Recommended Practices for Welding of, AWS D10.8

Heat Treatment of Welds in Piping and Tubing, Local, AWS D10.10

Plain Carbon Steel Pipe, Recommended Practices and Procedures for Welding, AWS D10.12

Root Pass Welding and Gas Purging, Recommended Practices for, AWS D10.11

Titanium Piping and Tubing, Gas Tungsten Arc Welding of, AWS D10.6

Sheet Metal

AWS D9.1, Specification for Welding of Sheet Metal covers non-structural fabrication and erection of sheet metal by welding for heating, ventilating, and air-conditioning systems; architectural usages; food-processing equipment; and similar applications. Where differential pressures of more than 120 inches of water (4 psig) or structural requirements are involved, other standards are to be used.

Structural

AWS D1.1, Structural Welding Code—Steel covers welding requirements applicable to welded structures of carbon and low alloy steels. It is to be used in conjunction with any complementary code or specification for the design and construction of steel structures. It is not intended to apply to pressure vessels, pressure piping, or base metals less than 1/8 in. thick. There are sections devoted exclusively to buildings (static loading), bridges (dynamic loading), and tubular structures.

AWS D1.2, Structural Welding Code—Aluminum addresses welding requirements for aluminum alloy structures. It is used in conjunction with appropriate complementary codes or specifications for materials, design, and construction. The structures covered are tubular designs, and static and dynamic non-tubular designs.

AWS D1.3, Specification for Welding Sheet Steel in Structures applies to the arc welding of sheet and strip steel, including cold-formed members, that are 0.18 in. or less in thickness. The welding may involve connections of sheet or strip steel to thicker supporting structural members. When sheet steel is welded to primary structural members, the provisions of *AWS D1.1, Structural Welding Code—Steel* also apply.

AWS D1.4, Structural Welding Code—Reinforcing Steel applies to the welding of concrete reinforcing steel for splices (prestressing steel excepted), steel connection devices, inserts, anchors, anchorage details, and other welding in reinforced concrete construction. Welding may be done in a fabrication shop or in the field. When welding reinforcing steel to primary structural members, the provisions of *AWS D1.1, Structural Welding Code—Steel* also apply.

Safety

ANSI/ASC Z49.1, Safety in Welding and Cutting was developed by the ANSI Accredited Standards Committee Z49, Safety in Welding and Cutting, and then published by AWS. The purpose of the Standard is the protection of persons from injury and illness, and the protection of property from damage by fire and explosions arising from welding, cutting, and allied processes. It specifically covers arc, oxyfuel gas, and resistance welding, and thermal cutting, but the requirements are generally applicable to other welding processes as well. The provisions of this standard are backed by the force of law since they are included in the General Industry Standards of the U.S. Department of Labor, Occupational Safety and Health Administration.

Other safety and health standards published by AWS include the following:

Electron Beam Welding and Cutting, Recommended Safe Practices for, AWS F2.1

Evaluating Contaminants in the Welding Environment, A Sampling Strategy Guide, AWS F1.3

Measuring Fume Generation Rates and Total Fume Emission for Welding and Allied Processes, AWS F1.2

Preparation for Welding and Cutting of Containers and Piping That Have Held Hazardous Substances, Recommended Safe Practices for, AWS F4.1

Sound Level Measurement of Manual Arc Welding and Cutting Processes, AWS F6.1

Thermal Spraying, Recommended Safe Practices for, AWS C2.1

ASSOCIATION OF AMERICAN RAILROADS

Manual of Standards and Recommended Practices

The primary source of welding information relating to the construction of new railway equipment is the *Manual of Standards and Recommended Practices* prepared by the Mechanical Division, Association of American Railroads (AAR). This manual includes specifications, standards, and recommended practices adopted by the Mechanical Division. The sections of the manual that relate to welding are summarized below.

Section C, Part II, Specifications For Design, Fabrication, And Construction of Freight Cars. This specification covers the general welding practices for freight car construction. Welding processes and procedures other than those listed in the document may be used. However, they must conform to established welding standards or proprietary carbuilder's specifications, and produce welds of quality consistent with design requirements and good manufacturing techniques. The welding requirements are similar to, though not so detailed as those in *AWS D1.1, Structural Welding Code—Steel.* In fact, the qualification of welders and welding operators must be done in accordance with the AWS Code.

Section C, Part III, Specification for Tank Cars. This specification covers the construction of railroad car tanks used for the transportation of hazardous and nonhazardous materials. The requirements for fusion welding of the tanks, and for qualifying welders and welding procedures to be used are described in one appendix. A second appendix describes the requirements for repairs, alterations, or conversions of car tanks. If welding is required, it must be performed by facilities certified by AAR in accordance with a third appendix.

The U.S. Department of Transportation issues regulations covering the transportation of explosives, radioactive materials, and other dangerous articles. Requirements for tank cars are set forth in the United States Code of Federal Regulations, Title 49, Sections 173.314, 173.316, and 179, which are included at the end of the AAR specifications.

Section D, Trucks And Truck Details. The procedures, workmanship, and qualification of welders employed in the fabrication of steel railroad truck frames are required to be in accordance with (1) the latest recommendations of the American Welding Society, (2) the *Specifications for Design, Fabrication, and Construction of Freight Cars,* and (3) the welding requirements of the *Specifications for Tank Cars.*

Field Manual of Association of American Railroads Interchange Rules

This manual covers the repair of existing railway equipment. The U.S. railway network is made up of numerous interconnecting systems, and it is often necessary for one system to make repairs on equipment of another system. The repair methods are detailed and specific so that they may be used as the basis for standard charges between the various railway companies.

CANADIAN STANDARDS ASSOCIATION

The Canadian Standards Association (CSA) is a voluntary membership organization engaged in standards development and also testing and certification. It is similar to ANSI in the United States. A CSA Certification Mark assures buyers that a product conforms to acceptable standards.

Examples of CSA welding documents are the following:

Aluminum Welding Qualification Code, CSA W47.2

Certification of Companies for Fusion Welding of Steel Structures, CSA W47.1

Code for Safety in Welding and Cutting (Requirements for Welding Operators), CSA W117.2

Qualification Code for Welding Inspection Organizations, CSA W178

Resistance Welding Qualification Code for Fabricators of Structural Members Used in Buildings, CSA W55.3

Welded Aluminum Design and Workmanship (Inert Gas Shielded Arc Processes), CSA S244

Welded Steel Construction (Metal Arc Welding), CSA W59

Welding Electrodes, CSA W48 Series

Welding of Reinforcing Bars in Reinforced Concrete Construction, CSA W186

COMPRESSED GAS ASSOCIATION

The purpose of the Compressed Gas Association (CGA) is to promote, develop, represent, and coordinate technical and standardization activities in the compressed gas industries, including end uses of products, in the interest of efficiency and public safety.

The Handbook of Compressed Gases, published by CGA, is a source of basic information about compressed gases, their transportation, uses, and safety considerations, and also the rules and regulations pertaining to them.

CGA C-3, Standards for Welding and Brazing on Thin Walled Containers is directly related to the use of welding and brazing in the manufacture of DOT compressed gas cylinders. It covers procedure and operator qualification, inspection, and container repair.

The following CGA publications contain information on the properties, manufacture, transportation, storage, handling, and use of gases commonly used in welding operations:

G-1, Acetylene

G-1.1, Commodity Specification for Acetylene

G-4, Oxygen

G-4.3, Commodity Specification for Oxygen

G-5, Hydrogen

G-5.3, Commodity Specification for Hydrogen

G-6, Carbon Dioxide

G-6.2, Commodity Specification for Carbon Dioxide

G-9.1, Commodity Specification for Helium

G-10.1, Commodity Specification for Nitrogen

G-11.1, Commodity Specification for Argon

P-9, The Inert Gases Argon, Nitrogen, and Helium

Safety considerations related to the gases commonly used in welding operations are discussed in the following CGA pamphlets:

P-1, Safe Handling of Compressed Gases in Containers

SB-2, Oxygen-Deficient Atmospheres

SB-4, Handling Acetylene Cylinders in Fire Situations

FEDERAL GOVERNMENT

Several departments of the Federal Government, including the General Services Administration, are responsible for developing welding standards or adopting existing welding standards, or both.

Consensus Standards

The U.S. Departments of Labor, Transportation, and Energy are primarily concerned with adopting existing national consensus standards, but they also make amendments to these standards or create separate standards, as necessary. For example, the Occupational Safety and Health Administration (OSHA) of the Department of Labor issues regulations covering occuptional safety and health protection. The welding portions of standards adopted or established by OSHA are published under Title 29 of the *United States Code of Federal Regulations.* Part 1910 covers general industry, while Part 1926 covers the construction industry. These regulations were derived primarily from national consensus standards of ANSI and of the National Fire Protection Association (NFPA).

Similarly, the U.S. Department of Transportation is responsible for regulating the transportation of hazardous materials, petroleum, and petroleum products by pipeline in interstate commerce. Its rules are published under Title 49 of the *United States Code of Federal Regulations,* Part 195. Typical of the many national consensus standards incorporated by reference in these regulations are API Standard 1104 and ASME B31.4, which were discussed previously.

The U.S. Department of Transportation is also responsible for regulating merchant ships of American registry. It is empowered to control the design, fabrication, and inspection of these ships by Title 46 of the *United States Code of Federal Regulations.*

The U.S. Coast Guard is responsible for performing the inspections of merchant ships. Its *Marine Engineering Regulations* incorporate references to national consensus standards, such as those published by ASME, ANSI, and ASTM. These rules cover repairs and alterations that must be performed with the cognizance of the local Coast Guard Marine Inspection Officer.

The U.S. Department of Energy is responsible for the development and use of standards by government and industry for the design, construction, and operation of safe, reliable, and economic nuclear energy facilities. National consensus standards, such as the *ASME Boiler and Pressure Vessel Code, Sections III and IX,* and *AWS D1.1, Structural Welding Code—Steel* are referred to in full or in part. These standards are supplemented by separate program standards, known as *RDT Standards.*

Military and Federal Specifications

Military specifications are prepared by the Department of Defense. They cover materials, products, or services specifically for military use, and commercial items modified to meet military requirements.

Military specifications have document designations beginning with the prefix *MIL.* They are issued as either coordinated or limited-coordination documents. Coordinated documents cover items or services required by more than one branch of the military. Limited-coordination documents cover items or services of interest to a single branch. If a document is of limited coordination, the branch of the military which uses the document will appear in parentheses in the document designation.

Two current military specifications cover the qualification of welding procedures or welder performance, or both. One is *MIL-STD-1595, Qualification of Aircraft, Missile, and Aerospace Fusion Welders.* The other, *MIL-STD-248, Welding and Brazing Procedure and Performance Qualification,* covers the requirements for the qualification of welding and brazing procedures; welders; brazers; and welding and brazing operators. It allows the fabricator to submit for approval certified records of qualification tests prepared in conformance with the standards of other government agencies, ABS, ASME, or other organizations. Its use is mandatory when referenced by other specifications or contractural documents.

MIL-STD-1595 establishes the procedure for qualifying welders and welding operators engaged in the fabrication of components for aircraft, missiles, and other aerospace equipment by fusion welding processes. This standard is applicable when required in the contracting documents, or when invoked in the absence of a specified welder qualification document.

MIL-STD-1595 superseded *MIL-T-5021, Tests; Aircraft and Missile Welding Operator's Qualification* which is obsolete. However, MIL-T-5021 is still referenced by other current government specifications and contract documents. When so referenced, a contractor has to perform the technically obsolete tests required by this Standard.

Federal Specifications are developed for materials, products, and services that are used by two or more Federal agencies, one of which is not a Defense agency. Federal Specifications are classified into broad categories. The *QQ* group, for example, covers metals and most welding specifications. Soldering and brazing fluxes are in the *O-F* group.

Some military and federal specifications include requirements for testing and approval of a material, process, or piece of equipment before its submission for use under the specification. In such cases, the testing procedure and acceptance criteria are given in the specification. If the acceptance tests pass the specification requirements, the material or equipment will be included in the applicable *Qualified Products List (QPL).*

In other specifications, the supplier is responsible for product conformance. This is often the case for welded fabrications. The supplier must show evidence that the welding procedures and the welders are qualified in accordance with the requirements of the specification, and must certify the test report.

The following Military and Federal Standards (currently listed in the Department of Defense Index) address welding, brazing, and soldering. Those standards covering base metals and welding equipment are not included.

Braze-Welding, Oxyacetylene, of Built-Up Metal Structures, MIL-B-12672

Brazing Alloy, Copper, Copper-Zinc, and Copper-Phosphorus, QQ-B-650

Brazing Alloy, Gold, QQ-B-653

Brazing Alloy, Silver, QQ-B-654

Brazing Alloy, 82 Gold-18 Nickel, Wire, Foil, Sheet and Strip, MIL-B-47043

Brazing Alloys, Aluminum and Magnesium, Filler Metal, QQ-B-655

Brazing and Annealing of Electromagnetic (Iron-Cobalt Alloy) Poles to Austenitic Stainless Steel, Process for, MIL-B-47291

Brazing of Steels, Copper, Copper Alloys, Nickel Alloys, Aluminum, and Aluminum Alloys, MIL-B-78838

Brazing Sheet, Aluminum Alloy, MIL-B-20148

Brazing, Aluminum, Process for, MIL-B-47292

Brazing, Nickel Alloy, General Specification for, MIL-B-9972

Brazing, Oxyacetylene, of Built-Up Metal Structures, MIL-B-12673

Electrode, Underwater Cutting, Tubular, Ceramic, MS-16857

Electrode, Welding Covered (Austenitic Chromium-Nickel Steel, for Corrosive and High Temperature Services), MIL-E-22200/2

Electrode, Welding, Bare, Aluminum Alloys, MIL-E-16053

Electrode, Welding, Bare, Copper and Copper Alloy, MIL-E-23765/3

Electrode, Welding, Bare, High Yield Steel, MIL-E-19822

Electrode, Welding, Bare, Solid, Nickel-Manganese-Chromium-Molybdenum Alloy Steel for Producing HY-130 Weldments for As-Welded Applications, MIL-E-24355

Electrode, Welding, Carbon Steel, and Alloy Steel, Bare, Coiled, MIL-E-18193

Electrode, Welding, Copper, Silicon-Deoxidized, Solid, Bare, MIL-E-45829

Electrode, Welding, Covered, Aluminum Bronze, MIL-E-278

Electrode, Welding, Covered, Austenitic Steel (19-9 Modified) for Armor Applications, MIL-E-13080

Electrode, Welding, Covered, Bronze, for General Use, MIL-E-13191

Electrode, Welding, Covered, Coated, Aluminum and Aluminum Alloy, MIL-E-15597

Electrode, Welding, Covered, Copper-Nickel Alloy, MIL-E-22200/4

Electrode, Welding, Covered, Low-Hydrogen, and Iron Powdered Low-Hydrogen, Chromium-Molybdenum Alloy Steel and Corrosion-Resisting Steel, MIL-E-22200/8

Electrode, Welding, Covered, Low Alloy Steel (Primarily for Aircraft and Weapons), MIL-E-6843

Electrode, Welding, Covered, Low-Hydrogen, Heat-Treatable Steel, MIL-E-8697

Electrode, Welding, Covered, Mild Steel,

QQ-E-450

Electrode, Welding, Covered, Molybdenum Alloy Steel Application, MIL-E-22200/7

Electrode, Welding, Covered, Nickel Base Alloy, and Cobalt Base Alloy, MIL-E-22200/3

Electrode, Welding, Mineral Covered, Iron-Powder, Low-Hydrogen Medium and High Tensile Steel, As-Welded or Stress-Relieved Weld Application, MIL-E-22200/1

Electrode, Welding, Mineral Covered, Iron-Powder, Low-Hydrogen, High Tensile Low Alloy Steel-Heat-Treatable Only, MIL-E-22200/5

Electrode, Welding, Mineral Covered, Low-Hydrogen, Chromium-Molybdenum Alloy Steel and Corrosion Resisting Steel, MIL-E-16589

Electrode, Welding, Mineral Covered, Low-Hydrogen, Medium and High Tensile Steel, MIL-E-22200/6

Electrode, Welding, Mineral Covered, Low-Hydrogen or Iron-Powder, Low-Hydrogen, Nickel-Manganese-Chromium-Molybdenum Alloy Steel for Producing HY-130 Weldments for As Welded Application, MIL-E-22200/9

Electrode, Welding, Surfacing, Iron Base Alloy, MIL-E-19141

Electrodes (Bare) and Fluxes (Granular), Submerged Arc Welding, High-Yield Low Alloy Steels, MIL-E-22749

Electrodes and Rods - Welding, Bare, Chromium and Chromium-Nickel Steels, MIL-E-19933

Electrodes and Rods - Welding, Bare, Solid, General Specification for, MIL-E-23765

Electrodes and Rods - Welding, Bare, Solid, Mild and Alloy Steel, MIL-E-23765/1

Electrodes and Rods - Welding, Bare, Solid, Low Alloy Steel, MIL-E-23765/2

Electrodes, Cutting and Welding, Carbon-Graphite, Uncoated and Copper-Coated, MIL-E-17777

Electrodes, Welding, Covered, General Specification for, MIL-E-22200

Electrodes, Welding, Mineral Covered, Iron-Powder, Low-Hydrogen-8 Nickel-Chromium-Molybdenum-Vanadium Alloy Steel for Producing HY-130 Weldments to be Heat Treated, MIL-E-22200/11

Electrodes, Welding, Mineral Covered, Low-Hydrogen, Iron-Powder for Producing HY-100 Steel Weldments for As-Welded Applications, MIL-E-22200/10

Fabrication Welding and Inspection of Hyperbaric Chambers and Other Critical Land Based Structures, MIL-STD-1693

Fabrication Welding and Inspection, and Casting Inspection and Repair for Machinery, Piping and Pressure Vessels In Ships of the United States Navy, MIL-STD-278

Fabrication, Welding, and Inspection of HY-130 Submarine Hull, MIL-STD-1681

Flux, Aluminum and Aluminum Alloy, Gas Welding, MIL-F-6939

Flux, Brazing, Silver Alloy, Low Melting Point, O-F-499

Flux, Galvanizing, MIL-F-19197

Flux, Soldering (Stearine Compound 1C-3), MIL-F-12784

Flux, Soldering, Liquid (Rosin Base), MIL-F-14256

Flux, Soldering, Paste and Liquid, O-F-506

Flux, Soldering, Rosin Base, General Purpose, MIL-F-20329

Flux, Welding (For Copper-Base and Copper-Nickel Alloys and Cast Iron), MIL-F-16136

Fluxes, Welding (Compositions), Submerged Arc Process With Type B, Electrodes, Carbon and Low Alloy Steel Application, MIL-F-19922

Fluxes, Welding, Submerged Arc Process Carbon and Low Alloy Steel Application, MIL-F-18251

Rod, Welding, Copper and Copper Alloy, MIL-R-19631

Rod, Welding, Copper and Nickel Alloy, QQ-R-571

Rod, Welding, High Strength, MIL-R-47191

Solder Bath Soldering of Printed Wiring Assemblies, MIL-S-46844

Solder, Lead-Tin Alloy, MIL-S-12204

Solder, Low-Melting Point, MIL-S-627

Solder, Tin Alloy, Lead-Tin Alloy and Lead Alloy, QQ-S-571

Soldering of Electrical Connections and Printed Wiring Assemblies, Procedure for, MIL-STD-1460

Soldering of Metallic Ribbon Lead Materials to Solder Coated Conductors, Process for Reflow, MIL-S-46880

Soldering Process, General Specification for, MIL-S-6872

Soldering, Manual Type, High Reliability, Electrical and Electronic Equipment, MIL-S-45743

Welded Joint Design, MIL-STD-22

Welded Joint Designs, Armored-Tank Type, MIL-STD-21

Welder Performance Qualification, Aerospace, MIL-STD-1595

Welding and Brazing Procedure and Performance Qualification, MIL-STD-248

Welding of Aluminum Alloys, Process for, MIL-W-8604

Welding of Armor, Metal-Arc, Manual, With Austenitic Electrodes, for Aircraft, MIL-W-41

Welding of Electronic Circuitry, Process for, MIL-W-46870

Welding of Homogeneous Armor by Metal Arc Processes, MIL-W-46086 (MR)

Welding of Magnesium Alloys, Gas and Arc, Manual and Machine Processes for, MIL-W-18326

Welding Procedures for Constructional Steels, MIL-STD-1261 (MR)

Welding Process and Welding Procedure Requirements for Manufacture of Equipment Utilizing Steels, MIL-W-52574

Welding Rod and Wire, Nickel Alloy High Permeability, Shielding Grade, MIL-W-47192

Welding Rods and Electrodes, Preparation for Delivery of, MIL-W-10430

Welding Symbols (ABCA-323), Q-STD-323

Welding Terms and Definitions (ABCA-324), Q-STD-324

Welding, Aluminum Alloy Armor, MIL-W-45206

Welding, Arc and Gas, for Fabricating Ground Equipment for Rockets and Guided Missiles, MIL-W-47091

Welding, Flash, Carbon and Alloy Steel, MIL-W-6873

Welding, Flash, Standard Low Carbon Steel, MIL-W-62160

Welding, Fusion, Electron Beam, Process for, MIL-W-46132

Welding, Gas Metal-Arc and Gas Tungsten-Arc, Aluminum Alloys, Readily Weldable for Structures, Excluding Armor, MIL-W-45205

Welding, Gas, Steels, Constructional, Readily Weldable, for Low Stressed Joints, MIL-STD-1183

Welding, High Hardness Armor, MIL-STD-1185

Welding, Metal Arc and Gas, Steels, and Corrosion and Heat Resistant Alloys, Process for, MIL-W-8611

Welding, Repair, of Readily Weldable Steel Castings (Other Than Armor) Metal-Arc, Manual, MIL-W-13773

Welding, Resistance, Electronic Circuit

Modules (Asg), MIL-W-8939

Welding, Resistance, Spot and Projection for Fabricating Assemblies of Carbon Steel Sheets, MIL-W-46154

Welding, Resistance, Spot and Seam, MIL-W-6858

Welding, Resistance, Spot, Seam, and Projection, for Fabricating Assemblies of Low-Carbon Steel, MIL-W-12332

Welding, Resistance, Spot, Weldable Aluminum Alloys, MIL-W-45210

Welding, Spot, Hardenable Steels, MIL-W-45223

Welding, Spot, Inert-Gas Shielded Arc, MIL-W-27664

Welding, Stud, Aluminum, MIL-W-45211

Weldment, Aluminum and Aluminum Alloy, MIL-W-22248

Weldment, Steel, Carbon and Low Alloy (Yield Strength 30,000-60,000 psi), MIL-W-21157

INTERNATIONAL ORGANIZATION FOR STANDARDIZATION

The International Organization for Standardization (ISO) promotes the development of standards to facilitate the international exchange of goods and services. It is comprised of the standards-writing bodies of more than 80 countries, and has adopted or developed over 4000 standards. Several of these standards are related to welding. The following are typical:

ISO 2401-1972, Covered electrodes - Determination of the efficiency, metal recovery and deposition coefficient

ISO 3041-1975, Welding requirements - Categories of service requirements for welded joints

ISO 3088-1975, Welding requirements - Factors to be considered in specifying requirements for fusion welded joints in steel (technical influencing factors)

ISO 3834-1978, Welding - Factors to be considered when assessing firms using welding as a prime means of fabrication

ISO 6520-1982, Classification of imperfections in metallic fusion welds, with explanations

The American National Standards Institute is the designated U.S. representative to ISO. ISO standards and publications are available from ANSI.

NATIONAL BOARD OF BOILER AND PRESSURE VESSEL INSPECTORS

The National Board of Boiler and Pressure Vessel Inspectors (NBBPVI), often referred to as the *National Board,* represents the enforcement agencies empowered to assure adherence to the *ASME Boiler and Pressure Vessel Code.* Its members are the chief inspectors or other jurisdictional authorities who administer the boiler and pressure vessel safety laws in the various jurisdictions of the United States and provinces of Canada.

The National Board is involved in the inspection of new boilers and pressure vessels. It maintains a registration system for use by manufacturers who desire or are required by law to register the boilers or pressure vessels that they have constructed. The National Board is also responsible for investigating possible violations of the *ASME Boiler and Pressure Vessel Code* by either commissioned inspectors or manufacturers.

The National Board publishes a number of pamphlets and forms concerning the manufacture and inspection of boilers, pressure vessels, and safety valves. It also publishes the *National Board Inspection Code* for the guidance of its members, commissioned inspectors, and others. The purpose of this code is to maintain the integrity of boilers and pressure vessels after they have been placed in service by providing rules and guidelines for inspection after installation, repair, alteration, or rerating. In addition, it provides inspection guidelines for authorized inspectors during fabrication of boilers and pressure vessels.

In some states, organizations that desire to repair boilers and pressure vessels, must obtain the National Board Repair (*R*) stamp by application to the National Board. The firm must qualify all welding procedures and welders in accordance with the *ASME Boiler and Pressure Vessel Code, Section IX,* and the results must be accepted by the inspection agency. The firm must also have and demonstrate a quality control system similar to, but not so comprehensive as that required for an ASME code symbol stamp.

NATIONAL FIRE PROTECTION ASSOCIATION

The mission of the National Fire Protection Association (NFPA) is the safeguarding of man and his environment from destructive fire through the use of scientific and engineering techniques and education. NFPA standards are widely used as the basis of legislation and regulation at all levels of government. Many are referenced in the regulations of the Occupational Safety and Health Administration (OSHA). The standards are also used by insurance authorities for risk evaluation and premium rating.

Installation of Gas Systems

NFPA publishes several standards that present general principles for the installation of gas supply systems and the storage and handling of gases commonly used in welding and cutting. These are:

NFPA 50, Bulk Oxygen Systems at Consumer Sites

NFPA 50A, Gaseous Hydrogen Systems at Consumer Sites

NFPA 51, Design and Installation of Oxygen-Fuel Gas Systems for Welding and Cutting and Allied Processes

NFPA 54, National Fuel Gas Code

NFPA 58, Storage and Handling of Liquefied Petroleum Gases

Users should check each standard to see if it applies to their particular situation. For example, NFPA 51 does not apply to a system comprised of a torch, regulators, hoses, and single cylinders of oxygen and fuel gas. Such a system is covered by *ANSI/ASC Z49.1, Safety in Welding and Cutting.*

Safety

NFPA publishes several standards which relate to the safe use of welding and cutting processes. These are:

NFPA 51B, Fire Prevention in Use of Cutting and Welding Processes

NFPA 91, Installation of Blower and Exhaust Systems for Dust, Stock, and Vapor Removal or Conveying

NFPA 306, Control of Gas Hazards on Vessels to be Repaired

NFPA 327, Cleaning Small Tanks and Containers

NFPA 410, Standard on Aircraft Maintenance

Again, the user should check the standards to determine those that apply to the particular situation.

PIPE FABRICATION INSTITUTE

The Pipe Fabrication Institute (PFI) publishes numerous documents for use by the piping industry. Some of the standards have mandatory status because they are referenced in one or more piping codes. The purpose of PFI standards is to promote uniformity of piping fabrication in areas not specifically covered by codes. Other PFI documents, such as technical bulletins, are not mandatory, but they aid the piping fabricator in meeting the requirements of codes. The following PFI standards relate directly to welding:

ES1, End Preparation and Machined Backing Rings for Butt Welds

ES7, Minimum Length and Spacing for Welded Nozzles

ES19, Preheat and Postheat Treatment of Welds

ES21, Manual Gas Tungsten Arc Root Pass Welding End Preparation and Dimensional Joint and Fit Up Tolerances

ES26, Welded Load Bearing Attachments to Pressure Retaining Piping Materials

ES27, Visual Examination - The Purpose, Meaning, and Limitation of the Term

ES28, Recommended Practice for Welding of Transition Joints Between Dissimilar Steel Combinations

SOCIETY OF AUTOMOTIVE ENGINEERS

The Society of Automotive Engineers (SAE) is concerned with the research, development, design, manufacture, and operation of all types of self-propelled machinery. Such machinery includes automobiles, trucks, buses, farm machines, construction equipment, airplanes, helicopters, and space vehicles. Related areas of interest to SAE are fuels, lubricants, and engineering materials.

Automotive Standards

Several SAE welding-related automotive standards are written in cooperation with AWS. These are:

HS J1156, Automotive Resistance Spot Welding Electrodes, Standard for, (AWS D8.6)

HS J1188, Automotive Weld Quality-Resistance Spot Welding, Specification for, (AWS D8.7)

HS J1196, Automotive Frame Weld Quality-Arc Welding, Specification for, (AWS D8.8)

Aerospace Material Specifications

Material specifications are published by SAE for use by the aerospace industry. The Aerospace Material Specifications (AMS) cover fabricated parts, tolerances, quality control procedures, and processes. Welding-related AMS specifications are listed below. The appropriate AWS filler metal classification or a common trade name follows some of the specifications, in parentheses, for clarification.

Processes

2664 *Brazing—Silver, for Use up to 800° F (425° C)*

2665 *Brazing—for Use Up to 400° F (205° C)*

2666 *Brazing—Silver, High Temperature*

2667 *Brazing—Silver, For Flexible Metal Hose—600° F (315° C) Max Operating Temperature*

2668 *Brazing—Silver, For Flexible Metal Hose—400° F (200° C) Max Operating Temperature*

2669 *Brazing-Silver, For Flexible Metal Hose—800° F (425° C) Max Operating Temperature*

2670 *Brazing—Copper Furnace, Carbon and Low Alloy Steels*

2671 *Brazing—Copper Furnace, Corrosion and Heat Resistant Steels and Alloys*

2672 *Brazing—Aluminum*

2673 *Brazing—Aluminum Molten Flux (Dip)*

2675 *Brazing—Nickel Alloy*

2680 *Electron Beam Welding for Fatigue Critical Applications*

2681 *Electron Beam Welding*

2685 *Welding, Metal Arc, Inert Gas, Nonconsumable Electrode (GTAW Method)*

2689 *Fusion Welding, Titanium and Titanium Alloys*

2690 *Welding (Parallel Gap) of Microelectric Interconnections to Thin Film Substrates*

2694 *Repair Welding of Aerospace Castings*

Flux

3410 *Flux—Brazing, Silver*

3411 *Flux—Brazing, Silver, High Temperature*

3412 *Flux—Brazing, Aluminum*

3414 *Flux—Welding, Aluminum*

3415 *Flux—Aluminum Dip Brazing, 1030° F Fusion Point*

3416 *Flux—Aluminum Dip Brazing, 1090° F Fusion Point*

3430 Paste, Copper Brazing—Water Thinning

Aluminum Alloys

4184 Wire, Brazing—10Si 4Cu (4145)

4185 Wire, Brazing—12Si (4047)

4188 Welding Wire

4188/1 Welding Wire, 4.5Cu 0.70Ag 0.30Mn 0.25Mg 0.25Ti (A201.0)

4188/2 Welding Wire, 4.6Cu 0.35Mn 0.25Mg 0.22Ti (A206.0)

4188/3 Welding Wire, 5.0Si 1.2Cu 0.50Mg (C355.0)

4188/4 Welding Wire, 7.0Si 0.30Mg (A356.0)

4188/5 Welding Wire, 7.0Si 0.52Mg (357.0)

4189 Wire, Welding—4.1Si 0.2Mg (4643)

4190 Wire, Welding—5.2Si (4043)

4191 Rod and Wire, Welding—6.3Cu 0.3Mn 0.18Zr 0.15Ti 0.10V (ER 2319)

Magnesium Alloys

4395 Wire, Welding—9Al 2Zn (ER AZ92A)

4396 Wire, Welding—3.3Ce 2.5Zn 0.7Zr (ER EZ33A)

Brazing and Soldering Filler Metals

4750 Solder—Tin-Lead 45Sn 55Pb

4751 Solder—Tin-Lead, Eutectic, 63Sn 37Pb

4755 Solder—Lead-Silver, 94Pb 5.5Ag

4756 Solder—97.5Pb 1.5Ag 1Sn

4764 Brazing Filler Metal—Copper, 52.5Cu 38Mn 9.5Ni, 1615°-1700° F (880°-925° C) Solidus-Liquidus Range

4765 Brazing Filler Metal, Silver—56Ag 42Cu 2.0Ni, 1420°-1640° F (770°-895° C) Solidus-Liquidus Range (BAg-13a)

4766 Brazing Filler Metal, Silver-85Ag 15Mn, 1760°-1780° F (960°-970° C)

4767 Brazing Filler Metal, Silver—92.5Ag 7.2Cu 0.22Li, 1435°-1635° F (780°-890° C) Solidus-Liquidus Range (BAg-19)

4768 Brazing Filler Metal, Silver—35Ag 26Cu 21Zn 18Cd, 1125°-1295° F (605°-700° C) Solidus-Liquidus Range (BAg-2)

4769 Brazing Filler Metal—Silver, 45Ag 24Cd 16Zn 15Cu, 1125°-1145° F (605°-620° C) Solidus-Liquidus Range (BAg-1)

4770 Brazing Filler Metal—Silver, 50Ag 18Cd 16.5Zn 15.5Cu, 1160°-1175° F (625°-635° C) Solidus-Liquidus Range (BAg-1A)

4771 Brazing Filler Metal—Silver, 50Ag 16Cd 15.5Zn 15.5Cu 3.0Ni, 1125°-1295° F (630°-690° C) Solidus-Liquidus Range (BAg-3)

4772 Brazing Filler Metal—Silver, 54Ag 40Cu 5.0Zn 1.0Ni, 1325°-1575° F (720°-855° C) Solidus-Liquidus Range (BAg-13)

4773 Brazing Filler Metal, Silver, 60Ag 30Cu 10Sn, 1115°-1325° F (600°-720° C) Solidus-Liquidus Range (BAg-18)

4774 Brazing Filler Metal, Silver, 63Ag 28.5Cu 6.0Sn 2.5Ni, 1275°-1475° F (690°-800° C) Solidus-Liquidus Range (BAg-21)

4775 Brazing Filler Metal, Nickel, 73Ni 4.5Si 14Cr 3.1B 4.5Fe, 1790°-1970° F (975°-1075° C) Solidus-Liquidus Range (BNi-1)

4776 Brazing Filler Metal, Nickel, 73Ni 4.5Si 14Cr 3.1B 4.5Fe, (Low Car-

bon) 1790°-1970° F (975°-1075° C) Solidus-Liquidus Range (BNi-1A)

4777 *Brazing Filler Metal, Nickel, 82Ni 4.5Si 7.0Cr 3.1B 3.0Fe, 1780°-1830° F (970°-1000° C) Solidus-Liquidus Range (BNi-2)*

4778 *Brazing Filler Metal, Nickel, 92Ni 4.5Si 3.1B, 1800°-1900° F (980°-1040° C) Solidus-Liquidus Range (BNi-3)*

4779 *Brazing Filler Metal, Nickel, 94Ni 3.5Si 1.8B, 1800°-1950° F (980°-1065° C) Solidus-Liquidus Range (BNi-4)*

4780 *Brazing Filler Metal, Manganese, 66Mn 16Ni 16Co 0.80B, 1770°-1875° F (965°-1025° C) Solidus-Liquidus Range*

4782 *Brazing Filler Metal, Nickel, 71Ni 10Si 19Cr, 1975°-2075° F (1080°-1135° C) Solidus-Liquidus Range (BNi-5)*

4783 *Brazing Filler Metal, High Temperature, 50Co 8.0Si 19Cr 17Ni 4.0W 0.80B, 2050°-2100° F (1120°-1150° C) Solidus-Liquidus Range (BCo-1)*

4784 *Brazing Filler Metal, High Temperature, 50Au 25Pd 25Ni, 2015°-2050° F (1100°-1120° C) Solidus-Liquidus Range*

4785 *Brazing Filler Metal, High Temperature, 30Au 34Pd 36Ni, 2075°-2130° F (1135°-1165° C) Solidus-Liquidus Range (BAu-5)*

4786 *Brazing Filler Metal, High Temperature, 70Au 8Pd 22Ni, 1845°-1915° F (1005°-1045° C) Solidus-Liquidus Range*

4787 *Brazing Filler Metal, High Temperature, 82Au 18Ni, 1740° F (950° C) Solidus-Liquidus Temperature (BAu-4)*

Titanium Alloys

4951 *Wire, Welding*

4953 *Wire, Welding—5Al-2.5Sn*

4954 *Wire, Welding—6Al-4V*

4955 *Wire, Welding—8Al-1Mo-1V*

4956 *Wire, Welding—6Al-4V, Extra Low Interstitial, Environment Controlled*

Carbon Steels

5027 *Wire, Welding—1.05Cr 0.55Ni 1.0Mo 0.07V (0.26-0.32C), Vacuum Melted, Environment Controlled, Packaged (D6AC)*

5028 *Wire, Welding—1.05Cr 0.55Ni 1.0Mo 0.07V (0.34-0.40C), Vacuum Melted, Environment Controlled, Packaged (D6AC)*

5029 *Wire, Welding—0.78Cr 1.8Ni 0.35Mo 0.20V (0.33-0.38C), Vacuum Melted, Environment Controlled, Packaged*

5030 *Wire, Welding—Low Carbon*

5031 *Welding Electrodes, Covered, Steel—0.07-0.15C (E6013)*

Corrosion and Heat Resistant Steels and Alloys

5675 *Wire, Welding—70Ni 15.5Cr 7Fe 3.0Ti 2.4Mn*

5676 *Wire, Welding—80Ni 20Cr*

5677 *Electrodes, Covered Welding—75Ni 19.5Cr 1.6 (Cb+Ta)*

5679 *Wire, Welding—73Ni 15.5Cr 8Fe 2.2 (Cb+Ta)*

5680 *Wire, Welding—18.5Cr 11Ni 0.40 (Cb+Ta)*

5681 *Electrodes, Covered Welding—19.5Cr 10.5Ni 0.60 (Cb+Ta) (E347)*

5683 *Wire, Welding—75Ni 15.5Cr 8Fe*

5684 *Electrodes, Covered, Alloy, Welding—72Ni 15Cr 9Fe 2.8 (Cb+Ta) (ENiCrFe-1)*

5691 *Electrodes, Covered, Steel, Welding—18Cr 12.5Ni 2.2Mo (E316)*

5694 *Wire, Welding—27Cr 21.5Ni (ER310)*

5695 Electrodes, Covered Welding—25Cr 20Ni (E310)

5774 Wire, Welding—16.5Cr 4.5Ni 2.9Mo 0.1N (AM350)

5775 Electrodes, Welding, Covered—16.5Cr 4.5Ni 2.9Mo 0.1N (AM-350)

5776 Wire, Welding—12.5Cr (410)

5777 Electrodes, Coated Welding Steel—12.5Cr (E410)

5778 Wire, Welding—72Ni 15.5Cr 2.4Ti 1 (Cb+Ta) 0.7Al 7Fe

5779 Electrodes, Covered Welding—75Ni 15Cr 1.5 (Cb+Ta) 1.9Ti 0.55Al 5.5Fe

5780 Wire, Welding—15.5Cr 4.5Ni 2.9Mo 0.10N (AM355)

5781 Electrodes, Welding Covered—15.5Cr 4.5Ni 2.9Mo 0.1N (AM-355)

5782 Wire, Welding—20.5Cr 9.0Ni 0.50Mo 1.5W 1.2 (Cb+Ta) 0.20Ti (19-9W/Mo)

5783 Electrodes, Covered Welding—19.5Cr 8.8Ni 0.50Mo 1.5W 1.0 (Cb+Ta) (E349)

5784 Wire Welding, Corrosion and Heat Resistant 29Cr 9.5Ni (29-9)

5785 Electrodes, Covered Welding—28.5Cr 9.5Ni (E312)

5786 Wire, Welding—62.5Ni 5.0Cr 24.5Mo 5.5Fe (Hastelloy W)

5787 Electrodes, Covered Welding—63Ni 5Cr 24.5Mo 5.5Fe (ENiMo-3)

5789 Wire, Welding—54Co 25.5Cr 10.5Ni 7.5W (Stellite 31)

5794 Wire, Welding—31Fe 20Cr 20Ni 20Co 3Mo 2.5W 1 (Cb+Ta) (N-135)

5795 Electrodes, Covered, Welding—31Fe 20Cr 20Ni 20Co 3.0Mo 2.5W 1.0 (Cb+Ta) (N-155)

5796 Wire, Welding, Alloy—52Co 20Cr 10Ni 15W (L-605)

5797 Electrodes, Covered, Welding—51.5Co 20Cr 10Ni 15W (L-605)

5798 Wire, Welding, Alloy—47.5Ni 22Cr 1.5Co 9.0Mo 0.60W 18.5Fe (Hastelloy X)

5799 Electrodes, Covered, Welding—48Ni 22Cr 1.50Co 9.0Mo 0.60W 18.5Fe (ENiCr-Mo-2)

5800 Wire, Welding—54Ni 19Cr 11Co 10Mo 3.2Ti 1.5Al 0.006B Vacuum Melted (Rene 41)

5801 Wire, Welding—39Co 22Cr 22Ni 14.5W 0.07La (Haynes 188)

5804 Wire, Welding—15Cr 25.5Ni 1.3Mo 2.2Ti 0.006B 0.30V (A-286)

5805 Wire, Welding—15Cr 25.5Ni 1.3Mo 2.2Ti 0.006B 0.30V Vacuum Induction Melted (A-286)

5812 Wire, Welding—15Cr 7.1Ni 2.4-Mo 1Al Vacuum Melted (WPH15-7Mo-VM)

5817 Wire, Welding—13Cr 2Ni 3W (Greek Ascoloy)

5821 Wire, Welding—12.5Cr, Ferrite Control Grade (410 Mod)

5823 Wire, Welding—11.8Cr 2.8Ni 1.6Co 1.8Mo 0.32V

5824 Wire, Welding—17Cr 7.1Ni 1Al (17-7PH)

5825 Wire, Welding—16.5Cr 4.8Ni 0.22 (Cb+Ta) 3.6Cu

5826 Wire, Welding—15Cr 5.1Ni 0.30 (Cb+Ta) 3.2Cu (17-4PH)

5827 Electrodes, Welding, Covered Steel—16.4Cr 4.8Ni 0.22 (Cb+Ta) 3.6Cu (17-4PH)

5828 Wire, Welding—57Ni 19.5Cr 13.5Co 4.2Mo 3.1Ti 1.4Al 0.006B Vacuum Induction Melted (Waspaloy)

5829 Wire, Welding—56Ni 19.5Cr 18Co 2.5Ti 1.5Al Vacuum Induc-

tion Melted (Nimonic 90)

5832 *Wire, Alloy, Welding—52.5Ni 19Cr 3.0Mo 5.1 (Cb+Ta) 0.90Ti 0.50Al 18Fe, Con. Elect. or Vac. Induct. Melted (718)*

5837 *Wire, Welding, Alloy—62Ni 21.5Cr 9.0Mo 3.7 (Cb+Ta) (625)*

5838 *Wire, Welding, Alloy—65Ni 16Cr 15Mo 0.30Al 0.06La (Hastelloy S)*

5840 *Wire, Welding—13Cr 8.0Ni 2.3Mo 1.1Al, Vacuum Melted (PH13-8Mo)*

Low Alloy Steels

6457 *Wire, Welding—0.95Cr 0.20Mo (0.28-0.33C) Vacuum Melted (4130)*

6458 *Wire, Welding—0.65Si 1.25Cr 0.50Mo 0.30V (0.28-0.33C), Vacuum Melted*

6459 *Wire, Welding—1.0Cr 1.0Mo 0.12V (0.18-0.23C) Vacuum Induction Melted*

6460 *Wire, Welding—0.75Si 0.62Cr 0.20Mo 0.10Zr (0.10-0.17C)*

6461 *Wire, Welding—0.95Cr 0.20V (0.28-0.33C) Vacuum Melted (6130)*

6462 *Wire, Welding—0.95Cr 0.20V (0.28-0.33C) (6130)*

6463 *Wire, Welding—18.5Ni 8.5Co 5.25Mo 0.72Ti 0.10Al, Vacuum Melted, Environment Controlled Packaging*

6464 *Electrodes, Covered Welding—1.5Mo 0.20V (0.06-0.12C) (E10013)*

6465 *Wire, Welding—2.0Cr 10Ni 8.0Co 1.0Mo 0.02Al 0.06V Vacuum Melted, Environment Controlled Packaging (HY180)*

6466 *Wire, Welding Corrosion Resistant 5.2Cr 0.55Mo (502)*

6467 *Electrodes, Covered Welding Steel—5Cr 0.55Mo (E502)*

6468 *Wire, Welding—1.0Cr 10Ni 3.8Co 0.45Mo 0.08V (0.14-0.17C) Vacuum Melted, Environment Controlled*

SAE Aerospace Recommended Practices of interest are:

ARP 1317, Electron Beam Welding

ARP 1330, Welding of Structures for Ground Support Equipment

ARP 1332, Wave Soldering Practice

ARP 1333, Nondestructive Testing of Electron Beam Welded Joints in Titanium Base Alloys

Unified Numbering System

The Unified Numbering System (UNS) provides a method for cross referencing the different numbering systems used to identify metals, alloys, and welding filler metals. With UNS, it is possible to correlate over 3500 metals and alloys used in a variety of specifications, regardless of the identifying number used by a society, trade association, producer, or user.

UNS is produced jointly by SAE and ASTM, and designated SAE HSJ1086/ASTM DS56. It cross references the metal and alloy designations of the following organizations and systems:

AA (Aluminum Association) Numbers

ACI (Steel Founders Society of America) Numbers

AISI (American Iron and Steel Institute)—SAE (Society of Automotive Engineers) Numbers

AMS (SAE Aerospace Materials Specifications) Numbers

ASME (American Society of Mechanical Engineers) Numbers

ASTM (American Society for Testing and Materials) Numbers

AWS (American Welding Society) Numbers

CDA (Copper Development Association) Numbers

Federal Specification Numbers

MIL (Military Specifications) Numbers

SAE (Society of Automotive Engineers) Numbers

Over 500 of the listed numbers are for welding and brazing filler metals. Numbers with the prefix *W* are assigned to welding filler metals that are classified by deposited metal composition.

MANUFACTURERS' ASSOCIATIONS

The following organizations publish literature which relates to welding. The committees that write the literature are comprised of representatives of equipment or material manufacturers. They do not generally include users of the products. Although some bias may exist, there is much useful information that can be obtained from this literature. The organization should be contacted for further information.

The Aluminum Association
818 Connecticut Avenue N.W.
Washington, D.C. 20006

American Iron and Steel Institute
1000 16th Street N.W.
Washington, D.C. 20036

Copper Development Association, Inc.
57th Floor, Chrysler Building
405 Lexington Avenue
New York, New York 10017

Electronic Industries Association
2001 Eye Street N.W.
Washington, D.C. 20006

National Electrical Manufacturers Association
2101 L Street N.W.
Washington, D.C. 20037

Resistance Welder Manufacturers Association
1900 Arch Street
Philadelphia, Pennsylvania 19103

Metric Conversion Factors

$t_C = 1.8 (t_F - 32)$

$1 \text{ lb} \cdot \text{f/in.}^2 = 6.894 \text{ kPa}$

$1 \text{ in.} = 25.4 \text{ mm}$

7
Qualification and Certification

Introduction *246*
Procedure Specifications *248*
Qualification of Procedure Specifications *256*
Performance Qualifications *261*
Standardization of Qualification Requirements *273*
Supplementary Reading List........... *274*

Chapter Committee

R.L. HARRIS, *Chairman*
R.L. Harris Associates, Incorporated

H.E. HELMBRECHT
Babcock and Wilcox Company

J.R. MCGUFFEY
Martin Marietta Energy Systems, Inc.

W.R. SMITH, SR.
Consultant

W.F. URBICK
The Boeing Company

Welding Handbook Committee Member

D.R. AMOS
Westinghouse Electric Corporation

7
Qualification and Certification

INTRODUCTION

Most fabricating codes and standards require qualification and certification of welding and brazing procedures and of welders, brazers, and operators who perform welding and brazing operations in accordance with the procedures.[1] Codes, standards, or contractual documents may also require that weldments or brazements be evaluated for acceptance by a qualified inspector. Nondestructive inspection of joints may be required. It should be done by qualified nondestructive testing personnel using standard testing procedures.

Technical societies, trade associations, and government agencies have defined qualification requirements for welded fabrications in codes or standards generally tailored for specific applications such as buildings, bridges, cranes, piping, boilers, and pressure vessels. Welding procedure qualification performed for one code or standard may qualify for another code or standard provided the qualification test results meet the requirements of the latter. The objectives of qualificaion are the same in nearly all cases. Some codes and standards permit acceptance of previous performance qualification by welders and welding operators having properly documented evidence.

A welding procedure specification (WPS) is a document that provides in detail required welding conditions for a specific application to assure repeatability by properly trained welders or welding operators. A specification normally identifies essential and non-essential variables. *Essential variables* are items in the welding procedure specification that, if changed beyond specified limits, require requalification of the welding procedure and either a revised or new welding procedure specification. *Non-essential variables* are items in the welding procedure specification which may be changed but do not affect the qualification status. Such changes, however, require a revision of the written welding procedure specification. Normally, a procedure specification must be qualified by demonstrating that joints made by the procedure can meet prescribed requirements. The actual welding conditions used to produce an acceptable test joint and the results of the qualification tests are recorded in the *procedure qualification record* (PQR).

Welders or welding operators are normally required to demonstrate their ability to produce welded joints that meet prescribed standards. This is known as *welder performance qualification.*

The results of welding procedure or perfor-

1. In the following discussion, the terms *weld*, *welder*, *welding*, and *welding operator* imply also *braze*, *brazer*, *brazing* and *brazing operator*, respectively, unless otherwise noted.

mance qualification must be certified by an authorized representative of the organization performing the tests. This is known as *certification.*

CODES AND STANDARDS

Typical codes and standards[2] that require welding qualification and sometimes brazing qualification are:

AWS D1.1, Structural Welding Code - Steel

AWS D1.2, Structural Welding Code - Aluminum

AWS D1.3, Structural Welding Code - Sheet Steel

AWS D1.4, Structural Welding Code - Reinforcing Steel

AWS D3.6, Specification for Underwater Welding

AWS D9.1, Specification for Welding Sheet Metal

AWS D14.1, Specification for Welding Industrial and Mill Cranes

AWS D14.2, Specification for Metal Cutting Tool Weldments

AWS D14.3, Specification for Earthmoving and Construction Equipment

AWS D14.4, Classification and Application of Welded Joints for Machinery and Equipment

AWS D14.5, Specification for Welding of Presses and Press Components

AWS D14.6, Specification for Rotating Elements of Equipment

ASME Boiler and Pressure Vessel Code

ANSI B31, Code for Pressure Piping

API STD 1104, Standard for Welding Pipelines and Related Facilities

The codes listed above generally include requirements for welding procedure and performance qualification. The *ASME Boiler and Pressure Vessel Code* specifically covers welding and brazing qualification in Section IX.

Other standards address only welding or brazing qualification. Examples are:

AWS B2.1, Welding Procedure and Performance Qualification

AWS B2.2, Brazing Procedure and Performance Qualification

AWS D10.9, Specification for Qualification of Welding Procedures and Welders for Piping and Tubing

MIL-STD-248, Welding and Brazing Procedure and Performance Qualification

MIL-STD-1595, Qualification of Aerocraft, Missile, and Aerospace Fusion Welders

PROCESSES

Qualification of procedures generally applies to all joining processes covered by a code or specification. The processes include shielded metal arc, gas metal arc, gas tunsten arc, plasma arc, oxyfuel gas, electron beam, resistance, electrogas, electroslag, and arc stud welding, and sometimes brazing. Performance qualification is normally required of personnel who will do manual, semiautomatic, and machine welding in production.

2. These and other codes and standards are discussed in Chapter 6.

PROCEDURE SPECIFICATIONS

ARC WELDING

Many factors contribute to the end result of a welding operation, whether it is manual shielded metal arc welding of plain carbon steel or gas shielded arc welding of exotic heat-resisting alloys. It is always desirable and often essential that the vital elements associated with the welding of joints are described in sufficient detail to permit reproduction, and to provide a clear understanding of the intended practices. Generally, proposed practices must be proved either by procedure qualification tests or by sufficient prior use and service experience to guarantee dependability. The purpose of a *welding procedure specification* (WPS) is, therefore, to define and document in detail the variables involved in welding a certain base metal. Typical variables are welding process, joint design, filler metal, welding position, and preheat. To fulfill this purpose efficiently, welding procedure specifications should be as concise and clear as possible, without extraneous detail.

Description and Details

Two different types of welding procedure specifications are in common use. One is a broad, general type that applies to all welding of a given kind on a specific base metal. The other is a narrower, more definitive type that spells out in detail the welding of a single size and type of joint in a specific base metal or part. Only the broader, more general type is usually required by codes, specifications, and contracts or by building, insurance, and other regulatory agencies.

The narrower, more definitive type is frequently used by manufacturers for control of repetitive in-plant welding operations or by purchasers desiring certain specific metallurgical, chemical, or mechanical properties. However, either type may be required by a customer or agency, depending upon the nature of the welding involved and the judgment of those in charge. In addition, the two types are sometimes combined to varying degrees, with addenda to show the exact details for specific joints attached to the broader, more general specification.

Arrangement and details of welding procedure specifications, as written, should be in accordance with the contract or purchase requirements and good industry practice. They should be sufficiently detailed to ensure welding that will satisfy the requirements of the applicable code, rules, or purchase specifications.

Codes and standards generally require that the manufacturer or fabricator prepare and qualify the welding procedure specifications. They should list all of the welding variables such as joint geometry, welding position, welding process, base metal, filler metal, preheat and interpass temperatures, welding current, arc voltage, shielding gas or flux, welding technique, and postweld heat treatment.

Some codes and standards are very specific in defining the content of information to be included in a WPS. they may list specific essential and nonessential variables that are to be addressed. Other codes refer to the welding variables of a specific process that affect qualification and leave it to the user to determine what needs to be included in the WPS.

Some fabrication codes permit the use of prequalified welding procedures. Under this system, the manufacturer or contractor prepares a welding procedure conforming to the specific requirements of that code for materials, joint design, welding technique, preheat, filler materials, etc. Weld qualification tests need not be made if the requirements are followed in detail. A contractor must accept responsibility for the use of prequalified welding procedures in construction when permitted to use them. Deviations from the requirements negate the prequalified status and require qualfication by testing. *AWS D1.1, Structural Welding Code - Steel* is an example of a code that permits prequalified procedures as an alternative to testing by each contractor or manufacturer.

Use of prequalified and qualified welding procedures does not guarantee satisfactory production welds. The quality of welds must be determined by the type and extent of nondestructive testing applied during and after welding.

Visual, magnetic particle, liquid penetrant, ultrasonic, and radiographic testing are commonly used on welded joints.

Most codes permit an organization other than the manufacturer or contractor to prepare test samples and to perform nondestructive or destructive tests, providing the manufacturer or contractor accepts responsibility for the tests. The welding operations must be under the direct supervision and control of the manufacturer or contractor.

Regardless of the differences in content and other requirements of codes and standards, the WPS provides direction to the welder or welding operator and is an important control document. It should have a specific reference number and an approval signature prior to release for production welding. Responsibility for the content, qualification status, and use of a WPS rests upon the manufacturer.

Typical subjects that may be listed in a WPS are discussed below. They do not necessarily apply to every process or application. Also, there may be important variables for certain welding processes that are not covered. The applicable code or standard should be consulted.

Scope. The welding processes, the applicable base metal, and the governing specification should be stated clearly.

Base Metal. The base metal(s) should be specified. This may be done either by giving the chemical composition or by referring to an applicable specification. If required, special treatment of the base metal before welding also should be indicated (e.g., heat treatment, cold work, or cleaning). In some cases, these factors are very important. A welding procedure that would provide excellent results with one base metal might not provide the same results with another, or even with the same base metal treated differently. Thus, the fabricator should identify the base metal, condition, and thickness.

Welding Process. The welding process that is to be used and type of operation should be clearly defined.

Filler Metal. Composition, identifying type, or classification of the filler metal should always be specified to ensure proper use. Filler metal marking is usually sufficient for identification. In addition, sizes of filler metal or electrodes that can be used when welding different thicknesses in different positions should be designated. For some applications, additional details are specified. These may include manufacturer, type, heat, lot, or batch of the welding consumables.

Marking does not always guarantee that the electrode is in the proper condition for welding. Certain types of filler metals and fluxes require special conditions during storage and handling for satisfactory service. For example, low hydrogen covered electrodes must be conditioned at elevated temperature to reduce moisture content. The welding procedure should state such requirements.

Type of Current and Range. Whenever welding involves the use of electric current, the type of current should be specified. Some covered welding electrodes operate with either ac or dc. If dc is specified, the proper polarity should be indicated. In addition, the current range for each electrode size when welding in different positions and for welding various thicknesses of base metal should be specified.

Arc Voltage and Travel Speed. For all arc welding processes, it is common practice to list an arc voltage range. Ranges for travel speed are mandatory for automatic welding processes, and are desirable many times for semiautomatic welding processes. If the properties of the base metal would be impaired by excessive heat input, permissible limits for travel speed or bead width are necessary.

Joint Design and Tolerances. Permissible joint design details should be indicated, as well as the sequence for welding. This may be done with cross-sectional sketches showing the thickness of the base metal and details of the joint, or by references to standard drawings or specifications. Tolerances should be indicated for all dimensions to avoid conditions that could prevent even an expert welder from producing an acceptable weld.

Joint and Surface Preparation. The methods that can be used to prepare joint faces and the degree of surface cleaning required should be designated in the procedure specification. They may include oxyfuel gas, air-carbon arc, or plasma arc cutting, with or without surface cleaning. Surface preparation may involve machining or grinding followed by vapor, ultrasonic, dip, or lint-free cloth cleaning. Methods and practices should conform to the application. In any case, the method or practice specified for production work should be used when qualifying the welding procedure, including the use of weld spatter-resisting compounds on the surfaces.

Tack Welding. Tack welding can affect weld soundness, hence details concerning tack welding procedures should be included in the welding procedure specification. Tack welders must use the designated procedures.

Welding Details. All details that influence weld quality, in terms of the specification requirements, should be clearly outlined. These usually include the appropriate sizes of electrodes for different portions of joints and for different positions; the arrangement of weld passes for filling the joints; and pass width or electrode weave limitations. Such details can influence the soundness of welds and the properties of finished joints.

Positions of Welding. A procedure specification should always designate the positions in which welding can be done. In addition, the manner in which the welding is to be done in each position should be designated (i.e., electrode or torch size; welding current range; shielding gas flow; number, thickness, and arrangement of weld passes, etc.). Positions of welding are shown in Figs. 7.1 through 7.4.

Preheat and Interpass Temperatures. Whenever preheat or interpass temperatures are significant factors in the production of sound joints or influence the properties of weld joints, the temperature limits should be specified. In many cases, the preheat and interpass temperatures must be kept within a well defined range to avoid degradation of the base metal heat-affected zone.

Peening. Indiscriminate use of peening should not be permitted. However, it is sometimes used to avoid cracking or to correct distor-

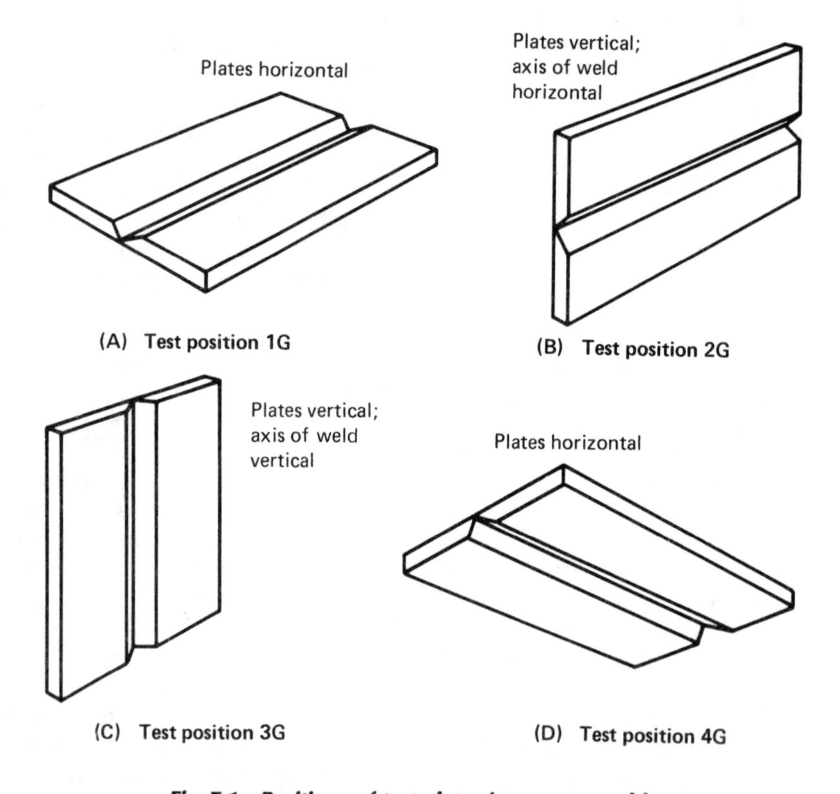

Fig. 7.1—Positions of test plates for groove welds

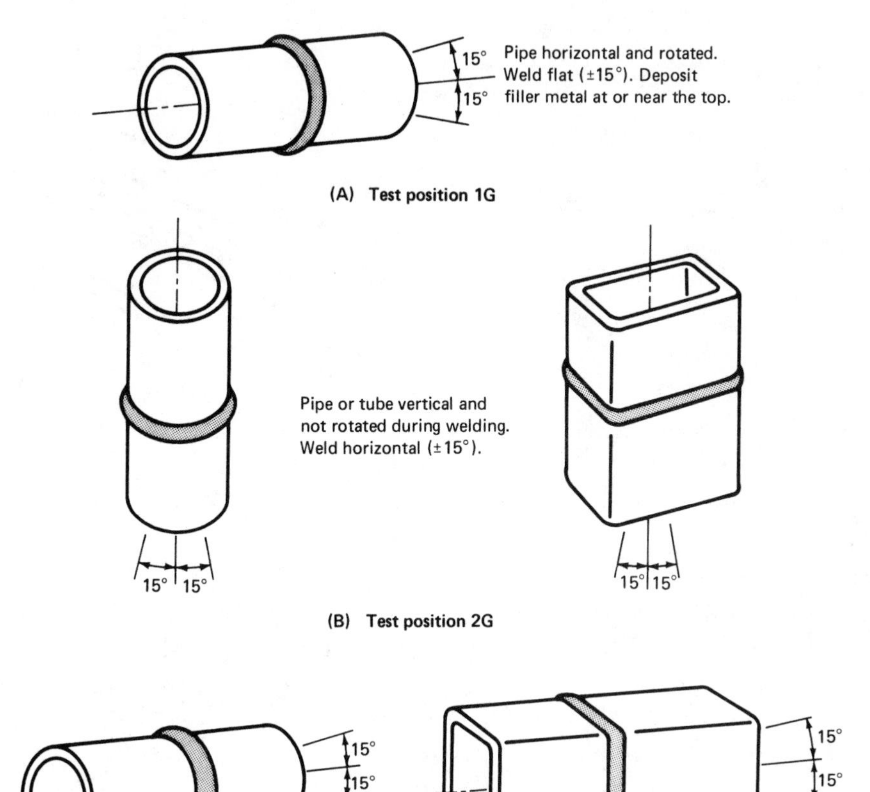

15°
15°
15°
15°

Pipe or tube horizontal fixed (±15°) and not rotated during welding. Weld flat, vertical, overhead.

(C) Test position 5G

Restriction ring
Test weld
45° ± 5°
45° ±5°

Pipe inclined fixed (45° ±5°) and not rotated during welding.

(D) Test position 6G

(E) Test position 6GR (T, K, or Y connections)

Fig. 7.2—Positions of test pipe or tubing for groove welds

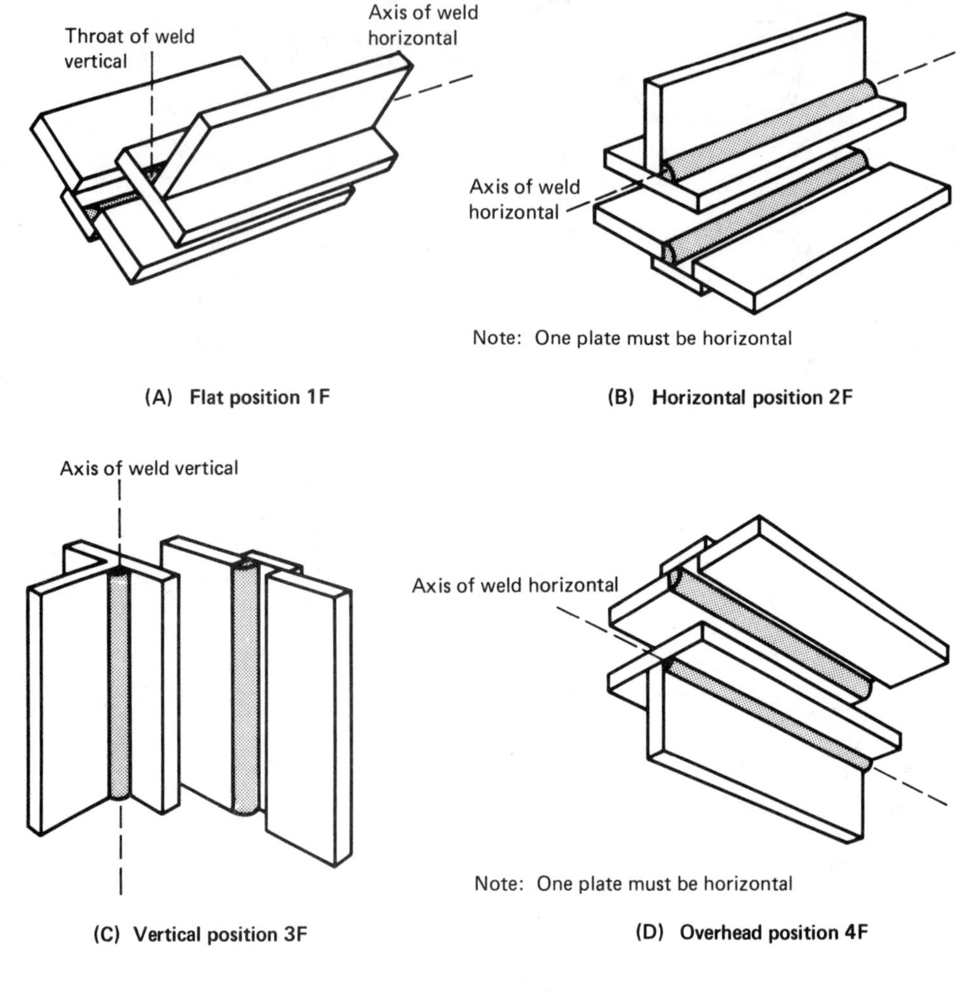

Fig. 7.3—Positions of test plates for fillet welds

tion of the weld or base metals. If peening is to be used, the details of its application and the appropriate tooling should be covered in the procedure specification.

Heat Input. Heat input during welding is usually of great importance when welding heat-treated steels and crack-sensitive ferrous and nonferrous alloys. Whenever heat input can influence final weld joint properties, details for its control should be prescribed in the procedure specification.

Second Side Preparation. When joints are to be welded from both sides, the methods that are to be used to prepare the second side should be delineated in the procedure specification. They may include chipping, grinding, and air-carbon arc or oxyfuel gas gouging of the root to sound metal. Frequently, this preparation is of primary importance in producing weld joints free from cracks and other unsound conditions.

Postheat Treatment. When welded joints or structures require heat treatment after welding to develop required properties, dimensional stability, or dependability, such treatment should be

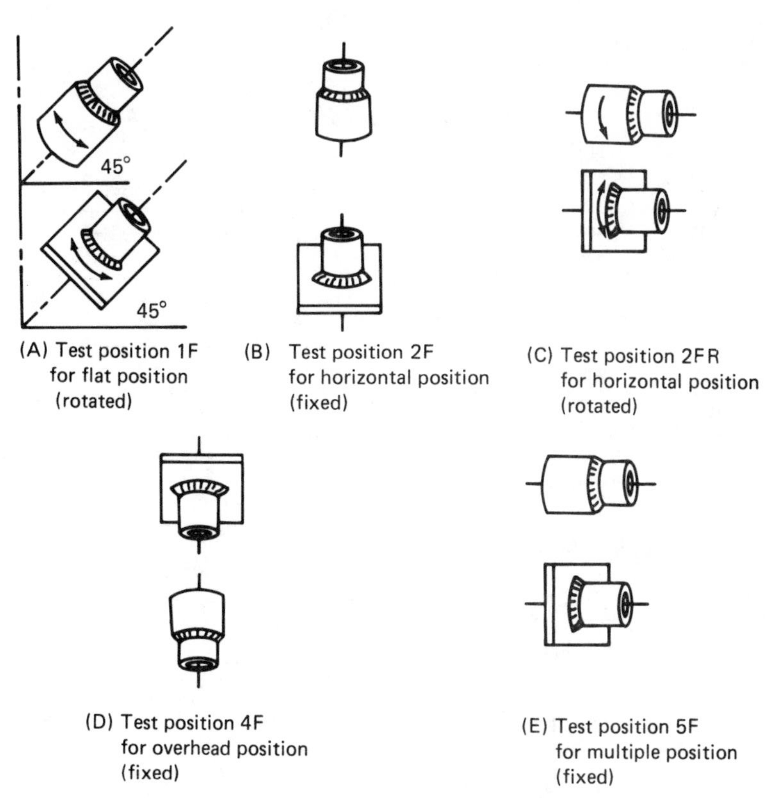

Fig. 7.4—Positions of test pipes for fillet welds

delineated in the WPS. The same heat treatment should be applied to all procedure qualification test welds. A full description of the heat treatment may appear in the WPS or in a separate fabrication document, such as a shop heat-treating procedure.

Records. If detailed records of the welding of joints are required, the specific requirements for these records may be included in the welding procedure specification.

Application

Welding procedure specifications are sometimes required by the purchaser to govern fabrication of a given product in a fabricator's shop. More often, however, the purchaser will specify the properties desired in the weldment in accordance with a code or specification. The fabricator then develops a welding procedure that will produce the specified results. In other cases, the purchaser will prescribe that the fabricator establish and test definite welding procedure specifications to prove that the resulting welds will meet certain requirements, and then follow those welding procedures in production.

Forms

Welding procedure specifications may be prepared in many different ways, and may be either quite brief or very long and detailed. Some codes have suggested forms which provide sufficient information but are not excessively complex. An example is shown in Fig. 7.5. Naturally, more complex and critical applications should have more detailed procedure specifications.

BRAZING

Brazing procedure specifications (BPS) are similar to those for arc welding except for the

WELDING PROCEDURE SPECIFICATION

Material specification ______
Welding process ______
Manual or machine ______
Position of welding ______
Filler metal specification ______
Filler metal classification ______
Flux ______
Weld metal grade* ______ Flow rate ______
Single or multiple pass ______
Single or multiple arc ______
Welding current ______
Polarity ______
Welding progression ______
Root treatment ______
Preheat and interpass temperature ______
Postheat treatment ______

*Applicable only when filler metal has no AWS classification.

WELDING PROCEDURE

Pass no.	Electrode size	Welding current		Travel speed	Joint detail
		Amperes	Volts		

This procedure may vary due to fabrication sequence, fit-up, pass size, etc., within the limitation of variables given in 4B, C, or D of AWS D1.1, (______ year) Structural Welding Code.

Procedure no. ______ Manufacturer or contractor ______

Revision no. ______ Authorized by ______

Date ______

Fig. 7.5—Sample form for a welding procedure specification

process data. Process information may include but not be restricted to the following:

(1) Type of brazing (torch, furnace, induction, etc.)
(2) Brazing filler metal and form
(3) Brazing temperature range
(4) Brazing flux or atmosphere
(5) Flow position
(6) Method of applying filler metal
(7) Time at brazing temperature
(8) Heating and cooling rates

Typical brazing flow positions are flat, vertical down, vertical up, and horizontal. In flat flow, the joint faces and capillary flow are horizontal. In vertical down and vertical up flow, the joint faces are vertical and capillary flow of filler metal is down and up respectively. With horizontal flow, the joint faces are also vertical, but capillary flow of filler metal is horizontal.

Typical codes and standards that address brazing qualification are *AWS B2.2, Standard for Brazing Procedure and Performance Qualification* and *Section IX, Welding and Brazing Qualification, ASME Boiler and Pressure Vessel Code.*

RESISTANCE WELDING

The following items are ordinarily covered in a resistance welding procedure specification. Others may be required depending on the welding process.

Welding Process

A WPS should specify the specific resistance welding process to be used (i.e., spot, seam, flash welding, etc.) because the various processes are distinctly different in many respects.

Composition and Condition of Base Metal

The base metal to be welded should be specified either by reference to a specification or by chemical composition. The permissible chemical composition range may include base metals covered by more than one specification if they can be welded by the same procedure. The condition (temper) should also be stated. The WPS should include any specific cleaning requirements. The above information is of great importance because a welding procedure that produces excellent results with one base metal may not be satisfactory for another, or for even the same metal in a different condition of heat treatment or cleanliness.

Joint Design

A WPS should specify all details of the joint design including contacting overlap, weld spacing, type and size of projection, and other similar factors.

Type and Size of Electrode

The specification should state the type of electrode to be used including alloy, contour, and size. If plates, dies, blocks, or other such devices are used, properties that would affect the quality of welding should be specified.

Machine Settings

The electrode force, squeeze time, weld time, hold time, off time, welding speed, upsetting time, and other such factors controlled by machine settings should be specifically prescribed in the WPS.

Weld Size

The size of each weld or the weld strength is generally an acceptance criterion and should be specified.

Surface Appearance

Factors such as indentation, discoloration, or amount of upset that affect the surface appearance of a weldment should be specified. If these factors are not considered too important, they may be covered by a general requirement rather than by detailed requirements.

Inspection Details

The properties to be checked - appearance, strength, tightness, etc. - should be specified as well as the method of testing, e.g., - shear test, pillow test, peel test, or workmanship sample.

QUALIFICATION OF PROCEDURE SPECIFICATIONS

The purpose of procedure qualification is to determine, by preparation and testing of standard specimens, that welding in accordance with the WPS will produce sound welds and adequate properties in a joint. The type and number of tests required by the code or standard are designated to provide sufficient information on strength, ductility, toughness, or other properties of the joint. The essential process variables are specified to maintain desired properties. When changes are made to the procedure specification that affect properties of the welded joint, requalification is required.

The mechanical and metallurgical properties of a welded joint may be altered by the WPS selected for the job. It is the responsibility of each manufacturer or contractor to conduct the tests required by the applicable codes and contractural documents. It is the duty of the engineer or inspector to review and evaluate the results of such qualification tests. These qualification activities must be completed prior to production to assure that the selected combination of materials and methods is capable of achieving the desired results. Qualification of welding procedures can be accomplished by any of the following concepts or a combination thereof.

(1) *Employment of prequalified procedures.* This concept is based on the reliability of certain proven procedures spelled out by the applicable code or specification. Any deviation outside specified limits voids prequalification.

(2) *Employment of standard qualification tests.* Standard tests may or may not simulate the actual conditions anticipated for a given project. Usually, such standard tests involve conventional butt joints on pipes or plates, or fillet welds between two plates. Base metals, welding consumables, and thermal treatments follow production welding plans within specific ranges. However, other variables, such as joint geometry, welding position, and accessibility may not be considered essential. (See the *ASME Boiler and Pressure Vessel Code, Section IX.*) Thus, certain variables may bear little resemblance to production conditions, but must be changed by revision of the welding procedure specification. Other documents, such as *API 1104, Standard for Welding Pipelines and Related Facilities,* require qualification tests that closely resemble production conditions.

(3) *Employment of mock-up tests.* Mock-up tests should simulate actual production conditions to the extent necessary to ascertain that a sound plan with proper tooling and inspection has been selected. Generally, welding codes do not require preparation of welded mock-ups or sample joints unless they are needed to demonstrate that the welding procedures will produce the specified joints. Preparation of mock-ups or sample joints may be required to satisfy contractual conditions or to avoid problems in production. For the latter purpose, mock-ups can indicate the expected quality levels under difficult or restricted welding conditions.

WELDING PROCEDURES

The basic steps in the qualification of a welding procedure are:

(1) Preparation and welding of suitable samples

(2) Testing of representative specimens

(3) Evaluation of overall preparation, welding, testing, and end results

(4) Possible changes in procedure

(5) Approval

Preparation of Sample Joints

Plate or pipe samples with a representative welding joint are usually prepared for procedure qualification testing. The size, type, and thickness of the sample are governed by the thickness and type of base metal to be welded in production, and by the type, size, and number of specimens to be removed for testing. The latter are usually prescribed by the applicable code or specification. The base and filler metals and other details associated with welding of sample joints should be in accordance with the particular welding procedure specification that is being qualified.

Tests of Procedure Qualification Welds

Test specimens are usually removed from the sample joints for examination to determine certain properties.[3] The type and number of specimens removed and the test details normally depend upon the requirements of the particular application or specification. Usually, the tests include tensile and guided bend specimens to determine strength, ductility, soundness, and adequacy of fusion. If only fillet welds are tested, tensile-shear tests and break or macroetch specimens are usually employed. Additional tests may be specified by the applicable codes or contract documents to meet specific needs. These may include:

(1) Impact tests to determine notch toughness of the weld and the heat-affected zone to minimize the risk of brittle fracture at specified temperatures. Charpy V-notch specimens are most commonly used for such tests, but many other tests, including drop weight and crack-opening-displacement (COD) tests, are sometimes employed.

(2) Nick-break tests to determine weld soundness at randomly selected locations.

(3) Free bend tests to determine the ductility of deposited weld metal.

(4) Shear tests to determine the shear strength of fillet welds or clad bonding.

(5) Hardness tests to determine adequacy of heat treatment and suitability for certain service conditions. Such tests may be performed on surfaces or on cross sections of welds.

(6) All-weld-metal tension tests to determine the mechanical properties of the deposited weld metal with minimum influence due to base metal dilution.

(7) Elevated temperature tests to determine mechanical properties at temperatures resembling service conditions.

(8) Restraint tests to determine crack susceptibility and the ability to achieve sound welds under highly rstrained conditions.

(9) Corrosion tests to determine the properties needed to withstand aggressive environments.

(10) Nondestructive inspection and macroetch or microetch tests to determine the soundness of a weld, and to evaluate the inspectability of production welds.

(11) Delayed cracking tests to detect resistance to hydrogen cracking in high strength, low alloy steels and some other alloys.

Recording Test Results

If all requirements are satisfactorily met, the details of the tests performed and the results of all tests and examinations are entered on a Procedure Qualification Record (PQR) (Refer to Fig. 7.6.) The PQR is certified to be accurate and to meet specified requirements of a particular code or standard by signature of a responsible individual within the organization performing the qualification. Since the PQR is a certified record of a qualification test, it should not be revised. If information needs to be added later, it can be in the form of a supplement or attachment; records should not be changed by revision.

In evaluating a welding procedure or the test results, applicable codes provide general guidance and some specific acceptance-rejection criteria. For instance, the minimum tensile strength and the maximum number of inclusions or other discontinuities are specified by many documents. The acceptability of other properties must be based on engineering judgment. In general, it is desirable that the weld match the mechanical and metallurgical properties of the base metal, but this is not always possible. Not only are weld metal and base metal different product forms, but they often have somewhat different chemical compositions and mechanical properties. It requires engineering judgment to select the most important properties for each individual application. This is especially important for service at high or low temperature, or under corrosive conditions.

Changes in a Qualified Procedure

If a fabricator has qualified a welding procedure and desires at some later date to make a change in that procedure, it may be necessary to conduct additional qualifying tests. Those tests should establish that the changed welding procedure will produce satisfactory results.

Such requalification tests are not usually required when only minor details of the original procedure are changed. They are required, how-

3. Standard welding tests are covered in the *Welding Handbook*, Vol. 1, 7th ed., 1976: 154-219, and *AWS B4.0, Standard Methods for Mechanical Testing of Welds.*

WELDING PROCEDURE QUALIFICATION TEST RECORD

PROCEDURE SPECIFICATION

Material specification ______
Welding process ______
Manual or machine ______
Position of welding ______
Filler metal specificatin ______
Filler metal classification ______
Weld metal grade* ______
Shielding gas ______ Flow rate ______
Single or multiple pass ______
Single or multiple arc ______
Welding current ______
Welding progression ______
Preheat temperature ______
Postheat treatment ______
Welder's name ______

*Applicable when filler metal has no AWS classification.

VISUAL INSPECTION

Appearance ______
Undercut ______
Piping porosity ______

Test date ______
Witnessed by ______

GROOVE WELD TEST RESULTS

Tensile strength, psi
1. ______
2. ______

Guided-bend tests (2 root-, 2 face-, or 4 side-bend)

Root	Face
1. ______	1. ______
2. ______	2. ______

Radiographic-ultrasonic examination

RT report no ______
UT report no. ______

FILLET WELD TEST RESULTS

Minimum size multiple pass Macroetch		Maximum size single pass Macroetch	
1. ______	3. ______	1. ______	3. ______
2. ______		2. ______	

All-weld-metal tension test

Tensile strength, psi ______
Yield point/strength, psi ______
Elongation in 2 in., % ______
Laboratory test no. ______

WELDING PROCEDURE

Pass no.	Electrode size	Welding power		Speed of travel	Joint detail
		Amperes	Volts		

We, the undersigned, certify that the statements in this record are correct and that the test welds were prepared, welded, and tested in accordance with the requirements of AWS D1.1. (______ year) Structural Welding Code — Steel.

Procedure no. ______
Revision no. ______

Manufacturer or contractor ______
Authorized by ______
Date ______

Fig. 7.6—Sample of a procedure qualification record form

ever, if the changes might alter the properties of the resulting welds. Reference should always be made to the governing code or specification to determine whether an essential variable has been changed. Typical procedure factors that may require requalification of the WPS are given in Table 7.1.

BRAZING PROCEDURES

Procedure specifications for brazing are qualified by brazing designated test specimens and evaluating the joints by appropriate tests. Typical tests, applicable joints, and purposes of the tests are given in Table 7.2. The results of the tests are recorded on the Brazing Procedure Qualification Record, which is similar to a Welding Procedure Qualification Record. It lists the information on the BPS and the results of the appropriate tests. The record is then certified by the manufacturer's representative who witnessed the tests.

RESISTANCE WELDING PROCEDURES

A proposed welding procedure must be evaluated by appropriate tests to determine whether joints can be welded consistently and

Table 7.1
Welding procedure and welder qualification variables that may require requalification

Procedure variable	Procedure requalification may be required when the welding practices are changed to the extent indicated below	Welder or welding operator requalification may be required for the practice changes indicated below
Type, composition, or process condition of the base material	When the base metal is changed to one not conforming to the type, specification, or process condition qualified. Some codes and specifications provide lists of materials which are approximately equivalent from the standpoint of weldability and which may be substituted without requalification.	Usually not required, unless a marked change is made in the type of filler metal used or covering thereon, e.g., ferritic to austenitic or non-ferrous metal; cellulosic to low-hydrogen type electrode covering, etc.
Thickness of base metal	When the thickness to be welded is outside the range qualified. The various codes and specifications may differ considerably in this respect. Most provide for qualification on one thickness within a reasonable range; e.g., 3/16 in. to 2T, 1/2T to 2T, 1/2T to 1.1T, etc. Some may require qualification on the exact thickness or on the minimum and maximum.	When the thickness to be welded is outside the range qualified.
Joint design	When established limits of root openings, root face, and included angle of groove joints are increased or decreased; i.e., basic dimensions plus tolerances. Some codes and specifications prescribe definite upper and lower limits for these dimensions, beyond which requalification is necessary. Others permit an increase in the included angle and root opening and a decrease in the root face without requalification. Requalification is also often required when a backing or spacer strip is added or removed or the basic type of material of a backing or spacer strip is changed.	When changing from a double-welded joint or a joint using backing material to an open root or to a consumable insert joint.

Table 7.1 (Continued)

Pipe diameter	Usually not required. In fact, some codes permit procedure qualification on plate to satisfy the requirements for welding to be performed on pipe.	When the diameter of piping or tubing is reduced below specified limits. It is generally recognized that smaller pipe diameters require more sophisticated techniques, equipment, and skills.
Type of current and polarity (if dc)	Usually not required for changes involving electrodes or welding materials adapted for the changed electrical characteristics, although sometimes required for change from ac to dc, or vice versa, or from one polarity to the other.	Usually not required for changes involving similar electrodes or welding materials adapted for the changed electrical characteristics.
Electrode classification and size	When electrode classification is changed or when the diameter is increased beyond that qualified.	When the electrode classification grouping is changed and sometimes when the electrode diameter is increased beyond specified limits.
Welding current	When the current is outside of the range qualified.	Usually not required.
Position of welding	Usually not required, but desirable.	When the change exceeds the limits of the position(s) qualified (only required by some codes).
Deposition of weld metal	When a marked change is made in the manner of weld deposition; e.g., from a small bead to a large bead or weave arrangement or from an annealing pass to a no annealing pass arrangement, or vice versa.	Usually not required.
Preparation of root of weld for second side welding	When method or extent is changed.	Usually not required.
Preheat and interpass temperatures	When the preheat or interpass temperature is outside the range for which qualified.	Usually not required.
Postheat treatment	When adding or deleting postheating or when the postheating temperature or time cycle is outside the range for which qualified.	Usually not required.
Use of spatter compound, paint, or other material on the surfaces to be welded.	Usually not required, but desirable when employing unproven materials.	Usually not required.

Note: This table is shown for illustration only to indicate the general nature of code requirements governing requalification of welding procedures, welders, and operators. It cannot be used by an inspector to determine whether requalification is required in a specific instance. For that information reference must be made to the particular code or specification applicable to the work being inspected.

Table 7.2
Brazing procedure qualification tests

Test	Applicable joints	Properties evaluated
Tension	Butt, scarf, lap, rabbet	Ultimate strength (tension or shear)
Guided bend	Butt, scarf	Soundness, ductility
Peel	Lap	Bond quality, soundness

will satisfactorily meet the service requirements to which they will be exposed. Because of the varied nature of resistance welded products, qualification of resistance welding procedures is also varied. Generally where the welded part is small in size, the procedure may be qualified by making a number of finished pieces and testing them to destruction under service conditions, either simulated or real. In other instances, welds can be made in test specimens that are tested in tension or shear, or inspected for other properties.[4]

There are three main steps in qualifying a resistance welding procedure.

Preparation of Weld Specimens

Sample weld specimens are prepared and welded in accordance with the welding procedure specifications to be qualified. The number, size, and type of samples are governed by the nature of the tests to be performed for qualification. Where qualification is in accordance with a standard, these requirements are usually specified.

Testing of Specimens

The nature of the tests to be performed will vary according to the service requirements of the completed parts. Test specimens may actually be subjected to real or simulated service conditions. More frequently, however, tests will be made to determine specific properties of the weld, such as tensile strength, shear strength, surface appearance, and soundness.

Evaluation of Test Results

After testing the welded specimens, the results must be reviewed to determine whether they meet the specified requirements. If all of the requirements are met, the welding procedure is considered qualified.

Changes in a Qualified Procedure

If changes become necessary in an established and qualified welding procedure, it may be necessary to conduct additional qualifying tests to determine whether the modified welding procedure will yield satisfactory results. Such qualification tests should not be required when only minor changes have been made in the original procedure, but should be required when the changes might alter the properties of the resulting welds. Where a governing code or specification exists, reference should be made to it to determine if requalification is required.

PERFORMANCE QUALIFICATION

WELDER PERFORMANCE

Welder, welding operator, and tack welder qualification tests determine the ability of the persons tested to produce acceptably sound welds with the process, materials, and procedure called for in the tests. Qualification tests are not intended to be used as a guide for welding during actual construction, but rather to assess whether an individual has a required minimum level of skill to produce sound welds. They do not tell what an individual normally will or can do in production. For this reason, complete reliance should not be placed on qualification testing of

4. Standard tests for resistance welds are given in *AWS C1.1, Recommended Practices for Resistance Welding.* Resistance welded butt joints can be evaluated by mechanical tests similar to those used for arc welded joints.

welders. The quality of all production welds should be determined by inspection during and following completion of the actual welding.

Various codes, specifications, and governing rules generally prescribe similar though frequently somewhat different methods or details for qualifying welders, welding operators, and tackers. The applicable code or specification should be consulted for specific details and requirements. The types of tests that are most frequently required are described below.

Qualification Requirements

The qualification requirements for welding pressure vessels, piping systems, and structures usually state that every welder or welding operator shall make one or more test welds using a qualified welding procedure. Each qualification weld is tested in a specific manner (e.g., radiography or bend tests).

Qualification requirements for welding pressure pipe differ from those for welding plate and structural members chiefly in the type of test assemblies used. The test positions may also differ to some extent. As a rule, the tests require use of pipe assemblies instead of flat plate. Space restrictions may also be included as a qualification factor if the production work involves welding in restricted spaces.

Limitation of Variables

Certain welding variables that will affect the ability of welders or welding operators to produce acceptable welds are considered essential. The specific variables are called out in the applicable code. Generally, a welder or welding operator must be requalified before commencing welding to a WPS that has one or more essential variables for which the welder or welding operator was not qualified by performance qualification. Following are examples of essential variables from *AWS D1.1, Structual Welding Code—Steel* that require requalification of welder performance if changed:

(1) *Welding Process.*

(2) *Filler metal.* Shielded metal arc welding electrodes are divided into groups and assigned "F" numbers. The grouping of electrodes is based essentially on those usability characteristics that fundamentally determine the ability of a welder to make satisfactory welds with all electrodes in a given group.

(3) *Welding position.* Positions are classified differently for plate and pipe. (See Figs. 7.1 through 7.4.) A change in position may require requalification. For pipe welding, a change from rotated to fixed position requires requalification. Positions of groove and fillet welds as established by the *American Welding Society* are shown in Figs. 7.7 and 7.8 respectively.

(4) *Joint detail.* A change in joint type, such as omission of backing on joints welded from one side only, requires requalification.

(5) *Plate thickness.* Test plates for groove welds are generally 3/8 or 1 in. in thickness. The 3/8-in. thickness is employed for qualifying a welder for limited thickness (3/4 in. max.) and the 1-in. test plate for unlimited thickness.

(6) *Technique.* In vertical position welding, for example, a change in direction of progression of welding requires requalification.

Codes and specifications usually require that welder qualification tests be made in one or more of the most difficult positions to be encountered in production (e.g., vertical, horizontal, or overhead) if the production work involves other than flat position welding. Qualification in a more difficult position may qualify for welding in less difficult positions (e.g., qualification in the vertical, horizontal, or overhead position is usually considered adequate for welding in the flat position).

Test Specimens

Typical groove weld qualification test plates for unlimited thickness are shown in Figs. 7.9 and 7.10. The groove weld plates for limited thickness qualification are essentially the same except the plate thickness is 3/8 in.

Joint detail for groove weld qualification tests for butt joints on pipe or tubing should be in accordance with a qualified welding procedure specification for a single-welded pipe butt joint. As an alternative, the joint details shown in Fig. 7.11 are frequently used. Consideration by some codes is also given to other joint designs such as T-, K-, or Y-connections in pipe, and also fillet and tack welds. A typical fillet weld test plate is shown in Fig. 7.12.

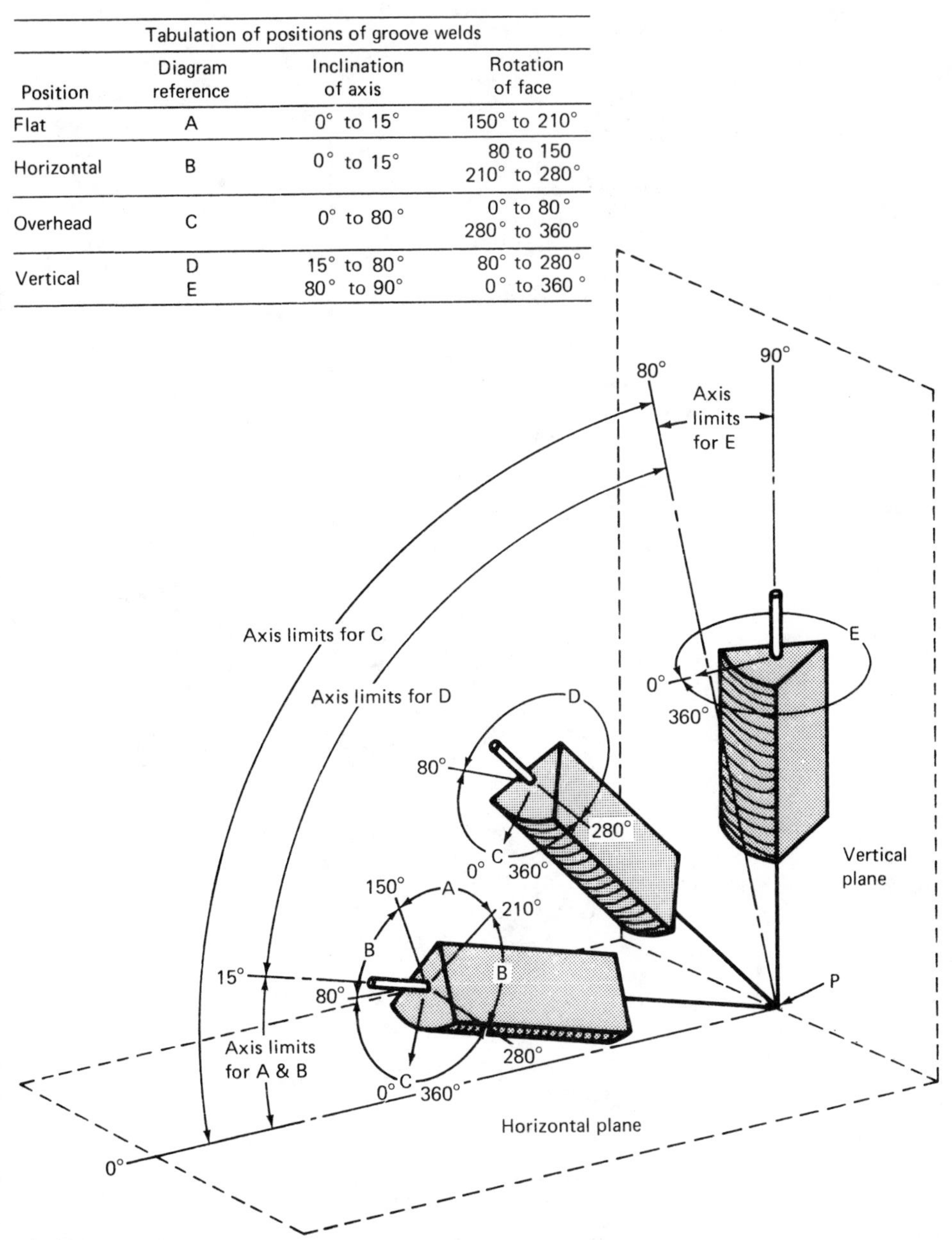

Tabulation of positions of groove welds			
Position	Diagram reference	Inclination of axis	Rotation of face
Flat	A	0° to 15°	150° to 210°
Horizontal	B	0° to 15°	80 to 150 210° to 280°
Overhead	C	0° to 80°	0° to 80° 280° to 360°
Vertical	D E	15° to 80° 80° to 90°	80° to 280° 0° to 360°

Notes:
1. The horizontal reference plane is always taken to lie below the weld under consideration.
2. The inclination of axis is measured from the horizontal reference plane toward the vertical reference plane.
3. The angle of rotation of the face is determined by a line perpendicular to the theoretical face of the weld which passes through the axis of the weld. The reference position (0°) of rotation of the face invariably points in the direction opposite to that in which the axis angle increases. When looking at point P, the angle of rotation of the face of the weld is measured in a clockwise direction from the reference position (0°).

Fig. 7.7—Positions of groove welds

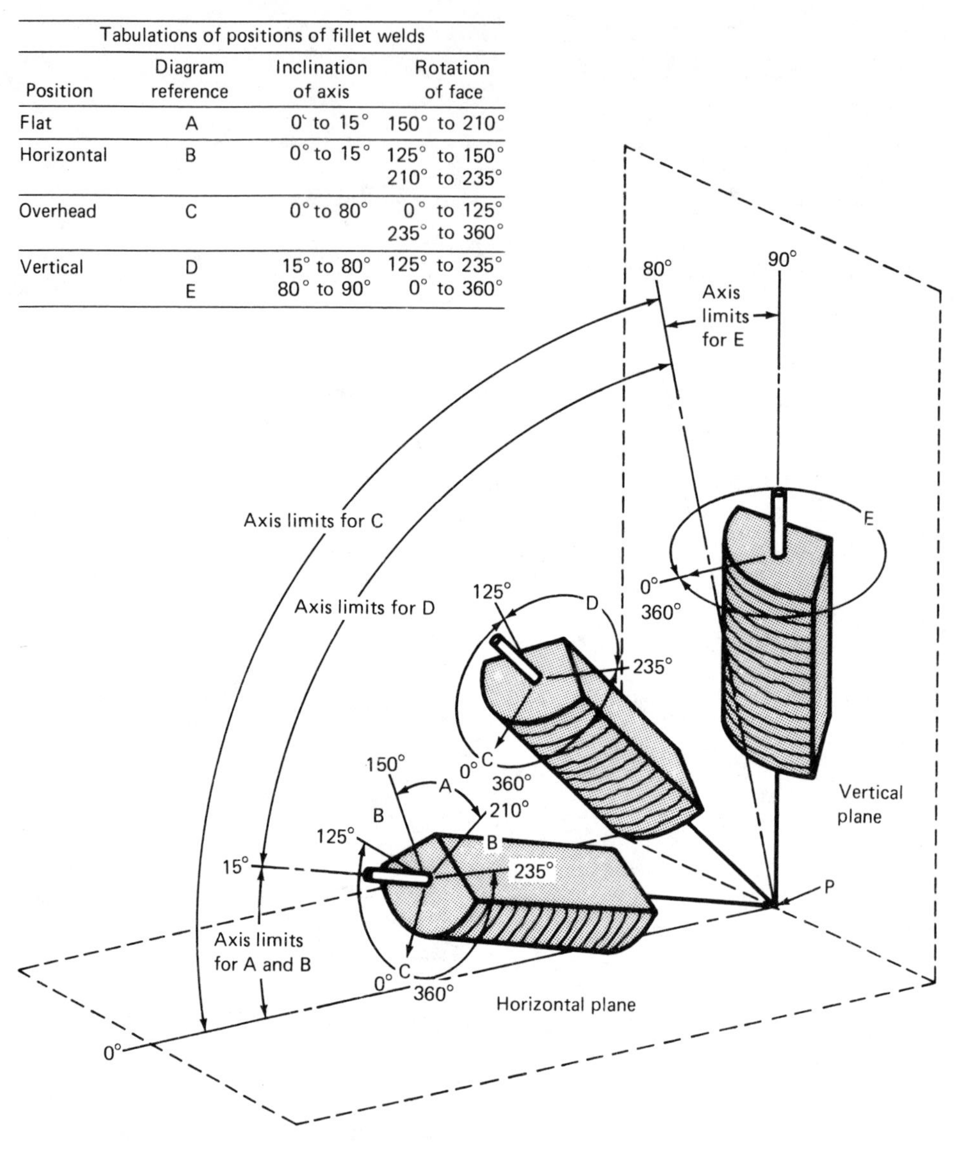

Tabulations of positions of fillet welds			
Position	Diagram reference	Inclination of axis	Rotation of face
Flat	A	0° to 15°	150° to 210°
Horizontal	B	0° to 15°	125° to 150° 210° to 235°
Overhead	C	0° to 80°	0° to 125° 235° to 360°
Vertical	D E	15° to 80° 80° to 90°	125° to 235° 0° to 360°

Note:
For groove welds in pipe the following definitions shall apply:
Horizontal Fixed Position: When the axis of the pipe does not deviate by more than 30° from the horizontal plane and the pipe is not rotated during welding.
Horizontal Rolled Position: When the axis of the pipe does not deviate by more than 30° from the horizontal plane, the pipe is rotated during welding, and the weld metal is deposited within an arc not to exceed 15° on either side of a vertical plane passing through the axis of the pipe.
Vertical Postiion: When the axis of the pipe does not deviate by more than 10° from the vertical position. (The pipe may or may not be rotated during welding.)*

*Positions in which the axis of the pipe deviates by more than 10° and less than 60° from the vertical shall be considered intermediate.

Fig. 7.8—Positions of fillet welds

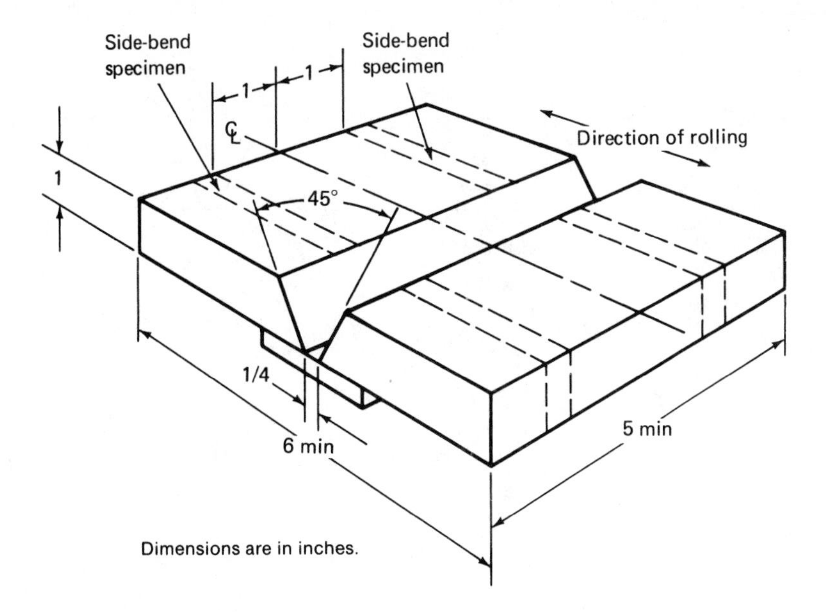

Fig. 7.9—Test plate for unlimited thickness—welder qualification

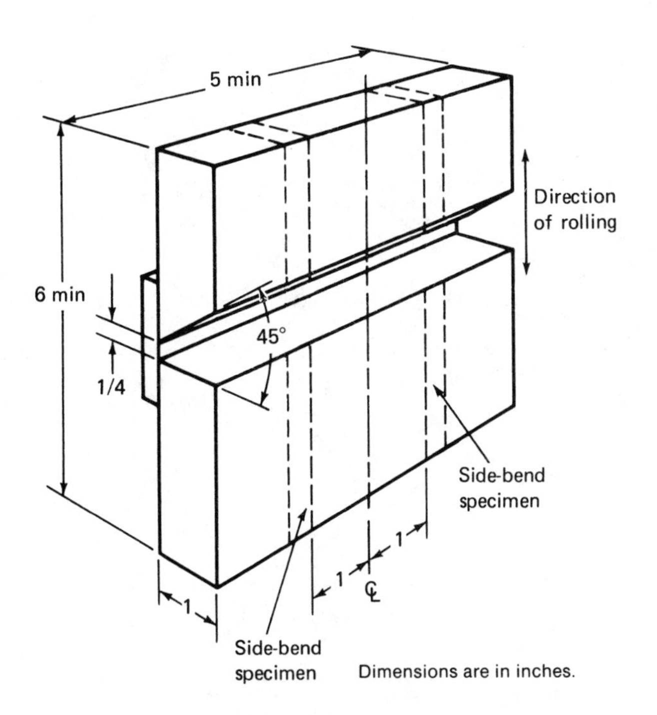

Fig. 7.10—Optional test plate for unlimited thickness—horizontal position—welder qualification

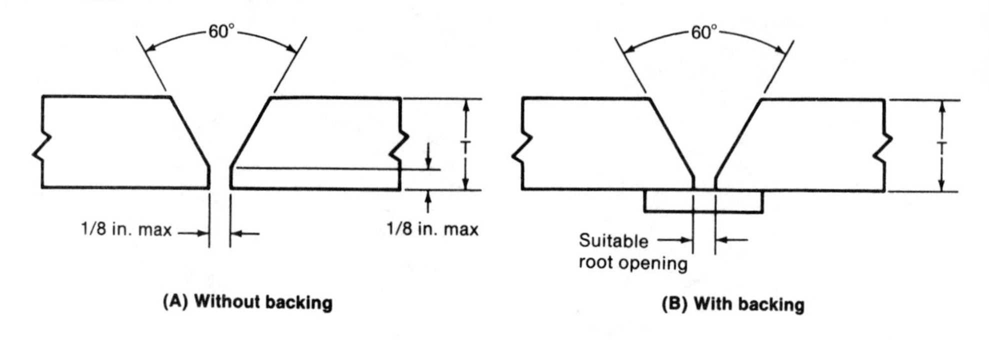

Fig. 7.11—Tubular butt joint—welder qualification

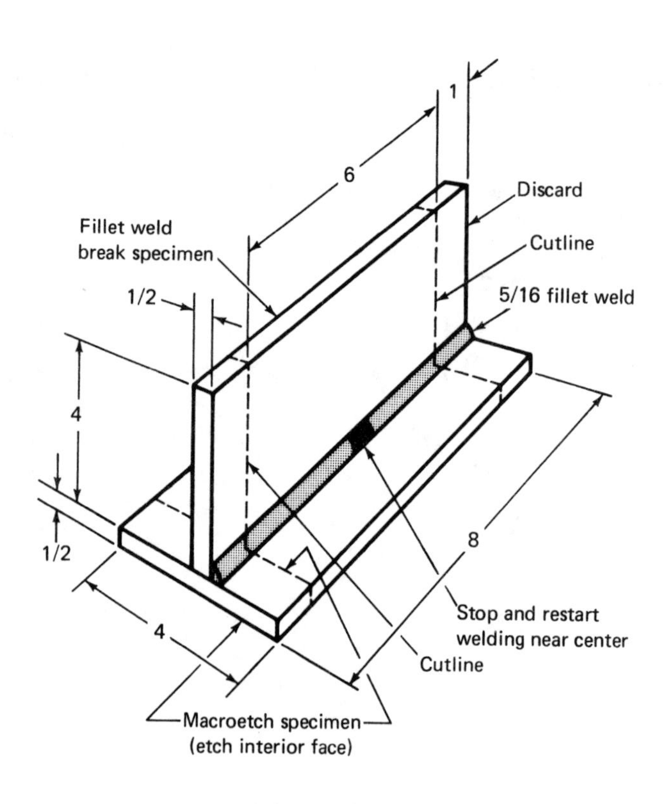

Fig. 7.12—Fillet-weld-break and macroetch test plate—welder qualification

Groove weld qualification usually qualifies the welder to weld both groove and fillet welds in the qualified positions. Fillet weld qualification limits the welder to fillet welding in only the position qualified and other specified positions of less difficulty. Common type and position limitations for welder qualification are shown in Table 7.3.

Testing of Qualification Welds

All codes and specifications have definite rules for testing qualification welds to determine compliance with requirements. For groove welds, guided bend test specimens are cut from specific locations in the welded plates and bent in specified jigs. Fillet welds do not readily lend themselves to guided bend tests. In most cases, fillet weld break tests or macroetch tests, or both, are required.

Radiographic testing is sometimes permitted as an alternative to mechanical or other tests. The primary requirement for qualification is that all test welds be sound and thoroughly fused to the base metal. If the test welds meet the prescribed requirements, the welder or welding operator is considered qualified to weld with processes, filler metals, and procedures similar to those used in the test but within prescribed limits.

Qualification Records

Responsibility for qualification of welders, as for qualification of welding procedures, is with

Table 7.3
Welder Qualification—type and position limitations

Qualification test		Type of weld and position of welding qualified*			
		Plate		Pipe	
Weld	Plate or pipe Positions**	Groove	Fillet	Groove	Fillet
Plate-groove	1G	F	F, H	F (Note 1)	F, H
	2G	F, H	F, H	F, H (Note 1)	F, H
	3G	F, H, V	F, H, V	F, H, V (Note 1)	F, H
	4G	F, OH	F, H, OH		F
	3G & 4G	All	All		F, H
Plate-fillet (Note 2)	1F		F		F
	2F		F, H		F, H
	3F		F, H, V		
	4F		F, H, OH		
	3F & 4F		All		
Pipe-groove	1G	F	F, H	F	F, H
	2G	F, H	F, H	F, H	F, H
	5G	F, V, OH	F, V, OH	F, V, OH	F, V, OH
	6G	Note 3	Note 3	Note 3	Note 3
	2G & 5G	Note 3	Note 3	Note 3	Note 3
	6GR	All	All	All	All

Notes:
1. Welders qualified to weld tubulars over 24 in. (600 mm) in diameter with backing or back gouging, for the test positions indicated.
2. Not applicable for fillet welds between parts having a dihedral angle (Ψ) of 60 degrees or less.
3. Qualified for all but groove welds for T-, Y-, and K-connections.

*Positions of welding: F = flat, H = horizontal, V = vertical, OH = overhead.
**See Figs. 7.1 through 7.4.

the manufacturer. After successful qualification of a welder or welding operator, the manufacturer is required to prepare a performance qualification record that provides the necessary details of qualification. The manufacturer then certifies the record by the affixed signature of a responsible individual within the organization. Hence, the term *certified welder* is sometimes referred to rather than *qualified welder*. An illustrative record form for welder or welding operator qualification testing is shown in Fig. 7.13.

Duration of Qualification

The duration of qualification for welders and welding operators varies from one code or specification to another. One code allows welders to be inactive with the welding process for three months and, if they have welded with another process, the period may be extended to six months. Other codes allow six months of inactivity with the welding process before requalification is required. However, qualification may be extended indefinitely if the welder or welding operator has welded with the process within the stated period, and this is documented in some way.

BRAZER PERFORMANCE

The purpose of brazer performance qualification tests is to determine the ability of brazers to make sound brazed joints following a brazing procedure specification (BPS) and under conditions that will be encountered in production applications. Brazing operators are tested to determine their ability to operate machine or automatic brazing equipment in accordance with a BPS.

Two standards that address brazing performance qualification are *AWS B2.2, Brazing Procedure and Performance Qualification* and *Section IX, ASME Boiler and Pressure Vessel Code.*

Acceptance Criteria

The brazer or brazing operator is required to make one or more test samples using a qualified brazing procedure. Acceptance of performance test brazements may be based on visual examination or specimen testing. Qualification by specimen testing qualifies the person to perform production brazing following a BPS that was qualified by either specimen testing or visual examination. When performance qualification is based on visual examination alone, the person is permitted to make only those production brazements that will be visually inspected for acceptance.

Qualification by Visual Examination

Qualification by visual examination is done with a workmanship test brazement representative of the design details of the joint to be brazed in production. Typical workmanship brazements are shown in Fig. 7.14. The completed brazement must meet the requirements of the standard.

Qualification by Specimen Testing

Either a standard test brazement or a workmanship test brazement is used. A standard test brazement may have a butt, scarf, lap, single- or double-spliced butt, or a rabbet joint in plate or pipe, as shown in Fig. 7.15.

The test joint normally is sectioned, the exposed surfaces are polished and etched, and the brazed joint is examined at low magnification for discontinuities. Peel tests may be used in place of macroetch tests, or vice versa, using lap joints or spliced butt joints. In the latter case, the splice plates are peeled from the specimens. The macroetched cross sections or the peeled surfaces must meet the acceptance criteria of the appropriate standard.

Qualification Variables

Typical brazing performance qualification variables that may require requalification when changed are as follows:

(1) Brazing process
(2) Base metal
(3) Base metal thickness
(4) Brazing filler metal composition
(5) Method of adding filler metal
(6) Brazing position
(7) Joint design

To minimize the number of brazing performance qualification tests, base metals are separated into groups that have similar brazeability. Brazing filler metals are grouped according to similarity of composition or melting range.

Performance Qualification Test Record

Brazing of the test brazement and the results of the acceptance tests are reviewed by the person

WELDER AND WELDING OPERATOR QUALIFICATION TEST RECORD

Welder or welding operator's name ________________ Identification no. ________
Welding process ________ Manual ________ Semiautomatic ________ Machine ________
Position ________________
(Flat, horizontal, overhead or vertical — if vertical, state whether upward or downward)
In accordance with procedure specification no ________________
Material specification ________________
Diameter and wall thickness (if pipe) — otherwise, joint thickness ________________
Thickess range this qualifies ________________

FILLER METAL

Specification no. ________ Classification ________ F no. ________
Describe filler metal (if not covered by AWS specification) ________________
____
Is backing strip used? ________________
Filler metal diameter and trade name ________________
Flux for submerged arc or gas for gas metal arc or flux cored arc welding ________________

VISUAL INSPECTION

Appearance ________ Undercut ________ Piping porosity ________

Guided Bent Test Results

Type	Result	Type	Result

Test conducted by ________________ Laboratory test no. ________
per ________________ Test date ________

Fillet Test Results

Appearance ________________ Fillet size ________
Fracture test root penetration ________________ Marcoetch ________
(Describe the location, nature, and size of any crack or tearing of the specimen.)
Test conducted by ________________ Laboratory test no. ________
per ________________ Test date ________

RADIOGRAPHIC TEST RESULTS

Film identification	Results	Remarks	Film identification	Results	Results

Test witnessed by ________________ Test no. ________
per ________________

We, the undersigned, certify that the statements in this record are correct and that the welds were prepared and tested in accordance with the requirements of AWS D1.1, (________) Structural Welding Code — Steel.
year

Manufacturer or contractor ________________
Authorized by ________________
Date ________________

Fig. 7.13—Sample performance qualification test record

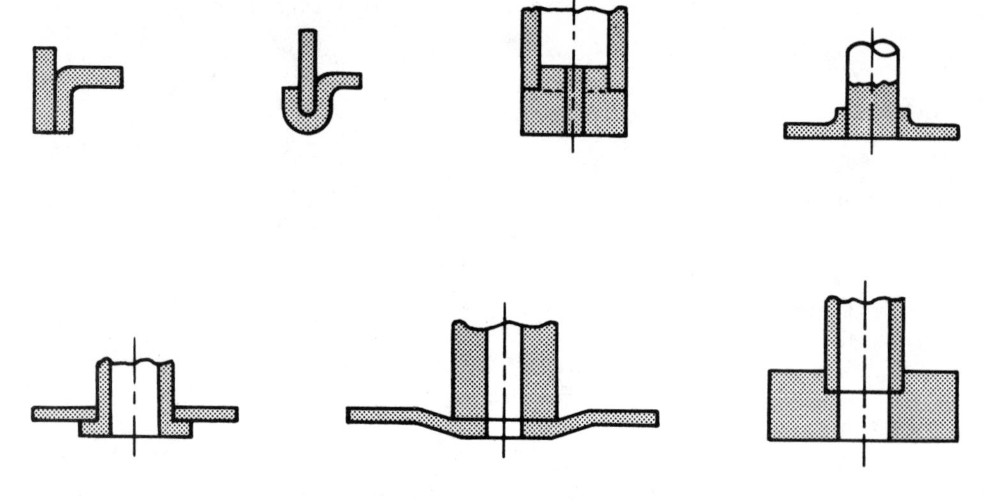

Fig. 7.14-Typical workmanship test brazements

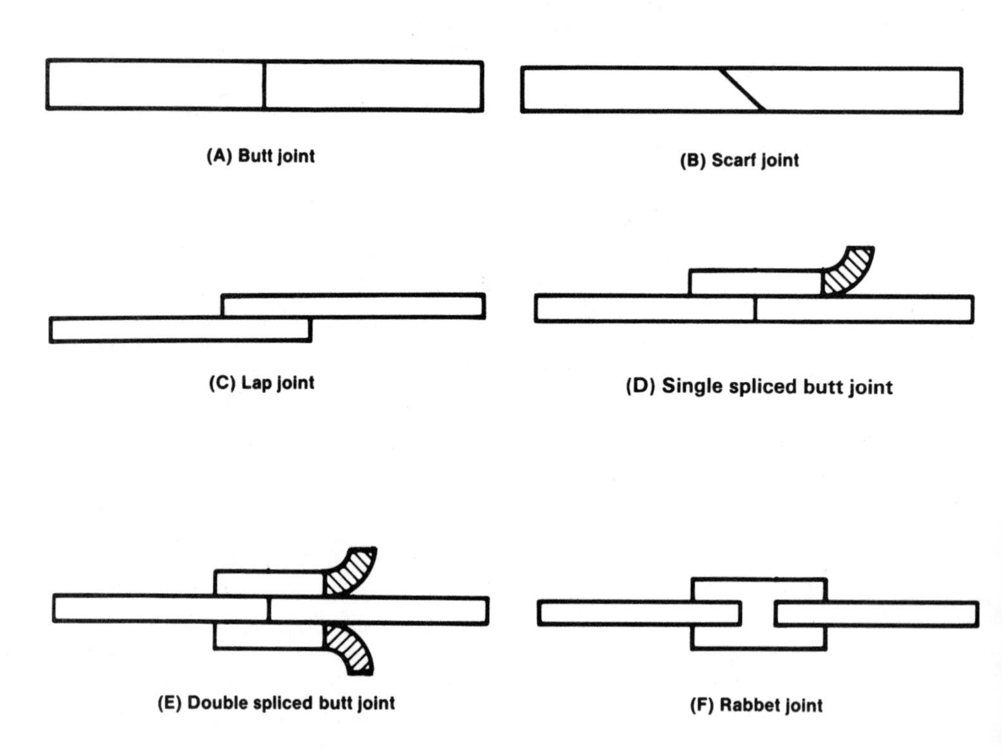

Fig. 7.15—Brazed joint designs for performance qualification

responsible for the test. The information is recorded on a Performance Qualification Test Record (see Fig. 7.16).

THERMAL SPRAY OPERATORS

Operators of equipment for applying thermal sprayed coatings may be qualified using the recommended procedures in *AWS C2.16, Guide for Thermal Spray Operator and Equipment Qualification.* They should be qualified for a specific coating process and method of application.

Initial operator qualification should consist of the following:

(1) A written test covering all aspects of the coating process and application method

(2) Demonstration of ability to operate the equipment

(3) Testing to determine knowledge of proper masking procedures for both surface preparation and spraying

(4) Surface preparation of the thermal spray test specimens

Test specimens should be prepared from alloys specified by the purchaser and agreed to by the certifying agent. The type and number of test specimens to be prepared should be specified in the contract documents. The testing conditions and results of the test should be recorded and certified by the representative that observed the tests.

NONDESTRUCTIVE TESTING PERSONNEL

Qualification of personnel is one of the most important aspects of nondestructive testing. As in welding, qualification assures that the person performing the test method has proper knowledge and experience to apply the test and interpret the results. Most nondestructive testing personnel are qualified in accordance with the American Society for Nondestructive Testing Recommended Practice No. SNT-TC-1A. This documetn establishes guidelines for education, training, experience, and testing requirements for various levels of competence. The levels and the functions that can be performed by each level are as follows:

NDT Level I. Level I personnel are qualified to properly perform specific calibrations, specific tests, and specific evaulations according to written instructions and to record the results. They receive the necessary guidance or supervision from a certified NDT Level II or III individual.

NDT Level II. Level II personnel are qualified to set up and calibrate equipment, and to interpret and evaluate results with respect to applicable codes, standards, and specifications. They must (1) be thoroughly familiar with the scope and limitations of the method; (2) exercise responsibility for on-the-job training and guidance of trainees and NDT Level I personnel; and (3) organize and report nondestructive testing investigations.

NDT Level III. Personnel certified for Level III are capable of and responsible for (1) establishing techniques; (2) interpreting codes, standards, and specifications; and (3) designating the particular test method and technique to be used. They are accountable for the complete NDT operation and evaluation of results in terms of existing codes and specifications. It is desirable that they have general familiarity with all commonly used NDT methods. They may be responsible for the training and examination of Level I and Level II personnel for certification.

WELDING INSPECTOR QUALIFICATION

The American Welding Society conducts examinations to determine the proficiency of welding inspectors and associate welding inspectors in accordance with *AWS QC1, Standard for Qualification and Certification of Welding Inspectors.* The objective of the welding inspector qualification and certification program is to verify the knowledge of welding inspectors who may be required to inspect weldments or welded products in accordance with codes or other specified requirements.

The examinations are designed to require a level of knowledge that can be determined by standardized tests, and do not necessarily reflect the level of competence required for each and every code or specification used in the welding fabrication field. Employers must recognize their

Performance Qualification Test Record

Name ______________________ Ident. ______________________

Date ______________________ BPS No. ______________________

Brazing Process ______________________ Brazer ☐ Operator ☐

Test Brazement

Base Metal Ident. ____________ BM No. ____________ BM Thickness ____________

Filler Metal Ident. ____________ FM No. ____________ FM Feed Method ____________

Test Position ____________ Joint Type ______________________

Other ______________________

Test Results

Visual	Pass	Fail

Macroetch or Peel

Specimen No.	Remarks	Pass	Fail

Qualified For

Brazing Process ______________________ Position ______________________

BM No. ______________________ BM Thickness ______________________

FM No. ______________________ BM Feed Method ______________________

Joint Type ______________________

The above named individual is qualified in accordance with the American Welding Society Standard for Brazing Procedure and Performance Qualification, AWS B2.2-84.

Date ______________________

Signed ______________________
Qualifier

Fig. 7.16—Typical brazer performance test record

responsibility for additional training and education of their welding inspectors when needed to satisfy current production needs and the requirements of new standards and specifications.

A certified welding inspector (CWI) must be able to perform inspections and verify that the work and records conform to the applicable codes and specifications. A certified associate welding inspector (CAWI) performs inspections under the direction of a CWI. However, the CWI is responsible for determining if weldments conform to workmanship and acceptance standards.

The applicants must have specified minimum visual, educational, and work qualifications directly related to weldments fabricated to a code or specification. The applicant's work experience must include one or more of the following:

(1) Design of weldments

(2) Production welding

(3) Inspection of welds

(4) Repair of welds

Candidates are also required to pass written tests covering the following:

(1) One of four specified codes and specifications

(2) Principles of welding, nondestructive testing, materials, heat treatment, and other fundamentals

(3) Practical application of welding inspection

Upon successful completion of the examination, the *American Welding Society* issues a certificate to the applicant. Certification must be renewed every three years. This may be done by re-examination or by meeting certain requirements for renewal without examination.

STANDARDIZATION OF QUALIFICATION REQUIREMENTS

The primary reason for standardization in any field is economics, in particular as it affects the consumer. Standardization in the field of welding qualification is an outstanding example of the savings that may be made through standardization. Increasing concern for the protection of consumer interest makes it imperative that standards be developed on a voluntary consensus basis with all affected interests represented.

Most welding qualification standards have only minor differences, but now require separate qualification for each. The resulting duplication of effort and cost in the fabrication and construction industries is highly undesireable, and results in significant increases in cost to the consumer with no related benefits.

The American Welding Society has responded in the field of welding qualification in several programs. The Welding Inspector Qualification and Certification program is one example of standardization of qualification that is receiving recognition and acceptance by industry and government agencies. The two general qualification standards, *AWS B2.1, Standard for Welding Procedure and Performance Qualification* and *AWS B2.2, Standard for Brazing Procedure and Performance qualification*, are designed to provide single standards to replace several different standards that contain similar rules for qualification of procedures and performance.

Metric Conversion Factor

1 in. = 25.4 mm

SUPPLEMENTARY READING LIST

Cary, H.B., *Modern Welding Technology*, Englewood Cliffs, NJ: Prentice-Hall, 1979.

Rules for Building and Classing Steel Vessels, New York: American Bureau of Shipping, latest edition.

The Procedure Handbook of Arc Welding, 12th Ed., Cleveland, OH: The Lincoln Electric Company, 1973.

Welding Inspection, 2nd Ed., Miami, FL: American Welding Society, 1980.

8
Weld Quality

Introduction *276*

Terminology *278*

Discontinuities in Fusion Welded Joints *278*

Causes and Remedies for Fusion Weld Discontinuities *293*

Discontinuities in Resistance and Solid State Welds *303*

Discontinuities in Brazed and Soldered Joints *306*

Significance of Weld Discontinuities ... *308*

Supplementary Reading List........... *316*

Chapter Committee

C.D. LUNDIN, *Chairman*
The University of Tennessee

A. LEMON
Welding Institute of Canada

J.W. LEE
AVCO Lycoming Division

A.E. PEARSON
Tucker Metal Manufacturing

A.W. PENSE
Lehigh University

Welding Handbook Committee Member

E.G. SIGNES
Bethlehem Steel Corporation

8
Weld Quality

INTRODUCTION

MEANING OF WELD QUALITY

To enable a weldment (or brazement) to have the required reliability throughout its life, it must have a sufficient level of *quality* or *fitness for purpose*. Quality includes design considerations, which means that each fabricated weldment should be

(1) Adequately designed to meet the intended service for the required life.

(2) Fabricated with materials and welds in accordance with the design concepts.

(3) Operated and maintained properly.

Quality is a relative term, and it is unnecessary and in fact often costly to have higher quality than is needed for the application. Thus, quality levels can be permitted to vary among different weldments and individual welds, depending on service requirements.

Weld quality often is taken narrowly as the level of geometric type imperfections in a weld, but it should also include other features such a hardness, composition, and toughness, all of which contribute to the fitness for purpose of a weld.

Selection of the required quality to give the desired reliability primarily depends on the possible modes of failure under the service conditions.

Weld quality relates directly to the integrity of a weldment. It underlies all of the fabrication and inspection steps necessary to ensure a welded product that will be capable of serving the intended function for the desired life. Both economics and safety influence weld quality considerations. Economics require that a product be competitive, and safety dictates the avoidance of potential injury to personnel or extensive damage to plant and equipment.

The majority of the codes and standards[1] governing welded fabrications define the quality requirements for welded construction to insure reasonably safe operation of the structure. The code or standard requirements are to be considered minimums, and the acceptance criteria for welds should not be encroached upon without sound engineering judgement. For critical applications, more stringent requirements than those specified in the code or standard may be necessary to ensure safety.

The criteria for weld acceptance revolve around examination of welds, generally by nondestructive means, for deviations from acceptable standards. All deviations are evaluated and the acceptance or rejection of a weld is usually based on well defined conditions. Repair of nonacceptable or defective conditions is normally permitted to bring the quality up to acceptance standards. Many codes and standards relating to weld quality do not govern product usage, and leave maintenance of the fabrication or product to the user. It is the user who must modify, amplify, or impose additional weld quality standards to insure that the product will exceed the minimum specified requirements.

Weld quality, while often difficult to define precisely, is often governed by codes, standards,

1. Codes and standards are discussed in Chapter 6 of this volume.

or regulations based on rational assessments of economics and safety. The documents may be modified by the user to reflect additional concerns of usage related to safety or economics, or both. In other cases, acceptable weld quality must be defined by design engineering or the customer. Welds are examined in regard to size, shape, contour, soundness, and other features. Thus, at the heart of weld quality is the understanding of the occurrence, inspection,[2] and elimination or repair of any defects.

SELECTION OF WELD QUALITY

Determination of the overall quality requirements for a weldment is a major consideration involving design teams and quality groups. Specifying excessive quality can lead to high costs with no benefits, while reduced quality in weldments can result in high maintenance costs and excessive loss of service. Thus, the aim is to specify features which lead to *fitness for purpose.*

Fortunately, valuable guidance is provided in codes and specifications which often indicate allowable stress levels, properties, and discontinuities in weldments. These standards are based on experience, and they have proven to be safe.

Quality selection also involves consideration of those major factors that can be analyzed by fracture mechanics (discussed later), and a number of other factors. The main factors are as follows:

(1) *Service Conditions*

(a) *Stress level.* Appropriate section sizes must be used to ensure that stress levels are not excessive.

(b) *Nature of stress.* Under dynamic or cyclic loading, particularly where the cycles exceed 10^4, fatigue must be considered.

(c) *Working temperature.* Low and high temperatures require design for brittle fracture and creep, respectively.

(d) *Corrosion and wear.*

(2) *Material Properties.* Material of suitable strength, toughness, corrosion resistance, and other relevant properties must be selected to suit the design. Also, the fabricator must use practices that do not lower material properties below acceptable levels, e.g., cold and hot forming, postweld heat treatment, excessive or insufficient heat inputs.

(3) *Geometric Imperfections.* The design must limit geometric imperfections to ensure that they do not cause cracking or leakage. The fabricator must use practices which avoid them in the weldment.

(4) *Risk of Defects Arising.* Possibility of defects is high when difficult-to-weld materials are welded in awkward positions or on site where quality control is usually less reliable.

(5) *Risk of Defects Not Being Detected.* Fillet welds are normally difficult to assess internally, as are other welds where inspection will be limited.

(6) *The Consequences of Failure.* Where the likelihood of failure is high, higher quality and increased inspection may be needed. Consequences increase with:

(a) *Size of weldment*

(b) *Stored energy* (high potential energy in tall towers, pressure vessels, dam gates, etc; high kinetic energy in moving trains, ships, etc.)

(c) *Location with respect to people*

(d) *Loss of production* (A simple boiler tube failure may require little repair effort but production loss is costly.)

Finally, the basis for selection of overall quality is a combination of design, fabrication, and testing which will provide the lowest *total* cost over the full life of the weldment. Least initial cost, least weight, least welding, least imperfections, etc., taken individually, should not be a basis for quality selection.

Optimum cost quality is based on (1) costs of design, materials, fabrication, quality assurance, and money; (2) costs of possible failure multiplied by the probability of failure; and (3) service costs (including maintenance). Thus, the purchaser should realize that the lowest fabrication costs may not represent the lowest total costs.

---

2. Inspection methods and procedures for detecting discrepancies in welded and brazed joints are discussed in Chapter 9 of this Volume.

TERMINOLOGY

The necessity for uniform understanding of quality and inspection requirements of weldments and brazements has resulted in the modification of standard definitions for certain terms to incorporate special meanings pertaining to welded joints.[3] Several important definitions follow.

Discontinuity. An interruption of the typical structure of a weldment such as a lack of homogeneity in the mechanical, metallurgical, or physical characteristics of the material or weldment. A discontinuity is not necessarily a defect.

Flaw. A near synonym for discontinuity, but with connotation of undesireability.

Defect. One or more discontinuities that by nature or accumulated effect render a part of product unable to meet minimum applicable acceptance standards or specifications. This term designates rejectability.

Arc strike. A discontinuity consisting of any localized remelted metal, heat-affected metal, or change in the surface profile of any part of a weld or base metal resulting from arc impingement.

Burn-through. A shop term indicating a hole through the weld metal in a single pass weld.

Crack. A fracture-type discontinuity characterized by a sharp tip and high ratio of length and width to opening displacement.

Fissure. A small crack-like discontinuity with only slight separation (opening displacement) of the fracture surfaces. The prefixes *macro* or *micro* may be used to indicate relative size.

Inadequate joint penetration. Joint penetration which is less than that specified.

Incomplete fusion. Fusion which is less than complete.

Overlap. The protrusion of weld metal beyond the toe, face, or root of the weld.

Porosity. Cavity type discontinuities formed by gas entrapment during weld solidification.

Shrinkage void. A cavity type discontinuity normally formed by shrinkage during solidification.

Slag inclusion. Nonmetallic solid material entrapped in weld metal or between weld metal and base metal.

Undercut. A groove melted into the base metal adjacent to the toe or root of a weld, and left unfilled by weld metal.

Underfill. A depression on the face of the weld or root surface extending below the surface of the adjacent base metal.

Defective weld. A weld containing one or more defects.

Acceptable weld. A weld that meets all the design and service requirements and the acceptance criteria prescribed by the welding code or specifications, or the contractual documents, or both.

DISCONTINUITIES IN FUSION WELDED JOINTS

CLASSIFICATION

Discontinuities in fusion welded joints may be generally classified into three major groups, namely process and procedures related, metallurgical, and design related. Table 8.1 lists those discontinuities usually considered in each of the three major groups. The discontinuities so listed should not be considered rigorously assigned to a group because they may have secondary origins in other groups. The process and procedure related and design related discontinuities reflect, for the most part, those that alter, raise, or amplify stresses in a weld or the heat-affected

3. Refer to AWS A3.0, *Welding Terms and Definitions,* latest edition, published by the American Welding Society.

zone. Some metallurgical discontinuities may vary the local stress distribution, but those connected with the microstructure may also change the properties of the metal and provide metallurgical notch effects.

Table 8.1
Classification of weld joint discontinuities

Welding Process and Procedures Related
- A. Geometric
  - Misalignment
  - Undercut
  - Concavity or convexity
  - Excessive reinforcement
  - Improper reinforcement
  - Overlap
  - Burn-through
  - Backing left on
  - Incomplete penetration
  - Lack of fusion
  - Shrinkage
  - Surface irregularity
- B. Other
  - Arc strikes
  - Slag inclusions
  - Tungsten inclusions
  - Oxide films
  - Spatter
  - Arc craters

Metallurgical
- A. Cracks or fissures
  - Hot
  - Cold or delayed
  - Reheat, stress-relief, or strain-age
  - Lamellar tearing
- B. Porosity
  - Spherical
  - Elongated
  - Worm-hole
- C. Heat-affected zone, microstructure alteration
- D. Weld metal and heat-affected zone segregation
- E. Base plate laminations

Design Related
- A. Changes in section and other stress concentrations
- B. Weld joint type

Discontinuities should not only be characterized as to their nature but also as to their shape, that is whether they are essentially planar or are three dimensional. Planar types, such as cracks, laminations, incomplete fusion, and inadequate joint penetration, generally have more pronounced stress amplification effects than three dimensional discontinuities. Therefore, the following characteristics of discontinuities should always be considered:

(1) Size

(2) Acuity or sharpness

(3) Orientation with respect to the principal working stress and residual stresses

(4) Location with respect to the weld and to the exterior surfaces of the joint

Discontinuities oriented or positioned where stresses tend to enlarge them are more detrimental than those that are not so positioned. Also, surface or near surface discontinuities may be more detrimental than similarly shaped, internal discontinuities.

MATERIAL AND PROCESS RELATED DISCONTINUITIES

Certain discontinuities found in welded joints are related to specific welding processes while others are related to a particular type of base metal or filler metal, or both. Slag inclusions are generally associated with shielded metal arc, flux cored arc, submerged arc, and electroslag welding. Tungsten inclusions result from improper gas tungsten arc welding practices. Hydrogen-induced cold cracking and lamellar tearing are only found in steel weldments. Hot cracking may take place in many types of weld and base metals.

Weld discontinuities that may be encountered with a specific welding process are discussed in the appropriate process chapter in Volumes 2 and 3 of this *Handbook* (7th Edition). Likewise, discontinuities associated with a specific base metal are covered in the appropriate chapter in Volume 4 of this *Handbook*.

LOCATION AND OCCURRENCE OF DISCONTINUITIES

Discontinuities may be found in the weld metal, the heat-affected zone, and the base metal of weldments. The common weld discontinuities, general locations, and specific occurrences are presented in Table 8.2

The discontinuities listed in Table 8.2 are depicted in butt, lap, corner, and T-joints in Figs. 8.1 through 8.6. Where the list indicates

Table 8.2
Common types of fusion weld discontinuities

Type of discontinuity	Depiction on Figs. 8.1-8.6	Location[a]	Remarks
Porosity		W	Weld only, as discussed herein
Uniformly scattered	1a		
Cluster	1b		
Linear	1c		
Piping	1d		
Inclusions		W	
Slag	2a		
Tungsten	2b		
Incomplete fusion	3	W	At joint boundaries or between passes
Inadequate joint penetration	4	W	Root of weld preparation
Undercut	5	BM	Junction of weld and base metal at surface
Underfill	6	W	Outer surface of joint preparation
Overlap	7	W	Junction of weld and base metal at surface
Laminations	8	BM	Base metal, generally near mid-thickness of section
Delamination	9	BM	Base metal, generally near mid-thickness of section
Seams and laps	10	BM	Base metal surface almost always longitudinal
Lamellar tears	11	BM	Base metal, near weld HAZ
Cracks (includes hot cracks and cold cracks)			
Longitudinal	12a	W, HAZ	Weld or base metal adjacent to weld fusion boundary
Transverse	12b	W, HAZ, BM	Weld (may propagate into HAZ and base metal)
Crater	12c	W	Weld, at point where arc is terminated
Throat	12d	W	Weld axis
Toe	12e	HAZ	Junction between face of weld and base metal
Root	12f	W	Weld metal, at root
Underbead and heat-affected zone	12g	HAZ	Base metal, in HAZ
Fissures		W	Weld metal

a. W—weld, BM—base metal, HAZ—heat-affected zone

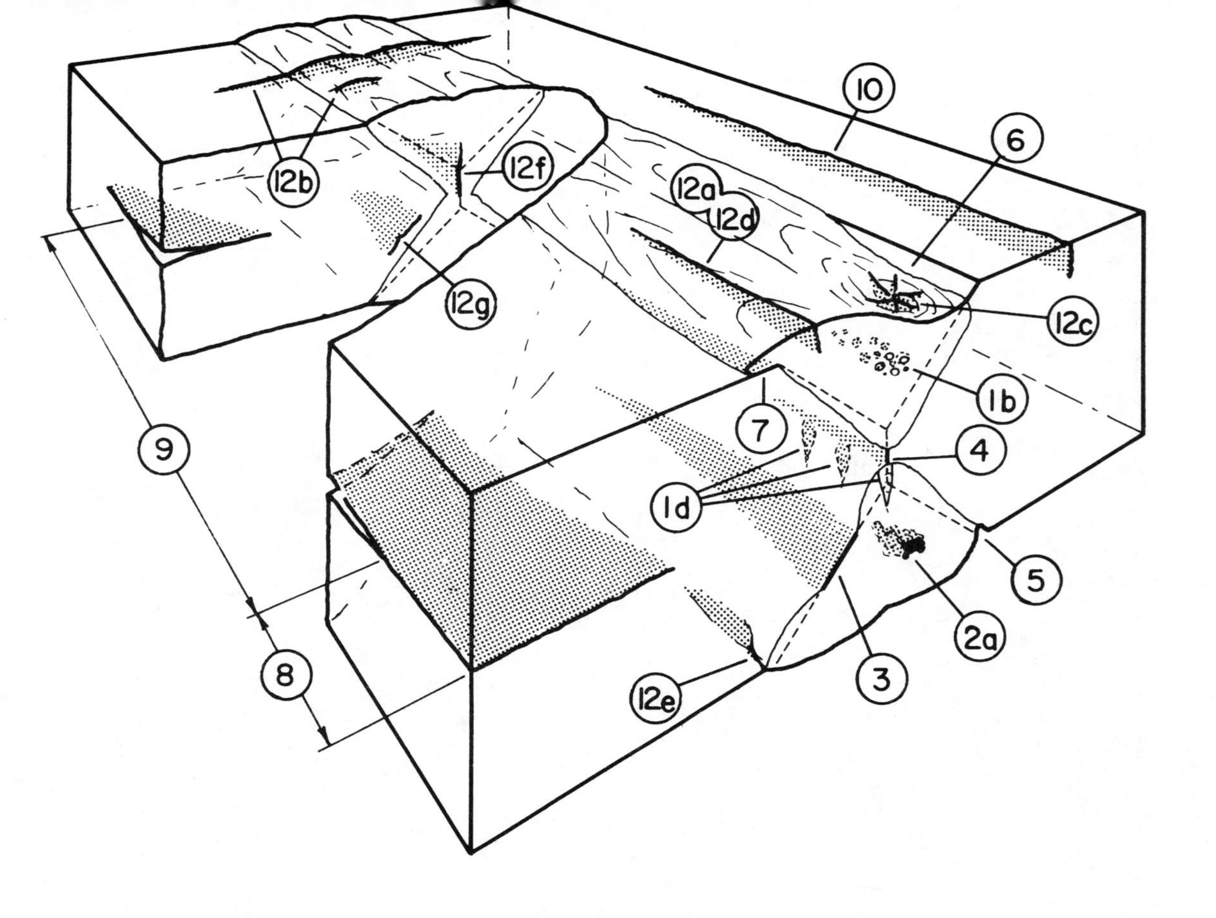

Fig. 8.1—Double-V-groove weld in butt joint

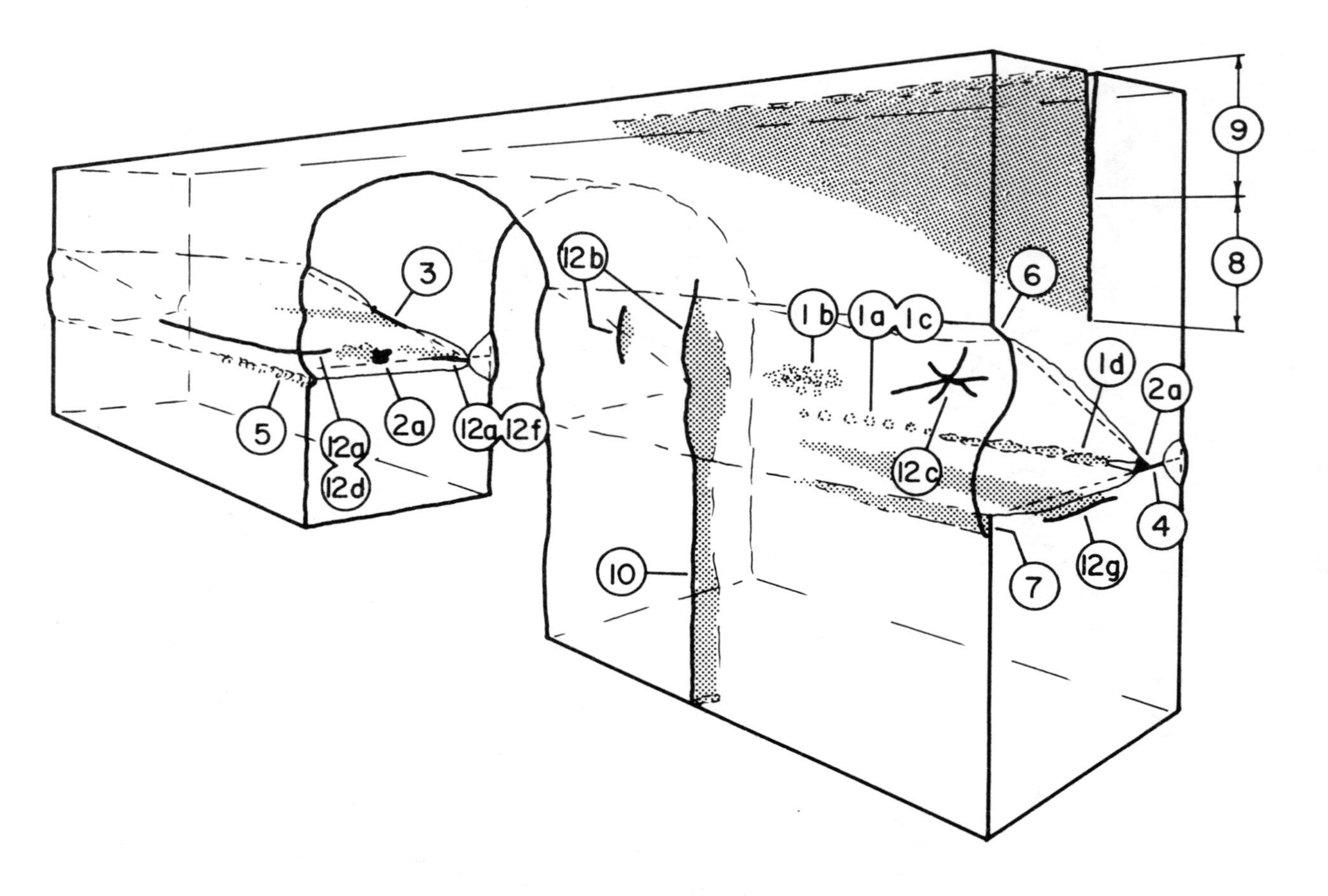

Fig. 8.2—Single-bevel-groove weld in butt joint

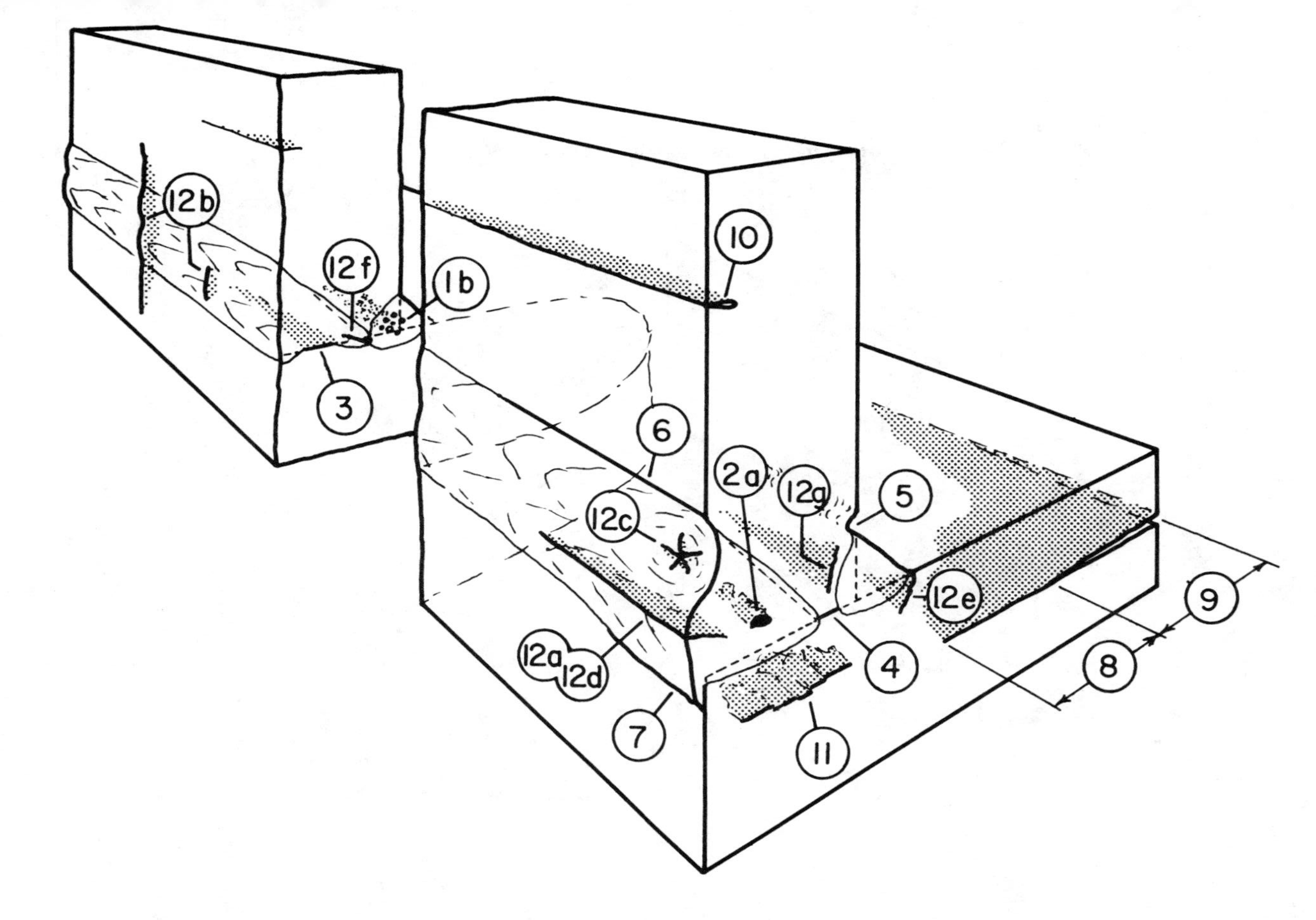

Fig. 8.3—Welds in corner joint

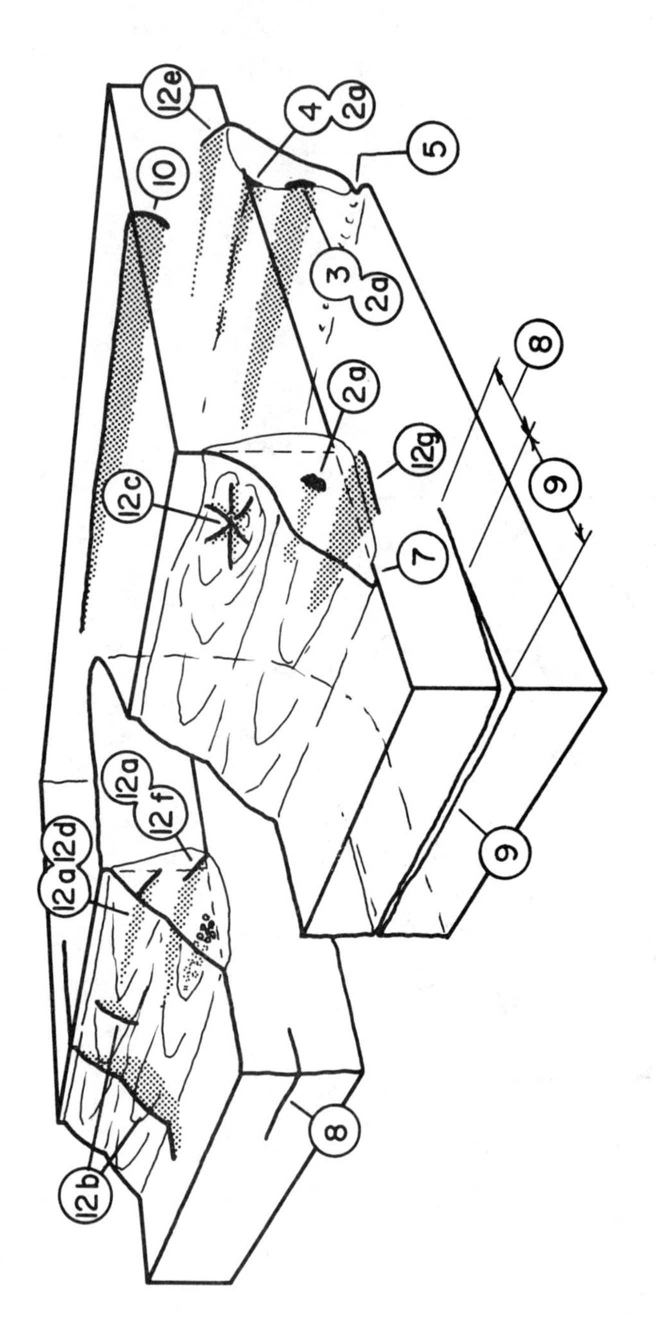

Fig. 8.4—Double fillet weld in lap joint

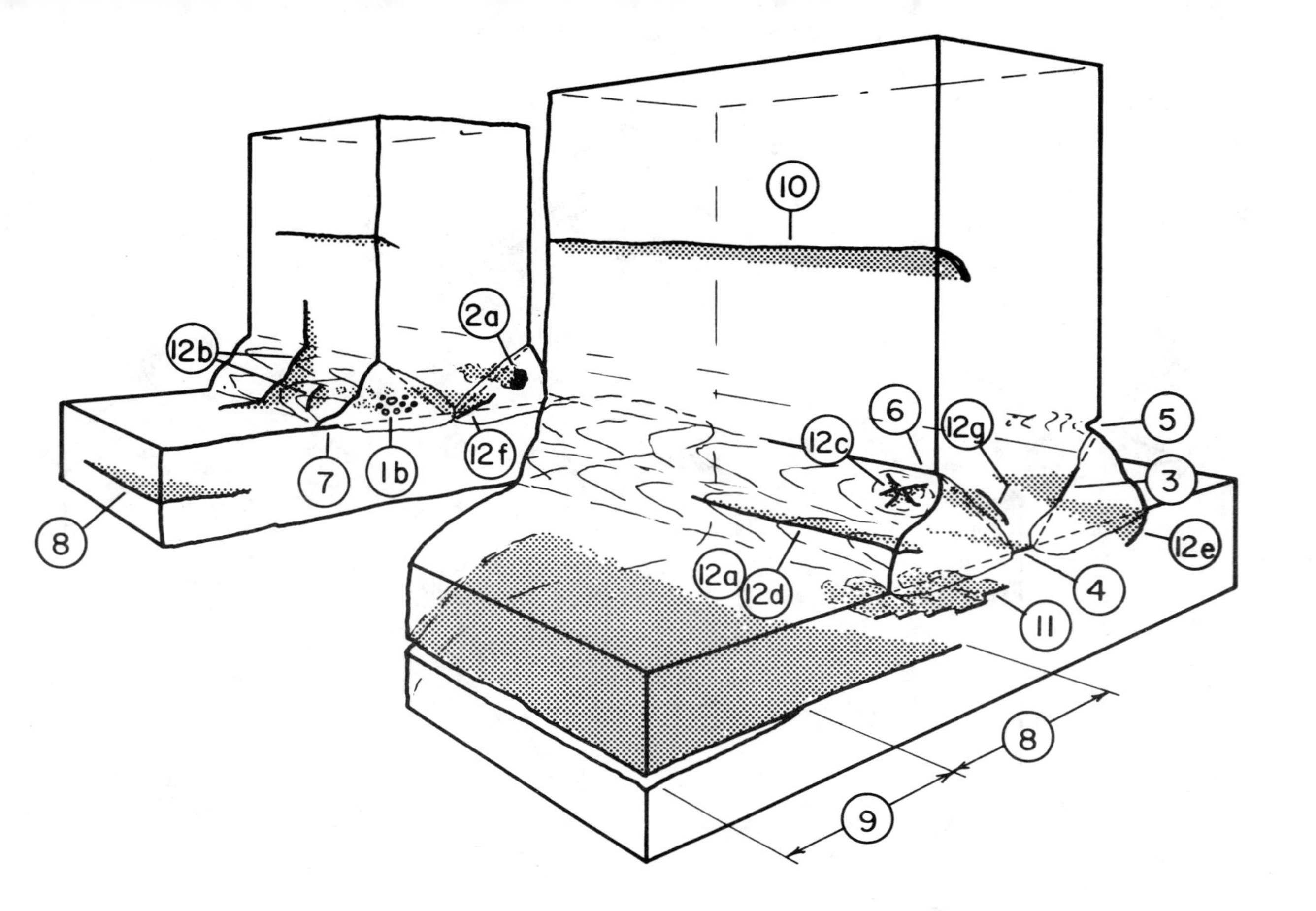

Fig. 8.5—Combined groove and fillet welds in T-joint

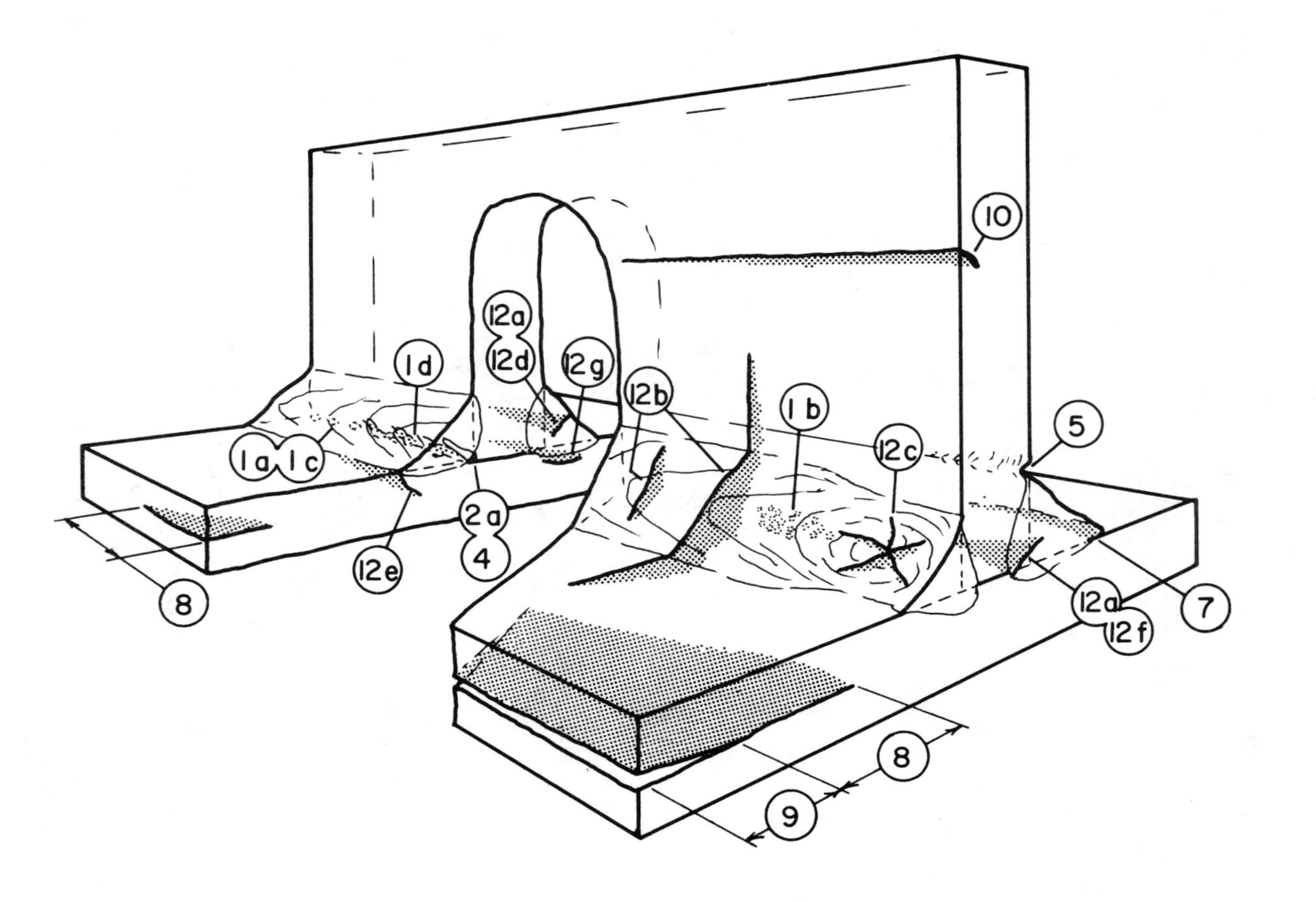

Fig. 8.6—Single pass fillet welds in a T-joint

that a discontinuity is generaly located in the weld, it may be expected to appear in almost any type of weld. However, there are exceptions. For example, tungsten inclusions are only found in welds made by the gas tungsten arc welding process, and macrofissures normally occur only in electroslag and electrogas welds.

Weld and base metal discontinuities of specific types are more common when certain welding processes and joint details are used. Table 8.3 indicates the types of discontinuities normally encountered with various welding processes. High restraint and limited access to portions of a weld joint may result in higher than normal incidence of weld and base metal discontinuities with a particular welding process.

Each general type of discontinuity is discussed below. The term *fusion type discontinuity* is sometimes used inclusively to describe slag inclusions, incomplete fusion, inadequate joint penetration, and similar generally elongated discontinuities in weld fusion. Many codes consider fusion type discontinuities less critical than cracks. However, some codes specifically prohibit cracks as well as incomplete fusion and partial joint penetation. Spherical discontinuities, almost

Table 8.3
Discontinuities commonly associated with welding processes

	Type of discontinuity						
Welding process	Porosity	Slag inclusions	Incomplete fusion	Inadequate joint penetration	Undercut	Overlap	Cracks
Stud welding			X				X
Plasma arc welding	X		X	X			X
Submerged arc welding	X	X	X	X	X	X	X
Gas tungsten arc welding	X		X	X			X
Gas metal arc welding	X	X	X	X	X	X	X
Flux cored arc welding	X	X	X	X	X	X	X
Shielded metal arc welding	X	X	X	X	X	X	X
Carbon arc welding	X	X	X	X	X	X	X
Resistance spot welding			X				X
Resistance seam welding			X				X
Projection welding			X				X
Flash welding			X				X
Upset welding			X				X
Percusion welding			X				X
Oxyacetylene welding	X		X	X			X
Oxyhydrogen welding	X		X	X			X
Pressure gas welding	X		X				X
Cold welding			X				X
Diffusion welding			X				X
Explosion welding			X				
Forge welding			X				X
Friction welding			X				
Ultrasonic welding			X				
Electron beam welding	X		X	X			X
Laser beam welding	X		X				X
Electroslag welding	X	X	X	X	X	X	X
Induction welding			X				X
Thermit welding	X	X	X				X

always gas pores, can occur anywhere within the weld. Elongated discontinuities may be encountered in any orientation. Specific joint types and welding procedures have an effect on the type, location, and incidence of discontinuities. When the welding process, joint details, restraint of the weldment, or a combination of these have an effect on the discontinuities to be expected, the controlling conditions are described below.

FUSION WELD DISCONTINUITIES

Porosity

Porosity is the result of gas being entrapped in solidifying weld metal. The discontinuity is generally spherical, but may be elongated. Porosity will be present in a weld if the welding technique, materials used, or condition of the weld joint preparation promote gas formation and entrapment. If the molten weld metal cools slowly and allows all gas to rise to the surface before solidification, the weld will be virtually free of porosity.

Uniformly scattered porosity may be distributed throughout single pass welds or throughout several passes of multiple pass welds. Whenever uniformly scattered porosity is encountered, the cause is generally faulty welding technique or defective materials, or both.

Cluster porosity is a localized grouping of pores that may result from improper initiation or termination of the welding arc.

Linear porosity may be aligned along (1) a weld interface, (2) the root of a weld, or (3) a boundary between weld beads. Linear porosity is caused by gas evolution from contaminants along a particular boundary.

Piping porosity is a term for elongated gas pores. Piping porosity in fillet welds normally extends from the root of the weld toward the face. When one or two pores are seen in the surface of the weld, it is likely that many subsurface piping pores are interspersed among the exposed pores. Much of the piping porosity found in welds does not extend to the surface. Piping porosity in electroslag welds may be relatively long.

Inclusions

Slag inclusions are nonmetallic solid material entrapped in the weld metal or between the weld metal and the base metal. They may be present in welds made by most arc welding processes. In general, slag inclusions result from faulty welding techniques, improper access to the joint for welding, or both. With proper welding techniques, molten slag will float to the surface of the molten weld metal. Sharp notches in joint boundaries or between weld passes often result in slag entrapment in the weld metal.

Tungsten inclusions are particles trapped in weld metal deposited with the gas tungsten arc welding process. These inclusions may be present in a weld if the tungsten electrode is dipped into the molten weld metal, or if the welding current is too high and causes melting and transfer of tungsten droplets into the molten weld metal. Tungsten inclusions appear as light areas on radiographs because tungsten is more dense than the surrounding metal and absorbs larger amounts of x-rays or gamma radiation. Almost all other weld discontinuities are indicated by dark areas on radiographs.

Incomplete Fusion

Incomplete fusion in a weld results from incorrect welding techniques, improper preparation of the materials for welding, or wrong joint design. Welding conditions that contribute to incomplete fusion include insufficient welding current and lack of access to all faces of the weld joint that should be fused during welding. With certain metals, the weld metal may fail to fuse with the base metal even though the joint faces have been melted because of the presence of tightly adhering oxides. This condition indicates improper preweld cleaning.

Inadequate Joint Penetration

When the joint penetration is less than the specified amount, it is considered inadequate for the application. Technically, this discontinuity may only be present when the welding procedure specification requires penetration of the weld metal beyond the original joint preparation. When the weld fails to penetrate an area of the weld joint that depends on penetration for fusion, the unpenetrated and unfused area is a discontinuity described as inadequate joint penetration. Inadequate joint penetration may result from insufficient welding heat, improper joint design (too much metal for the welding arc to penetrate), or poor control of the welding arc. Some welding processes have much greater penetrating ability than others. Many designs call for backgouging of the first weld of double groove welds to insure that there are no areas of inadequate joint penetration.

Undercut

Visible undercut is generally associated with either improper welding techniques or excessive welding currents, or both. It is generally located parallel to the junction of weld metal and base metal at the toe or root of the weld. Undercut discontinuities create a mechanical notch at the weld interface. If examined carefully, all welds have some undercut. Often the undercut may only be seen in metallographic tests where etched weld cross sections are examined under magnification. When undercut is controlled within the limits of the specifications and does not constitute a sharp or deep notch, it is usually not considered a weld defect.

The term *undercut* is sometimes used in the shop to describe melting away of the groove face of a joint at the edge of a layer or bead of weld metal. It forms a recess in the joint face where the next layer or bead of weld metal must fuse to the base metal. If the depth of fusion at this location is too shallow when the next layer of weld metal is applied, voids may be left in the fusion zone. These voids would be identified as incomplete fusion. This type of undercut is usually associated with incorrect manipulation of the welding electrode while depositing a weld bead or layer next to the joint face.

Underfill

Underfill results simply from the failure of the welder or welding operator to fill the joint with weld metal as called out in the welding procedure specification or on the design drawing. Normally, the condition is corrected by adding one or more additional layers of weld metal in the joint prior to subsequent processing.

Overlap

Overlap is usually caused by incorrect welding procedures, wrong selection of welding materials, or improper preparation of the base metal prior to welding. If tightly adhering oxides on the base metal interfere with fusion, overlap will result along the toe, face, or root of the weld. Overlap is a surface-connected discontinuity that forms a severe mechanical notch parallel to the weld axis.

Cracks

Cracks will occur in weld metal and base metal when localized stresses exceed the ultimate strength of the metal. Cracking is generally associated with stress amplification near discontinuities in welds and base metal, or near mechanical notches associated with the weldment design. Hydrogen embrittlement often contributes to crack formation in steel. Plastic deformation at crack edges is very limited.

Cracks can be classified as either *hot* or *cold* cracks. Hot cracks develop at elevated temperatures. They commonly form during solidification of the metal at temperatures near the melting point. Cold cracks develop after solidification of a fusion weld as a result of stresses. Such cracks in steel are sometimes called *delayed cracks* when they are associated with hydrogen embrittlement. Hot cracks propagate between the grains while cold cracks propagate both between grains and through grains.

Cracks may be longitudinal or transverse depending on their orientation with respect to the weld axis. Longitudinal cracks are parallel to the axis of the weld regardless of location in the weld metal or the heat-affected zone.

Throat cracks run longitudinally in the face of the weld extending toward the root of the weld. They are generally, but not always hot cracks. *Root cracks* also run longitudinally but in the root of the weld. They are generally hot cracks.

Longitudinal cracks in automatic submerged arc welds are commonly associated with high welding speeds, and are sometimes related to internal porosity. Longitudinal cracks in small welds between heavy sections are often the result of high cooling rates and high restraint.

Transverse cracks are nearly perpendicular to the axis of the weld. They may be limited in size and completely within the weld metal, or they may propagate from the weld metal into the adjacent heat-affected zone, but not farther into the base metal. Transverse cracks are generally the result of longitudinal shrinkage strains acting on weld metal of low ductility.

Crater cracks are formed by improper termination of a welding arc. They are usually shallow hot cracks and sometimes are referred to as star cracks when they form a star-like cluster.

Toe cracks are generally cold cracks that initiate approximately normal to the base material surface and then propagate from the toe of the weld where residual stresses are higher. These cracks are generally the result of thermal shrinkage strains acting on a weld heat-affected zone that has been embrittled. Toe cracks sometimes occur when the base metal cannot accommodate the shrinkage strains that are imposed by welding.

Underbead cracks are generally cold cracks that form in the heat-affected zone. They are generally short and discontinuous, but may extend to form a continuous crack. Underbead cracking can occur when three elements are present: (1) hydrogen in steel, (2) a relatively brittle microstructure, and (3) high residual stresses. These cracks are usually found at regular intervals under the weld metal, and do not extend to the surface. Therefore, they cannot be detected by visual methods of inspection, and may be difficult to locate by other inspection methods.

Fissures are crack-like separations of small or moderate size along internal grain boundaries. They can occur in fusion welds made by most welding processes, but they are easier to see in electroslag welds because of the larger grain sizes in the weld. The separations can be either hot or cold cracks. Their effects on the performance of welded joints are the same as cracks of similar sizes in the same locations and orientations.

Surface Irregularities

Surface Pores. Occcasionally, welding conditions cause pores to form in the face of the weld bead. The pattern can vary from a single pore every few inches to many pores per inch. The gases shielding the droplets of molten filler metal in the arc and the molten weld pool are an important factor in the quality of the weld metal produced. Any change that varies the gas shielding will affect the weld. An example of such a change is a difference between the shielding at the arc when a weld pass is being deposited at the bottom of a narrow groove, compared to that during deposition of the top layer of the same weld. Improvement in weld bead appearance is usually obtained by changing the welding conditions, such as polarity or arc length.

It is important to eliminate surface pores because they can result in slag entrapment or other discontinuities. When sound welds are the goal, it is not safe to assume that pores will be eliminated when the next layer of weld metal is deposited. Surface pores should be removed by chipping or grinding before deposition of succeeding layers.

Other Surface Irregularities. Varying width of the surface layers, depressions, variations in height of reinforcement, nonuniformity of weld ripples, and other surface irregularities may not be classified as weld discontinuities. However, they are a part of the weld. Surface irregularities may not affect the integrity of a completed weld, but they frequently are covered by specification requirements and subject to inspection.

Magnetic disturbances, poor welding technique, and improper electrical conditions can account for certain surface irregularities. Such conditions might be caused by lack of experience, inaccessibility to the weld joint, or other factors peculiar to a specific job. Generally speaking, surface appearance reflects the ability and experience of the welder, and presence of surface irregularities may be deleterious. Welds with uniform surfaces are desireable for structural as well as visual reasons.

A fillet weld with poor surface appearance is shown in Fig. 8.7. A satisfactory weld is shown in Fig. 8.8. Acceptable weld surfaces are best judged with workmanship samples.

BASE METAL FLAWS

Not all discontinuities are a result of improper welding procedures. Many difficulties with weld quality may be attributed to the base metal. Base metal requirements may be defined by applicable specifications or codes. Departure

Fig. 8.7—Single-pass horizontal fillet weld with poor surface appearance caused by improper welding technique

Fig. 8.8—Single-pass horizontal fillet weld produced with the proper welding technique

from these requirements should be considered cause for rejection.

Base metal properties that may not meet prescribed requirements include chemical composition, cleanliness, laminations, stringers, surface conditions (scale, paint, oil), mechanical properties, and dimensions. The inspector should keep factors such as these in mind when trying to determine the causes of indications in welded joints that have no apparent cause. Several flaws that may be found in base metal are also shown in Figs. 8.1 through 8.6.

Laminations

Laminations in plate and other mill shapes are flat and generally elongated discontinuties found in the central zone of wrought products. Laminations may be completely internal and only detectable nondestructively by ultrasonic tests, or they may extend to an edge or end where they may be visible at the surface. They may also be found when cutting through or machining edges or ends of components.

Laminations are formed when gas voids or shrinkage cavities in an ingot are flattened during hot working operations. They generally run parallel to the surfaces of rolled products, and are most commonly found in mill shapes and plates. Some laminations are partially welded together during hot rolling operations. Tight laminations will sometimes conduct sound across the interface, and therefore may not be detectable by ultrasonic tests. Metals containing laminations often cannot be relied upon to carry tensile stress in the through-thickness direction.

Delamination in the base metal may occur when laminations are subject to transverse stresses. The stresses may be residual from welding or they may result from external loading. Delamination may be detected visually at the edges of pieces or by ultrasonic testing with longitudinal waves through the thickness. Delaminated metal should not be used to transmit tensile loads.

Lamellar Tears

Some rolled structural shapes and plates are susceptible to a cracking defect known as *lamellar tearing*. Lamellar tears are generally terrace-like separations in base metal typically caused by thermally-induced shrinkage stresses resulting from welding. Lamellar tearing, a form of fracture resulting from high stress in the through-thickness direction, may extend over long distances. The tears take place roughly parallel to the surface of rolled products. They generally initiate either in regions having a high incidence of coplanar, stringer-like, nonmetallic inclusions or in areas subject to high residual (restraint) stresses, or both. The fracture usually propagates from one lamellar place to another by shear along planes that are nearly normal to the rolled surface.

Laps and Seams

Laps and *seams* are longitudinal flaws at the surface of the base metal that may be found in hot-rolled mill products. When the flaw is parallel to the principal stress, it is not generally a critical defect; if perpendicular to applied or residual stresses, or both, it will often propagate as a crack. Seams and laps are surface connected discontinuties. However, their presence may be masked by manufacturing proceses that have subsequently modified the surface of a mill product. Welding over seams and laps can cause cracking. Seams and laps are harmful for applications involving welding, heat treating, or upsetting, and in certain components that will be subjected to cyclic loading. Mill products can be produced with special procedures to control the presence of laps and seams. Open laps and seams can be detected by penetrant and ultrasonic inspection methods, while those tightly closed may not be revealed during inspection.

Arc Strikes

Striking an arc on base metal that will not be fused into the weld metal should be avoided. A small volume of base metal may be momentarily melted when the arc is initiated. The molten metal may crack from quenching, or a small surface pore may form in the solidified metal. These discontinuities may lead to extensive cracking in service. Any cracks or blemishes caused by arc strikes should be ground to a smooth contour and reinspected for soundness.

DIMENSIONAL DISCREPANCIES

The production of satisfactory weldments depends upon, among other things, the maintenance of specified dimensions, whether these be the size and shape of welds or finished dimensions of an assembly. Requirements of this nature will be found in the drawings and specifications. Departure from the requirements in any respect should be regarded as dimensional discrepancies that, unless a waiver is obtained, must be corrected before acceptance of a weldment. Dimensional discrepancies can be largely avoided if proper controls are exercised in the welding procedure.

Warpage

Warpage or distortion is generally controllable by the use of suitable jigs, welding sequences, or presetting of joints prior to welding. The exact method employed will be dictated by the size and shape of the parts as well as by the thickness of the metal.

Incorrect Joint Preparation

Established welding practices require proper dimensions for each type of joint geometry consistent with the base metal composition, the welding process, and the thickness of the base metal. Departure from the required joint geometry may increase the tendency to produce weld discontinuities. Therefore, actual joint preparation should be that described in the applicable specifications, and be within specified limits.

Joint Misalignment

The term *misalignment* is often used to denote the amount of offset or mismatch across a butt joint between members of equal thickness. Many codes and specifications limit the amount of allowable offset because misalignment can result in stress raisers at the toe and root of the weld. Excessive misalignment is a result of improper fit-up, fixturing, or tack welding, or a combination of them.

Incorrect Size of Weld

Fillet Welds. The required size of fillet welds should be specified in welding procedure specifications. Fillet weld size can be measured with gages designed for this purpose. Undersize fillet welds can be corrected by adding one or more weld passes. Oversize fillet welds are not harmful if they do not interfere with subsequent assembly. However, they are uneconomical and can cause excessive distortion.

Groove Welds. The size of a groove weld is the joint penetration, which is equal to the depth of the joint preparation plus root penetration. Underfilled groove welds can be repaired by adding additional passes.

Repair of inadequate joint penetration depends upon the requirements of the joint design. When the weld is required to extend completely through the joint, repair may be made by backgouging to sound metal from the back side and applying a second weld. When only partial joint penetration is required or the back side of the weld is inaccessible, the weld must be removed and the joint rewelded using a modified welding procedure that will provide the required size of weld.

Incorrect Weld Profile

The profile of a finished weld may considerably affect performance of the joint under load. Moreover, the profile of a single pass or layer of a multiple pass weld may contribute to the formation of discontinuities such as incomplete fusion or slag inclusions when subsequent layers are deposited. Requirements concerning discontinuities of this nature in finished welds are usually included in the specifications and drawings involved. Nonconformity with the requirements constitutes a weld defect. Figure 8.9 illustrates various types of acceptable and defective weld profiles.

CAUSES AND REMEDIES FOR FUSION WELD DISCONTINUITIES

The common causes of porosity, inclusions, and cracks in welds and suggested methods of preventing them are summarized in Table 8.4.

POROSITY IN FUSION WELDS

Causes of Porosity

Dissolved gases are always present in molten weld metal during welding. Porosity is formed in weld metal as it solidifies when these dissolved gases are present in amounts greater than their solid solubility limits. The causes of porosity in weld metal are related to the welding process and the welding procedure, and in some instances, to the base metal type and chemistry. The welding

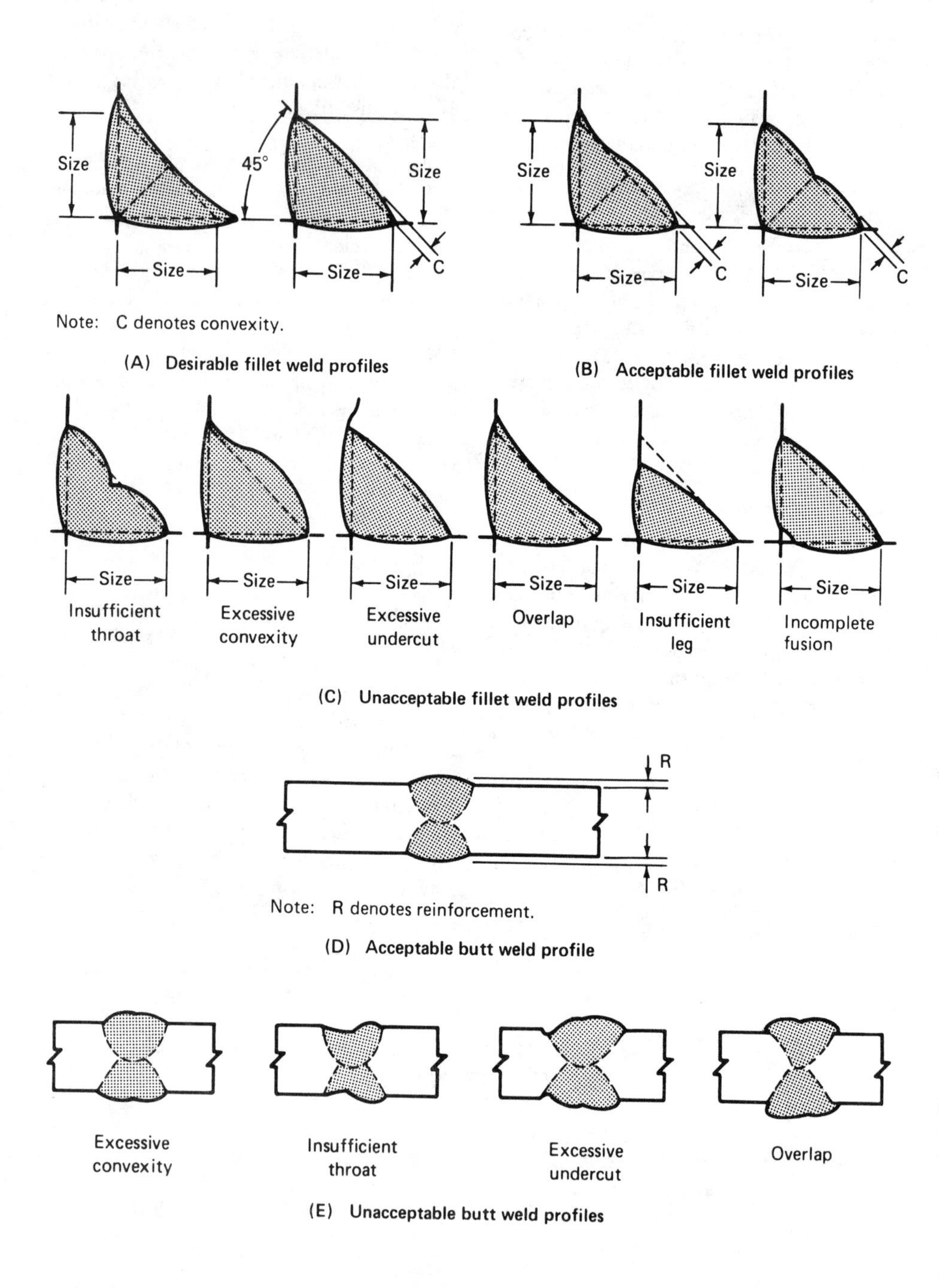

Fig. 8.9—Acceptable and unacceptable weld profiles

Table 8.4
Common causes and remedies of weld discontinuities

Cause	Remedies
	Porosity
Excessive hydrogen, nitrogen, or oxygen in welding atmosphere	Use low-hydrogen welding process; filler metals high in deoxidizers; increase shielding gas flow
High solidification rate	Use preheat or increase heat input
Dirty base metal	Clean joint faces and adjacent surfaces
Dirty filler wire	Use specially cleaned and packaged filler wire, and store it in clean area
Improper arc length, welding current, or electrode manipulation	Change welding conditions and techniques
Volatization of zinc from brass	Use copper-silicon filler metal; reduce heat input
Galvanized steel	Use E7010 electrodes and manipulate the arc heat to volatilize the galvanized (zinc) ahead of the molten weld pool
Excessive moisture in electrode covering or on joint surfaces	Use recommended procedures for baking and storing electrodes
	Preheat the base metal
High sulfur base metal	Use electrodes with basic slagging reactions
	Inclusions
Failure to remove slag	Clean surface and previous weld bead
Entrapment of refractory oxides	Power wire brush the previous weld bead
Tungsten in the weld metal	Avoid contact between the electrode and the work; use larger electrode
Improper joint design	Increase groove angle of joint
Oxide inclusions	Provide proper gas shielding
	Weld metal cracking
Highly rigid joint	Preheat
	Relieve residual stresses mechanically
	Minimize shrinkage stresses using backstep or block welding sequence
Excessive dilution	Change welding current and travel speed
	Weld with covered electrode negative; butter the joint faces prior to welding

Table 8.4 (Continued)

Cause	Remedies
Weld metal cracking	
Defective electrodes	Change to new electrode; bake electrodes to remove moisture
Poor fit-up	Reduce root opening; build up the edges with weld metal
Small weld bead	Increase electrode size; raise welding current; reduce travel speed
High sulfur base metal	Use filler metal low in sulfur
Angular distortion	Change to balanced welding on both sides of joint
Crater cracking	Fill crater before extinguishing the arc; use a welding current decay device when terminating the weld bead
Base metal cracking	
Hydrogen in welding atmosphere	Use low-hydrogen welding process; preheat and hold for 2 h after welding or postweld heat treat immediately
Hot cracking	Use low heat input; deposit thin layers; change base metal
Low ductility	Use preheat; anneal the base metal
High residual stresses	Redesign the weldment; change welding sequence; apply intermediate stress-relief heat treatment
High hardenability	Preheat; increase heat input; heat treat without cooling to room temperature
Brittle phases in the microstructure	Solution heat treat prior to welding

process, welding procedure, base metal type, and process used to produce the base metal directly affect the quantities and types of gases that are present in the molten weld pool. The welding process and welding procedure control the solidification rate; the latter is the overriding factor causing weld metal porosity. Proper welding procedures for a given welding process and base metal should produce welds that are essentially free of porosity.

Dissolved Gases in Weld Metal

The gases which may be present in the molten weld pool include the following:

(1) Hydrogen
(2) Oxygen
(3) Nitrogen
(4) Carbon monoxide
(5) Carbon dioxide
(6) Water vapor
(7) Hydrogen sulphide
(8) Argon
(9) Helium

Of these, only hydrogen, oxygen, and nitrogen are considered soluble to any significant extent in the molten weld pool. Their solubilities in the solidified metal are significantly less.

Hydrogen is considered to be the major cause of porosity in the welding of metals. It may enter the molten weld pool from many sources. For example, it may be present in the gas atmosphere surrounding the arc zone or in hydrogen-forming constituents, such as cellulose in the flux or electrode covering. Hydrogen may also be introduced into the molten weld pool by the dissociation of water. Moisture may be present in flux, electrode coverings, and the atmosphere, or on the base metal surfaces. Hydrogen dissolved in the base metal itself may also be introduced into the weld during welding. Filler metals may also contain dissolved hydrogen. Sulfur or selenium in the base metal may combine with hydrogen to form a gas.

Nitrogen may be a cause of porosity in steel welds and in nickel alloys. This gas may enter the molten weld pool from the atmosphere or from contaminated shielding gas. It may also be present in the base metal or filler metal in the form of dissolved nitrogen or nitrides.

Oxygen dissolved in the molten weld metal may also cause porosity. When present in molten steel, oxygen will react with carbon to form CO or CO_2. Oxygen may enter the molten weld pool as oxides on filler wire or on the base metal, in the form of compounds in flux or electrode covering, and from the atmosphere. Insufficient amounts of deoxidizers in steel base metals, filler metals, flux, or electrode coverings may result in incomplete deoxidation of the molten weld pool.

Significance to Weld Integrity

Porosity has been extensively evaluated as a discontinuity in welds. Tests have been conducted to determine the effects of porosity on both static and the dynamic behaviors of welded joints, and for virtually all types of base metals. At levels of 2 to 3 percent, porosity has an insignificant effect on static strength. This level is generally higher than that permitted by industry codes and standards. The effect on yield strength follows the same pattern as with tensile strength.

The effect of porosity on ductility is slightly more pronounced. The higher the tensile or yield strength of the materials, the greater is the effect on porosity ductility.

In addition, the gas or contaminant causing porosity may influence the properties of the weld metal by dissolving in it. Also, the gas in the pores or cavities may influence the metal surrounding the pore to the extent that the pore acts as a crack when loaded in service. (H_2 has this effect in steel, but other gases may not.)

The influence of porosity on the dynamic toughness of weld metal is less certain. The designer should investigate the effects of porosity on welds subjected to the expected type of loading before specifying acceptable porosity limits for the weldment.

In ferrous materials which have been postweld heat treated, the effect of porosity is mitigated to a great extent. The simple loss of cross-sectional area does not affect behavior until porosity levels exceed about 3 percent by volume.

In alloys that have a face-centered cubic crystal structure (Al, Cu, Ni), the influence of porosity is generally considered small. For all metals at high temperatures (creep range), the effect is again only related to the loss in cross-sectional area.

The most significant work on porosity has been done in terms of its effect on the fatigue properties of fusion welded butt joints with and without weld reinforcement. In any amount, the effect of porosity on the fatigue strengths of reinforced welds was overshadowed by stress concentrations on the surfaces. However, when the weld reinforcement was removed, exposed porosity contributed to failure by fatigue.

For fillet welds, the stress concentration effects of the weld toes and the start and stop locations are great, and they override all porosity considerations. Porosity in fillet welds does not appear to be a factor influencing behavior when held to reasonable limits.

The effect of surface porosity in butt and fillet welds is slightly different. Surface porosity is considered more injurious than buried or internal porosity, but certainly no worse than a crack. The appearance of surface porosity may reduce the effective throat of the weld needed to support the desired load. Surface porosity also often indicates that the process is out-of-control, and thus, other factors may act in concert to alter fatigue behavior.

SLAG INCLUSIONS

Causes and Remedies

Enrapped slag discontinuities typically occur with the flux shielded welding processes: shielded metal arc, flux cored arc, submerged arc, and electroslag welding. The slag produced during welding results from chemical reactions, in the molten weld pool, of elements present in the electrode coverings or welding fluxes, and in the filler and base metals. Some products of these reactions are nonmetallic compounds soluble only to a slight degree in the molten weld metals. Normally, the slag is found at the surface of the molten weld metal (unless it is physically restricted) because of lower specific gravity and interfacial energy considerations.

During the welding process, slag is formed and forced below the surface of the molten weld metal by the stirring action of the arc. Slag may also flow ahead of the arc, and metal may be deposited over it. In any case, slag tends to rise to the surface because of its lower density.

A number of factors may prevent the release of slag and result in the slag being trapped in the weld metal. Some of these factors are:

(1) High viscosity of weld metal

(2) Rapid solidification

(3) Too low a temperature

(4) Improper manipulation of the electrode

(5) Undercut on previous passes

Geometric factors such as poor bead profile, sharp undercuts, or improper groove geometry also may cause entrapment of slag by providing places where it may accumulate beneath the weld bead. In making a root pass, if the electrode is too large and the arc impinges on the groove faces instead of the root faces, the slag may roll down into the root opening, and it may be trapped under the weld metal.

The factors that contribute to slag entrapment may be controlled by welding technique, which is ultimately the cause of or cure for entrapped slag. The causes and remedies for entrapped slag are summarized as follows:

Causes

(1) Poor electrode manipulative technique

(2) Slag flooding in advance of the arc because of improper positioning of the joint

(3) Incomplete removal of solidified slag from weld beads or layers of a multiple pass weld

(4) Poor bead profile in multiple pass welds

(5) Presence of heavy mill scale or rust on the base metal

(6) Inclusions of pieces of unfused covering from electrodes with damaged coatings

Remedies

(1) The work should be positioned to prevent loss of slag control.

(2) The electrode or flux may be changed to improve control of molten slag.

(3) Slag removal between weld passes should be thorough.

(4) If the weld surface is rough and likely to entrap slag, it should be dressed smooth.

(5) Heavy mill scale or rust on weld preparations or faces should be removed.

(6) Avoid using electrodes with damaged covering.

Significance to Weld Integrity

The influence of slag inclusions on weld behavior is similar to that of porosity as described previously. The effect of slag inclusions on static tensile properties is considered to be significant principally to the extent it influences the cross-sectional area available to support the load. The toughness of the weld metal is generally to be considered unaffected by isolated slag with volumes of 4 percent or less of the weld zone. In weld metals of less than 75 ksi tensile strength, ductililty is generaly unaffected. As the tensile strength increases, however, ductility drops in proportion to the slag present.

Slag inclusions may be elongated with tails that may act as stress raisers. Therefore, slag can influence the fatigue behavior of welds, particularly when the weld reinforcement is removed and the weld is not postweld heat treated. As with porosity, slag at or very near to the weld surface (face or root) influences fatigue behavior to a considerably greater extent than similarly constituted slag buried within the weld metal. Slag together with hydrogen dissolved in the weld metal may influence fatigue strength by reducing the critical slag particle size for the initiation of a fatigue crack.

INCOMPLETE FUSION

Incomplete fusion results from the failure to fuse the weld metal to the base metal, or to fuse adjacent beads or layers of weld metal to each other. Failure to obtain fusion may occur at any point in a groove or fillet weld, including the root of the weld.

Causes and Remedies

Incomplete fusion almost always occurs as a result of improper welding techniques for a given joint geometry and welding process. It may be caused by failure to raise the temperature of the base metal or the previously deposited weld metal to the melting point. The presence of oxides or other foreign materials, such as slag, on the surfaces of the metals to be joined may also result in incomplete fusion if they are not removed either by fluxing or mechanical means. The causes and remedies for incomplete fusion are summarized as follows:

Causes

(1) Insufficient heat input as a result of low welding current or high travel speeds

(2) Incorrect electrode position

(3) Molten metal flooding ahead of the arc because of the position of the joint

(4) Excessive inductance in the welding circuit during short-circuiting transfer with gas metal arc welding

(5) Failure to remove oxides or slag from groove faces or previously deposited beads

(6) Wrong type or size of electrode

(7) Improper joint design

(8) Inadequate gas shielding

Remedies

(1) The welding conditions for complete fusion should be verified by welding tests to establish minimum heat input requirements.

(2) Proper electrode position should be used and maintained.

(3) Work position or welding conditions should be changed to prevent molten weld metal from flooding ahead of the arc.

(4) Excessive inductance in short-circuiting transfer with gas metal arc welding should be reduced, even at the expense of increased spatter.

(5) Oxides and slag should be removed either by chemical or mechanical means.

Significance to Weld Integrity

Incomplete fusion discontinuities affect weld joint integrity in much the same manner as porosity and slag inclusions. The degree to which incomplete fusion can be tolerated in a welded joint for various types of loading is similar to those for porosity and slag inclusions.

INADEQUATE JOINT PENETRATION

Inadequate joint penetration is generally associated with groove welds. In some weldment designs, complete joint penetration is not always required in all welded joints. Some joints are designed with partial-joint-penetration welds. However, such welds can have inadequate joint penetration when the effective throat of the weld is less than that specified in the welding specifications. The occurrence of inadequate joint penetration in welds is a function of groove geometry as well as welding procedure. The causes and cures of inadequate joint penetration are as follows:

Causes and Remedies

Causes

(1) Excessively thick root face or insufficient root opening

(2) Use of improper welding progression

(3) Insufficient heat input as a result of too low welding current or excessive travel speed

(4) Slag flooding ahead of the arc due to improper electrode or work position

(5) Electrode diameter too large

(6) Excessive inductance in the welding circuit during short-circuiting transfer with gas metal arc welding

(7) Misalignment of second side weld

(8) Failure to gouge to sound metal when back gouging a weld

(9) Bridging of root openings in joints

Remedies

(1) Reduce the thickness of the root face or increase the root opening.

(2) Use the prescribed welding procedures.

(3) Adjust the welding conditions and verify the minimum heat input needed to achieve specified penetration.

(4) Use the proper electrode or work position to minimize slag flooding.

(5) Select a smaller electrode when the groove angle is too small for larger electrodes.

(6) Reduce the inductance in the welding circuit, even at the expense of increased spatter, with short-circuiting transfer.

(7) Improve the welder's visibility of the weld groove.

(8) Re-assess back gouging procedure to insure gouging to sound metal.

(9) Use a wider root opening or substitute a deeply penetrating welding process.

Significance to Welding Integrity

Inadequate joint penetration is undesirable in single-groove welds, particularly if the root of the weld is subject to either tension or bending stresses. The unfused area permits stress concentrations that could cause failure without appreciable deformation. Even though the service stresses in the structure may not involve tension or bending at this point, shrinkage stresses and consequent distortion of the parts during welding will frequently cause a crack to initiate at the unfused area. Such cracks may progress, as successive beads are deposited, until they extend through or nearly through the entire thickness of the weld.

Inadequate joint penetration is undesirable in any groove weld that will be subjected to cyclic tension loading in service. The discontinuity can initiate a crack that may propagate and result in catastropic failure.

CRACKS

Cracking in welded joints results from localized stresses that exceed the ultimate strength of the metal. When cracks occur during or as a result of welding, little deformation is usually apparent.

Weld metal or base metal that has considerable ductility under uniaxial stress may fail without appreciable deformation when subjected to biaxial or triaxial stresses. Shrinkage caused by welding operations frequently sets up multidirectional stress systems. If a joint or any portion of it (such as the heat-affected zone) is unable to take appreciable deformation without failure because of such stresses, additional stresses imposed on the joint may cause it to fail.

An unfused area at the root of a weld may result in cracks without appreciable deformation if this area is subjected to tensile or bending stresses. When welding two plates together, the root of the weld is subjected to tensile stress as successive layers are deposited, and, as already stated, a partially fused root will frequently permit a crack to start and progress through practically the entire thickness of the weld.

After a welded joint has cooled, cracking is more likely to occur if the weld metal or heat-affected zone is either hard or brittle. A ductile metal, by localized yielding, may withstand stress concentrations that might cause a hard or brittle metal to fail.

Weld Metal Cracking

The ability of weld metal to remain intact under a stress system imposed during a welding operation is a function of the composition and structure of the weld metal. In multiple layer welds, cracking is most likely to occur in the first layer (root pass) of weld metal. Unless such cracks are repaired, they will often propagate through subsequent layers as deposited. When cracking of the weld metal is encountered, improvement may be obtained by one or more of the following modifications:

(1) Change the electrode manipulation or electrical conditions to improve the contour or composition of the deposit.

(2) Decrease the travel speed to increase the thickness of the deposit and provide more weld metal to resist the stresses.

(3) Use preheat to reduce thermal stresses.

(4) Use low hydrogen electrodes.

(5) Sequence welds to balance shrinkage stresses.

(6) Avoid rapid cooling conditions.

Three types of cracks that can occur in weld metal are as follows:

Transverse weld cracks. These cracks are perpendicular to the axis of the weld and, in some cases, extend beyond the weld into the base metal. This type of crack is more common in joints that have a high degree of restraint.

Longitudinal weld cracks. These cracks are found mostly within the weld metal, and are usually confined to the center of the weld. Such cracks may occur as the extension of cracks formed at the end of the weld. They may also be the extension, through successive layers, of a crack that started in the first layer.

Crater cracks. Whenever the welding operation is interrupted, there is a tendency for a crack to form in the crater. These cracks are usually star-shaped and progress only to the edge of the crater. However, these may be starting points for longitudinal weld cracks, particularly when cracks occur in the crater formed at the end of the weld.

Crater cracks are found most frequently in materials with high coefficients of thermal expansion, such as austenitic stainless steel. However, the occurrence of such cracks can be minimized or prevented by filling craters to a slightly convex shape prior to breaking the welding arc.

Base Metal Cracking

Cracking in the base metal is usually longitudinal in nature, takes place within the heat-affected zone, and is almost always associated with hardenable base metals. Hardness and low ductility in the heat-affected zone of welded joints are metallurgical effects that result from the thermal cycle of welding, and are among the principal factors that tend to cause cracking.

In the case of low carbon, medium carbon, and low alloy steels, hardness and abililty to deform without rupture depend upon the group to which the steel belongs, and also upon the rate of cooling from elevated temperatures following the welding operation. The rate of cooling will depend upon a number of physical factors such as:

(1) The temperature produced by the welding operation

(2) The temperature of the unaffected base metal

(3) The thickness and thermal conductivity of the base metal

(4) The heat input per unit time at a given section of the weld

(5) The ambient temperature

With a given cooling rate, low carbon steels will harden considerably less than medium carbon steels. Low alloy steels exhibit a wider variation in hardening characteristics; some of them may be similar to low carbon steel while others will react like medium carbon steel.

High alloy steels include the austenitic and ferritic stainless steels. The latter behave similarly to medium carbon and low alloy steels, except that they harden to a greater degree with a given cooling rate. Austenitic stainless steels and ferritic stainless steels will not harden upon quenching from elevated tempertures. In general, ferritic stainless steels are rendered brittle (but not hard) by welding operations.

The metallurgical characteristics of metals are of prime importance. Base metal cracking is associated with lack of ductility in the heat-affected zone. It has been established that different heats of the same hardenable steel can vary appreciably in cracking tendency. Furthermore, in shielded metal arc welding, the characteristics of the electrode as determined by the covering considerably affect the susceptibility to heat-affected zone cracking.

When base metal cracking is encountered in hardenable steels, the condition can be improved in the following ways:

(1) Use of preheat to control cooling rate

(2) Controlled heat input

(3) Use of the correct electrode

(4) Proper control of welding materials

Another problem that may be encountered when welding many steels under certain conditions is hydrogen induced cracking.[4] Such cracking is known by various other names, including underbead, cold, and delayed cracking. It generally occurs at some temperature below 200°F immediately upon cooling or after a period of several hours. The time delay depends upon the type of steel, the magnitude of the welding stresses, and the hydrogen content of the weld and heat-affected zones. Delayed cracking normally takes place after the weldment has cooled to ambient temperture. In any case, it is caused by dissolved hydrogen entrapped in small voids or dislocations in the metal. Sometimes, the weld metal may crack, although this seldom occurs when its yield strength is below about 90 ksi. Diffusion of hydrogen into the heat-affected zone from the weld metal during welding contributes to cracking in this zone. Microstructure of the steel is also a contributing factor.

---

4. Hydrogen induced cracking is discussed in detail in the *Welding Handbook*, Volume 4, 7th Ed., 1982; 3-6.

Hydrogen-induced cracking can be controlled using (1) a welding process or an electrode that produces little or no hydrogen, (2) a combination of welding and thermal treatments that drive off the hydrogen or produce a microstructure that is insensitive to it, or (3) welding procedures that result in low welding stresses.

Significance to Weld Integrity

Cracking, in all of its forms, constitutes the weld discontinuity considered most detrimental to performance. A crack, by its very nature, is sharp at its extremities, and thus acts as a stress concentrator. The stress concentration effect produced by cracks is greater than that of other discontinuities. Therefore, cracks are not normally permitted, regardless of size, in weldments governed by most fabrication codes. Cracks must be removed from the weld area, and the weld repaired.

Cracks and crack-like discontinuities can be evaluated with a fracture mechanics approach. When cracks are assessed on a service performance basis, they can often be tolerated for a given size, orientation, and location.

SURFACE IRREGULARITIES

The following surface irregularties may be observed on welds:

(1) Badly-shaped, irregular surface ripples

(2) Excessive spatter

(3) Craters

(4) Protrusions (an overfilled crater)

(5) Arc strikes

The welder is usually directly responsible for these discontinuities as a result of incorrect welding technique or improper machine settings. Sound welds finished in a poor manner should not be accepted, even though the joint is adequate for the intended service. The ability and integrity of the welder must be questioned.

In some cases, faulty or wet electrodes and unsuitable base metal chemistry may cause discontinuities and unsatisfactory weld appearance.

Gross bead irregularities are discontinuities inasmuch as they constitute an abrupt change of section. Such abrupt changes of section are potential causes of high stress concentration, and should be carefully evaluated with respect to service requirements.

Spatter in itself is not necessarily a defect, but is quite likely indicative of improper welding technique and of the likelihood of other associated faults.

Arc strikes with either the electrode or the holder can initiate failure in bending or cyclic loading. They can create a hard and brittle condition in alloy steels, and are inadvisable even on mild steel when high static or normal fatigue stresses may be encountered. The repair of such damage may be difficult and costly, involving chipping and possible preheating in the case of low alloy steels.

INADEQUATE WELD JOINT PROPERTIES

Specific mechanical or chemical properties, or both, are required of all welds in a weldment. These requirements depend on the codes or specifications covering the weldment, and departure from specified requirements may be considered a defect. The required properties are generally determined with specially prepared test plates but may be made on sample weldments taken from production. Where test plates are used, the inspector should see that specified procedures are followed, otherwise the results obtained will not necessarily indicate the properties of production weldments.

Mechanical properties which may be inadequate are tensile strength, yield strength, ductility, hardness, and toughness. Chemical properties may be deficient because of either incorrect filler metal composition or excessive dilution. Both may result in lack of corrosion resistance.

DISCONTINUITIES IN RESISTANCE AND SOLID STATE WELDS

WELDING PROCESSES

Certain resistance and solid state welding processes are done at elevated temperatures essentially below the melting point of the base metal being joined without the addition of filler metal. When the joint reaches the desired welding temperature, force is applied to the joint to consummate the weld. As welding is being acccomplished, some molten metal may be generated between the faces being joined, but it is expelled as a result of pressure on the interface from the force applied during the welding cycle. Typical resistance and solid state welding processes of this nature are flash welding, friction welding, high frequency resistance welding, and upset welding.

The resulting welds are all characterized by a flat or linear weld interface. They are not volumetric as are fusion welds and thus, the discontinuities of importance are those associated with a flat weld interface. Further, some inspection techniques applicable to fusion welds, such as radiographic and ultrasonic testing, are difficult to apply to these welds because the results are hard to interpret.

WELD DISCONTINUITIES

Normally, resistance and solid state welding processes are completely automated with good reproducibility. However, discontinuities are encountered in solid state welded joints. The weld discontinuities may be classified by origin as either mechanical or metallurgical discontinuities.

The most probable causes for rejection of these types of welds are related to mechanical discontinuities, such as misalignment of the workpieces. The discrepancies are easily found by visual inspection, and can usually be corrected by simple machine adjustments. A typical example of a mechanical discontinuity is an inferior weld caused by misalignment or offset of the workpieces, as indicated in Fig. 8.10.

In the case of flash welds and upset welds, the shape and contour of the upset metal is a good indicator of weld quality. The upset geometry of the weld in Fig. 8.11(A) indicates proper heat distribution as well as proper upset. Insufficient upset, as in Fig. 8.11(B), may indicate trapped oxides or flat spots at the weld interface and possibly poor interface bonding.

The condition and shape of the flash on friction welds is an indicator of possible discontinuities along the weld interface. Figure 8.12 shows the effect of axial shortening on weld quality. These inertia friction welds were made with the same speed and inertial mass but with a decreasing heating pressure from left to right. Two of the welds exhibited center discontinuities

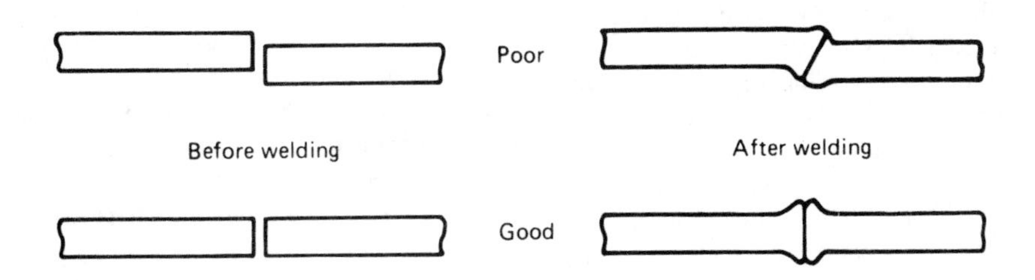

Fig. 8.10—Effect of part alignment on joint geometry

because insufficient axial shortening (pressure) was used.

Lack of center bonding may also occur in continuous drive friction welds when inadequate speed, heating time, or pressure is used. Depressions in the faying surfaces that prevent uniform contact during the early stages of friction welding often limit center heating and also entrap oxides.

Metallurgical discontinuities are usually associated with material defects or heterogeneities. Such discontinuities are much harder to find by nondestructive inspection techniques because they are usually internal defects.

The various types of metallurgical discontinuities found in resistance and solid state welds are as follows:

(1) Cracking

(2) Intergranular oxidation

(3) Decarburization

(4) Voids

(5) Oxides and other inclusions

(6) Flat spots

(7) Cast metal at the interface (flash welds only)

(8) Out-turned fibrous metallurgical structure in the weld

Cracking in welds can be divided into two categories depending upon the temperature of formation: (1) cold cracking and (2) hot cracking. Cold cracking, as indicated in Fig. 8.11(C), can be caused by a combination of excessive upset and an improper temperature profile. Excessive cooling rates in hardenable steels can cause cold cracking from intolerable strains acting upon martensitic structures. The most common form of hot cracking in upset welds occurs as microfissures in the heated zone, and is known as *break-up*.

In the case of flash welds, a form of intergranular oxidation known as *die-burn* can occur at clamp locations. This discontinuity is caused by localized overheating of the workpiece where it is held in the clamping die. Cleaning of the workpiece surfaces in the clamping area to bright metal will usually eliminate this problem. Too large an initial die opening can result in excessive metal temperatures near the faying surfaces during flashing. This may lead to intergranular oxidation as well as nonuniform upsetting and joint misalignment.

Occasionally, another type of solid state welding discontinuity results from elemental redistribution during welding. In carbon steel, this may be manifested as decarburization. This discontinuity appears as a bright band on a polished and etched surface of an upset welded steel specimen which is cut transverse to the weld interface. Chemical heterogeneity affects the mechanical properties of a welded joint. In particular, a variation in carbon concentration greatly affects hardness, ductility, and strength.

Oxides and other inclusions, voids, and cast metal along the interfaces of weld joints are related by the fact that they can usually be eliminated by increasing the upset distance. In the case of flash welds, craters are formed on the faying interfaces by the expulsion of molten metal during flashing. If the flashing voltage is too high or the platen motion is incorrect, violent flashing causes deep craters in the faying surfaces. Molten metal and oxides may be trapped in these deep craters and not expelled during upsetting.

The cause and prevention of flat spots in welds are not well understood. In fact, the term itself is vague and is often used to describe several different features. Flat features may be the result of a number of metallurgical phenomena having inherently different mechanisms. The smooth, irregular areas indicated by arrows on the fractured surfaces of a flash weld in Fig. 8.13 are typical flat spots. Flat spots (low ductility zones along the weld interface) in flash welds and friction welds apparently are associated with base metal characteristics as well as welding process variables.

The inherent fibrous structure of wrought mill products may cause anisotropic mechanical behavior. An out-turned fibrous structure at the weld interface often results in some decrease in mechanical properties as compared to the base metal, particularly in ductility. The decrease in ductility is not normally significant unless one or both of the following conditions are present:

(1) The base metal is extremely inhomogeneous. Examples are severely banded steels, alloys with excessive stringer type inclusions, and mill products with seams and cold shuts produced during the fabrication process.

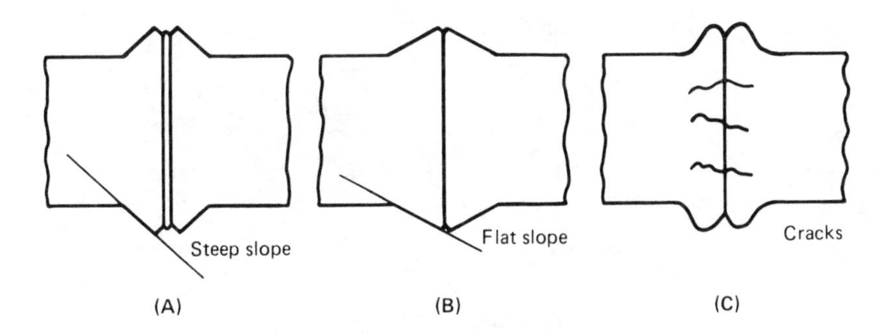

Fig. 8.11—Visual indications of flash weld quality: (A) satisfactory heat and upset, (B) insufficient heat or upset or both, (C) cracks due to insufficient heat

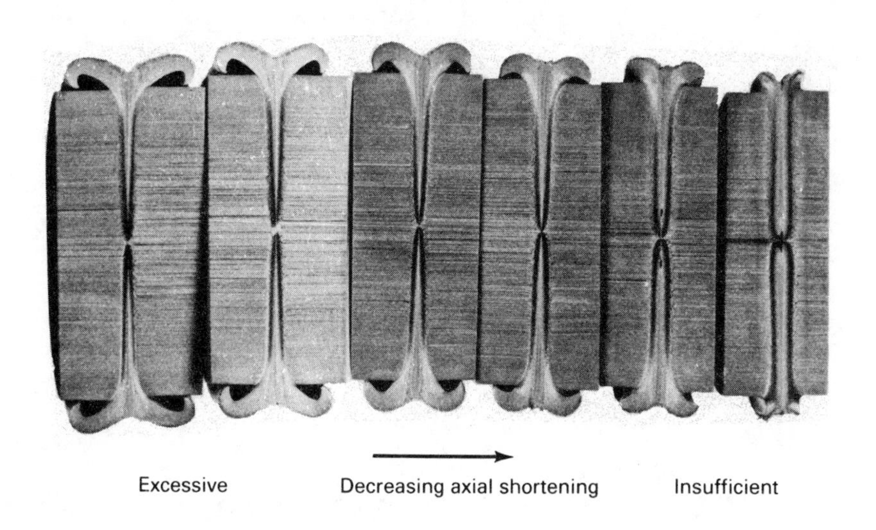

Fig. 8.12—The effect of axial shortening on the bonding and flash of friction welds

Fig. 8.13—Flat spots, indicated by arrows, on the mating fractured surfaces of a flash weld (X1.5)

(2) The upset distance is excessive. When excessive upset distance is employed, the fibrous structure may be completely reoriented transverse to the original structure.

DISCONTINUITIES IN BRAZED AND SOLDERED JOINTS

Brazing and soldering are heterogeneous joining processes that rely on capillary action to draw liquid filler metal into a controlled joint clearance and also on wetting of the faying surfaces by the liquid filler metal. The two processes differ only in the filler metals used and temperatures employed to obtain the desired joint properties, e.g., strength and corrosion resistance.

Discontinuities found in soldered joints and their causes and remedies are similar to those for brazed joints. Therefore, the discussion will be limited to brazed joints.[5]

COMMON DISCONTINUITIES IN BRAZED JOINTS

Lack of Fill

Lack of fill in the form of voids and porosity can be the result of improper cleaning, excessive joint clearance, insufficient filler metal, entrapped gas, and movement of the mating parts, caused by improper fixturing, while the filler metal is in the liquid or partially liquid state. This discontinuity reduces the strength of the joint by reducing the load-carrying area, and it may provide a path for leakage.

Flux Entrapment

Entrapped flux may be found in a brazed joint where a flux is used to prevent and remove oxidation during the heating cycle. Entrapped flux prevents flow of filler metal into that particular area, thus reducing the joint strength. It may also cause false leak or proof test acceptance. The entrapped flux, if corrosive, may reduce service life.

Noncontinuous Fillets

Segments of joints void of filleted filler metal are usually found by visual inspection. They may or may not be acceptable depending upon the specification requirements of the brazed joint.

Base Metal Erosion

Erosion of the base metal by the filler metal is caused by the filler metal alloying with the base metal during brazing. It may cause undercut or the disappearance of the mating surface, and reduce the strength of the joint by changing the filler metal composition and by reducing the base metal cross-sectional area.

Unsatisfactory Surface Appearance

Excessive flow of brazing filler metal onto the base metal, surface roughness, and excessive filler metal may be detrimental for several reasons. In addition to aesthetic considerations, these may act as stress concentrations, corrosion sites, or may interfere with inspection of the brazement.

Cracks

Cracks reduce both the strength and service life expectancy. They may act as stress raisers, causing premature fatigue failure as well as lowering the mechanical strength of the brazement.

Acceptance Limits

When defining the acceptance limit for a particular type of brazing discontinuity, the following should be considered: its shape, orientation, and location in the brazement, including whether surface or subsurface, and its relationship to other imperfections.

Judgments for disposition of components containing flaws should be made by persons competent in the fields of brazing metallurgy and quality assurance who fully understand the function of the component. Such dispositions should be documented.

TYPICAL DISCONTINUITIES AND THEIR CAUSES

Discontinuities found in brazed joints are an indication that the brazing procedure is out of control or that improper techniques were used. Several typical discontinuities are shown here so that they may be recognized when they occur in production. The possible cause is given in each case.

Figure 8.14 is a section of a brazed copper lap joint in which a large void in the fillet is evident. The flawed joint was detected by subjecting the assembly to pressurized air and submerging it in water. The probable causes of the void were underheating or improper fluxing procedures, or both. This is borne out by the irregular flow of the filler metal into the joint.

Figure 8.15 shows a section through a brazed joint in which severe erosion of the base metal occurred. This erosion resulted from over-

---

5. Additional information on discontinuities in soldered joints and inspection procedures may be found in the *Soldering Manual*, 2nd Edition, Revised, 1978 published by the American Welding Society.

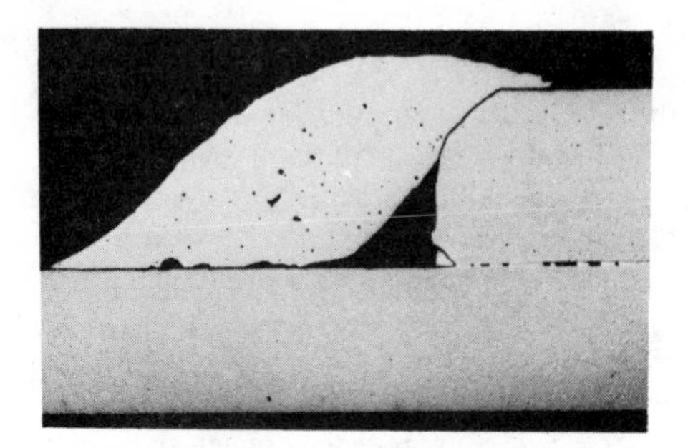

Fig. 8.14—Void under the filler metal fillet in a brazed copper lap joint

Fig. 8.16—Brazed joint with essentially no base metal erosion

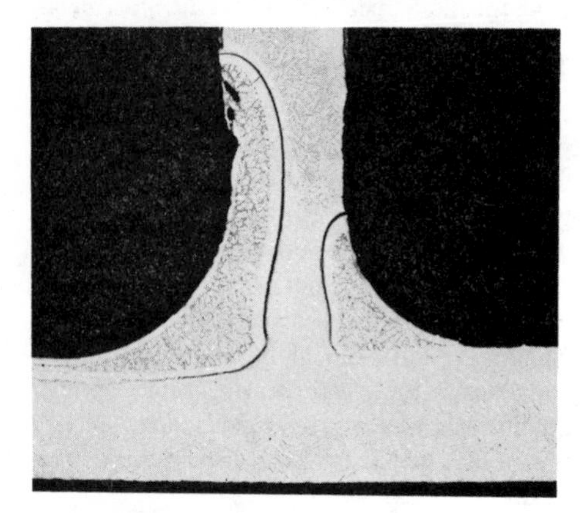

Fig. 8.15—Severe alloying and erosion of the base metal in a brazed joint

heating of the joint during brazing. Such erosion may not be serious where thick sections are joined, but cannot be permitted where relatively thin sections are used. A joint with essentially no erosion is shown in Fig. 8.16.

Figure 8.17(A) shows a brazed copper socket joint. A radiograph of the joint, Fig. 8.17(B) shows that large areas of the joint are void of filler metal. Figure 8.17(C) is a macrograph of cross section through the braze fillet showing an extensive void in the capillary of the joint. The voids throughout the joint were caused by insufficient heating. If a joint contains leaks from voids in the braze, these are sometimes revealed by a pressure test.

Figure 8.18 shows a brazed lap joint, a radiograph of the joint, and sections through the joint. The assembly shown in Fig. 8.18(A) is of two flat pieces of low carbon steel brazed with a silver base filler metal. The radiograph taken through the joint, Fig. 8.18(B), shows large voids as dark areas. Cross sections *1-1* and *2-2* through the joint are shown in Fig. 8.18(C). The voids in the joint capillary are evident.

SIGNIFICANCE OF WELD DISCONTINUITIES

Increasing design demands are resulting in more critical inspection methods and acceptance standards. Ideally, these acceptance standards should represent the minimum weld quality that can be tolerated to assure satisfactory performance of the welded part or component. They should be based on tests of welded specimens containing the particular discontinuity under consideration. Correlation of these test results with allowable results gives a basis for acceptance

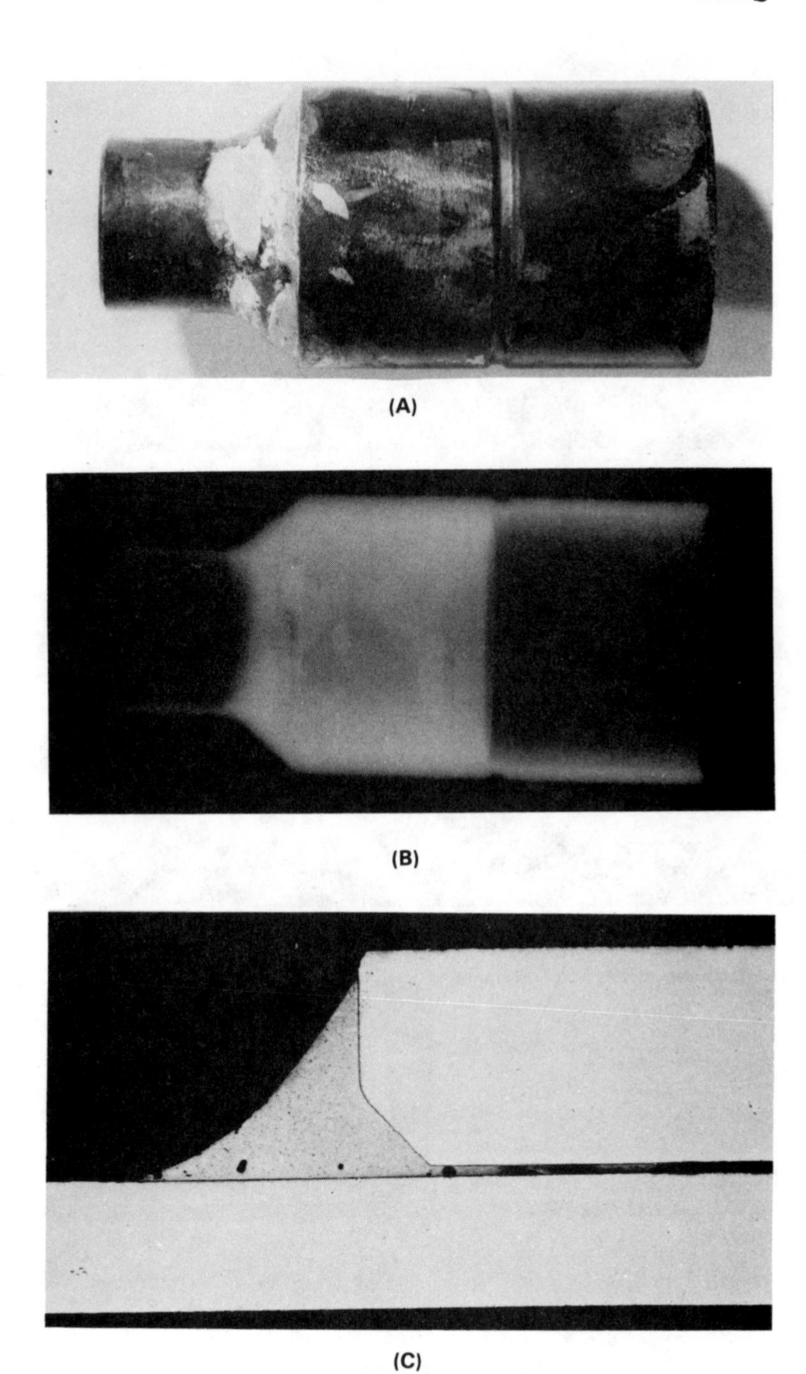

Fig. 8.17—(A) Brazed copper socket joint, (B) Radiograph of the joint, (C) Cross section through the braze fillet and capillary (about X20)

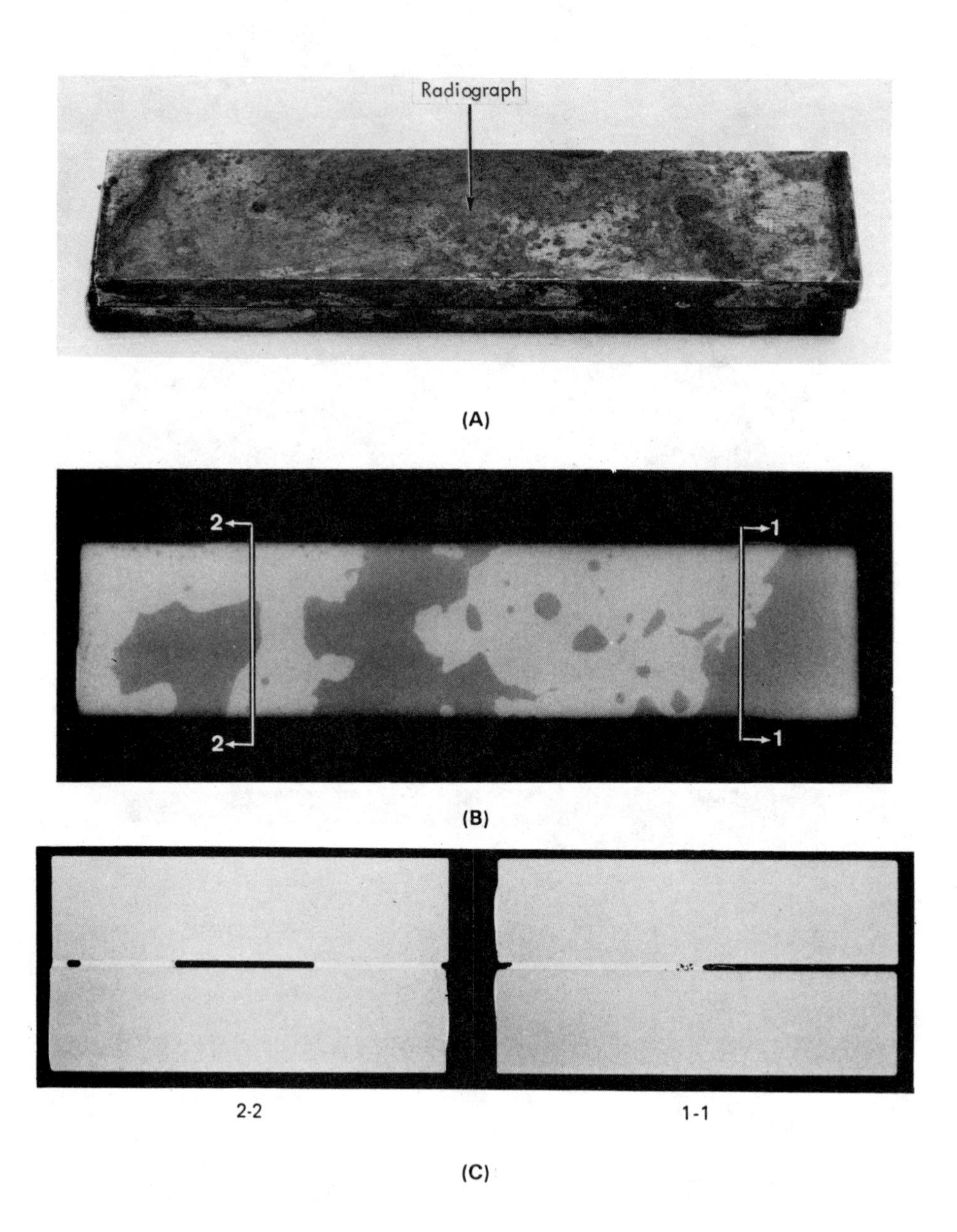

Fig. 8.18—(A) Brazed lap joint, (B) Radiograph showing voids (dark areas) in the joint, (C) Cross sections through the joint showing voids in the capillary

of these particular discontinuities. Usually, a safety factor will be added to yield the final acceptance standard.

However, too frequently acceptance standards are arbitrary, bearing little or no relationship to actual performance, or are merely an attempt to define the limits for practical welding and good workmanship. Such standards can be either inadequate or overly conservative, thus increasing production costs unnecessarily.

DISCONTINUITY – MATERIAL RELATIONSHIPS

The effect of a particular weld discontinuity on structural integrity, economics, and safety depends to a certain extent on the materials being welded. For example, the effect of cracks in austenitic stainless steel is generally less significant than in ferritic steels. Various carbon steels have a wide range in tolerance for discontinuities before fracture can initiate. The interrelationship between the material and discontinuities is not addressed in current codes and standards. Separate and distinct engineering analytical techniques are required to take this into account. The ability to assess the significance of a discontinuity in terms of material properties falls under the fracture mechanics approach to weld discontinuities.

DISCONTINUITY – MECHANICAL PROPERTY RELATIONSHIPS

The significance of a particular discontinuity in a weld depends on the function that the weld must perform. Certain weld discontinuities which have no effect on mechanical properties can, nevertheless, become significant by virture of corrosion considerations; for example, inclusions of heavy metals at or near the weld surface cannot be tolerated in certain environments. It is important to note that a repair of a harmless discontinuity can result in a more damaging, less readily detectable discontinuity.

The heat-affected zone plays a major role in the determination of mechanical properties of the joint, and can render the presence of weld discontinuities less significant. The heat-affected zone itself can be considered a discontinuity because it almost invariably experiences modifications of mechanical properties. For example, strain-hardened base metals with mechanical properties enhanced by prior cold work experience both recrystallization and grain growth in the heat-affected zone. In such metals, both the strength and hardness of the heat-affected zone are invariably diminished. A drastic loss in tensile strength also occurs in the heat-affected zone of precipitation-hardened alloys because some region in the heat-affected zone usually experiences overaging during welding.

The effects of discontinuities on mechanical properties are governed by the shape, size, quantity, interspacing, distribution, and orientation of the discontinuities within the weld. Due to the number of variables, any qualitative correlation between a specific discontinuity class and mechanical properties is extremely difficult. Therefore, trends rather than direct relationships are presented here.

As the strength of the base metal is increased, so is its sensitivity to discontinuities. Ultra-high strength steels, as well as hard heat-affected zones in relatively mild steels, may be extremely sensitive to very small discontinuities. When welding very strong steels, it is also difficult to match the strength of the weld to that of the base metal. Therefore, the protection afforded to discontinuities in mild steels by the overmatching strength of the weld metal is no longer available.

Tensile Strength

For the purpose of characterizing static tensile performance, welded joints fall into two categories: (1) those having weld metal strengths that closely match or undermatch the base metal strength and (2) those having weld metal strengths that overmatch the base metal strength.

In general, the welded joints in the first category are weakened more by weld discontinuities than those in the second category because there is no reserve of additional weld metal strength to counteract the decrease in cross-sectional area. Because of the additional strength of overmatching weld metal, a certain amount of discontinuity can be tolerated before transverse tensile properties are adversely affected. The loss in transverse tensile strength is roughly proportional to the loss in cross-sectional area. Additional cross-sectional area provided by any weld reinforcement compensates for some of this loss.

In welded joints of the overmatching type, the transverse strength degradation is usually accompanied by a change in location of the fracture from base metal to weld metal when the number of discontinuities present changes from a few to a significant quantity. Usually, the ductility will be reduced roughly proportionally to the volume of the discontinuities present. The yield strength is not significantly affected.

Fatigue

Fatigue failure at normal working stresses is invariably associated with stress concentrations around weld or design discontinuities. Fatigue is probably the most common cause of failure in welded construction.[6] The discontinuities most significant in promoting fatigue failures except for the obvious effect of gross cracking or extensive incomplete fusion, are those which affect the weld surface. The combination of excessive weld reinforcement, Fig. 8.19(A), and slight undercutting is one of the most serious discontinuities causing fatigue failure. Fabrication codes usually specify the maximum permissible height of the reinforcement. However, many welding fabricators merely grind off the excess without altering the reinforcement contact angle as shown in Fig. 8.19(B). This practice does not improve the fatigue life, and therefore probably would not in itself provide a finished weld meeting the required quality level. Most codes also specify that the reinforcement shall blend smoothly into the base metal at the edges of the weld as shown in Fig. 8.19(C).

Figure 8.20 illustrates the influence of weld reinforcement contact angle on fatigue strength under repeated tension loading for an endurance of 2 x 10^6 cycles. These data indicate why design defects are often responsible for service failures. The use of a load-carrying lap joint with a fillet weld in place of a butt welded joint, for example, has the effect of reducing the strength by a factor of up to three.

6. The fatigue properties of welded structural joints are discussed in the *Welding Handbook*, Vol. 1, 7th Ed., 1976: 186-201.

Tests with both porosity and weld reinforcement have shown that fatigue cracks initiate at the toe of the weld, and that porosity has little effect on the fatigue life. However, when the reinforcement is removed, the position and orientation of the porosity rather than the magnitude is the important factor, with the fatigue strength being critically dependent on the presence of pores breaking the surface or just under the surface.

Tests with weld reinforcement intact and very large tungsten inclusions in the weld metal have also shown that fatigue cracks initiate at the toe of the weld. When the reinforcement was removed, fatigue failure did not result from the tungsten inclusions but from small oxide inclusions associated with them.

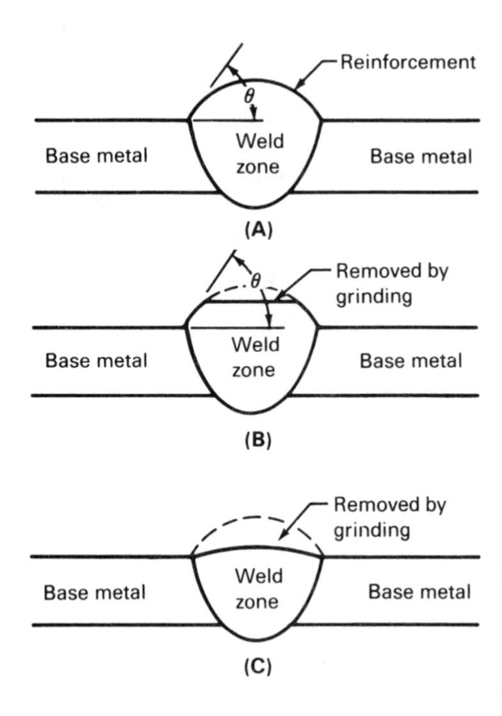

Fig. 8.19—(A) Weld with excessive reinforcement, (B) Improper treatment of weld reinforcement, (C) Acceptable weld reinforcement profile for fatigue applications

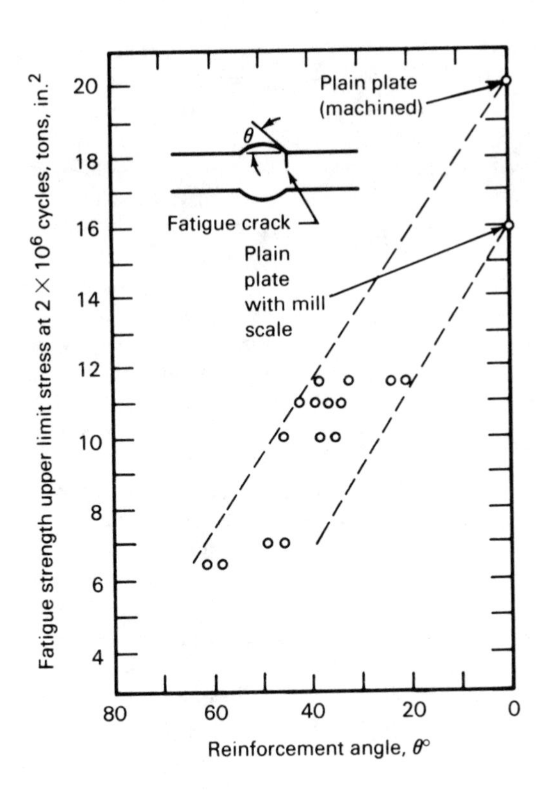

Fig. 8.20—Effect of weld reinforcement angle on fatigue strength of steel

Increasing slag inclusion length initially leads to decreasing fatigue strength. However, when the discontinuity length becomes large relative to its depth through the thickness, no further reduction in fatigue strength occurs.

Discontinuities which are located in the middle of the weld can be blanketed by compressive residual stress so that other smaller discontinuities nearer to the surface control the fatigue strength. Such discontinuities may be below the size detectable by radiography.

Brittle Fracture

A fracture mechanics approach permits the determination of a relationship between stresses and the critical size of a discontinuity that will propagate. This critical discontinuity size decreases in proportion to the square of the increase in applied stress whereas, in general, the fracture toughness of metals decreases as the yield strength increases.

For a given weld structure, an appropriate equation must be developed that relates the fracture toughness to the applied stress and discontinuity size.[7] This equation should account for the location, dimension, and shape of the discontinuity. The location and shape (depth-to-length ratio) of the discontinuity must be known or accurately predicted, particularly in those instances where discontinuities can grow by fatigue

7. Fracture toughness is discussed in the *Welding Handbook*, Vol. 1, 7th Ed., 1976: 170-85.

or stress corrosion. With this information and a valid plane strain fracture toughness value, one can estimate the combination of stress and discontinuity size at which a structure can be operated safely.

The effectiveness of a discontinuity as an initiator of fracture in a given weld metal depends on the degree of stress concentration and triaxility at the discontinuity tip. Therefore, it is clear that the edge radius of discontinuities will have an effect on performance. Thus, sharp natural cracks are considered the severest discontinuity for brittle fracture considerations. In a similar manner, incomplete fusion can be the next most severe discontinuity. Inadequate joint penetration normally would not be as severe as incomplete fusion because the edges of the former normally do not have so sharp a radius. Slag inclusions and porosity would be relatively harmless as initiating discontinuities unless there is cracking associated with them.

Results of precracked Charpy impact tests have shown that the resistance of weld metal to crack propagation under impact loading is not significantly affected by porosity. Porosity should be considered a risk only when the weld metal is too brittle to be used safely or when fatigue loading initiates cracks.

FRACTURE MECHANICS EVALUATION OF DISCONTINUITIES

The traditional approach to the significance of discontinuities in welds by the majority of current codes and standards is to establish their sizes and extent limits from a workmanship standpoint. The evolution of the current discontinuity standards has had no rigorous engineering approach, and thus the criteria for acceptance are without a true engineering base.

Another approach to the prevention of failures on a material and discontinuity control basis is fracture mechanics.[8] The fracture mechanics evaluation of materials, discontinuities, and stress state has led to significant advances in the methods for assessment of discontinuity acceptance levels. The most advanced area of fracture mechanics as far as applications to structures is concerned is Linear Elastic Fracture Mechanics.

There are two published documents which apply a linear elastic fracture mechanics analysis to weld discontinuities: the ASME Boiler and Pressure Vessel Code, Section XI, *Inservice Inspection*, and the British Standards Institute Document PD 6493, *Guidance on Some Methods for the Derivation of Acceptance Levels for Defects in Fusion Welded Joints*. These documents are being employed to advance the fitness for service concept which applies reasonably based acceptance standards to structures of known service usage.

In the fracture mechanics approach, the most probable failure modes must first be defined. They are as follows:

(1) Elastic—plastic instability

(2) Overload—excessive deformation, ductile rupture, or leakage

(3) General corrosion

(4) Stress corrosion

(5) Brittle fracture

(6) Fatigue—high and low cycle corrosion fatigue

(7) Creep

The first three failure modes are basic design considerations; stress corrosion is treated from the environment control standpoint. Creep failure, being a high temperature phenomenon and time dependent, does not normally lead to sudden unexpected failures. It is difficult, if not impossible, to treat it from elastic behavior standpoints. The brittle fracture and fatigue aspects of failure are the most amenable modes to assess on a fracture mechanics basis. Brittle fracture has received the greatest attention, and the basic concepts will be outlined here.

In assessing the importance of discontinuities, a ranking is usually established for the severity of a given sized discontinuity. A ranking in decreasing importance as shown below would not be uncommon.

(1) Cracking
(2) Undercut
(3) Lack of fusion

8. For additional information, refer to the *Welding Handbook*, Vol. 1, 7th Ed., 1976; 174-79, and Welding Research Council Bulletin 295 (see Supplementary Reading List).

(4) Lack of penetration
(5) Slag
(6) Porosity

Cracks always head the list as the most potentially injurious (to service behavior) discontinuity, and porosity normally is relegated to a low position in the ranking because porosity tends to be rounded with a minor stress concentration effect.

Fracture mechanics generally treat the worst case discontinuity (cracks) and then makes an assessment of other discontinuities in proportion to their relative stress concentration influence. By treating discontinuities as crack-like, a conservative assessment of all similar sized discontinuities is obtained. In determining the effect of crack-like discontinuities (planar discontinuities) the critical sizes can be defined. It is this definition of flaw size that is at the heart of the fitness for service approach.

If a crack can be characterized as to size and acuity or sharpness by nondestructive testing methods, and if the material properties and the working and welding residual stresses are known as well, it is possible to determine whether the crack will cause failure.

From the material and stress state viewpoint, the following additional aspects must be considered:

(1) Material type
(2) Material thickness
(3) Yield strength
(4) Microstructure in the vicinity of the tip of the discontinuity
(5) Magnitude of the applied and residual stress and strain
(6) The rate of application of strain (loading rate)
(7) The environment

With the details of the foregoing considerations defined and evaluated, the mathematic relationships of discontinuity-material interactions can be applied. The end result of such determinations is the employment of a relatively simple equation for acceptable-unacceptable discontinuity assessment. The equation normally takes the form:

$$K_I = C\sigma (\pi a)^{1/2}$$

where

K_I is a plane strain stress intensity factor

σ = the applied or residual stress magnitude acting on a discontinuity

C = a constant depending on discontinuity size and shape

a = the discontinuity size or depth

π = 3.1416

The plane strain stress intensity factor is in reality a property of a given material, and when assessed at its minimum value, it is known as K_{Ic} (critical value of fracture toughness). The two documents mentioned above provide charts or tables, or both, for evaluation of the constants in the appropriate equations. The fracture toughness can be measured directly by defined tests or estimated from Charpy V-notch impact data.

Metric Conversion Factors

1 ksi = 6.89 MPa
1 ton/in.^{2} = 13.79 MPa

SUPPLEMENTARY READING LIST

Boulton, C.F., Acceptance levels of weld defects for fatigue service, *Welding Journal*, 56(1): 13s-22s; 1977 Jan.

Burdekin, F.M., Some defects do - some defects don't (lead to the failure of welded structures), *Metal Construction*, 14(2): 91-94; 1982 Feb.

Cox, E.P., and Lamba, E.P., Cluster porosity effects on transverse fillet weld strength, *Welding Journal*, 63(1): 1s-7s; 1984 Jan.

Guidance on Some Methods for the Derivation of Acceptance Levels or Defects in Fusion Welded Joints, London: British Standards Institute, 1980; PD6493.

Lundin, C.D., Fundamentals of Weld Discontinuities and Their Significance, New York: Welding Research Council, 1984 June; Bulletin 295.

Lundin, C.D., Review of Worldwide Discontinuity Acceptance Standards, New York: Welding Research Council, 1981 June; Bulletin 268.

Lundin, C.D., The Significance of Weld Discontinuities - A Review of Current Literature, New York: Welding Research Council, 1976 Dec: Bulletin 222.

Lundin, C.D., and Pawel, S.J., An Annotated Bibliography on the Significance, Origin, and Nature of Discontinuities in Welds, 1975-1980, New York: Welding Research Council, 1980 Nov; Bulletin 263.

Masubuchi, K., *Analysis of Welded Structures - Residual Stresses, Distortion, and Their Consequences*, Oxford, England: Pergamon Press, 1980.

Pellini, W.S., Principles of fracture-safe design, *Welding Journal*, 50(1): 91s-109s; 1971 Mar., 50(2): 147s-162s; 1971 Apr.

Pfluger, A.R. and Lewis, R.E., eds., *Weld Imperfections*, Addison-Wesley Co., 1968.

Reed, R.P., McHenry, H.I., and Kasan, M.B., A Fracture Mechanics Evaluation of Flaws in Pipeline Girth Welds, New York: Welding Research Council, 1979 Jan: Bulletin 245.

Tsai, C.L. and Tsai, M.J., Significance of weld undercut in design of fillet welded T-joints, *Welding Journal*, 63(2): 64s-70s; 1984 Feb.

Wells, A.A., Fitness for purpose and the concept of defect tolerance, *Metal Construction*, 13(11): 677-81; 1981 Nov.

Will, W., Technical basis for acceptance standards for weld discontinuities, *Naval Engineers Journal*, 60-70; 1979 April.

Wilkowski, G.M., and Eiber, R.J., Review of Fracture Mechanics Approaches to Defining Critical Size Girth Weld Discontinuities, New York: Welding Research Council, 1978 July; Bulletin 239.

9
Inspection

Welding Inspectors *318*
Inspection Plan *319*
Nondestructive Testing *321*
Destructive Testing *370*
Proof Testing *374*
Brazed Joints *376*
Supplementary Reading List *377*

Chapter Committee

D.R. AMOS, *Chairman*
Westinghouse Electric Corporation

H.W. EBERT
Exxon Research and Engineering Company

R.L. HOLDREN
Welding Consultants

C.D. LUNDIN
University of Tennessee

W.C. MINTON
Southwest Research Institute

J.R. TAGLIEBER
Midwest Testing Laboratories

Welding Handbook Committee Member

J.R. HANNAHS
Midwest Testing Laboratories

9
Inspection

WELDING INSPECTORS

Inspectors of weldments should know and understand (1) the requirements and the duties of welding inspectors, engineers, and others, and (2) the principles of those tests employed to evaluate the adequacies of weldments and their compliance with welding procedures, codes, and specifications.[1] In general, the information about the inspection of weldments applies to brazements also. However, some inspection methods are not suitable for capillary brazed joints because of the thinness of the filler metal in a joint.

Welding inspectors must be familiar with all phases of the fabrication activities that apply to their product line. This includes a working knowledge of applicable codes, specifications, and laws governing the quality of specific components. Often this will include requirements for welding procedure specifications, qualification testing, and the application of mechanical, proof, and nondestructive tests. In this regard, it is beneficial if such inspectors are certified under the Welding Inspector Qualification and Certification Program of the American Welding Society.[2]

REQUIREMENTS FOR INSPECTORS

In conjunction with their duties, welding inspectors may represent the manufacturer, the purchaser, or the insurer of welded components, or a public interest or governmental organization that has no commercial involvement. The represented organization can subject inspectors to various pressures and influences. Therefore, it is essential to maintain highly ethical standards. Such pressures can be minimized by active management efforts to reduce conflicts and to support the inspectors in their work. For instance, it is usually desirable that inspectors report to design engineers or welding engineers who establish quality requirements rather than to the manufacturing managers who are responsible for meeting production requirements and costs.

Inspectors must be physically fit to observe the welding and related activities. In some cases, this may involve ability to climb around or enter specific components. In all cases, the inspector's duties involve good vision. Some inspection tools also require adequate color vision for maximum effectiveness.

Good interpersonal skill is another nontechnical trait associated with effective inspectors. Serving as a link between different departments with differing interests and as persons who at times must make unpopular decisions, inspectors heed positive attitudes toward the job and other persons. To be successful, cooperation and respect must be earned by impartial, consistent, and technically correct decisions.

Naturally, inspectors must be knowledgeable in welding and welding related activities. Usually this includes several welding processes, welding consumables, and both preweld and

1. Additional information may be found in *Welding Inspection, 2nd Ed.*, 1980, and *ANSI/AWS B1.0, Guide for the Nondestructive Inspection of Welds*, (latest edition) published by the American Welding Society.

2. Refer to Chapter 7 for information on this program.

postweld activities. In addition to having demonstrated ability to interpret drawings, specifications, and fabrication and testing procedures, they must be able to communicate orally and in writing. They must know testing methods and be able to perform or adjudicate such tests, or both. Finally, they must be able to maintain records documenting their findings, acceptances, and rejections. The ability to analyze these findings and recommend corrective actions is highly desirable. These skills can be learned through formal and on-the-job training and can be verified by the certification program offered by the American Welding Society.

DUTIES OF INSPECTORS

Inspectors must be familiar with the product, engineering drawings, specification requirements, and manufacturing and testing procedures. This includes the handling and disposition of deviations from requirements or procedures.

The writing and qualifying of welding procedure specifications are usually an engineering function, while verification of welder and welding operator qualifications and the surveillance of performance records are inspection functions. Various codes and contractural documents have different testing and acceptance criteria. Decisions must be based on the detailed knowledge of these requirements. The same also applies when performing or monitoring applicable mechanical, nondestructive, and proof testing operations.

In many cases, actual inspection activities should not be limited to acceptance or rejection of the final product. Once a component has been completed, it may be difficult to evaluate its total quality, and to take corrective actions that could improve the quality. Thus, certain activities are appropriate prior to and during fabrication. Typically, these may include:

(1) Inspection prior to welding
- (a) Procedures and qualifications
- (b) Fabrication and testing plans
- (c) Base metal specifications and quality
- (d) Welding equipment and welding consumables
- (e) Joint designs and joint preparations

(2) Inspection during welding
- (a) Conformity to welding procedures and fabrication plans
- (b) Preheat and interpass temperature requirements and measurement methods
- (c) Filler metal control and handling
- (d) Use of welders qualified for specific operations
- (e) Interpass and final cleaning
- (f) Visual and, if required, nondestructive inspection

(3) Inspection after welding
- (a) Conformity to drawings and specifications
- (b) Cleaning and visual inspection
- (c) Nondestructive, proof, and mechanical testing, if required
- (d) Repair activities
- (e) Postweld heat treatment, if required
- (f) Documentation of fabrication and inspection activities

When the above activities are applied to specific industries and products, the duties and requirements of welding inspectors must be tailored to meet specific needs. At times, these details are left to the discretion of individual inspectors; in other cases, they are documented as part of a detailed quality assurance plan.

INSPECTION PLAN

In some cases, 100 percent inspection of all production may be required. In other cases, sampling procedures are specified. *Sampling* is the selection of a representative portion of the production for inspection or test purposes. Conclusions can be drawn about the overall quality of an entire production run based on the statistical basis for the sampling procedure and the results of inspections or tests on the samples. The test specifications or the procedures, or both, usually detail the sampling methods for inspection.

SAMPLING METHODS

Some of the terms and methods found in test procedures for sampling are *partial, specified,* and *random.*

Partial Sampling

Partial sampling involves the inspection of a certain number of weldments which is less than the total number in the lot or production run. The method of selection of the weldments to be inspected and the type of inspection should be prescribed. The rejection criteria and disposition routine for any substandard weldment found by the sampling inspection should also be specified on the procedure.

Specified Partial Sampling. Specified partial sampling occurs when a particular frequency or sequence of sample selection from the lot or production run is prescribed on the drawing or specification. An example of specified partial sampling is the selection of every fifth unit for inspection, starting with the fifth unit. This type of sampling is usually selected for fully automated operations where process deterioration is monitored.

Random Partial Sampling. Random partial sampling occurs when the units to be inspected are selected in a random manner. For example, one out of every five units from a production run is to be inspected, and the specific selection is made by the inspector in a random manner.

Since it is not generally known in advance which units will be selected for inspection, equal care must be taken in the production of all units and a more uniform product quality should result.

Statistical Sampling

Statistical sampling uses one procedure to select probability samples, and another to summarize the test results so that specific inferences can be drawn and the risks calculated by probability. This method of sampling can miss some defective products. The percentage of defective products accepted is reduced by increasing the size of the sample. This method of selecting samples is used primarily for high-speed mass production.

Statistical sampling will function most economically after statistical control has been attained as indicated by a control chart. Control charts are the usual criteria for statistical control. When it is desirable to take full advantage of the various plans for sampling, a statistical control chart must first be established.

Plans for statistical inspection are designed to maximize acceptance of good material and rejection of poor material, and to minimize acceptance of poor material. It is never possible to estimate the quality of a production lot from a relatively small sample. The higher the percentage of production sampled, the lower the probability of error.

PLAN SELECTION

When the sampling plan is given in a procedure specification, the inspector need only follow the procedure.

Complete inspection is used when weldments of the highest quality are required for critical services. One or more methods of nondestructive testing, along with visual inspection, may be specified for critical joints.

For the average welding job, inspection generally involves a combination of complete examination and random partial sampling. All of the welds are inspected visually; random partial sampling is then applied using one or more of the other methods of nondestructive testing.

NONDESTRUCTIVE TESTING

DEFINITIONS AND GENERAL DESCRIPTION

Nondestructive testing (NDT) is a term used to designate those inspection methods that allow materials to be examined without changing or destroying their usefulness. *Nondestructive evaluation* (NDE) and *nondestructive inspection* (NDI) are terms sometimes used interchangeably with NDT and are generally considered synonymous. Nondestructive tests are used on weldments for the following reasons:

(1) Improved product reliability

(2) Accident prevention by eliminating faulty products

(3) Determination of acceptability in accordance with a code or specification

(4) Information for repair decisions

(5) Reduction of costs by eliminating further processing of unacceptable components

Nondestructive testing methods, other than visual inspection, include the following elements:

(1) A source of probing energy

(2) A component that is compatible with the energy source

(3) A detection device for measuring differences in or effects upon the probing energy

(4) A means to display for record the results of the test

(5) A trained operator

(6) A procedure for conducting the tests

(7) A system for reporting the results

There is considerable overlap in the application of destructive and nondestructive tests. For example, destructive tests or proof tests are frequently used to supplement, confirm, or establish the limits of nondestructive tests, and to provide supporting information.

The forms of energy that can be used for nondestructive testing of a particular weldment depend on the physical properties of the base and weld metals and the joint designs. Thorough knowledge of each NDT method is needed for proper selection of the appropriate methods for each application. The commonly used NDT methods that are applicable to the inspection of weldments are:

(1) Visual inspection, with or without optical aids (VT)

(2) Liquid penetrant (PT)

(3) Magnetic particle (MT)

(4) Radiography (RT)

(5) Eddy current (ET)

(6) Ultrasonic (UT)

(7) Acoustic emission (AET)

There are other NDT methods, such as heat transfer and ferrite testing, that are used for special cases. They are not covered here. Table 9.1 summarizes the considerations generally used in selecting an NDT method for weld inspection.

With respect to welding inspection, the meaning of the terms *discontinuity, flaw,* and *defect* are as follows:

A *discontinuity* is an interruption in the typical structure of a weldment. It may consist of a lack of homogeneity in the mechanical, metallurgical, or physical characteristics of the base metal or the weld metal. A discontinuity is not necessarily a defect.

A *flaw* is nearly synonymous with a discontinuity but has a connotation of undesirability.

A *defect* is a discontinuity which by nature or effect renders a weldment unable to meet specifications or acceptance standards. The term designates a rejectable condition.

VISUAL INSPECTION

Visual inspection is the one nondestructive testing method most extensively used to evaluate the quality of weldments. It is easily done, relatively inexpensive, does not use special equipment, and gives important information about conformity to specifications. One requirement for this method of inspection is that the inspector have good vision.

Visual inspection should be the primary evaluation method of any quality control program. It can, in addition to flaw detection, discover signs of possible fabrication problems in subsequent operations, and can be utilized in process control programs. Prompt detection and correction of flaws or process deviations can result in significant cost savings. Conscientious visual

Table 9.1
Nondestructive testing methods

Equipment needs	Applications	Advantages	Limitations
		Visual	
Magnifiers, color enhancement, projectors, other measurement equipment, i.e., rulers, micrometers, optical comparators, light source.	Welds which have discontinuities on the surface.	Economical, expedient, requires relatively little training and relatively little equipment for many applications.	Limited to external or surface conditions only. Limited to the visual acuity of the observer/inspector.
		Radiography (Gamma)	
Gamma ray sources, gamma ray camera projectors, film holders, films, lead screens, film processing equipment, film viewers, exposure facilities, radiation monitoring equipment.	Most weld discontinuities including cracks, porosity, lack of fusion, incomplete penetration, slag, as well as corrosion and fit-up defects, wall thickness, dimensional evaluations.	Permanent record—enables review by parties at a later date. Gamma sources may be positioned inside of accessible objects, i.e., pipes, etc., for unusual technique radiographs. Energy efficient source requires no electrical energy for production of gamma rays.	Radiation is a safety hazard—requires special facilities or areas where radiation will be used and requires special monitoring of exposure levels and dosages to personnel. Sources (gamma) decay over their half-lives and must be periodically replaced. Gamma sources have a constant energy of output (wavelength) and cannot be adjusted. Gamma source and related licensing requirenents are expensive. Radiography requires highly skilled operating and interpretive personnel.
		Radiography (X-Rays)	
X-ray sources (machines), electrical power source, same general equipment as used with gamma sources (above).	Same applications as above.	Adjustable energy levels, generally produces higher quality radiographs than gamma sources. Offers permanent record as with gamma radiography (above).	High initial cost of x-ray equipment. Not generally considered portable, radiation hazard as with gamma sources, skilled operational and interpretive personnel required.

Table 9.1 (continued)

Equipment needs	Applications	Advantages	Limitations
Ultrasonic			
Pulse-echo instrument capable of exciting a piezoelectric mat'l and generating ultrasonic energy within a test piece, and a suitable cathode ray tube scope capable of displaying the magnitudes of received sound energy. Calibration standards, liquid couplant.	Most weld discontinuities including cracks, slag, lack of fusion, lack of bond, thickness. Poisson's ratio may be obtained by determining the modulus of elasticity.	Most sensitive to planar type discontinuities. Test results known immediately. Portable. Most ultrasonic flaw detectors do not require an electrical power outlet. High penetration capability.	Surface condition must be suitable for coupling to transducer. Couplant (liquid) required. Small, thin welds may be difficult to inspect. Reference standards are required. Requires a relatively skilled operator/inspector. No record of results in most cases.
Magnetic Particle			
Prods, yokes, coils suitable for inducing magnetism into the test piece. Power source (electrical). Magnetic powders, some applications require special facilities and ultraviolet lights.	Most weld discontinuities open to the surface—some large voids slightly subsurface. Most suitable for cracks.	Relatively economical and expedient. Inspection equipment is considered portable. Unlike dye penetrants, magnetic particle can detect some near surface discontinuities. Indications may be preserved on transparent tape.	Must be applied to ferromagnetic materials. Parts must be clean before and after inspection. Thick coatings may mask rejectable indications. Some applications require parts to be demagnetized after inspection. Magnetic particle inspection requires use of electrical energy for most applications.
Liquid Penetrant			
Fluorescent or dye penetrant, developers, cleaners (solvents, emulsifiers, etc.). Suitable cleaning gear. Ultraviolet light source if fluorescent dye is used.	Weld discontinuities open to surface, i.e., cracks, porosity, seams.	May be used on all non-porous materials. Portable, relatively inexpensive equipment. Expedient inspection results. Results are easily interpreted. Requires no electrical energy except for light source. Indications may be further examined visually.	Surface films such as coatings, scale, smeared metal mask or hide rejectable defects. Bleed out from porous surfaces can also mask indications. Parts must be cleaned before and after inspection.

Table 9.1 (continued)

Equipment needs	Applications	Advantages	Limitations
Eddy Current			
An instrument capable of inducing electromagnetic fields within a test piece and sensing the resulting electrical currents (eddy) so induced with a suitable probe or detector. Calibration standards.	Weld discontinuities open to the surface (i.e., cracks, porosity, lack of fusion) as well as some subsurface inclusions. Alloy content, heat treatment variations, wall thickness.	Relatively expedient, low cost. Automation possible for symmetrical parts. No couplant required. Probe does not have to be in intimate contact with test piece.	Limited to conductive materials. Shallow depth of penetration. Some indications may be masked by part geometry due to sensitivity variations. Reference standard required.
Acoustic Emission			
Emission sensors, amplifying electronics, signal processing electronics including frequency gates, filters. A suitable output system for evaluating the acoustic signal (audio monitor, visual monitor, counters, tape recorders, X-Y recorder).	Internal cracking in welds during cooling, crack initiation and growth rates.	Real time and continuous surveillance inspection. May be inspected remotely. Portability of inspection apparatus.	Requires the use of transducers coupled on the test part surface. Part must be in “use” or stressed. More ductile materials yield low amplitude emissions. Noise must be filtered out of the inspection system.

inspection before, during, and after welding can detect about 80 to 90 percent of the discontinuities before they are found by more expensive nondestructive testing methods.

Equipment

Visual aids and gages are sometimes used to make detection of flaws easier and to measure the sizes of welds or flaws in the joint. Lighting of the welded joint must be sufficient for good visibility. Auxiliary lighting may be needed. If the area to be inspected is not readily visible, the inspector may use mirrors, borescopes, flashlights, or other aids. Low power magnifiers are helpful for detecting minute flaws. However, care must be taken with magnifiers to avoid improper judgement of flaw size.

Inspection of welds usually includes quantitative as well as qualitative assessment of the joint. Numerous standard measuring tools are available to make various measurements, such as joint geometry and fit-up; weld size and reinforcement; misalignment; and depth of undercut. Typical weld measuring gages are shown in Fig. 9.1.

Some situations require special inspection gages to assure that specifications are met. Indicators, such as contact pyrometers and crayons, should be used to verify correct preheat and interpass temperatures. Proper usage of visual aids and gages requires adequate inspector training.

Prior to Welding

Examination of the base metal prior to fabrication can detect conditions that tend to cause weld defects. Scabs, seams, scale, or other harmful surface conditions may be found by visual examination. Plate laminations may be observed on cut edges. Dimensions should be confirmed by measurements. Base metal should be identified by type and grade. Corrections should be made before work proceeds.

After the parts are assembled in position for welding, the inspector should check the weld joint for root opening, edge preparation, and other features that might affect the quality of the weld. Specifically, the inspector should check the following conditions for conformity to the applicable specifications:

(1) Joint preparation, dimensions, and cleanliness

(2) Clearance dimensions of backing strips, rings, or consumable inserts

(3) Alignment and fit-up of the pieces being welded

(4) Welding process and consumables

(5) Welding procedures and machine settings

(6) Specified preheat temperature

(7) Tack weld quality

Sometimes, examination of the joint fit-up reveals irregularities within code limitations. These may be of concern, and should be watched carefully during later steps. For example, if the fit-up for a fillet weld exhibits a root opening when none is specified, the adjacent leg of the fillet weld needs to be increased by the amount of the root opening.

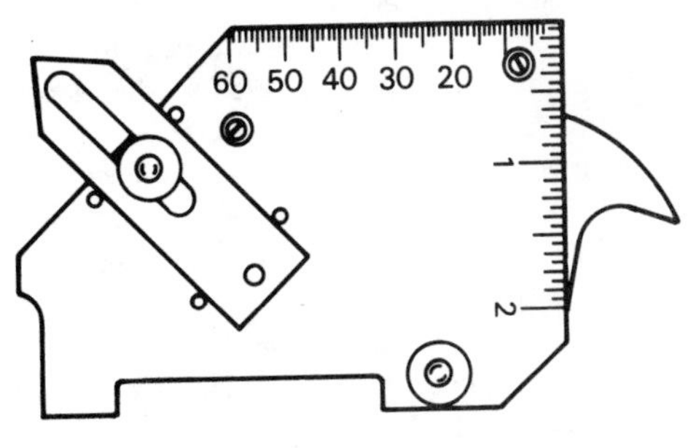

(A) Combination gage

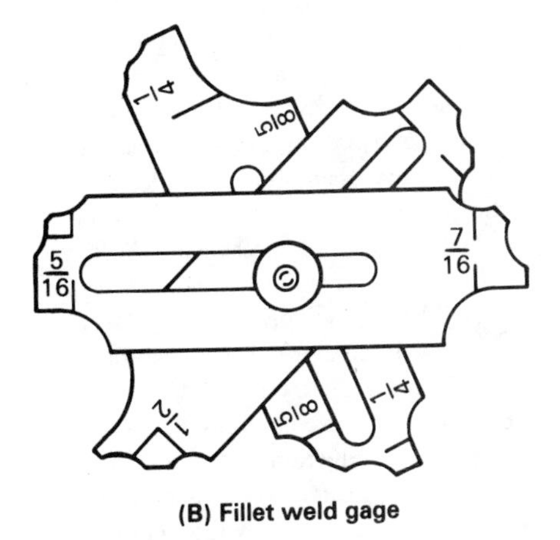

(B) Fillet weld gage

Fig. 9.1—Typical weld measurement gages

During Welding

During welding, visual inspection is the primary method for controlling quality. Some of the aspects of fabrication that can be checked include the following:

(1) Treatment of tack welds

(2) Quality of the root pass and the succeeding weld layers

(3) Proper preheat and interpass temperatures

(4) Sequence of weld passes

(5) Interpass cleaning

(6) Root preparation prior to welding a second side

(7) Conformance with the applicable procedure

The most critical part of any weld is the root pass because many weld discontinuities are associated with the root area. Competant visual inspection of the root pass will help to avoid discontinuities in the completed weld. Another critical root condition exists when second side treatment is required for a double welded joint. This includes removal of slag and other irregularities by chipping, arc gouging, or grinding to sound metal.

The root opening should be monitored as welding of the root pass progresses. Special emphasis should be placed on the adequacy of tack welds, clamps, or braces designed to maintain the specified root opening to assure proper joint penetration and alignment.

Inspection of successive layers of weld metal usually concentrates on bead shape and interpass cleaning. Sometimes, it is carried out with the assistance of workmanship standards. Examples of such standards are shown in Fig. 9.2. They are sections of joints similar to those in manufacture in which portions of successive weld layers are shown. Each layer of the production weld may be compared with the corresponding layer of the workmanship standard. A workmanship sample can only cover ideal conditions, and may not represent actual job conditions. Allowances must be made for production tolerances.

When preheat and interpass temperatures are specified, they should be monitored at the proper times with a suitable temperature measuring device (e.g., crayons or pyrometer). The amount of heat input and also the sequence and placement of each weld pass may be specified to maintain mechanical properties or limit distortion, or both.

To ensure the quality of the weld as work progresses, each weld layer should be visually checked by the welder for surface irregularities and adequate interpass cleaning to avoid slag inclusions or porosity.

After Welding

Items that are checked by visual inspection after welding include the following:

(1) Final weld appearance

(2) Final weld size

(3) Extent of welding

(4) Dimensional accuracy

(5) Amount of distortion

(6) Postweld heat treatment

Most codes and specifications describe the type and size of discontinuities which can be accepted. Many of the following discontinuities on the surface of a completed weld can be found by visual inspection:

(1) Cracks

(2) Undercut

(3) Overlap

(4) Exposed porosity and slag inclusions

(5) Unacceptable weld profile

(6) Roughness of the weld faces

For accurate detection of such discontinuities, the weld surface should be thoroughly cleaned of oxide and slag. The cleaning operation must be carried out carefully to avoid masking discontinuities from view. For example, if a chipping hammer is used to remove slag, the hammer marks could mask fine cracks. Shot blasting may peen the surface of relatively soft weld metal and hide possible flaws.

Dimensional accuracy of weldments is determined by conventional measuring methods. The conformity of weld size and contour may be determined by the use of a suitable weld gage. The size of a fillet weld in joints whose members are at right angles, or nearly so, is defined in terms of the length of the legs. The gage should determine whether the leg size is within allowable limits, and whether there is excessive concavity or

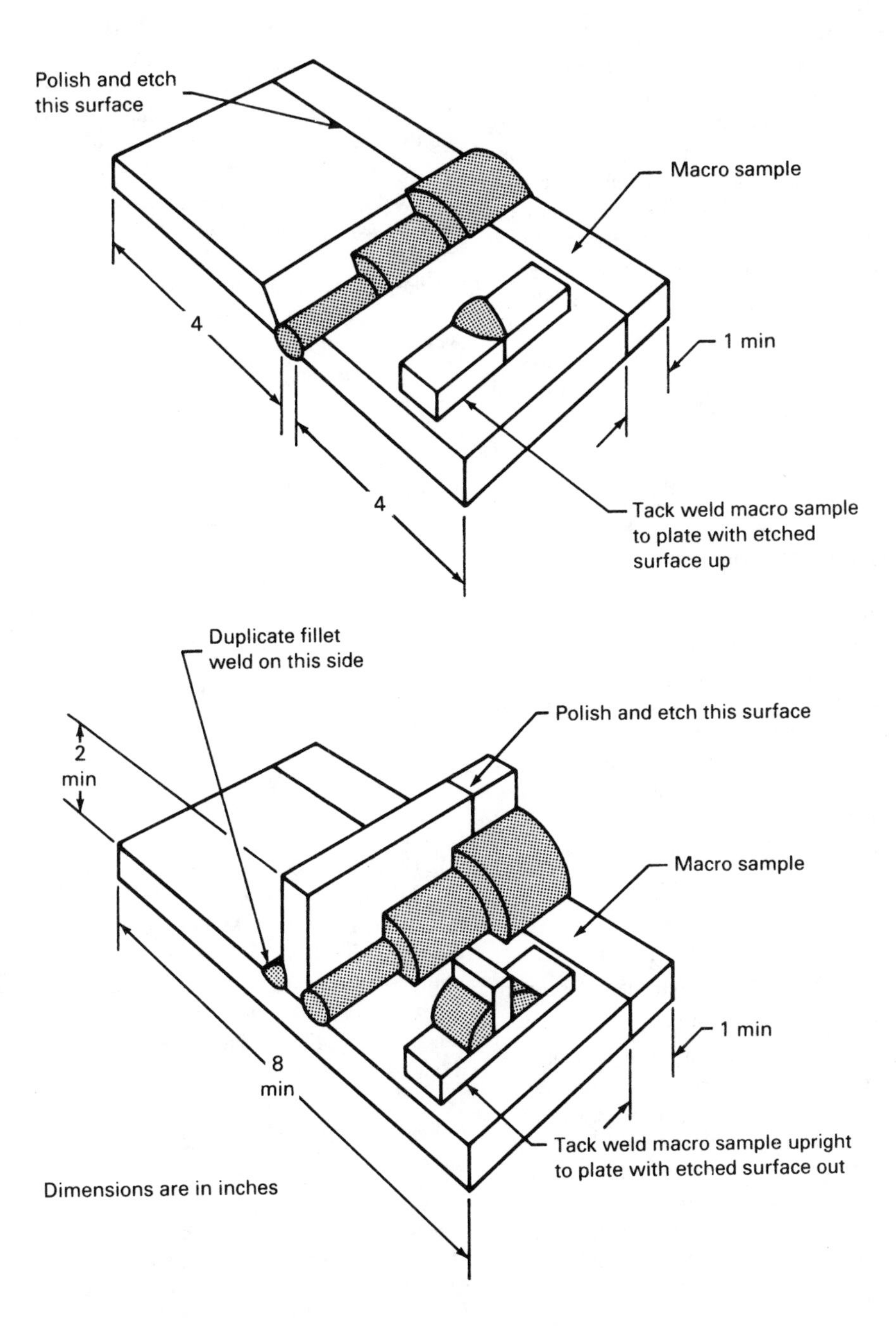

Fig. 9.2—Examples of weld workmanship standards for groove welds (top) and fillet welds (bottom)

convexity. Special gages may be made for use where the members are at acute or obtuse angles.

For groove welds, the height of reinforcement should be consistent with specified requirements. Where not specified, the inspector may need to rely on judgment, guided by what is considered good welding practice. Requirements as to surface appearance differ widely. In general, the weld surface should be as specified in the code or specifications. Visual standards or sample weldments submitted by the fabricator and agreed to by the purchaser can be used as guides to appearance. Sometimes a smooth weld, strictly uniform in size, is required because the weld is part of the exposed surface of the product, and good appearance is desirable.

A code may permit limited amounts of undercut, undersize, and piping porosity, but cracks, incomplete fusion, and unfilled craters are generally not acceptable. Undercut, overlap, and improper weld profile act as stress raisers under load, and cracks may develop at these locations under cyclic loading.

Some steels, such as ASTM A514 and A517, are susceptible to delayed cracking.[3] The applicable code may specify a delay before visual inspection of welds in crack-sensitive steels.

When a postweld heat treatment is specified, the operation should be monitored and documented by an inspector. Items of importance in heat treatment may include the following:

(1) Area to be heated

(2) Heating and cooling rates

(3) Holding temperature and time

(4) Temperature measurement and location

(5) Equipment calibration

Care should be taken when judging the quality of a weld from the visible appearance alone. Acceptable surface appearance does not prove careful workmanship, and is not a reliable indication of subsurface weld integrity. However, proper visual inspection procedures before and during fabrication can increase product reliability over that based only on final inspection.

LIQUID PENETRANT TESTING

Liquid penetrant testing (PT) is a method that reveals open discontinuities by bleedout of a liquid penetrant medium against a contrasting background developer.[4] The technique is based on the ability of a penetrating liquid to wet the surface opening of a discontinuity and to be drawn into it. If the flaw is significant, penetrant will be held in the cavity when the excess is removed from the surface. Upon application of a liquid-propelled or dry powder developer, blotter action draws the penetrant from the flaw to provide a contrasting indication on the surface.

The basic steps involved in the application of a liquid penetrant test are relatively simple. A novice individual can probably conduct a satisfactory test simply by reading the instructions printed on many of the various types of penetrant containers.

The following sequence is normally used in the application of a typical penetrant test. When the order is changed or short cuts are taken, validity of the test can be lost.

(1) Clean the test surface.

(2) Apply the penetrant.

(3) Wait for the prescribed dwell time.

(4) Remove the excess penetrant.

(5) Apply the developer.

(6) Examine the surface for indications and record results.

(7) Clean, if necessary, to remove the residue.

Liquid penetrant methods can be divided into two major groups: Method A—fluorescent penetrant testing, and Method B—visible penetrant testing. The major differences between the two types of tests is that for Method A, the penetrating medium is fluorescent meaning that it glows when illuminated by ultraviolet or "black" light. This is illustrated in Fig. 9.3. Method B utilizes visible penetrant, usually red in

3. Refer to Vol. 4, *Metals and Their Weldability, Welding Handbook,* 7th Ed., 1982, 3-7.

4. Refer to *ASTM E165, Standard Recommended Practice for Liquid Penetrant Inspection* for more information on the subject. *ASTM E433, Standard Reference Photographs for Liquid Penetrant Inspection* may be useful for classifying indications. These standards are available from the American Society for Testing and Materials.

color; that produces a contrasting indication against the white background of a developer, as shown in Fig. 9.4.

The sensitivity may be greater using the fluorescent method; however, both offer extremely good sensitivity when properly applied. The difference in sensitivity is primarily due to the fact that the eye can discern the contrast of a fluorescent indication under black light more readily than a color contrast under white light. In the latter case, the area must be viewed with adequate white light.

Fig. 9.3—Fluorescent penetrant indication of a discontinuity in a weld

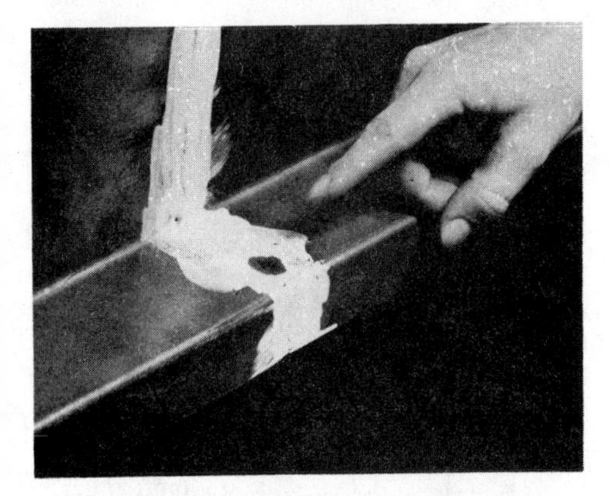

Fig. 9.4—Dye penetrant indication of a discontinuity in a weld

Applications

For the inspection of welds, the liquid penetrant test is reliable when properly performed. Except for visual inspection, it is perhaps the most commonly utilized nondestructive test for the surface inspection of nonmagnetic parts. While the test can be performed on as-welded surfaces, the presence of weld bead ripples and other irregularities can hinder interpretation of indications. In the examination of welds joining cast metals with this method, the inherent surface imperfections in castings may also cause problems with interpretation. If the surface condition causes an excessive amount of irrelevant indications, it may be necessary to remove troublesome imperfections by light grinding prior to inspection. The visible penetrant test method is commonly used for field testing applications because of its simplicity.

Another common use of penetrant inspection is to check the accuracy of results obtained by magnetic particle testing of ferromagnetic weldments.

Equipment

Unlike many of the other types of nondestructive testing, liquid penetrant testing requires little equipment. Most of the equipment consists of containers and applicators for the various liquids and solutions which are utilized. The testing materials are available in convenient aerosol spray cans, which can be purchased separately or in kits. Fluorescent penetrant testing requires a high intensity ultraviolet light source and facilities for reduction or elimination of outside lighting. Other related equipment, such as additional lighting, magnifiers, drying apparatus, rags, and paper towels, might be needed depending on the specific test application.

Materials

Liquid penetrant inspection materials consist of fluorescent and visible penetrants, emulsifiers, solvent base removers, and developers.[5] Intermixing of materials of different methods and manufacturers is not recommended. The

5. These materials can be flammable or emit hazardous and toxic vapors. Observe all manufacturer's instructions and precautionary statements.

inspection materials used should not adversely affect the serviceability of the parts tested.

Penetrants. Water-washable penetrants are designed to be directly water-washable from the surface of the test part, after a suitable penetrant (dwell) time. It is extremely important to avoid overwashing because the penetrants can be washed out of discontinuities if the rinsing step is too long or too vigorous.

Post-emulsifiable penetrants are insoluble in water and cannot be removed with water rinsing alone. They are selectively removed from the surface of a part using a separate emulsifier. The emulsifier combines with the penetrant to form a water-washable mixture, which can be rinsed from the surface of the part.

Solvent-removable penetrants can be removed by first wiping with clean, lint-free material, and repeating the operation until most traces of penetrant have been removed. The remaining traces are removed by wiping the surface with clean, lint-free material lightly moistened with a solvent remover. This type is intended primarily for portability and for localized areas of inspection. To minimize removal of penetrant from discontinuities, care shall be taken to avoid the use of excess solvent.

Emulsifiers. These are liquids used to emulsify the excess oily penetrant on the surface of the part, rendering it water washable.

Developers. Dry powder developers are free-flowing and noncaking as supplied. Care should be taken not to contaminate the developer with penetrant, as the specks can appear as indications.

Aqueous wet developers are normally supplied as dry powders to be suspended or dissolved in water, depending on the type of aqueous wet developer.

Nonaqueous suspendible developers are supplied as suspensions of developer particles in nonaqueous solvent carriers ready for use as supplied. They are applied to the area by conventional or electrostatic spray guns or by aerosol spray cans after the excess penetrant has been removed and the part has dried. Nonaqueous wet developers form a white coating on the surface of the part when dried, which serves as a contrasting background for visible penetrants and developing media for fluorescent penetrants.

Liquid film developers are solutions or colloidal suspensions of resins in a suitable carrier. These developers will form a transparent or translucent coating on the surface. Certain types of film developer may be stripped from the part and retained for record purposes.

Procedures

The following general processing procedures, as outlined in Fig. 9.5, apply to both the fluorescent and visible penetrant inspection methods. The temperature of the penetrant materials and the surface of the weldment should be between 60° and 125° F.

Satisfactory results can usually be obtained on surfaces in the as-welded condition. However, surface preparation by grinding or machining is necessary when surface irregularities might mask the indications of unacceptable discontinuities, or otherwise interfere with the effectiveness of the examination.

Precleaning. The success of any penetrant inspection procedure is greatly dependent upon the freedom of both the weld area and any discontinuities from contaminants (soils) that might interfere with the penetrant process. All parts or areas of parts to be inspected must be clean and dry before the penetrant is applied. "Clean" is intended to mean that the surface must be free of any rust, scale, welding flux, spatter, grease, paint, oily films, dirts, etc., that might interfere with penetration. All of these contaminants can prevent the penetrant from entering discontinuities. If only a section containing a weld is to be inspected, the weld and adjacent areas within 1 in. of the weld must also be cleaned.

Drying After Cleaning. It is essential that the parts be thoroughly dry after cleaning because any liquid residue will hinder the entrance of the penetrant. Drying may be accomplished by warming the parts in drying ovens, with infrared lamps, forced hot air, or exposure to ambient temperature. Part temperatures must not exceed 125° F prior to application of penetrant.

Penetrant Application. After the part has been cleaned, dried, and cooled to approximate ambient temperatures (125° F maximum), the penetrant is applied to the surface to be inspected

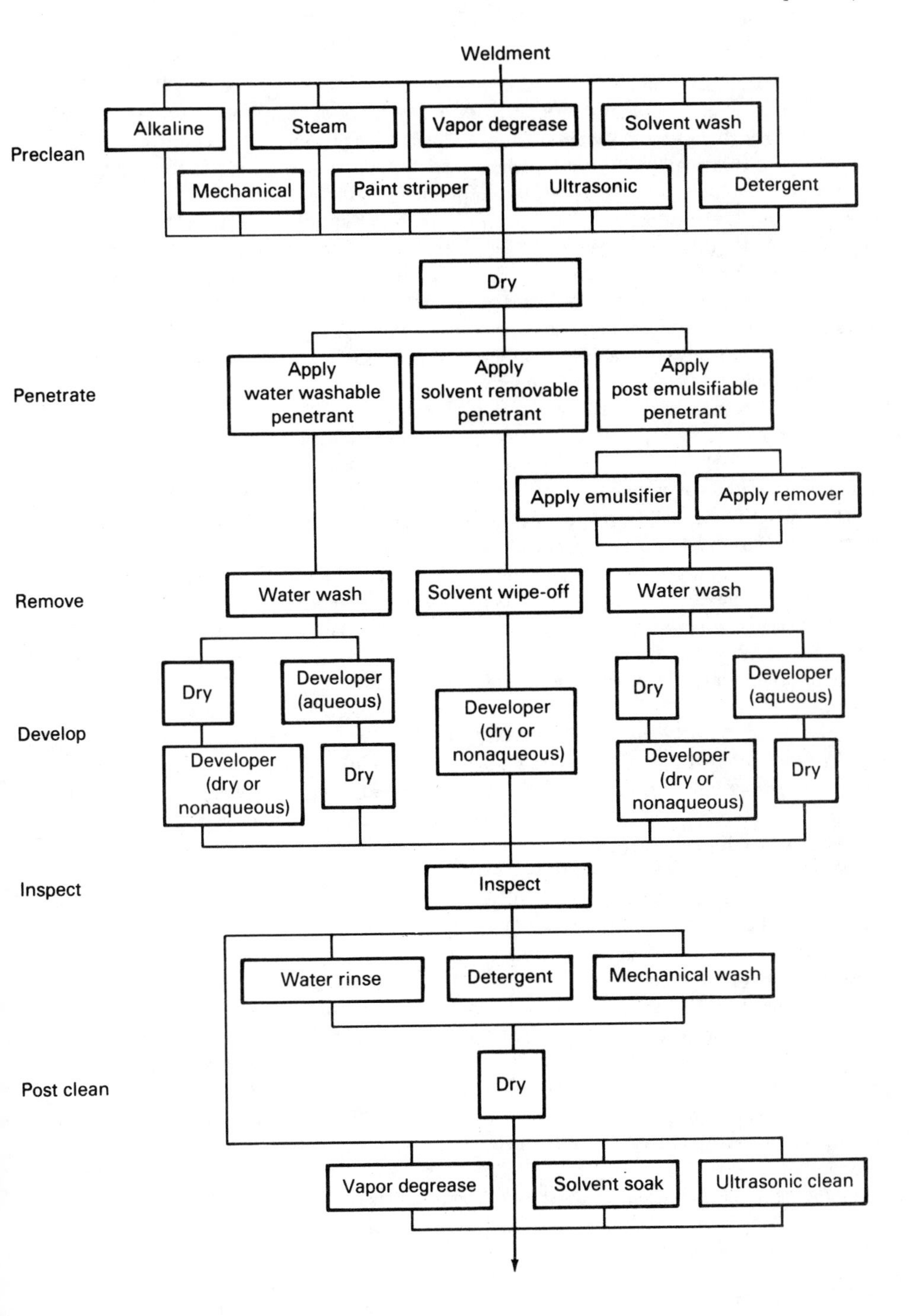

Fig. 9.5—Fluorescent and visible penetrant inspection processing flowsheet

so that the entire part or the weld area is completely covered with penetrant.

There are various methods for effective application of penetrant such as dipping, brushing, flooding, or spraying. Small parts are quite often placed in suitable baskets and dipped into a tank of penetrant. On larger parts, and those with complex geometries, penetrant can be applied effectively by brushing or spraying. Aerosol sprays are also a very effective and convenient means of application. With spray applications, it is important that there is proper ventilation. This is generally accomplished through the use of a properly designed spray booth and exhaust system.

After application, excess penetrant is drained from the part. Care should be taken to prevent pools of penetrant on the part.

The length of time that the penetrant must remain on the part to allow proper penetration should be as recommended by the penetrant manufacturer. If penetrant characteristics are materially affected by a prolonged dwell time, as evidenced by difficulty in removing the excess, reapply the penetrant for the original prescribed dwell time.

Removal of Excess Penetrant. After the required dwell time, the excess penetrant is removed. The procedure depends on the type of penetrant.

Water-washable penetrants can be removed directly from the part surface using a water spray or immersion equipment. Most water-washable penetrants can be removed effectively within a temperature range from 60° to 110° F, but for consistent results, the temperature recommended by the penetrant supplier should be used.

Excessive washing can cause penetrant to be washed out of discontinuities. The rinsing operation for fluorescent penetrants may be done under black light so that it can be determined when the surface penetrant has been adequately removed.

In special applications, where water rinse facilities are not available, penetrant removal may be performed by wiping the surface with a clean, absorbent material dampened with water until the excess surface penetrant is removed.

Post-emulsifiable penetrants are not directly water-washable; they require the use of an emulsifier (oil or water base). After the required penetration time, the excess penetrant on the part is emulsified by dipping, flooding, or spraying with the required emulsifier. After application of the emulsifier, the part should be drained to avoid pooling of the emulsifier.

Emulsification dwell time begins as soon as the emulsifier has been applied. Nominal emulsification time should be as recommended by the manufacturer.

Effective rinsing of the emulsified penetrant from the surface of the part can be accomplished in the same manner as for water-washable penetrants.

With solvent-removable penetrants, excess penetrant is removed, insofar as possible, by using wipes of clean, lint-free material, repeating the operation until most traces of penetrant have been removed. Then, a lint-free material is lightly moistened with solvent and wiped over the surface until all remaining traces of excess penetrant have been removed. To minimize removal of penetrant from discontinuities, avoid the use of excess solvent.

For some critical applications, flushing of the surface with solvent cleaner should be prohibited. When cleaning as previously described cannot be done, flushing the surface with cleaner may be necessary, but it will jeopardize the accuracy of the test.

The solvent may be trichorethylene, perchlorethylene, acetone, or a volatile petroleum distillate. The former two are somewhat toxic; the latter two are flammable. Proper safety precautions should be observed. Producers of the inspection materials market cleaners for their penetrant products, and their use is recommended.

Drying of Parts. During the preparation of parts for inspection, drying is necessary following either application of aqueous wet developer or water rinsing, before the application of dry or nonaqueous developers. Parts can be dried in a hot-air recirculating oven, with a hot-air blast, or by exposure to dry air at ambient temperature. Part temperature should never exceed 125° F.

Excessive drying time can cause evaporation of the penetrant, which may impair the sensitivity of the inspection. Drying time will vary with the size, nature, and number of parts under inspection. Where excess penetrant is removed with the solvent wipe-off technique, drying should be by normal evaporation.

Developing Indications. Indications are developed by bringing the penetrant back out of any discontinuities, through blotting action, which spreads it on the surface to increase its visibility to the eye. Developers are used either dry or suspended in an aqueous or nonaqueous solvent that is evaporated to dryness before inspection. They should be applied immediately after the excess penetrant has been removed from the part surface, prior to drying in the case of aqueous developers, and immediately after the part has been dried for all other developer forms.

There are various methods for effective application of the various types of developers, such as dipping, immersing, flooding, spraying, or dusting. The size, configuration, surface condition, and number of parts to be processed will influence the choice of developer.

Dry powder developers should be applied in a manner that assures complete coverage of the area being inspected. Excess powder may be removed by shaking or tapping the part gently, or by blowing with low-pressure (5 to 10 psi) dry, clean, compressed air.

Aqueous developers should be applied immediately after the excess penetrant has been removed from the part, but prior to drying. After drying, the developer appears as a white coating on the part. Aqueous developers should be prepared and maintained in accordance with the manufacturer's instructions and applied in a manner to assure complete and even coverage. The parts are then dried as described previously.

Nonaqueous wet developers are applied to the part by spraying, as recommended by the manufacturer, after the excess penetrant has been removed and the part has been dried. This type of developer evaporates very rapidly at normal room temperature, and does not require the use of a dryer. It should be used, however, with proper ventilation. The developer must be sprayed in a manner that assures complete coverage with a thin, even film.

The length of time before the coated area is visually observed for indications should not be less than about 7 minutes, or as recommended by the manufacturer. Developing time begins immediately after the appplication of dry powder developer or as soon as a wet (aqueous or nonaqueous) developer coating is dry. If bleedout does not alter the inspection results, development periods of over 30 minutes are permitted.

Inspection. Although examination of parts is done after the appropriate development time, it is good practice to observe the surface while applying the developer as an aid in evaluating indications. Inspection for fluorescent penetrant indications is done in a dark area. Maximum ambient light of about 3 footcandles is allowed for critical inspection. Higher levels may be used for noncritical inspections. Black light intensity should be a minimum of 800 $\mu W/cm^2$ on the surface of the part being inspected.

Visible penetrant indications can be inspected in either natural or artificial white light. A minimum light intensity at the inspection site of 32.5 footcandles is recommended.

Post Cleaning. Post cleaning is necessary in those cases where residual penetrant or developer could interfere with subsequent processing or with service requirements. It is particularly important where residual inspection materials might combine with other materials in service to produce corrosion products. A suitable technique, such as simple water rinsing, machine washing, vapor degreasing, solvent soaking, or ultrasonic cleaning may be employed. In the case of developers, it is recommended that cleaning be carried out within a short time after inspection so that the developer does not adhere to the part. Developers should be removed prior to vapor degreasing because that can bake the developer onto the parts.

Interpretation of Indications

Probably the most common defects found using this process are surface cracks. Due to the critical nature of surface cracks for most applications, this test capability is a valuable one. An indication of a crack is very sharp and well defined. Most cracks exhibit an irregular shape, and the penetrant indication will appear identical. The width of the bleedout is a relative measure of the depth of a crack. A very deep crack will continue to produce an indication even after recleaning and redeveloping several times.

Surface porosity, metallic oxides, and slag will also hold penetrant and cause an indication. Depending on the exact shape of the pore, oxide, or slag pocket, the indication will be more or less circular. In any case, the length to width ratio usually will be far less than that of a crack.

Other discontinuities such as inadequate joint penetration and incomplete fusion can also be detected by penetrant inspection if they are open to the surface. Undercut and overlap are not easily detected by this type of testing; they can be evaluated more effectively using visual inspection.

MAGNETIC PARTICLE TESTING

Principles

Magnetic particle testing (MT) is a nondestructive method used to detect surface or near surface discontinuities in magnetic materials.[6] The method is based on the principle that magnetic lines of force, when present in a ferromagnetic material, will be distorted by an interruption in material continuity, such as a discontinuity or a sharp dimensional change. If a discontinuity in a magnetized material is open to or close to the surface, the magnetic flux lines will be distorted at the surface, a condition termed *flux leakage.* When fine magnetic particles are distributed over the area of the discontinuity while the flux leakage exists, they will accumulate at the discontinuity and be held in place. This principle is illustrated in Fig. 9.6. The accumulation of particles will be visible under proper lighting conditions. While there are variations in the magnetic particle test method, they all are dependent on the principle that magnetic particles will be retained at the locations of magnetic flux leakage.

There are three essential requirements of the test method:

(1) The part must be magnetized.

(2) Magnetic particles must be applied while the part is magnetized.

(3) Any accumulation of magnetic particles must be observed and interpreted.

6. Additional information is contained in *ASTM E709, Standard Recommended Practice for Magnetic Particle Examination,* and *ASTM E125, Magnetic Particle Indications on Ferrous Castings,* available from the American Society for Testing and Materials.

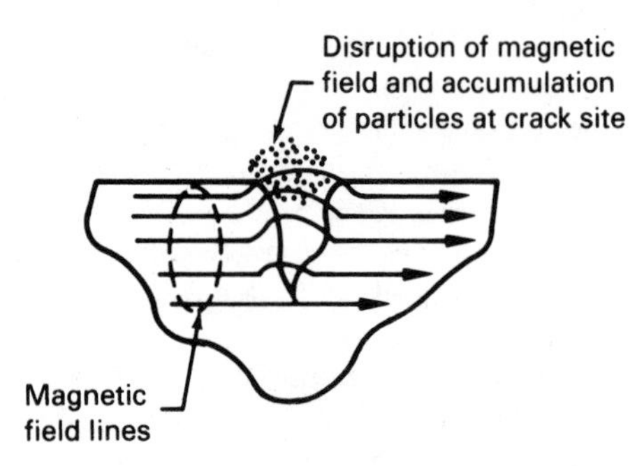

Fig. 9.6—Accumulation of magnetic particles at a surface flaw

A ferromagnetic material can be magnetized either by passing an electric current through the material or by placing the material within a magnetic field originated by an external source. The entire component or a portion of it can be magnetized as dictated by size and equipment capacity or by need. As previously noted, the discontinuity must interrupt the normal path of the lines of force. If a discontinuity is open to the surface, the flux leakage will be a maximum for a given size and shape of discontinuity. When a discontinuity is below the surface, flux leakage will be lower. Practically, discontinuities must be open to the surface or must be in the near subsurface to create flux leakage of sufficient strength to accumulate magnetic particles.

If a discontinuity is oriented parallel to the lines of force, it will be essentially undetectable. Because discontinuities may occur in any orientation, it is usually necessary to magnetize the part at least twice to induce magnetic lines of force in different directions to perform an adequate examination.

The lines of force must be of sufficient strength to indicate those discontinuities which are unacceptable, yet not so strong that an excess of particles is accumulated locally thereby masking relevant indications.

Applications

Magnetic particle inspection has wide usage in the field of welding. The process can assist in determining the resultant quality of welds in magnetic steels and cast irons. Considerable magnetic particle testing is performed on finished weldments. Moreover, when utilized at prescribed intervals in the completion of a multiple pass weld, it can be an extremely valuable process control tool. When used in this manner, flaws are found while they are easily correctable rather than later when the difficulty and cost of repair are greater. As an example, one usage is in the examination of the root of a back gouge or a repair groove prior to welding to assure that all defects have been satisfactorily removed. Magnetic particle inspection is also applied to weldments following, and sometimes prior to, stress relief. Most defects that might occur during this treatment would be surface-related.

The aerospace industry sometimes uses magnetic particle inspection on light structural components. With fatigue being a prime design consideration for some components, the surface quality becomes extremely critical. With thin cross sections, many subsurface flaws are also detectable. Magnetic particle inspection may be performed during routine maintenance. It provides a good check for potential structural problems.

Magnetic particle inspection is frequently applied to plate edges prior to welding to detect cracks, laminations, inclusions, and segregations. It will reveal only those discontinuities that are near or extend to the edge being inspected. Not all discontinuities found on plate edges are objectionable. However, it is necessary to remove those that would affect either the soundness of the welded joint or the ability of the base metal to assume design loads under service conditions.

Magnetic particle inspection can be applied in conjunction with repair work or rework procedures on both new parts and parts that may have developed cracks in service. This applies not only to the repair of weldments but also to rework done by welding in the repair or salvage of castings and forgings. It may be advisable to check the completed repair for cracks or other objectionable discontinuities in the weld or in the adjacent metal, if they exist. In general, the same inspection procedures should be used in connection with repair or rework procedures as would be used on the original parts.

Limitations

The magnetic method of inspection is applicable only to ferromagnetic metals in which the deposited weld metal is also ferromagnetic. It cannot be used to inspect nonferrous metals or austentic steel. Difficulties may arise with weldments where the magnetic characteristics of the weld metal are appreciably different from those of the base metal. Joints between metals of dissimilar magnetic characteristics create magnetic discontinuities that may produce indications even though the joints themselves are sound.

Subsurface porosity and slag inclusions produce powder patterns that are not clearly defined. The degree of sensitivity in this method depends upon certain factors. Sensitivity decreases with a decrease in size of the discontinuity, and also with an increase in depth below the surface. A decrease in sensitivity is evident when discontinuities are rounded or spherical, rather than cracklike.

A discontinuity must be sufficiently large to interrupt or distort the magnetic field to cause external leakage. Fine, elongated discontinuities, such as seams, inclusions, or fine cracks, will not interrupt a magnetic field that is parallel to the direction of the discontinuity. In this case, no indication will be apparent.

Surface conditions also influence the sensitivity of the inspection process. The surface of the weld and surrounding areas should be clean, dry, and free from oil, water, excessive slag, or other accumulations that would interfere with magnetic particle movement. A rough surface decreases the sensitivity and tends to distort the magnetic field. It also interferes mechanically with the formation of powder patterns. Light grinding may be used to smooth rough weld beads, using care to avoid smearing the surface.

Orientation of the Magnetic Field

The orientation of the magnetic field has a great influence on the validity and performance of the test. If testing is done on a weld using only a single orientation of the magnetic field, there may be discontinuities present that are undetected because they are aligned with the flux path. The direction of the magnetic field must be known so that it can be shifted to provide the necessary coverage. The best results are obtained when the magnetic field is perpendicular to the length of existing discontinuities.

Circular Magnetization

A magnetic field can be produced by passing an electrical current through a conductor. This method is referred to as circular magnetization. Most magnetic particle testing utilizes this principle to produce a magnetic field within the part. An electrically-induced magnetic field is highly directional. Also, the intensity of the field is proportional to the strength of the current. The direction of the magnetic lines of force for circular magnetism is shown in Fig. 9.7.

When current is passed through a conductor, a magnetic field is present on its surface as well as around it. However, when current is passed through a ferromagnetic conductor, such as carbon steel, most of the field is confined within the conductor itself. This behavior is illustrated in Fig. 9.8.

If current is made to flow through a uniform steel section, the magnetic field will be uniform as well. Upon application of fine magnetic particles to the surface while the current is flowing, the particles will be uniformly distributed over the surface. If a discontinuity is present on the surface, the particles will tend to build up across the discontinuity because of flux leakage.

When the excess magnetic powder is removed, the outline of the discontinuity, with regard to dimensions and orientation, is rather well defined by the powder that remains, providing the discontinuity is perpendicular to the flux path. This is illustrated in Fig. 9.9. If the flux path is parallel to the discontinuity, it is probable that no indication will appear. For a very large flaw parallel to the flux path, an indication might be visible, but it would likely be very weak and indefinite.

If circular magnetization is to be utilized to detect longitudinal discontinuities lying on the inner surface of a hollow part, a slightly different technique is required because the bore of a hollow part is not magnetized when current passes directly through the part. To accomplish this inspection, a conductor is placed through the opening or hole in the hollow part. When current

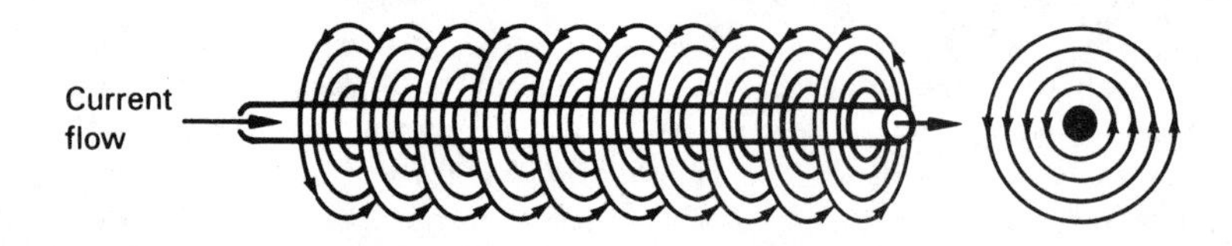

Fig. 9.7—Magnetic field around a conductor carrying current

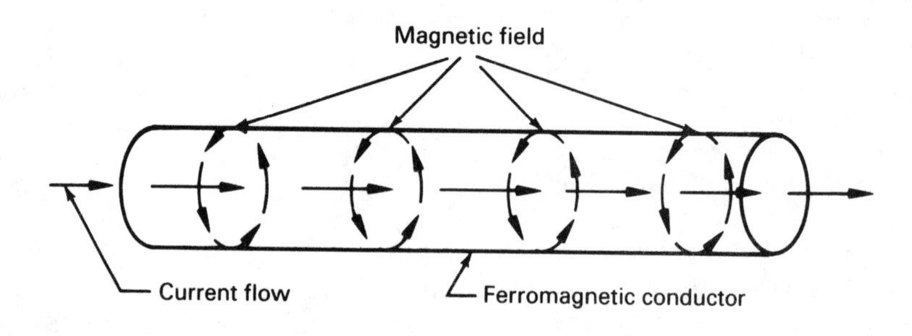

Fig. 9.8—Circular magnetic field within a ferromagnetic electric conductor

is passed through the conductor, circular magnetic fields are induced at both the inner and outer surfaces of the hollow part.

Longitudinal Magnetization

Sometimes, discontinuities are oriented such that they would be parallel to circular magnetic flux in a steel part. The detection of such discontinuities requires a somewhat different approach. The conductor is coiled, and the part to be tested is placed within the coil such that it becomes the core of a solenoid. This produces a magnetic field in line with the axis of the coil. Two or more poles are produced, usually at the ends of the part. This technique is referred to as *longitudinal* or *bipolar magnetization.* With longitudinal magnetization, it is possible to reveal those discontinuities lying transverse to the long axis of the part, Fig. 9.10. Flaws oriented at 45 degrees may be detected by circular or longitudinal magnetization. Sometimes in an emergency, longitudinal magnetization can be induced in parts, such as shafts, pipes, and beams, simply by coiling a length of welding cable around the area to be tested and applying current.

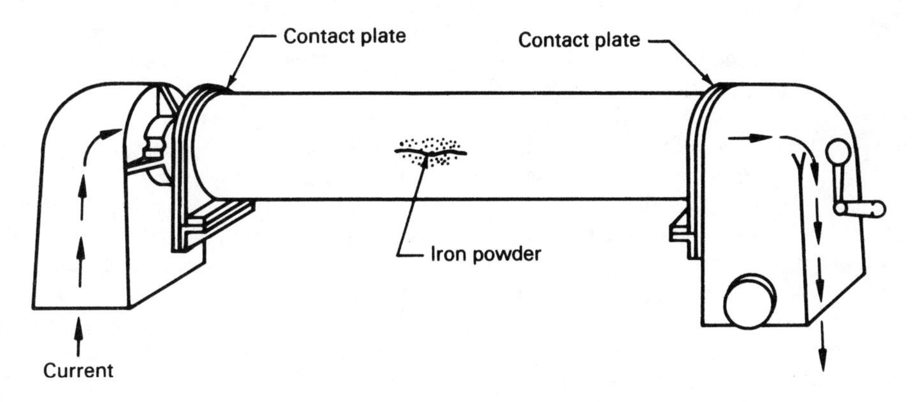

Fig. 9.9—Detection of a flaw with circular magnetization

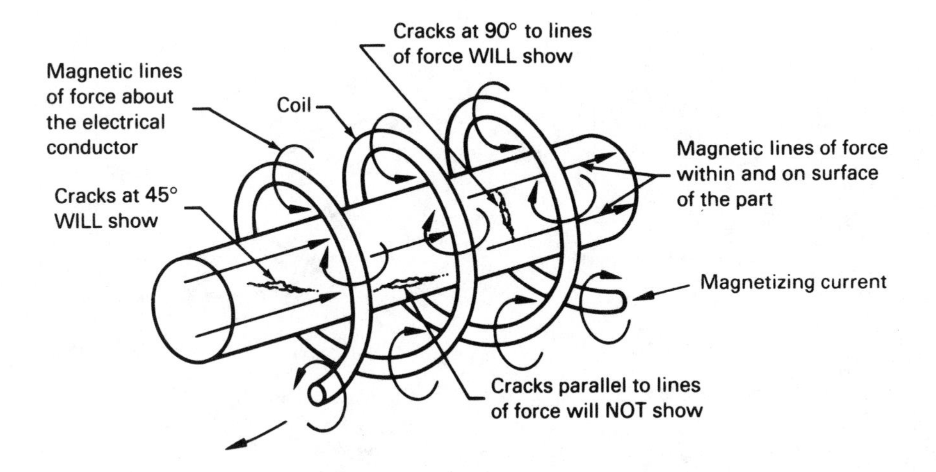

Fig. 9.10—Longitudinal magnetization

Localized Magnetization

For large parts, there are two basic types of equipment available that will provide a magnetic field in a localized area. Both can be utilized as portable methods so that the equipment can be taken to the component to be inspected on location.

The first of these two techniques is referred to as *prod magnetization*. Using this method, a localized area can be magnetized by passing current through the part by means of hand-held contacts or prods, as shown in Fig. 9.11. Manual clamps or magnetic leeches may be used in place of prods.

The current creates local, circular magnetic fields in the area between the contact points. This method is used extensively for localized inspection of weldments where the area of interest is confined to the weld zone. The prods must be securely held in contact with the part to avoid arcing at the contact points. For this reason, a low open-circuit voltage is utilized (2 to 16 V). This method provides only a unidirectional magnetic field. Therefore, it is necessary to reorient the contacts at about 90 degrees and remagnetize for complete inspection of the area.

Another way to induce a localized magnetic field is with a solenoid having flexible extensions of the core. When the extensions make contact with the part and the coil is energized, the magnetic field of the solenoid is concentrated in the part between the contact points. Referred to as the *yoke* method, the equipment can be relatively small and lightweight. Another desirable feature of this technique is that electric current is not transferred to the part, as with the prod method. Thus, there is no tendency to arc or burn the part.

Types of Magnetizing Current

Alternating or direct current may be used for magnetizing the parts. High amperage, low voltage power is usually employed.

Portable equipment that makes use of electromagnets or permanent magnets is occasionally used. These magnets are satisfactory for the detection of surface cracks only.

Alternating Current. Only the surface of the metal is magnetized by alternating current. The method is effective for locating discontinuities that extend to the surface, such as cracks, but deeper discontinuities or incomplete fusion would not be detected. It may be used to inspect welds where subsurface evaluation is not required.

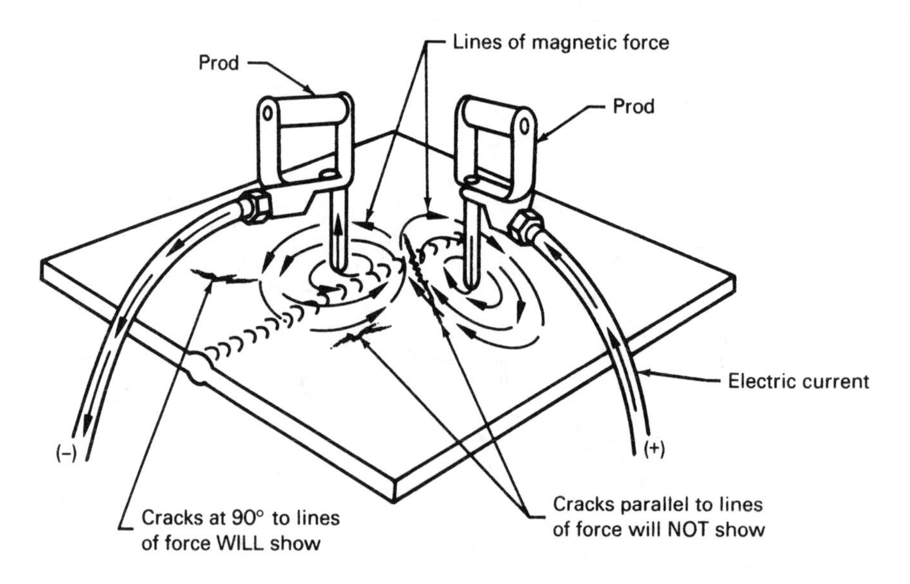

Fig. 9.11—Prod technique for local magnetization with magnetic particle inspection

Direct Current. Direct current produces a magnetic field that penetrates throughout the part and is, therefore, more sensitive than alternating current for the detection of subsurface discontinuities. Full-wave, three-phase rectified current produces results essentially comparable to direct current obtained from generator or batteries.

Half-wave rectified, single-phase current provides maximum sensitivity. The pulsating field increases particle mobility and enables particles to line up more readily in weak leakage fields.

With suitable equipment, two different types of current can be applied to a part. The characteristics of the same indications with different currents could provide useful information for more accurate interpretation of a flaw. For example, using both ac and dc magnetization, the relative amounts of particle buildup could help to determine the depth of the flaw. If the particle buildup with dc magnetization is considerably greater than that with ac, it is highly probable that the flaw has a significant depth, and is not just a surface irregularity.

Amount of Magnetizing Current

The magnetizing current should be of sufficient strength to indicate all detectable discontinuities that might affect the performance of the weldment in service. Excessive magnetizing currents should be avoided because they produce irrelevant patterns. Magnetizing currents should be specified in the test procedures or specifications. If these are unavailable, current requirements may be determined by experience or experiment. The applied voltage has no effect on the magnetic fields, and should be kept low to prevent arcing and overheating.

The approximate amperage ranges for the various magnetizing methods are as follows:

(1) Longitudinal magnetization: 3,000 to 10,000 ampere-turns depending on the ratio of the coil and part diameters.

(2) Overall circular magnetization: 100 to 1,000 amperes per inch of part diameter.

(3) Prod magnetization: 90 to 125 amperes per inch of prod spacing, depending on metal thickness.

(4) Yoke magnetization: The magnetizing current must be sufficient to lift 40 lb with dc magnetization and 10 lb with ac magnetization.

Inspection Media

With the availability of magnetic particles of various colors, mobility, and luminescence, a type can be selected that will provide for the greatest visual sensitivity for each specific test situation. The condition of the test surface as well as the types of flaws suspected are additional criteria to be considered in this decision.

Dry Method. With this technique, finely divided ferromagnetic particles in dry powder form are coated to provide for enhanced mobility, and then dyed various colors to create a distinct contrast with the background. They are applied uniformly to the part by means of a particle dispenser, an atomizer, or a spray gun. The dry method provides for the greatest portability of inspection. Dry powder is more satisfactory on rough surfaces. (Application of a wet medium would result in an increase in the extent of irrelevant indications, making interpretation more difficult.)

The powder should be applied in the form of a low velocity cloud with just enough motive force to direct the particles to the area of interest. This permits the particles to line up in indicating patterns as they are drawn to locations of leakage flux. The excess powder should be removed using a stream of air just sufficient to carry the unwanted powder away while not disturbing lightly held powder patterns.

Wet Method. Besides the obvious difference between the wet and dry methods, the magnetic particles for wet magnetic particle testing are smaller in size, and are suspended in a liquid bath of light petroleum distillate. These factors make wet magnetic particle testing better suited for the discovery of fine surface flaws on smooth surfaces. Conversely, it is less likely to reveal a subsurface flaw than is the dry method.

The magnetic particles for liquid suspension are available in either paste or concentrate forms for use with either an oil or water bath. The proportion used in the bath should be in compliance with the manufacturer's recommendation. Aerosol cans containing premixed particle-bearing suspensions are commercially available. Continuous agitation of the solution is necessary to prevent the indicating particles from settling out, which would result in a reduction of test sensitivity.

While both oil-based and water suspensions provide nearly equal sensitivity, the fire hazard due to arcing is avoided with water suspension. However, the presence of water near electrical apparatus creates a shock hazard that must be guarded against. Operators should be made aware of any potential hazards, and be informed how to prevent them.

During inspection using the wet technique, the magnetic particle solution is either flowed over or sprayed on the local area or the part is immersed in a tank containing the liquid bath. The smaller particle size increases the test sensitivity so that very fine discontinuities can be revealed consistently. The particles are available in two colors: red and black. The red particles are better for dark surfaces. Particles coated with a dye that fluoresces brilliantly under ultraviolet (black) light increase the sensitivity of the test. Fluorescent particles can indicate very small or fine discontinuities, and permit rapid inspection of irregular or dark surfaces.

Sequence of Operation

The sequence of operation in magnetic particle testing involves timing and application of the particles and the magnetizing current. Two basic sequences, continuous and residual, are commonly employed.

Continuous Magnetization. This method of operation with either wet or dry particles is employed for most applications. The sequence of operation differs for wet and dry continuous magnetization techniques. The wet technique is generally used for those parts processed on a horizontal, wet-type testing unit. In practice, it involves (1) bathing the part with the inspection medium to provide an abundant source of suspended particles on the surface, and (2) terminating the bath application simultaneously with the initiation of the magnetizing current. The duration of the magnetizing current is typically 0.5 seconds.

With the dry technique, the particles lose mobility when they contact the surface of a part. Therefore, it is imperative that the part be under the influence of the applied magnetic field while the particles are still airborne and free to migrate to leakage fields. This dictates that the flow of magnetizing current be initiated prior to the application of dry magnetic particles, and continued until the application of powder has been completed and any excess blown off. Half-wave rectified or alternating current provides additional particle mobility on the surface of the part, which can be an asset. Examination with dry particles is usually carried out in conjunction with prod type, localized magnetization.

Residual Magnetization. In this technique, the examination medium is applied after the magnetizing current has been discontinued. It can be used only if the weldment being tested has relatively high retentivity so that the residual magnetic field will be of sufficient strength to produce and hold the indications. Unless experiments with typical parts show that the residual field has sufficient strength to produce satisfactory indications, it should not be used.

Equipment for residual magnetization must be designed to provide a consistent, quick break of the magnetizing current. Half-wave rectifying equipment generally cannot be used for this purpose because it cannot provide this quick-break magnetizing current.

Equipment

The basic equipment for magnetic particle testing is relatively simple. It includes facilities for setting up magnetic fields of proper strengths and directions. Means are provided for adjusting current. An ammeter should be provided so that the inspector will know that the correct magnetizing force is being used for each test.

Most commercially available equipment is capable of providing certain amounts of versatility. However, no single piece of magnetic particle equipment can perform all variations of testing in the most effective and economical manner. Therefore, the following factors should be considered when selecting the type of equipment for a specific task:

(1) Type of magnetizing current
(2) Size of part or weldment

(3) Specific purpose of test or the type of defects anticipated

(4) Test media to be used

(5) Portable or stationary equipment

(6) Area to be examined and its location on the part

(7) Number of pieces to be tested

Recording of Results

Little is gained by any inspection if there is no system for consistently and accurately recording indications. Sometimes, dimensions from reference locators can be recorded so that the exact location of the discontinuity can be determined later. If repair is to take place immediately following inspection, the powder build-up may remain to positively identify the affected area.

However, many cases arise where a permanent, positive test record is required. In such an instance, it is possible to preserve the actual powder build-up using the following technique. Upon discovery of a magnetic particle indication, a piece of transparent pressure-sensitive tape, sufficiently large to cover the entire area of interest, can be carefully applied to the surface, over the indication. For location determination, it may be helpful to extend the tape to also cover other nearby reference locators such as holes, keyways, etc. Upon removal of the tape from the surface of the part, the magnetic particle indication will remain on the tape to provide an accurate record of its shape, extent, and location. This tape can then be applied to a piece of contrasting white paper. Additional sketches may be necessary to further clarify the exact location of the occurrence.

Also, various photographic techniques are used extensively for the permanent recording of magnetic particle indications. It is sometimes beneficial to provide better contrast of the indication on the part. Many times this can be accomplished by spraying the part with penetrant developer prior to testing, which will provide a white background for a dark magnetic particle indication.

Demagnetization

Ferromagnetic steels exhibit various degrees of residual magnetism once exposed. In some situations, a residual magnetic field remaining in a component could be detrimental in service. In these cases, demagnetization is necessary. An example would be a component that will be located close to instruments which might be affected by a magnetic field, such as a compass.

Demagnetization of small parts can be accomplished by inserting each one into the magnetic field of a strong alternating current solenoid, and gradually withdrawing it from the field. An alternative technique is to subject each part to an alternating current field that is gradually reduced in intensity.

When a massive structure is encountered, alternating current does not work because it fails to penetrate sufficiently to accomplish demagnetization. In such cases, direct current magnetization should be used, and the field should be gradually reduced to zero while undergoing cyclic reversals. Hammering or rotating in the field will sometimes assist demagnetization.

Annealing or stress relief heat treatment will demagnetize steel weldments, and total demagnetization is always accomplished when the weldment is heated above the Curie temperature (1414° F for carbon steel).

Indications

Magnetic particle powder indications must be evaluated to judge their compliance with the governing code or standard. Indications will have a variety of configurations based on the flux leakage field caused by the discontinuity. Such indication characteristics as height, width, shape, and sharpness of detail provide information as to the type and extent of the discontinuity. Certain discontinuities exhibit characteristic powder patterns that can be identified by a skilled operator. Some of the typical discontinuity indications are discussed below.

Surface Cracks. The indication exhibited by a surface crack is well-defined and tightly held, with heavy powder build-up. The amount of powder build-up is a relative measure of the depth of the crack.

Subsurface Cracks. Cracks that have not broken to the surface exhibit indications different from those of surface cracks. The powder build-up is slightly wider and less well-defined.

Incomplete Fusion. The indication will be fairly well-defined, but it will normally occur at the edge of a weld pass. Like a crack indication, when incomplete fusion is below the surface, its indication is less distinct.

Slag Inclusions and Porosity. Subsurface slag inclusions and porosity can be found, but the indications will be very vague unless the extent is severe. Powder build-up will not be clearly defined, but can be distinguished from surface indications.

Inadequate Joint Penetration. When test conditions permit, inadequate joint penetration can be located with magnetic particle inspection. Its powder indication will be wide and fuzzy, like a subsurface crack, but the pattern should be a straight line.

Laminations. When plate edges are inspected, especially in the case of weld preparations prior to welding, it is possible to reveal plate rolling laminations. These indications will be significant and tightly held. They may be continuous or intermittent.

Seams. These indications are straight, sharp, fine, and often intermittent. Powder build-up is less significant. A magnetizing current greater than that required for the detection of cracks may be necessary.

Undercut. This surface indication will occur at the toe of a groove or fillet weld. The indication will be slightly less well-defined than that from inadequate penetration. Visual inspection is probably a better method for evaluating this discontinuity.

When performing magnetic particle inspection, certain test conditions present will result in the occurrence of irrelevant or false indications. While these indications are real, the flux leakage fields are caused by discontinuities or phenomena that are not necessarily significant to weld soundness or ability to perform the intended function. In such cases, the operator must make a judgment as to the relevancy of the indication. Many times, this will require further grinding or cleaning to make an absolute determination of the indication type. Use of other nondestructive test methods to assist in the interpretation is also helpful. Some of the more commonly encountered irrelevant indications are discussed below:

Surface Finish. When a weld has a rough or irregular surface, it is highly likely that the powder will build up and give false indications. Grinding the weld smooth and retesting should determine the relevancy of any indications.

Differences in Magnetic Characteristics. Differences in magnetic characteristics in a welded joint can occur by several mechanisms, but all result in irrelevant indications. Consequently, some cases may warrant the use of a different nondestructive test to check the weld quality. A change in magnetic characteristics can take place in the heat-affected zone. The resulting indications will run along the edge of the weld and be fuzzy. The pattern could be mistaken for undercut, but is less tightly held. Postweld thermal stress relief will eliminate this phenomenon.

Another change in magnetic characteristics is noted when two metals of differing magnetic properties are joined, or when the filler metal and the base metal have different magnetic properties. The most common example is the joining of carbon steel (magnetic) and austenitic stainless steel (nonmagnetic). Another common occurrence is when austenitic stainless steel filler metal is used to make a repair in carbon steel. A false indication will form at the junction of the two, sharp and well-defined, much like a crack.

Banding. A magnetic phenomenon occurs when the magnetizing current is too high for the volume of metal subjected to the magnetic field. The pattern will be shaped like the lines of force in the field. When encountered, the current should be reduced or the prod spacing increased to prevent the possible masking of real flaws.

Residual Magnetism. Lifting of steel with electromagnets or contact with permanent magnets may give false indications from residual magnetism in the steel.

Cold Working. Sometimes, cold working of a magnetic steel will change the local magnetic characteristics such that the zone will hold powder. This can occur when the surface of a steel weld is marred by a blunt object. The track will appear as a fuzzy, lightly-held powder build-up.

Interpretation of Discontinuity Indications

When magnetic particle inspection is used, there is normally a code or standard that governs both the methodology as well as the acceptance-rejection criteria of indications. Most standards will dictate the permissible limits of discontinuities found using magnetic particle testing. Although size and type of discontinuities are prime criteria, many times there are other conditions such as location and direction, which may greatly influence the effect of discontinuities on the structural integrity of the weld. Such decisions normally require input from engineering personnel.

RADIOGRAPHIC TESTING

Description

Radiographic testing of weldments or brazements employs x-rays or gamma rays, or both, to penetrate an object and detect any discontinuities by the resulting image on a recording or a viewing medium. The medium can be a photographic film, sensitized paper, a fluorescent screen, or an electronic radiation detector. Photographic film is normally used to obtain a permanent record of the test.

When a test object or welded joint is exposed to a penetrating radiation, some of the radiation will be absorbed, some scattered, and some transmitted through the metal to a recording medium. The variations in amount of radiation transmitted through the weld depend upon (1) the relative densities of the metal and any inclusions, (2) thru-thickness variations, and (3) the characteristics of the radiation itself. Nonmetallic inclusions, pores, aligned cracks, and other discontinuities result in more or less radiation reaching the recording or viewing medium. The variations in transmitted radiation produce optically contrasting areas on the recording medium. This concept is illustrated in Fig. 9.12[7]

The essential elements of radiographic testing are:

(1) A source of penetrating radiation, such as an x-ray machine or a radioactive isotope

(2) The object to be radiographed, such as a weldment

(3) A recording or viewing device, usually photographic (x-ray) film enclosed in a light-tight holder

(4) A qualified radiographer, trained to produce a satisfactory exposure

(5) A means to process exposed film or operate other recording media

(6) A person skilled in the interpretation of radiographs

Sources

X-rays most suitable for welding inspection are produced by high-voltage x-ray machines. The wavelengths of the X-radiation are determined by the voltage applied between elements in the x-ray tube. Higher voltages produce x-rays of shorter wavelengths and increased intensities, resulting in greater penetrating capability. Typical applications of x-ray machines for various thicknesses of steel are shown in Table 9.2. With other metals, the penetrating ability of the machines may be greater or lesser depending upon the x-ray absorption properties of the particular metal. X-ray absorption properties are generally related to metal density.

Gamma rays are emitted from the disintegrating nuclei of radioactive substances known as radioisotopes. Although the wavelength of the radiation produced can be quite different, both X- and gamma radiations behave similarly for radiographic purposes. The three radioisotopes in common use are cobalt-60, cesium-137, and iridium-192, named in order of decreasing energy level (penetrating ability). Cobalt-60 and iridum-192 are more widely used than cesium-137. The appropriate thickness limitations of steel for these radioisotopes are given in Table 9.3.

Each source of radiation has advantages and limitations as noted in Table 9.4. The most significant aspect of a radiation source is usually related to its image quality producing aspects. However, other important considerations in selecting a source include its portability and costs.

All radiation producing sources are hazardous and special precautionary measures should be taken when entering or approaching a radiographic area. This subject is discussed later.

7. Details are given in *ASTM E94, Standard Recommended Practice for Radiographic Testing*, available from the American Society for Testing and Materials.

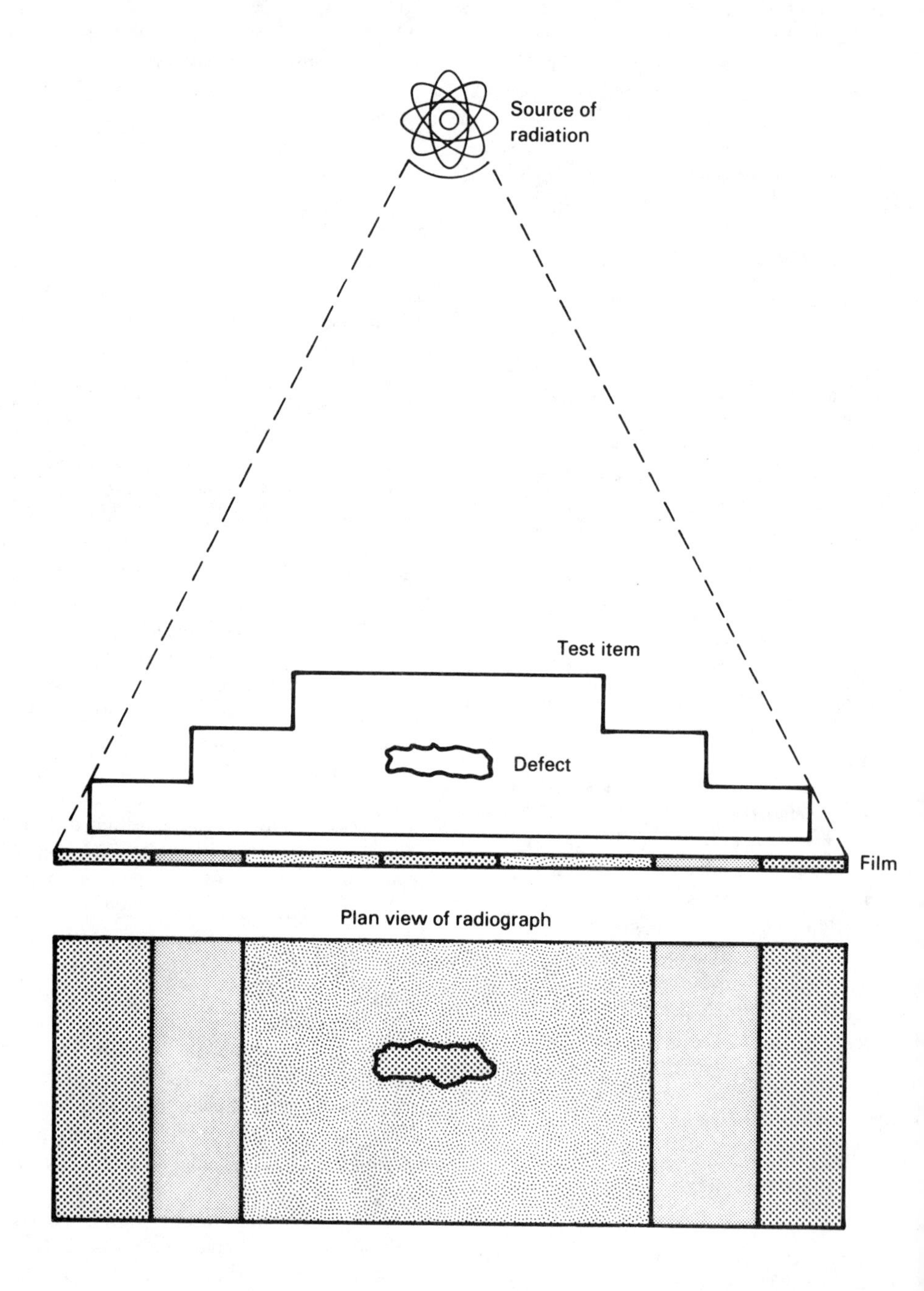

Fig. 9.12—Concept of radiographic testing

Table 9.2
Approximate thickness limitations of steel for x-ray machines

Max voltage, kVp	Approx. max thickness, in.
100	0.33
150	0.75
200	1
250	2
400	3
1000	5
2000	8

Table 9.3
Approximate thickness limitations of steel for radioisotopes

Radioisotope	Approx. equivalent x-ray machine kVp	Useful thickness range, in.
Iridium-192	800	0.5 – 2.5
Cesium-137	1000	0.5 – 3.5
Cobalt-60	2000	2 – 8

Table 9.4
Advantages and limitations of radiation sources

Radioisotopes	X-ray machines
Advantages	
(1) Small and portable	(1) Radiation can be shut off
(2) No electric power required	(2) Penetrating power (kV) is adjustable
(3) No electrical hazards	(3) Can be used on all metals
(4) Rugged	(4) Provides radiographs with good contrast and sensitivity
(5) Low initial cost	
(6) High penetrating power	
(7) Access into small cavities	
(8) Low maintenance costs	
Limitations	
(1) Radiation emitted continuously by the isotope	(1) High initial cost
(2) Radiation hazard if improperly handled	(2) Requires source of electrical power
(3) Penetrating power cannot be adjusted	(3) Equipment comparatively fragile
(4) Radioisotope decays in strength requiring recalibration and replacement	(4) Less portable
(5) Radiographic contrast generally lower than with x-rays	(5) Tube head usually large in size
(6) Cannot be used on all metals, e.g., aluminum	(6) Electrical hazard from high voltage
	(7) Radiation hazard during operation

Test Object

Radiographic testing depends upon the differential absorption of the radiation as it passes through the test object. The rate of absorption is determined by (1) the penetrating power of the source, (2) the densities of the materials subject to radiation, and (3) relative thicknesses of materials in the radiation path.

Various discontinuities found in welds may absorb more or less radiation than the surrounding metal depending on the density of the discontinuity and its thickness parallel to the radiation. Most discontinuities, such as slag inclusions, pores, and cracks are filled with material of relatively low density. They may appear as dark regions on the recording medium (film). Lighter regions usually indicate areas of greater thickness or density such as weld reinforcement, spatter, and tungsten inclusions. (See Fig. 9.12.)

Recording or Viewing Means

The commonly used recording means is radiographic film made expressly for the purpose. An industrial radiographic film is a thin transparent flexible plastic base on which a gelatin coating containing microscopic crystals of silver bromide has been deposited. Some film is coated on one side, others are coated on both sides with this gelatin emulsion. The emulsion is sensitive to both penetrating radiation and to visible light, and must be loaded in a darkroom into light-tight film cassettes (holders). Radiographic films are classified on the basis of speed, contrast, and grain size. Film selection depends on the nature of the inspection, the thickness and type of metal, required exposure time, and desired sensitivity.

Fluorescent screens or image amplifier systems may be used for direct viewing by the radiographer. Electronic devices can enhance the image, or convert it to electrical signals for further processing, display, or recording. However, most radiography of weldments is currently done with film.

Qualified Radiographer

The radiographer is a key factor in successful radiographic testing. The relative positioning of the source and film with the test object or weld affects the sharpness, density, and contrast of the radiograph. Many decisions must be made in choosing the procedure variables for specific test conditions. The radiographer must select the proper film type, intensifying screens, and filters.

Proper safety procedures must be followed to prevent dangerous levels of radiation exposure in the area. Applicable federal and local safety regulations must be followed during handling and use of radiographic equipment.

Many fabrication codes and specifications require that radiographers be trained, examined, and certified to certain proficiency levels. The American Society for Nondestructive Testing publishes recommended procedures for certification of radiographers.[8]

Film Processing

Many times, the processing of exposed radiographic film can determine the success or failure of the method. Radiographs are only as good as the developing process. The processing is essentially the same as that for black and white photographic film. During film handling and processing steps, cleanliness and care are essential. Dust, oily residues, fingerprints, droplets of water, and rough handling can produce false indications or mask real ones.

Skilled Interpreter

The finished radiograph must be evaluated or "read" to determine (1) the quality of the exposure, (2) the type and number of discontinuities present, and (3) the freedom of the weldment from unacceptable indications. This work requires a skilled film interpreter who can determine radiographic quality and knows the requirements of the applicable codes or specifications. The skills for film interpretation are acquired by a combination of training and experience, and require a knowledge of (1) weld and related discontinuities associated with various metals and alloys, (2) methods of fabrication, and (3) radiographic techniques. A code or specification may require that the film interpreter meet min-

---

8. Refer to *SNT-TC-1A, Personnel Qualification and Certification in Nondestructive Testing*, Columbus, OH: American Society for Nondestructive Testing.

imum levels of training, education, and experience, in conjunction with certification for technical competence.

Radiographic Image Quality

The radiographic image must provide useful information regarding the internal soundness of the weld. Image quality is governed by radiographic contrast and definition. The variables that affect contrast and definition are shown in Fig. 9.13.[9] Control of these variables, except for the weld to be tested, is primarily with the radiographer and the film processor.

Image Quality Indicators

There are a number of variables that affect the image quality of a radiograph. Consequently, some assurance is needed that adequate radiographic procedures are used.[10] The tool used to demonstrate technique is an image quality indicator (IQI) or penetrameter. Typical penetrameters are shown in Fig. 9.14. They consist of a piece of metal of simple geometric shape that has absorption characteristics similar to the weld under investigation.

When a weldment is to be radiographed, a penetrameter with a thickness equal to 2 percent

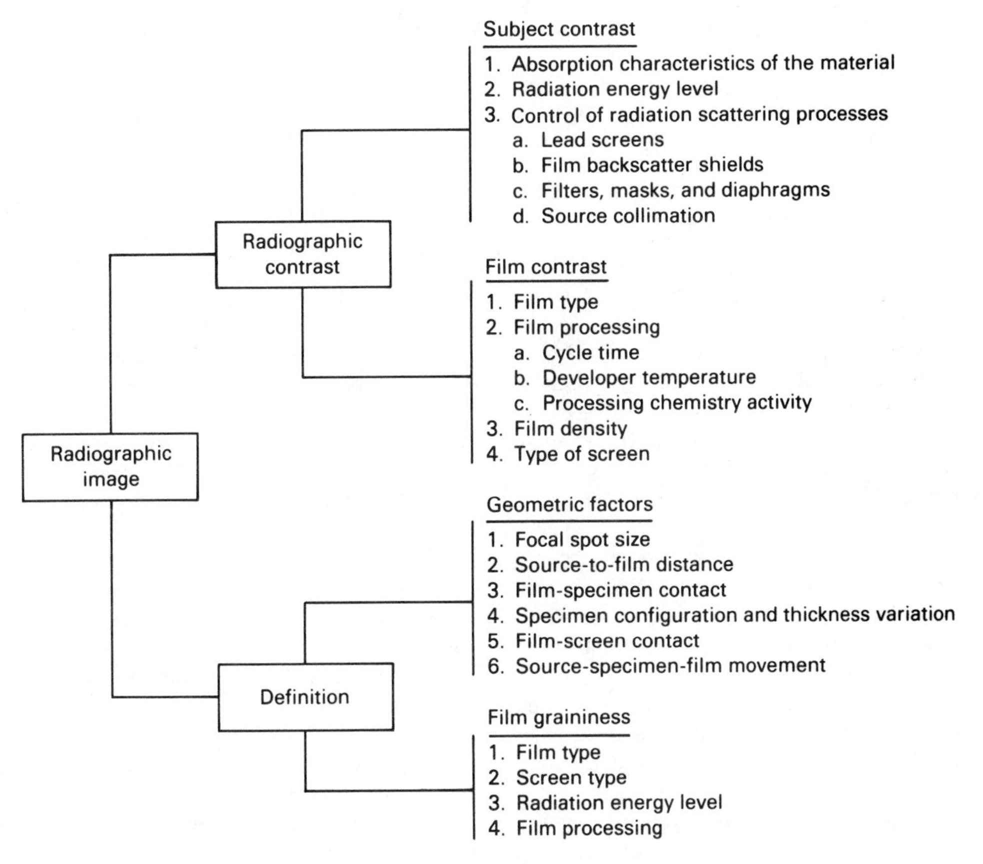

Fig. 9.13—Factors affecting quality of radiographic image

9. The factors affecting radiographic image quality are discussed in *Welding Inspection*, 2nd Ed., 1980: 128-34, published by the American Welding Society.

10. Refer to *ASTM E142, Standard Method for Controlling Quality of Radiographic Testing* for more detailed information.

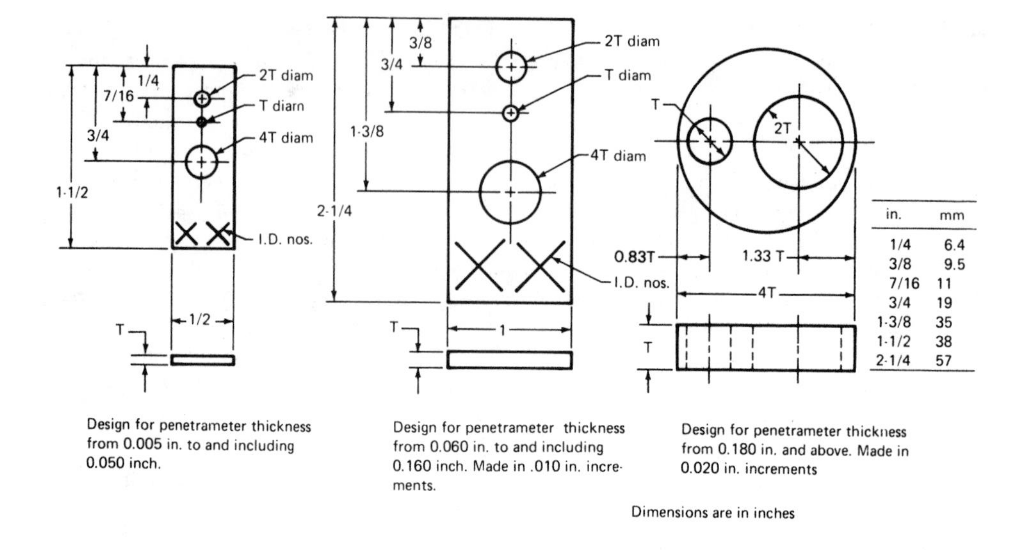

in.	mm
1/4	6.4
3/8	9.5
7/16	11
3/4	19
1-3/8	35
1-1/2	38
2-1/4	57

Design for penetrameter thickness from 0.005 in. to and including 0.050 inch.

Design for penetrameter thickness from 0.060 in. to and including 0.160 inch. Made in .010 in. increments.

Design for penetrameter thickness from 0.180 in. and above. Made in 0.020 in. increments

Dimensions are in inches

Fig. 9.14—Typical penetrameter designs

of the weld thickness is generally selected. A lead identification number at one end shows the thickness of the penetrameter in thousandths of an inch or in millimeters.

Conventional penetrameters usually contain three holes, the diameters of which vary in size as multiples of the thickness. Most specifications and codes call for 1T, 2T, and 4T diameter holes, where T is the penetrameter thickness. Penetrameters are manufactured in standard sizes and increments of thickness. When an exact thickness penetrameter is not available for a particular weld thickness, the next closest one is normally used.

In addition, conventional penetrameters are usually manufactured in material groupings. Most codes and specifications organize commonly used metals and alloys into a minimum of five absorption categories, ranging from light to heavy metals. A penetrameter of the appropriate grade is used when radiographic testing a weldment made of an alloy in the group.

Most radiographic image quality requirements are expressed in terms of penetrameter thickness and desired hole size. For example, the requirement might be a *2-2T* level of sensitivity. The first *2* requires the penetrameter thickness to be 2 percent of the thickness of the specimen; the symbol *2T* requires that the hole having a diameter twice the penetrameter thickness be visible on the radiograph. This image quality level is commonly specified for routine radiography. For more sensitive radiography, a *1-2T* or *1-1T* could be required. More relaxed image quality requirements would include *2-4T* and *4-4T*.

In most instances, one or more penetrameters will be placed on the source side of the specimen near the weld. This is the position of least favorable geometry. However, this will not always be practical; in radiography of a circumferential weld in a long pipe section, for example, the pipe is inaccessible from the inside. In this case, a special penetrameter may be located on the film side of the weldment.

The appearance of the penetrameter image on the radiograph will indicate the quality of the radiographic technique. Even though a certain hole in a penetrameter may be visible on the radiograph, a discontinuity of the same approximate diameter and depth as the penetrameter may not be visible. Penetrameter holes have sharp boundaries and abrupt changes in dimensions, whereas voids or discontinuities may have gradual changes in dimension and shape. A penetrameter, therefore, is not used for measuring the size of a discontinuity or the minimum detectable flaw size.

Identification Markers

Radiograph identification markers of lead alloy are usually in the form of a coded series of letters and numbers. The markers are placed on the test piece at marked locations adjacent to the welded joint during setup. When a welded joint is radiographed, a distinct, clear image of the identification markers should be produced at the same time. Identification markers must be located so that their projected shadows do not coincide with the shadows of any regions of interest in the weldment.

The view identification and the test-piece identification almost always appear in coded form. View identification is usually a simple code (such as A, B, C, or 1, 2, 3) that relates some inherent feature of the weldment or some specific location on the weldment to the view used. The location of the view markers is handwritten in chalk or crayon directly on the piece so that correlation of the radiographic image with the test piece itself can be made during interpretation and evaluation of the radiograph. As a minimum requirement, the identification code for a weld must enable each radiograph to be traced to a particular test piece or section of a test piece. The pertinent data concerning weld and test-piece identification should be recorded in a logbook opposite the corresponding identification number.

Exposure Techniques

Radiographic film exposures are performed in a number of different arrangements; typical ones are shown in Fig. 9.15. The exposure arrangement is chosen with consideration of the following factors:

(1) The best coverage of the weld and the best image quality.

(2) The shortest exposure time.

(3) Optimum image of discontinuities that are most likely to be present in a particular type of weld.

(4) The use of either multiple exposures or one or more exposures at some angle to fully cover all areas of interest.

(5) Radiation safety considerations.

(6) With a pipe weld, whether single- or double-wall exposures should be used.

Typical exposure arrangements for radiography of pipe welds are shown in Figs. 9.15(B) through (F). The appropriate arrangement depends upon the pipe diameter and wall thickness.

Interpretation of Radiographs

The essential steps of radiograph interpretation are as follows:

(1) Determine the accuracy of the identification of the radiograph.

(2) Determine the weld joint design and the welding procedure.

(3) Verify the radiographic setup and procedure.

(4) Review the film under good viewing conditions.

(5) Determine if any false or irrelevant indications are present on the film. (Re-radiograph if necessary.)

(6) Identify any surface irregularities, and verify their type and presence by visual or other NDT methods.

(7) Evaluate relevance of discontinuities with code or specification requirements.

(8) Prepare radiographic report.

Viewing Conditions

Film viewing equipment must be located in a space with subdued lighting to reduce interfering glare. There should be a masking arrangement so that only the film itself is illuminated, and the light is shielded from the viewer's eyes. Variable intensity lighting is usually desirable to accommodate film of various average densities.

Discontinuities in Welds

Radiographic testing can produce a visible image of weld discontinuities, either surface or subsurface, when they have sufficient differences in density from the base metal and in thickness parallel to the radiation. The process will not reveal very narrow discontinuities, such as cracks, laps, and laminations that are not nearly aligned with the radiation beam. Surface discontinuities

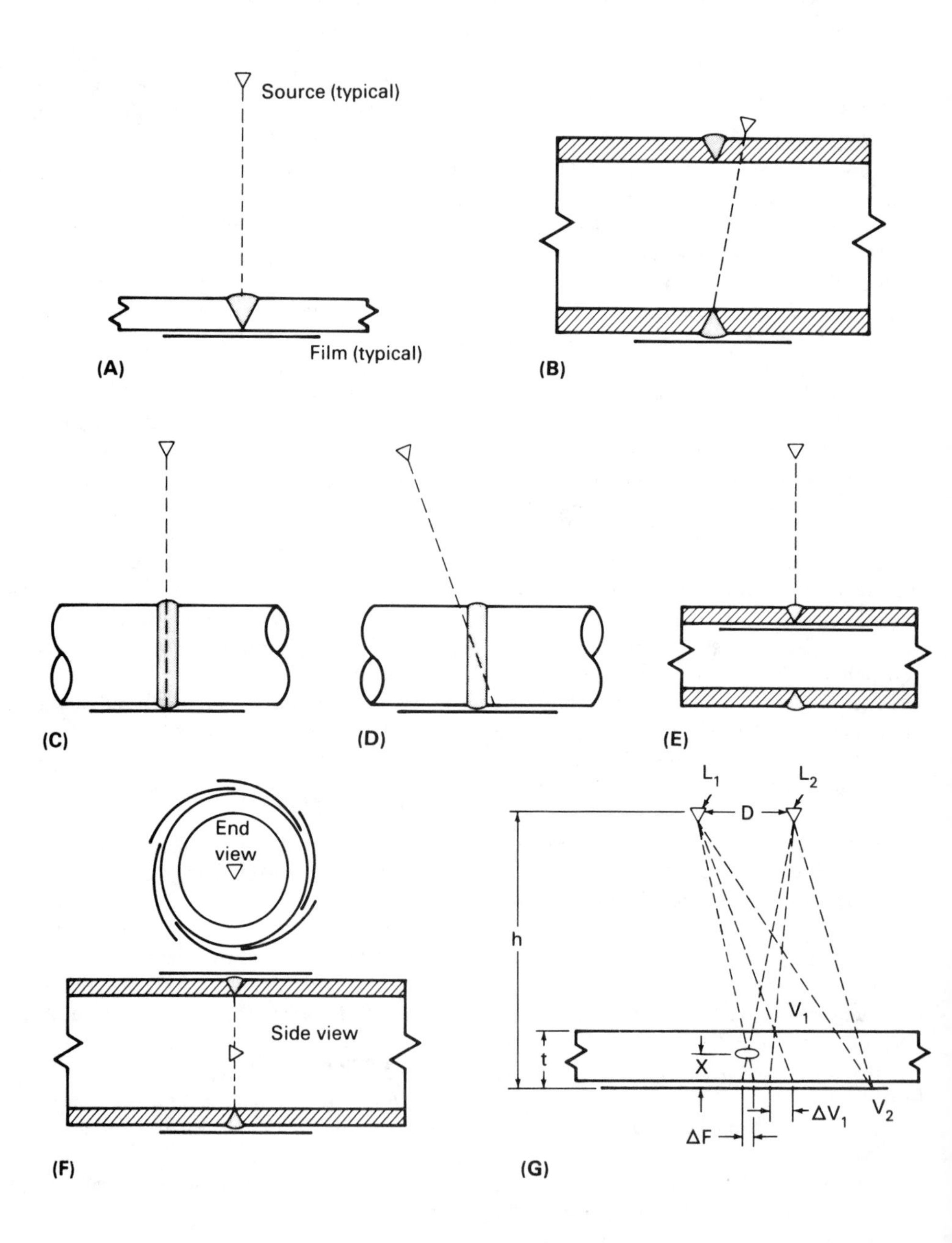

Fig. 9.14—Typical radiographic exposure arrangements

are better identified by visual, penetrant, or magnetic particle testing unless the face and root of the weld are not accessible for examination.

Slag inclusions are usually irregularly shaped dark areas, and appear to have some width. Inclusions are most frequently found at the weld interfaces, as illustrated in Fig. 9.16. In location, slag inclusions often are found between passes. Tungsten inclusions will appear as white spots.

Porosity (gas holes) appear as nearly round voids that are readily recognizable as dark spots with the radiographic contrast varying directly with diameter. These voids may be randomly dispersed, in clusters, or may even be aligned along the centerline of the fusion zone. An example of porosity is shown in Fig. 9.17. A worm hole appears as a dark rectangle if the long axis is perpendicular to the beam, and as concentric circles if the long axis is parallel to the beam.

Cracks are frequently missed if they are very small or are not aligned with the radiation beam. When present in a radiograph, cracks often appear as fine dark lines of considerable length, but without great width. Even some fine crater

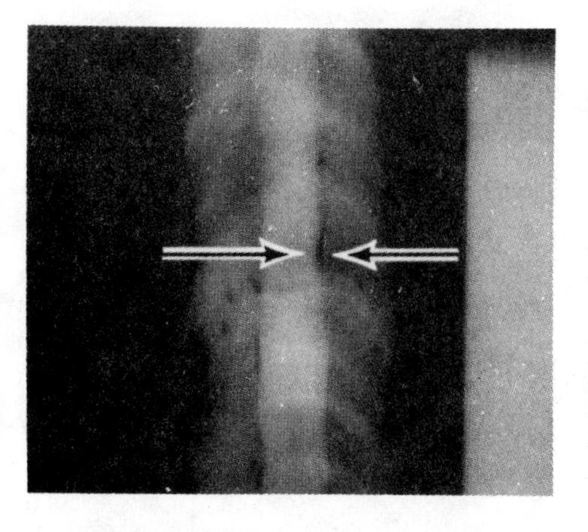

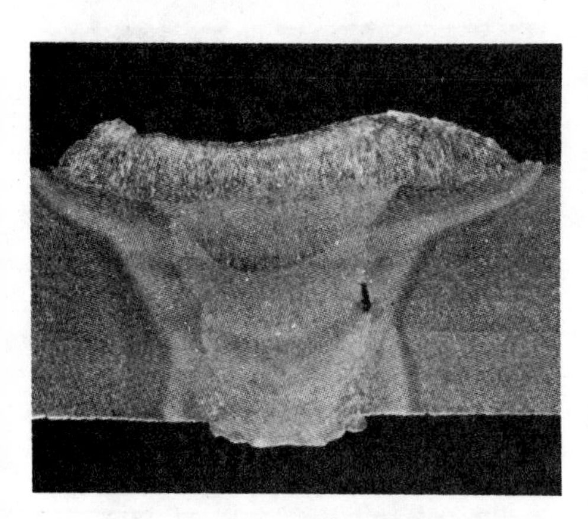

Fig. 9.16—Elongated slag at the weld interface between passes. Top: radiograph; bottom: metallographic section

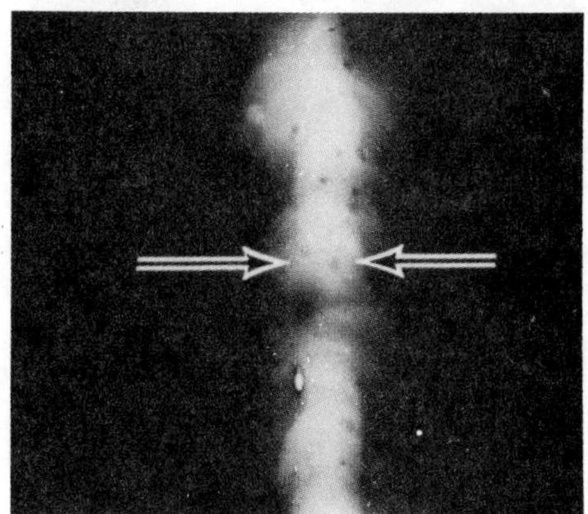

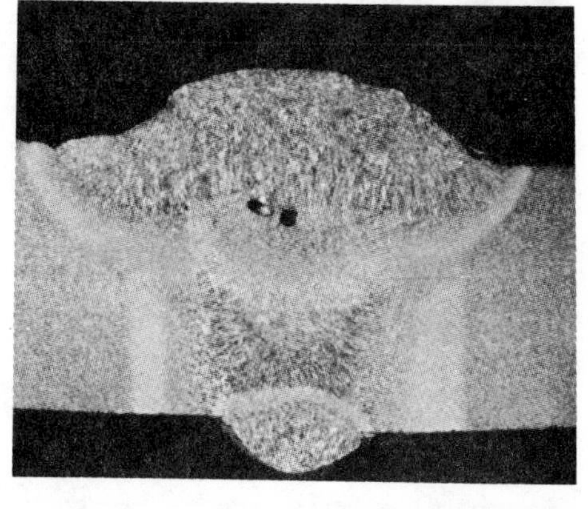

Fig. 9.17—Weld metal porosity. Top: radiograph; bottom: metallographic section

cracks are readily detected. Cracks in weldments may be transverse or longitudinal, and may be either in the fusion zone or in the heat-affected zone of the base metal. See Figs. 9.18 and 9.19.

Incomplete fusion appears as an elongated dark line or band. It sometimes appears very similar to a crack or an inclusion, and could even be interpreted as such. Incomplete fusion occurs between weld and base metal and also between successive beads in multiple-pass welds.

Inadequate joint penetration shows on a radiograph as a very narrow dark line near the center of the weld, as shown in Fig. 9.20. The narrowness can be caused by drawing together of the plates being welded. Slag inclusions and gas holes sometimes are found in connection with incomplete penetration, and cause the line to appear broad and irregular.

Undercutting appears as a dark zone of varying width along the edge of the fusion zone, as shown in Fig. 9.21. The darkness of density or the line is an indicator of the depth of the undercut.

Concavity at the weld root occurs only in joints that are welded from one side, such as pipe joints. It appears on the radiograph as a region darker than the base metal, and running along the center of the weld.

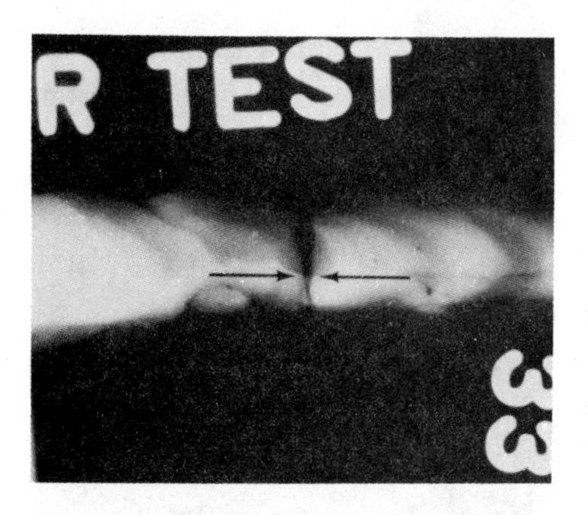

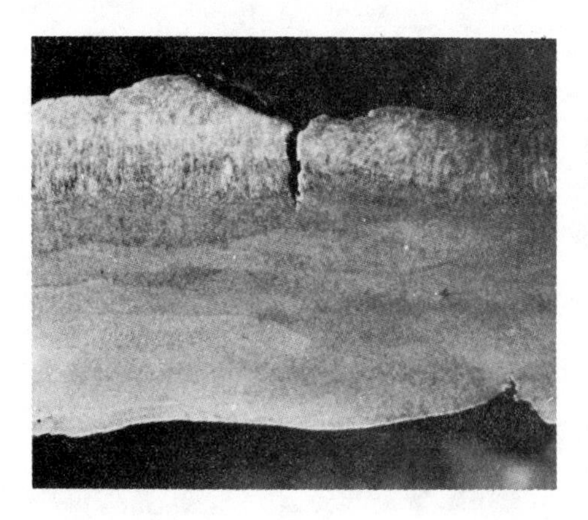

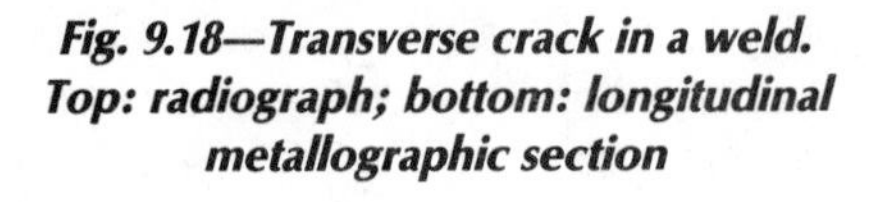

Fig. 9.18—Transverse crack in a weld. Top: radiograph; bottom: longitudinal metallographic section

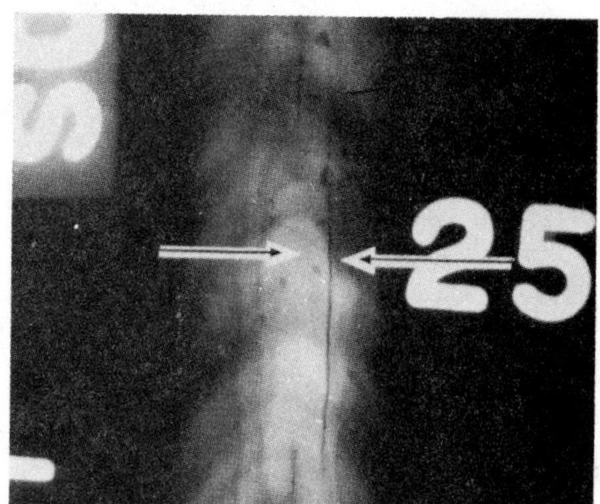

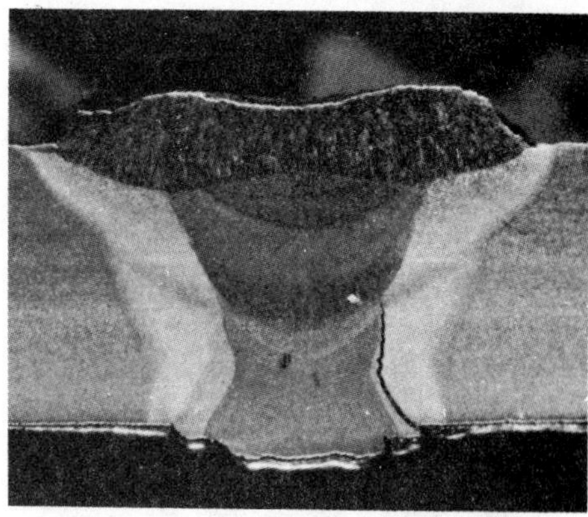

Fig. 9.19—Longitudinal crack in the heat-affected zone. Top: radiograph; bottom: metallographic section

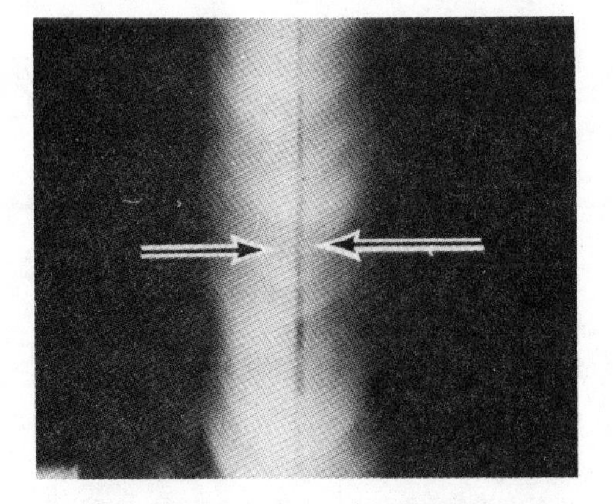

Fig. 9.20—Incomplete penetration in a weld. Top: radiograph; bottom: metallographic section

If weld reinforcement is too high, the radiograph shows a lighter zone along the weld seam. There is a sharp change in image density where the reinforcement meets the base metal, and the edge of the reinforcement image is usually irregular.

As joint thickness increases, radiography becomes less sensitive as an inspection method. Thus, for thick welds, other nondestructive inspection methods are preferred before, during, and after welding for both the base metal and weld metal.

Safe Practices

Radiation Protection. Federal, state, and local governments issue licenses for the operation of radiographic facilities. The federal licensing program is concerned mainly with those companies that use radioactive isotopes as sources. However, in most localities, state and local agencies exercise similar regulatory prerogatives. To become licensed under any of these programs, a facility or operator must show that certain min-

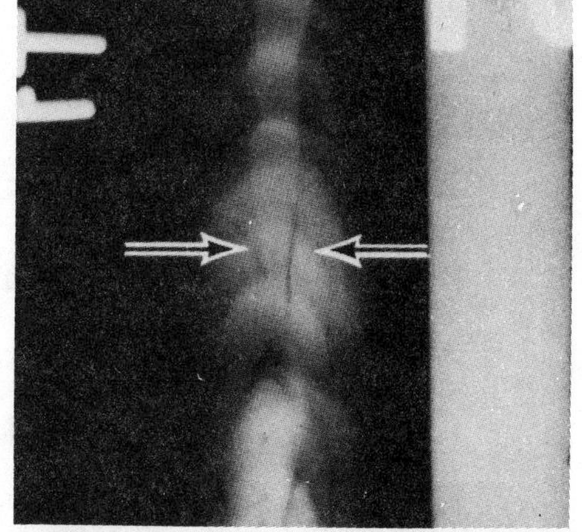

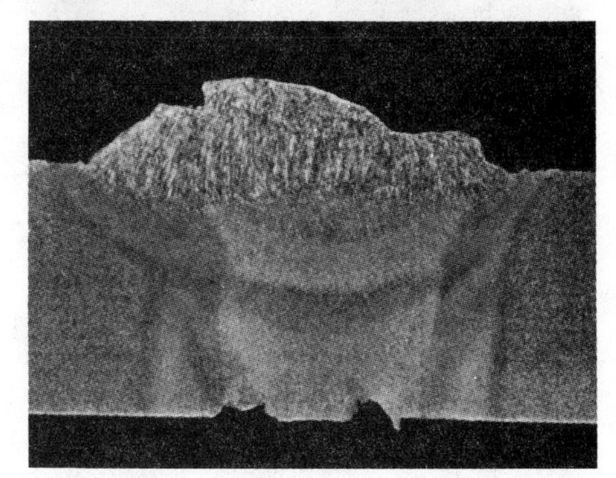

Fig. 9.21—Undercut at the root of a weld. Top: radiograph; bottom: metallographic section

imum requirements for protection of both operating personnel and the general public from excessive levels of radiation have been met. Although local regulations may vary in the degree and type of protection afforded, certain general principles apply to all.

The amount of radiation that is allowed to escape from the area over which the licensee has direct and exclusive control is limited to an amount that is safe for continuous exposure. In most instances, 2 millirems (mrem) per hour, 100 mrem in seven consecutive days, and 500 mrem in a calendar year can be considered as safe.

Radiation Monitoring. A radiation safety program must be controlled to ensure that both the facility itself and all personnel subject to radiation exposure are monitored. Facility monitoring generally is accomplished by periodically taking readings of radiation leakage during operation of each source under various conditions. Calibrated instruments can be used to measure radiation dose rates at various points within the restricted area, and at various points around the perimeter of the restricted area.

To guard against inadvertent leakage of large amounts of radiation from a shielded work area, interlocks and alarms are often required. Basically, an interlock disconnects power to an x-ray tube if an access door is opened, or prevents any door from being opened if the unit is turned on. Alarms are connected to a separate power source, and activate visible or audible signals, or both, whenever the radiation level exceeds a preset value.

All personnel within the restricted area must be monitored to assure that no one absorbs excessive amounts of radiation. Devices such as pocket dosimeters and film badges are the usual means of monitoring. Often both are worn. Pocket dosimeters may be direct reading or remote reading.

Access Control. Permanent facilities are usually separated from unrestricted areas by shielded walls. Sometimes, particularly in on-site radiographic inspection, access barriers may be only ropes or sawhorses. In such instances, the entire perimeter around the work area should be under continual surveillance by radiographic personnel. Signs that carry a symbol designated by the U.S. Government must be posted around any high-radiation area. This helps to inform casual bystanders of the potential hazard, but should never be assumed to prevent unauthorized entry into the danger zone. In fact, no interlock, no radiation alarm, and no other safety device should be considered a substitute for constant vigilance on the part of radiographic personnel.

EDDY CURRENT TESTING

General Description

Eddy current testing (ET) is an electromagnetic nondestructive testing method in which eddy current flow is induced in the test piece. Changes in the flow of the eddy currents caused by variations in the test piece or discontinuities in a weld are detected by a nearby coil or coils and measured by suitable instruments. The method can be used to detect certain discontinuities in welds. However, the results can be affected by variations in dimensions of the test piece or the testing arrangement, and by variations in the physical and metallurgical properties of the test piece.

Applications

Eddy current testing is primarily used for continuous inspection of seamless and welded piping and tubing during production. Testing of ferromagnetic steel, austenitic stainless steel, copper alloy, and nickel alloy tubular products are covered by ASTM specifications.

Eddy Currents

Eddy currents are circulating alternating electrical currents in a conducting material induced by an adjacent alternating magnetic field. Generally, the currents are induced in the component by making it the core of an ac induction coil, as shown in Fig. 9.22. A crack in a welded seam can disrupt the eddy current flow and the magnetic field produced by that current.

There are two ways of measuring changes that occur in the magnitude and distribution of these currents. Either the resistive component or the inductive component of impedance of the exciting (or of a secondary) coil can be measured. Electronic equipment is available for measuring either or both of these components.

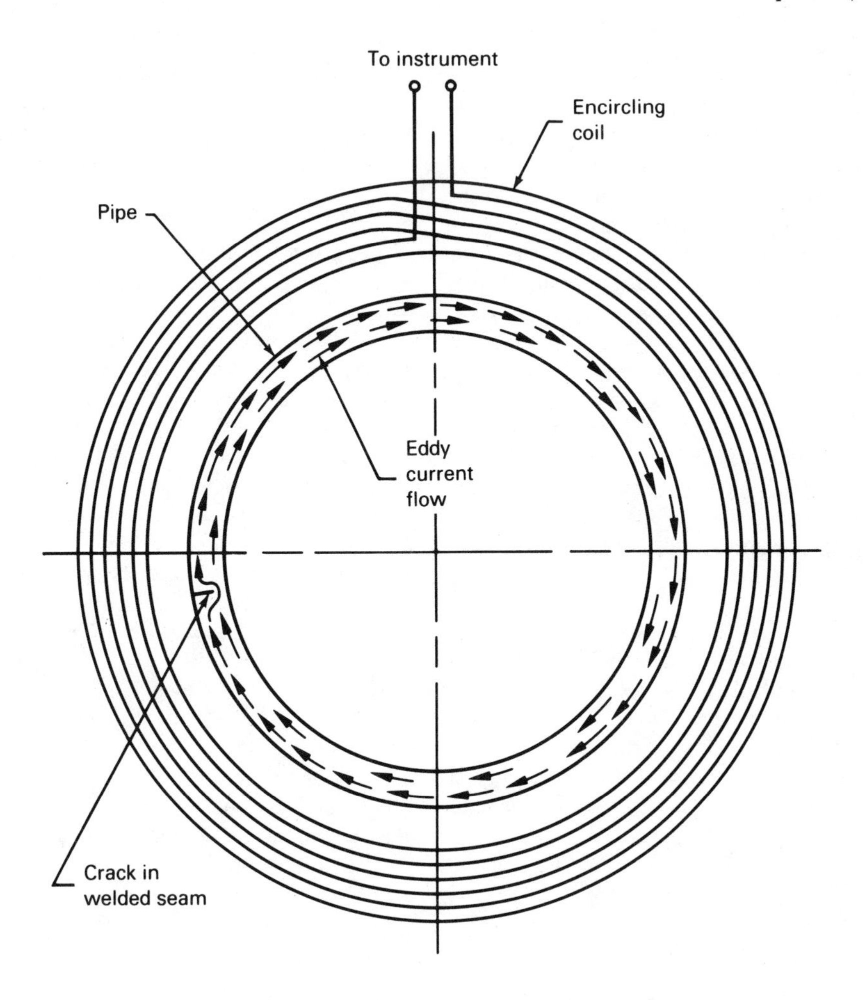

Fig. 9.22—Effect of a crack in eddy current flow during eddy current testing

The current distribution within a test piece may be changed by the presence of inhomogeneities. If a component is homogeneous, free from discontinuities, and has an undistorted grain structure, the mean free path of an electron passing through will be maximum in length. A crack, inclusion, cavity, or other conditions in an otherwise homogeneous material will cause a back scattering of electron flow, and thereby shorten their mean free paths. If an electric wave is considered instead of electrons, the same factors impeding the flow of a single electron will impede the passage of a wave front, causing it to be totally or partially reflected, or absorbed, or both. The various relationships between conductivity and such factors as impurities, cracks, grain size, hardness, strength, etc., have been investigated and reported in the literature.

Electromagnetic Properties of Coils

When an energized coil is brought near a metal test piece, eddy currents are induced into the piece. Those currents set up magnetic fields that act in opposition to the original magnetic field. The impedance, Z, of the exciting coil or any coil in close proximity to the specimen is affected by the presence of the induced eddy currents in the test piece. When the path of the eddy currents in the test piece is distorted by the presence of flaws or other inhomogeneties (Fig. 9.22), the apparent impedance of the coil is altered. This change in impedance can be measured, and is useful in giving indications of flaws or other differences in structure.

The test coil used to induce the eddy currents determines, to a large extent, the information which can be obtained during the test. The basic electrical variables of an eddy current testing system are the ac voltage, E, the current flow. .g through the coil, I, and the coil impedance, Z, that are related by the equation:

$$I = E/Z$$

The impedance of the coil is affected by its magnetic field. Any changes in that field will affect the current flow in the coil. Also, the distance between the coil and the test piece affects the impedance of the coil and the eddy current flow. This distance must be held constant to detect flaws in a moving test piece.

Impedance changes also affect the phase relationships between the voltage across the coil and the current through the coil. This provides a basis for sorting out the effects of variations in spacing and other variables. Phase changes can be observed by means of a cathode ray tube or other display.

Properties of Eddy Currents

In generating eddy currents, the test piece is brought into the field of a coil carrying alternating current. The coil may encircle the part, may be in the form of a probe, or in the case of tubular shapes, may be wound to fit inside a tube or pipe (see Fig. 9.23).

The eddy current in the metal test piece also sets up its own magnetic field, which opposes the original magnetic field. The impedance of the exciting coil, or of a second coil coupled to the first, in close proximity to the test piece, is affected by the presence of the induced eddy currents. This second coil is often used as a convenience, and is called a *sensing* or *pickup coil.* The path of the eddy current is distorted by the presence of a flaw or other inhomogeneity. In the case of a crack or an unwelded seam, the flaw must be oriented nearly normal to the eddy current flow to disturb it. The change in coil impedance can be measured, and is used to give an indication of the extent of defects.

Subsurface discontinuities may also be detected, but eddy currents decrease with depth.

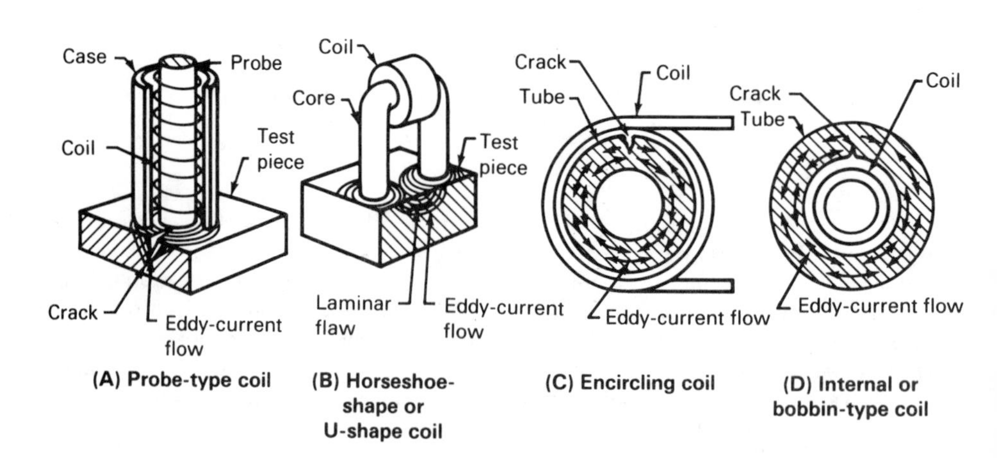

Fig. 9.23—Types and applications of coils used in eddy-current inspection

Alternating Current Saturation

A high ac magnetizing force may be used to simultaneously saturate (magnetically) a test piece and create an eddy current signal for flaw detection equipment. This increases the penetration of the eddy currents and suppresses the influence of certain disturbing magnetic variables.

All ferromagnetic materials that have been magnetically saturated will retain a certain amount of magnetization, called the residual field, when the external magnetizing force has been removed. This residual field may be large or small, depending upon the nature of the magnetizing force applied and material retentivity.

Demagnetization is necessary whenever the residual field (1) may affect the operation or accuracy of instruments when placed in service, (2) interferes with inspection of the part at low field strengths or with proper functioning of the part, and (3) might cause particles to be attracted and held to the surface of moving parts, particularly parts running in oil, thereby causing undue wear. There are many ways to demagnetize an object, the most common being to pass current directly through the test piece. The selected method should give the required degree of removal of the residual field.

Skin Effects

Eddy currents are strongest near the surface of the test piece. This effect is shown in Fig. 9.24. The term *standard depth* refers to that depth where the eddy current density is approximately 37% of that at the surface. Depth of current penetration varies inversely with the electrical conductivity and magnetic permeability of the metal and with the frequency of the alternating eddy currents. Table 9.5 shows typical standard depths of penetration for several metals and magnetizing current frequencies. Normally, a part being inspected must have a thickness of at least two or three standard depths before thickness ceases to have an effect on eddy current response.

Electromagnetic Testing

Electromagnetic testing consists of observing the interaction between electromagnetic fields and metal test pieces. The three things required for an electromagnetic test are:

(1) A coil or coils carrying an alternating current

(2) A means of measuring the electrical properties of the coil or coils

(3) A metal part to be tested

The test coils, being specialized sensing elements, are in some way similar to lenses in an optical system, and their design is a fundamental consideration depending upon the nature of the test. Probe coils that are brought up against the surface to be tested are used in testing a variety of metallic shapes for flaws. Annular coils encircle the part and are used especially for inspecting tubing, rods, bars, wires, and small parts.

Electromagnetic testing involves (1) interaction between applied and induced electromagnetic fields, and (2) imparting of energy into the test piece much like the transmission of x-rays, heat, or ultrasound. Upon entering the test piece, a portion of the electromagnetic energy produced by the test coil is absorbed and converted into heat through the action of resistivity and, if the conductor is magnetic, hysteresis. The rest of the energy is stored in the electromagnetic field. As a result, the electrical properties of the test coil are altered by the properties of the part under test. Hence, the current flowing in the coil reflects certain information about the part, namely dimensions, mechanical, metallurgical, and chemical properties, and presence of discontinuities.

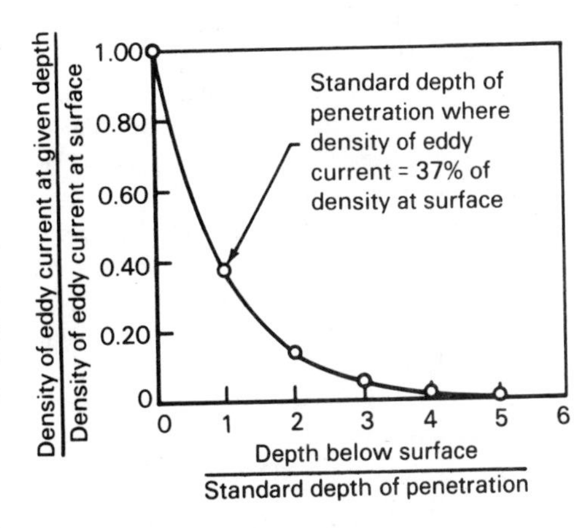

Fig. 9.24—Variation in density of eddy current as a function of depth below the surface of a conductor

Table 9.5
Typical depths of eddy current penetration with magnetizing current frequency

Metal	Standard (37%) depth of penetration, inch					
	1 kHz	4 kHz	16 kHz	64 kHz	250 kHz	1 MHz
Aluminum – 6061 T6	0.126	0.063	0.032	0.016	0.008	0.0040
Aluminum – 7075 T6	0.144	0.072	0.036	0.018	0.009	0.0046
Copper	0.082	0.041	0.021	0.010	0.005	0.0026
Lead	0.292	0.146	0.073	0.037	0.018	0.0092
Magnesium	0.134	0.066	0.033	0.017	0.008	0.0042
Stainless steel – Type 304	0.516	0.257	0.130	0.065	0.031	0.0165
Zirconium	0.445	0.222	0.112	0.056	0.028	0.0141
High alloy steel[a]	0.020	0.0095	0.0049	0.0025	0.001	0.0006

a. The standard depths are for tests without magnetic saturation. When saturated, the values for ferromagnetic steels are appoximately the same as those for austenitic stainless steel.

The character of the interaction between the applied and induced electromagnetic fields is determined by two basically distinct phenomena within the test part:

(1) The induction of eddy currents in the metal by the applied field

(2) The action of the applied field upon the magnetic domains, if any, of the part

Obviously, only the first phenomenon can act in the case of nonferromagnetic metals. In the case of ferromagnetic metals, both phenomena are present; however, the second usually has the stronger influence. This accounts for the basic difference in principle between the testing of ferromagnetic and nonferromagnetic metals.

Among the physical and metallurgical variables that affect electromagnetic tests in metals are the following:

(1) Physical shape, external dimensions, and thickness of the part

(2) Distance between the part and the electromagnetic coil

(3) Plating or coating thickness, if any

(4) Chemical composition

(5) Distribution of alloying or impurity elements (influenced by heat treatment of the part)

(6) Lattice dislocations caused by mechanical working

(7) Temperature

(8) Inhomogeneities and most types of discontinuities

(9) In ferromagnetic metals, residual and applied stresses

In practice, many, and sometimes all, of the above factors may vary simultaneously. It is difficult under such conditions to obtain a meaningful response from the magnetic flux set up within the test piece because several variables may have affected the test signal. The resulting voltage, which is the variable usually sensed by electromagnetic testing devices, must be very carefully analyzed to isolate the pertinent effects from the extraneous effects.

Associated with any electromagnetic test signal are three important attributes: amplitude, phase, and frequency. The test signal may contain either a single frequency (that selected for the test), or a multitude of frequencies (harmonics of the test signal frequency). In the latter case, the test signal frequency is referred to as the "fundamental frequency." In addition, there are amplitude and phase factors associated with each harmonic frequency. The engineer has available a number of techniques that make use of all this information, thereby permitting discrimination between test variables. The important techniques are amplitude discrimination, phase discrimination, harmonic analysis, coil design, choice of test frequency, and magnetic saturation.

Equipment Calibration and Quality Assurance Standards

In using electromagnetic test methods for inspection of metals, it is essential that adequate standards are available to (1) make sure that the equipment is functioning properly and is picking up discontinuities, and (2) ascertain whether the discontinuities are cause for rejection of the part.

It is not the discontinuity itself that is detected by the test equipment, but rather the effect that it has on the eddy currents in the piece being inspected. It is necessary, therefore, to correlate the change in eddy currents with the cause of the change. For this reason, standards must be available when calibrating an electromagnetic testing unit. The standards must contain either natural or artificial imperfections that can accurately reproduce the exact change in electromagnetic characteristics expected when production items are tested. Such standards are usually considered equipment calibration standards; that is, they demonstrate that the equipment is picking up any discontinuities for which the piece is being inspected. These standards are not only used to facilitate the initial adjustment or calibration of the test instrument, but also to check periodically on the reproducibility of the measurements.

It is not enough just to be able to locate discontinuities in a test piece; the inspector must be able to determine if any discontinuity is severe enough to be cause for rejection. For this purpose, quality assurance standards are required against which the test instrument can be calibrated to show the limits of acceptability or rejectability for any type of discontinuity. Quality assurance standards may be either actual production items representing the limits of acceptability or prepared samples containing artificial discontinuities to serve the same purpose.

The types of reference discontinuities that must be used for a particular application are usually given in the product specification.

Several discontinuities that have been used for reference standards include filed transverse notches, milled or electrical discharge machined longitudinal and transverse notches, and drilled holes.

Detectable Discontinuities

Basically, any discontinuity that appreciably alters the normal flow of eddy currents can be detected by eddy-current testing. With encircling-coil inspection of either solid cylinders or tubes, surface discontinuities having a combination of predominantly longitudinal and radial dimensional components are readily detected. When discontinuities of the same size are located beneath the surface of the part being inspected at progressively greater depths, they become increasingly difficult to detect, and can be detected at depths greater than 1/2 in. only with special equipment designed for this purpose.

On the other hand, laminar discontinuities, such as may be found in welded tubes, may not alter the flow of the eddy currents enough to be detected unless the discontinuity breaks either the outside or inside surfaces, or unless a discontinuity results in a weld with outward bent fibers caused by upsetting during welding. A similar difficulty could arise for the detection of a thin planar discontinuity that is oriented substantially perpendicular to the axis of the cylinder.

Regardless of the limitations, a majority of objectionable discontinuities can be detected by eddy-current inspection at high travel speed and at low cost. Some of the discontinuities that are readily detected are seams, laps, cracks, slivers, scabs, pits, slugs, open welds, missed welds, misaligned welds, black or gray oxide weld penetrators, and pinholes.

ULTRASONIC TESTING

General Description

Ultrasonic testing (UT) is a nondestructive method in which beams of high frequency sound waves are introduced into a test object to detect and locate surface and internal discontinuities. A sound beam is directed into the test object on a predictable path, and is reflected at interfaces or other interruptions in material continuity. The reflected beam is detected and analyzed to define the presence and location of discontinuities.

The detection, location, and evaluation of discontinuities is possible because (1) the velocity of sound through a given material is nearly constant, making distance measurements possible, and (2) the amplitude of a reflected sound pulse is nearly proportional to the size of the reflector.

Applications

Ultrasonic testing can be used to detect cracks, laminations, shrinkage cavities, pores, slag inclusions, incomplete fusion or bonding, inadequate joint penetration, and other discontinuities in weldments and brazements. With proper techniques, the approximate position and depth of the discontinuity can be determined, and in some cases, the approximate size of the discontinuity.

Advantages

The principle advantages of UT compared to other NDT methods for weldments are as follows:

(1) Good penetrating characteristics for detection of discontinuities in thick sections.

(2) Relatively high sensitivity to small discontinuities.

(3) Ability to determine position of internal discontinuities and to estimate their size and shape.

(4) Accessibility to one side is normally adequate.

(5) Portable equipment can be used at the job site.

(6) Nonhazardous to personnel or other equipment.

Limitations

The principal limitations of UT are as follows:

(1) Setup and operation requires trained and experienced technicians.

(2) Weldments that are rough, irregular in shape, very small, or thin are difficult to inspect; this includes fillet welds.

(3) Discontinuities at the surface may not be detected.

(4) Couplants are needed between the transducers and weldments to transmit the ultrasonic wave energy.

(5) Reference standards are required to calibrate the equipment and to evaluate flaw size.

Basic Equipment

Most ultrasonic testing systems utilize the following basic components:

(1) An electronic signal generator (pulser) that produces bursts of alternating voltage.

(2) A sending transducer that emits a beam of ultrasonic waves when alternating voltage is applied.

(3) A couplant to transmit the ultrasonic energy from the transducer to the test piece and vice versa.

(4) A receiving transducer to convert the sound waves to alternating voltage. This transducer may be combined with the sending transducer.

(5) An electronic device to amplify and demodulate or otherwise change the signal from the receiving transducer.

(6) A display or indicating device to characterize or record the output from the test piece.

(7) An electronic timer to control the operation.

The basic components are shown in block form in Fig. 9.25.

Equipment operating in the pulse-echo method with video presentation is most commonly used for hand scanning of welds. The pulse-echo equipment produces repeated bursts of high frequency sound with a rest time interval between bursts to allow for return signals from the far side of the test piece and from any discontinuities in the weld or base metal. The time rate between bursts, called the pulse rate, usually occurs between 100 and 5000 pulses per second.

In the video presentation, the time base line is located horizontally along the bottom of a cathode ray tube (CRT) screen, with a vertical initial pulse indication at the left side of the base line. An *A* scan indicates that the time lapse between pulses is represented by the horizontal direction, and the relative amplitude of the returning signal is represented by the degree of vertical deflection on the CRT screen. The screen is usually graduated in both horizontal and verti-

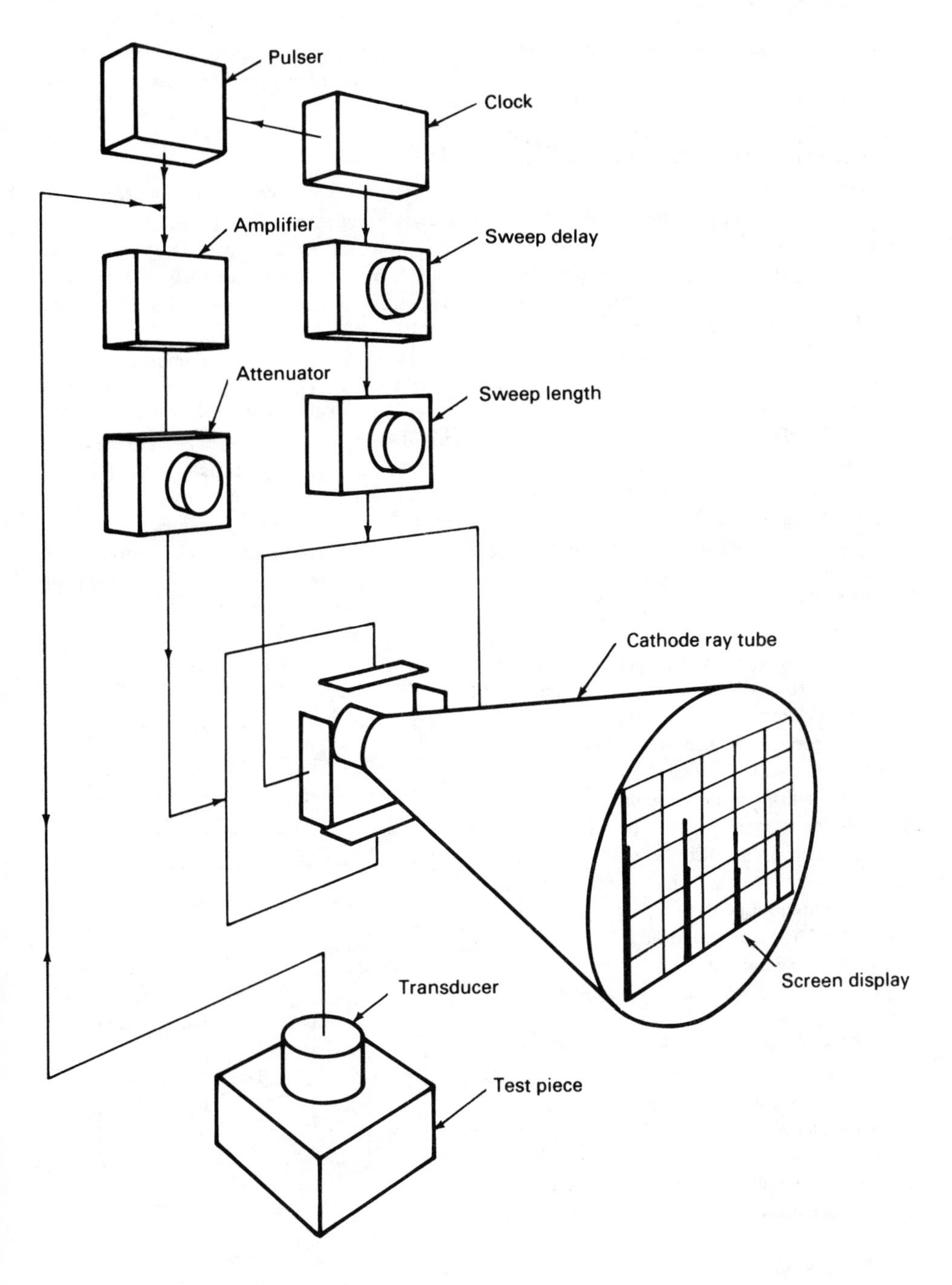

Fig. 9.25—Block diagram, pulse-echo flaw detector

cal directions to facilitate measurement of pulse displays.

A search unit is used to direct a sound beam into the test object. The search unit consists of a holder and a transducer. The transducer element is usually a piezoelectric crystalline substance. When excited with high frequency electrical energy, the transducer produces mechanical vibrations at a natural frequency. A transducer also can receive physical vibrations and transform them into low energy electrical impulses. In the pulse-echo mode, the ultrasonic unit senses reflected impulses, amplifies them, and presents them as pips on the CRT screen. The horizontal location of a reflector pip on the screen is proportional to the distance the sound has traveled in the test piece, making it possible to determine location of reflectors by using horizontal screen graduations as a distance measuring ruler.

Sound Behavior

Sound passes through most metals in a fairly well-defined beam. Initially, the sound beam has a cross section approximately the size of the transducer element. It propagates with slight divergence in a fairly straight line. As the sound beam travels through the metal, there is some attenuation or decrease in energy. The beam will continue to propagate until it reaches a boundary within the object being tested. Either partial or complete reflection of the sound beam takes place at a boundary.

The behavior of sound at UT frequencies resembles visible light in several ways:

(1) Divergence of the beam can be controlled by focusing.

(2) The beam will reflect predictably from surfaces of different densities.

(3) The beam will refract at an interface between materials of different density.

On the other hand, the behavior differs from light in that different vibrational modes and velocities can occur in the same medium.

Test Frequency

The sound wave frequencies used in weld inspection are between 1 and 6 MHz, beyond the audible range. Most weld testing is performed at 2.25 MHz. Higher frequencies, i.e., 5 MHz, will produce small sharp sound beams useful in locating and evaluating discontinuities in thin wall weldments.

Wave Form

There are three basic modes of propagating sound through metals: (1) longitudinal, (2) transverse, and (3) surface waves. In the first two modes, waves are propagated by the displacement of successive atoms or molecules in the metal.

Longitudinal waves, sometimes called straight or compressional, represent the simplest wave mode. This wave form exists when the motions of the particles are parallel to the direction of sound beam propagation, as shown in Fig. 9.26(A). Longitudinal waves have relatively high velocity and relatively short wave length. As a result, the energy can be focused into a sharp beam with a minimum of divergence. They are readily propagated in water.

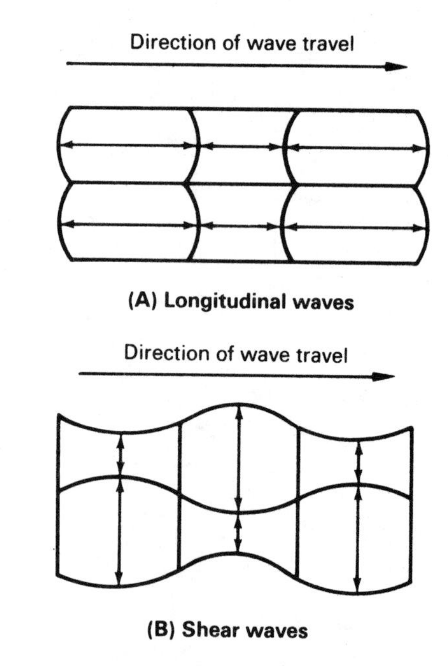

Fig. 9.26—Internal particle movement with longitudinal and shear waves

Another mode is the transverse wave, also called shear wave, in which the principal particle motion is perpendicular to the direction of sound beam propagation, as shown in Fig. 9.26(B). The velocity of these waves is approximately half that of longitudinal waves. Advantages of this wave form are (1) lower velocities allow easier electronic timing, and (2) greater sensitivity to small indications. On the other hand, these waves are more easily dispersed, and cannot be propagated in a liquid medium (water).

The third mode is surface waves, sometimes referred to as Rayleigh waves. In this mode, ultrasonic waves are propagated along the metal surface. This is similar to waves on the surface of water. These surface waves have little movement below the surface into a metal. Therefore, they are not used for examination of welded and brazed joints.

Longitudinal wave ultrasound is generally limited in use to detecting inclusions and lamellar type discontinuities in base metal. Shear waves are most valuable in the detection of weld discontinuities because of their ability to furnish three dimensional coordinates for discontinuity locations, orientations, and characteristics. The sensitivity of shear waves is also about double that of longitudinal waves for the same frequency and search unit size.

It is advisable to pretest the zones in the base metal adjacent to a weld with longitudinal waves to ensure that the base metal does not contain discontinuities that would interfere with shear wave evaluation of the weld.

Coupling

A liquid or hydraulic material is used for transmission of ultrasonic waves into the test object. Some of the more common couplants are water, light oil, glycerine, and cellulose gum powder mixed with water. Couplants generally are a housekeeping problem, so care must be used to prevent accidents. Also, couplants, or the various solvents used to remove them, can be detrimental to repair welding or subsequent operations. Several proprietary couplants are available that minimize these problems.

A weldment must be smooth and flat to allow intimate coupling. Weld spatter, slag, and other irregularities should be removed. Weld reinforcements may or may not be removed, depending on the technique of testing.

Calibration

Ultrasonic testing is basically a comparative evaluation. Both the horizontal (time) and the vertical (amplitude) dimensions on the CRT screen of the test unit are related to distance and size, respectively. It is necessary to establish a zero starting point for these variables, and essential that an ultrasonic unit be calibrated to some basic standard before use.

Various test blocks are used to assist in calibration of the equipment.[11] Known reflecting areas can simulate typical discontinuities. Notches substitute for surface cracks, side-drilled holes for slag inclusions or internal cracks, and angulated flat-bottomed holes for small areas of incomplete fusion. The test block material must be similar in acoustic qualities to the metal being tested.

The International Institute of Welding (IIW) test block is widely used as a calibration block for ultrasonic testing of steel welds. This and other test blocks are used to calibrate an instrument for sensitivity, resolution, linearity, angle of sound propagation, and distance and gain calibrations.

Test Procedures

Most ultrasonic testing of welds is done following a specific code or procedure. An example of such a procedure is that contained in *AWS D1.1, Structural Welding Code—Steel*[12] for testing groove welds in structures.

ASTM E164, Standard Practice for Ultrasonic Contact Examination of Weldments covers examination of specific weld configurations in wrought ferrous and aluminum alloys to detect weld discontinuities. Recommended procedures for testing butt, corner, and T-joints are given for weld thicknesses from 0.5 to 8 inches. Procedures for calibrating the equipment and appropriate

11. Standard test blocks are shown in *ASTM E164, Standard Practice for Ultrasonic Contact Examination of Weldments*, available from the American Society for Testing and Materials.

12. Available from the American Welding Society.

calibration blocks are included. Other ASTM Standards cover testing procedures with various ultrasonic inspection methods for inspection of pipe and tubing.

Procedures for UT of boiler and pressure vessel components are given in the *ASME Boiler and Pressure Vessel Code, Section V, Nondestructive Examination.*[13]

To properly ultrasonic test a welded joint, the search unit must be manipulated in one or more specific patterns to adequately cover the thru-thickness and length of the joint. In most cases, the joint must be scanned from two or more directions to ensure that the beam will intercept any discontinuities that exist in the joint. A typical search pattern for testing butt joints is shown in Fig. 9.27. Similar procedures for butt, corner, and T-joints are illustrated in ASTM E164.

Evaluation of Weld Discontinuities

The reliability of ultrasonic examination depends greatly upon the interpretive ability of the ultrasonic testing technician. With the use of proper inspection techniques, significant information concerning a discontinuity can be learned from the signal response and display on the CRT screen.

There are six basic items of information available through ultrasonic testing which describe a weld discontinuity, depending upon the sensitivity of the test. These are as follows:

(1) The returned signal amplitude is a measure of the reflecting area. (See Fig. 9.28.)

(2) Discontinuity length is determined by search unit travel in lengthwise direction. (See Fig. 9.29.)

(3) The location of the discontinuity in the weld cross section can be accurately measured. (See Fig. 9.30.)

(4) The orientation, and to some degree, the shape of the discontinuity can be determined by comparing signal sizes when tested from both sides of the weld. (See Fig. 9.31.)

(5) The reflected pulse shape and sharpness can be used as an indicator of discontinuity type. (See Fig. 9.32.)

(6) The height of the discontinuity in the weld can be measured by the coordination of the travel distance of the search unit to and from the weld with the rise and fall of the signal. (See Fig. 9.33.)

These six items of information concerning or describing a weld discontinuity can be used in the following manner. The first two items, the returned signal size and length, can be used as a basis for accepting or rejecting a single discontinuity in a weld. The third item, the location of the discontinuity within the cross section of the weld, is useful information when making a repair to a weld. Each of these first three items of information is essential in the proper inspection of welding for acceptance or repair.

The latter three items of information, namely orientation, pulse shape, and dynamic envelope, can be used to more accurately determine the nature of the discontinuity. This information is of value in determining if the welding procedures are under control.

Operator Qualifications

The operator is the key to successful ultrasonic testing. Generally speaking, UT requires more training and experience than the other nondestructive testing methods, with the possible exception of radiographic testing. Many critical variables are controlled by the operators. Therefore, the accuracy of the test depends largely on their knowledge and ability. For this reason, all codes and governing agencies require qualification of operators to ASNT-TC-1A. Experience in welding and other nondestructive testing methods is helpful.

Reporting

Careful tabulation of information on a report form, similar to that in *AWS D1.1, Structural Welding Code—Steel,* is necessary for a meaningful test. The welding inspector should be familiar with the kinds of data that must be recorded and evaluated so that a satisfactory determination of weld acceptability can be obtained.

---

13. Available from the American Society of Mechanical Engineers.

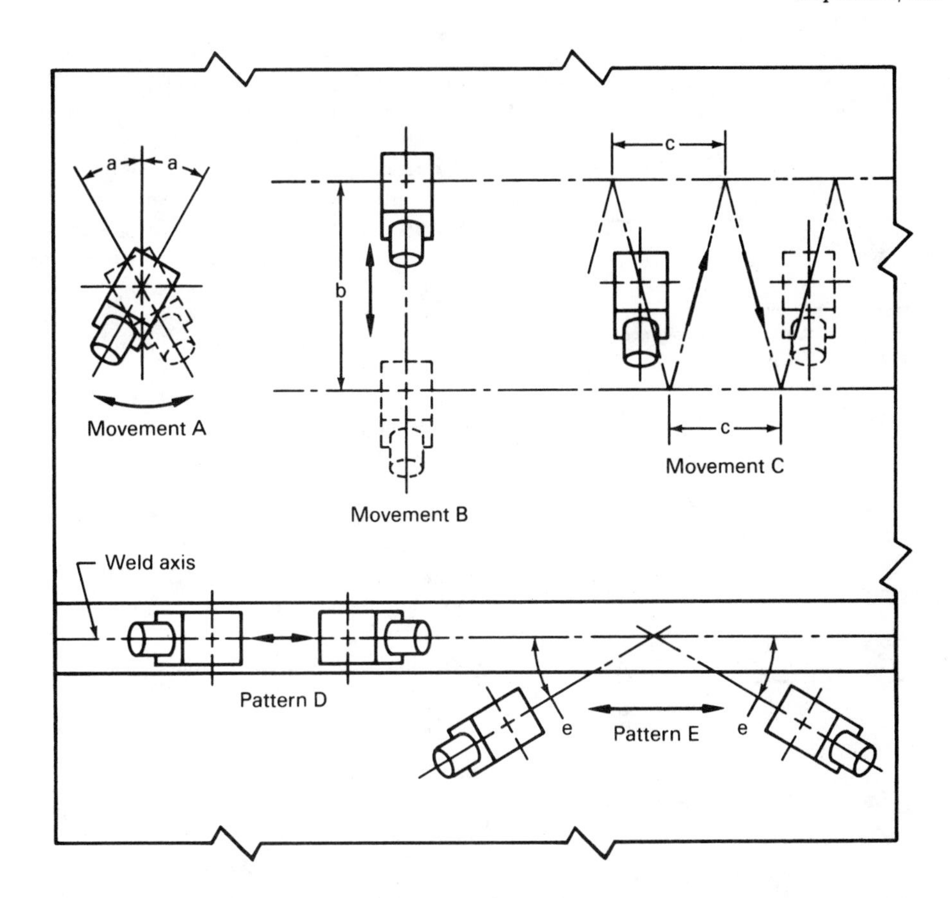

Scanning Patterns

Longitudinal Discontinuities

Movements A, B, and C are combined as one scanning pattern.

Scanning Movement A. Rotation angle a = 10 degrees.

Scanning Movement B. Scanning distance b shall be such that the section of weld being tested is covered.

Scanning Movement C. Progression distance c shall be approximately one-half the transducer width.

Transverse Discontinuities

Scanning pattern D (when welds are ground flush).

Scanning pattern E (when weld reinforcement is not ground flush).

Scanning angle e = 15 degrees max.

Notes:

1. Testing patterns are all symmetrical around the weld axis with the exception of pattern D which is conducted directly over the weld axis.
2. Testing from both sides of the weld axis is to be made wherever mechanically possible.

Fig. 9.27—Typical ultrasonic testing pattern for inspecting welded butt joints

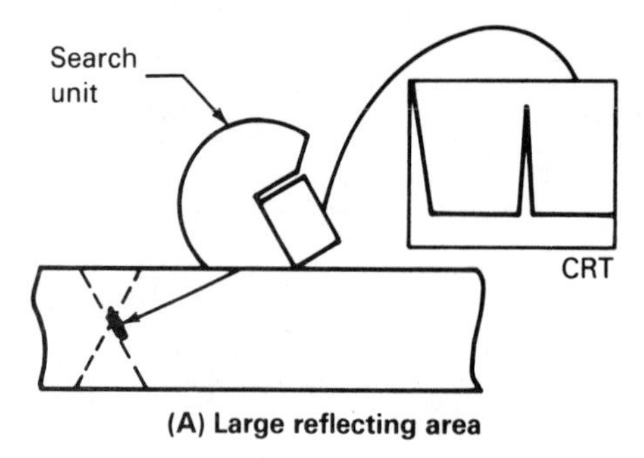

(A) Large reflecting area

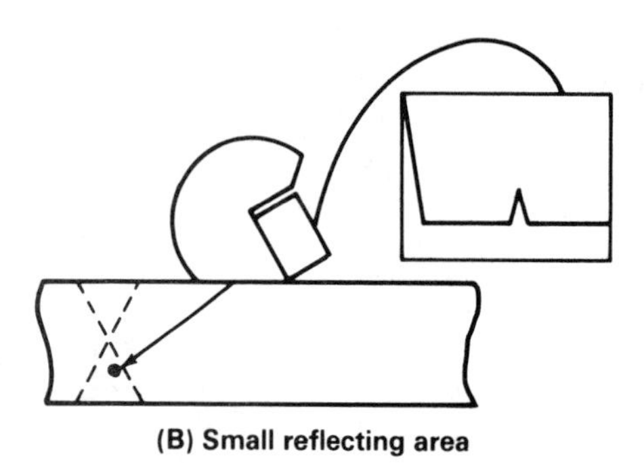
(B) Small reflecting area

Fig. 9.28—Size of reflecting area

ACOUSTIC EMISSION TESTING

General Description

Acoustic emission testing (AET) consists of the detection of acoustic signals produced by plastic deformation or crack formation during loading or thermal stressing of metals. These signals are present in a wide frequency spectrum along with ambient noise from many other sources. Transducers, strategically placed on a structure, are activated by arriving acoustic signals. Ambient noise in the composite signal is significantly reduced by suitable filtering methods, and any source of significant signals is located by triangulation based on the arrival times of these signals at the different transducers.

Acoustic emissions can be used to assess weld quality by monitoring during or after welding, or both. In weldments, regions having incomplete penetration, cracking, porosity, inclusions, or other flaws can be identified by detecting acoustic emissions originating at these regions. During welding processes, acoustic emissions are caused by many things, including plastic deformation, melting, friction, solidification, solid phase transformation, and cracking. Monitoring of acoustic emissions during welding can even include, in some instances, automatic feedback control of the welding process. In large-scale automatic welding, the readout equipment can be conveniently located near the welding controls or in a quality-monitoring area.

Location of Acoustic Emissions

Positions of acoustic sources along a welded joint can be presented in a variety of ways. One technique displays the number of events versus distance along the weld on an oscillosope screen or an X-Y plotter. Another technique uses a digital-line printer that gives the time of the event, its location, and its intensity. This information facilitates appraisal of the severity of each source. After the acoustic emission sources are graded, other nondestructive testing methods can be used to study the flaw in detail.

Applications

Monitoring During Continuous Welding. Detection and location of acoustic emission sources in weldments during fabrication may provide information related to the integrity of the weld.[14] Such information may be used to direct repair procedures on the weld or as a guide for application of other NDT methods. A major attribute of AET for in-process monitoring of welds is the ability of the method to provide immediate real-time information on weld integrity. This feature makes the method useful for lower welding costs by repairing of defects at the most convenient point in the production process. Acoustic emission activity from discontinuities in the weldment is stimulated by the thermal stresses from the welding process. The resulting activity from this stimulation is detected by sensors in the

14. For detailed information, refer to *ASTM E749, Standard Practice for Acoustic Emission Monitoring During Continuous Welding,* available from the American Society for Testing and Materials.

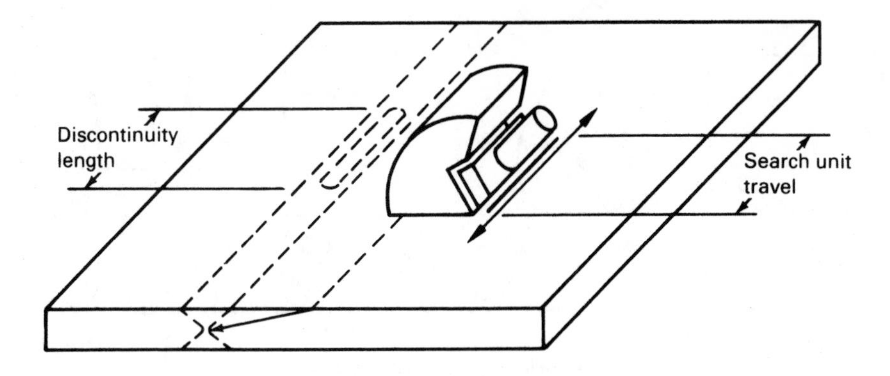

Fig. 9.29—Discontinuity length

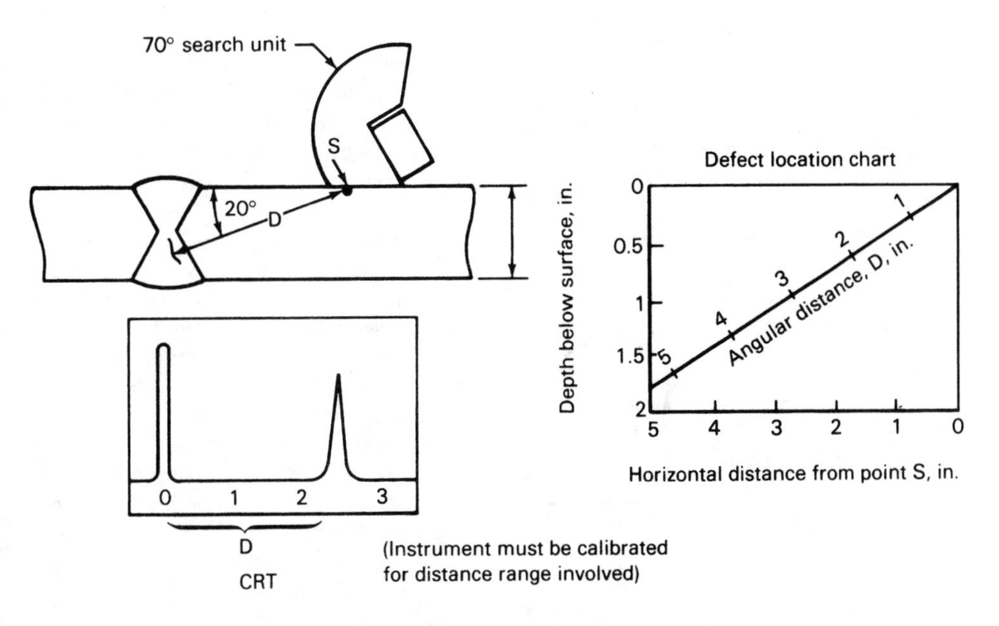

Fig. 9.30—Location of the discontinuity in the weld cross section

vicinity of the weld that convert the acoustic signals into electronic signals.

Acoustic emission monitoring for the evaluation of quality and the control of a welding process requires preliminary studies for each application to establish operating conditions such as the number, location, and mounting of sensors; gain settings; filtering; data presentation; and data interpretation. These studies normally include correlation with other nondestructive and destructive methods of inspection.

The monitoring equipment normally consists of sensors, electronic instrumentation, and recording devices. Acoustic emission monitoring during welding may place specialized requirements on the apparatus because of severe environmental factors and interfering noise sources.

Acoustic emission data may be accumulated during the welding process. Because of the delay between weld fusion and acoustic emission

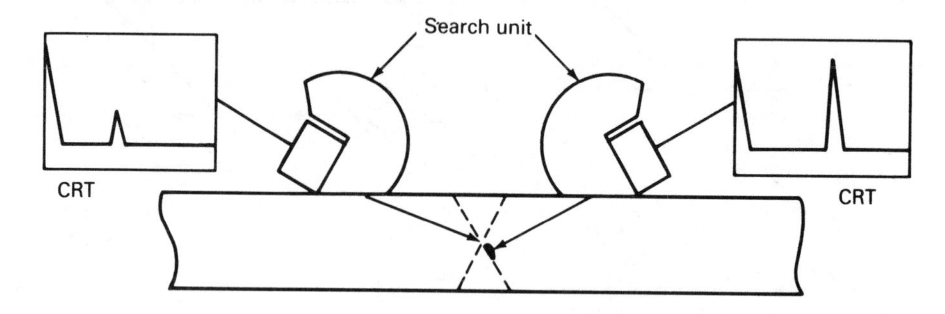

Fig. 9.31—Orientation of the discontinuity

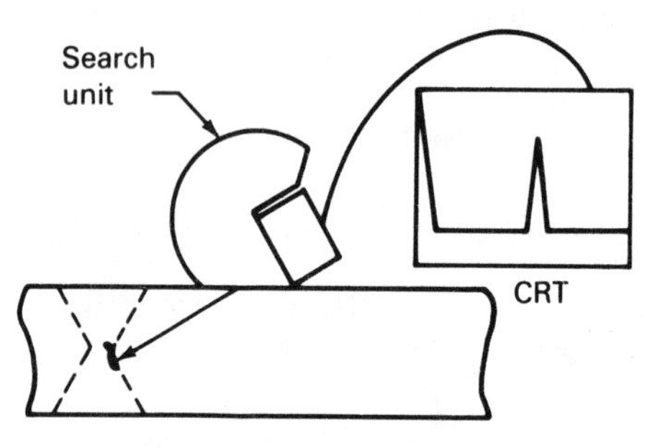

(A) Single flat discontinuity

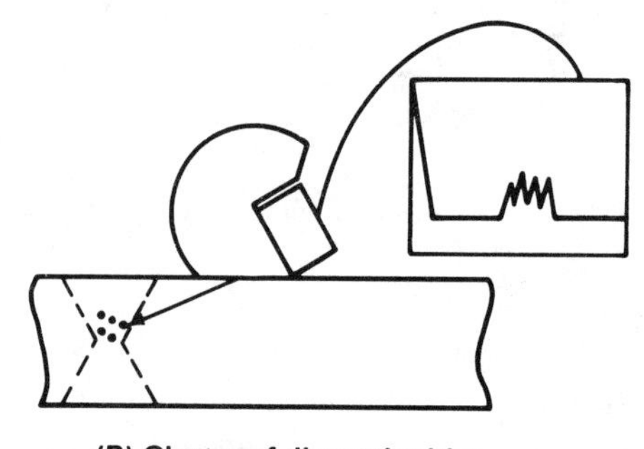

(B) Cluster of discontinuities

Fig. 9.32—Type of discontinuity

activity, monitoring must continue for a time period following welding to acquire all significant data. The monitoring time after welding increases with increasing weld heat input, ranging from 10 seconds for manual gas tungsten arc welding (approximately 100A) to more than 2 minutes for submerged arc welding (600 to 800A). The time should be established during developmental monitoring of trial welds.

Observable conditions that occur in conjunction with unusual acoustic emission activity are recorded to aid in later interpretation of the data. This would include clean up or chipping and grinding by the welder, for example.

In general, acoustic emission weld data must be evaluated against a base line obtained from known acceptable welds of a given type using the specific acoustic emission system, and from signals from the same weld type known to contain defects. Significant weld discontinuities may be characterized by increases in the acoustic emission event count, the rate of events, the intensity, or the peak amplitude.

Monitoring During Resistance Spot Welding. The acoustic emissions produced during the making of a spot weld can be related to weld quality characteristics such as the strength and size of the nugget, the amount of expulsion, and the amount of cracking. Therefore, in-process acoustic emission monitoring can be used both as an examination method, and as a means for providing feedback control.[15]

The resistance spot welding process consists of several stages. These are the set-down of the electrodes, squeeze time, weld time, forging, hold time, and lift-off. Many acoustic emission signals

---

15. Additional information may be found in *ASTM E751, Standard Practice for Acoustic Emission Monitoring During Resistance Spot Welding,* available from the American Society for Testing and Materials.

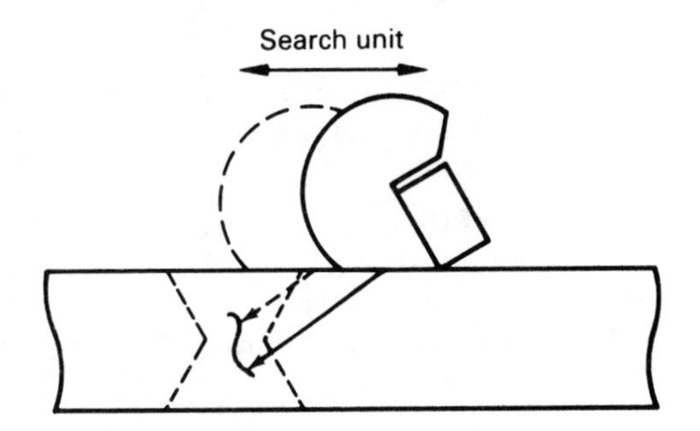

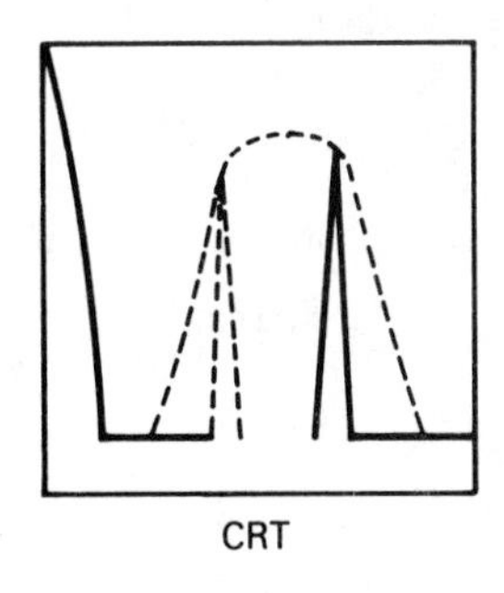

Fig. 9.33—Height of discontinuity

are produced during each of these stages. Often, these signals can be identified with respect to the nature of their source.

Most of the emission signal features can be related to factors of weld quality. The emissions occurring during set-down and squeeze time can often be related to the condition of the electrodes and the surface of the parts. The large but brief signal at current initiation can be related to the initial contact resistance and the cleanliness of the parts.

During welding current flow, signals are produced by plastic deformation, nugget expansion, friction, melting, and expulsion. Those signals caused by expulsion (spitting or flashing, or both), generally have large amplitudes, and can be distinguished from the rest of the acoustic emission associated with nugget formation.

Following termination of the welding current, some metals exhibit appreciable noise during solidification, which can be related to nugget size and inclusions. As the nugget cools during the hold period, acoustic emission can result from solid-phase transformations and cracking.

During the electrode lift-off stage, separation of the electrode from the part produces signals that can be related to the condition of the electrode as well as the cosmetic condition of the weld. The acoustic emission response corresponding to each stage can be separately detected and analyzed.

A measure of the cumulative acoustic emissions during resistance spot welding cannot be expected to relate clearly to weld quality. On the other hand, by using both time discrimination and multiple detection levels, the various segments of acoustic emissions can be separately measured and related to various indicators of quality. Commercial instrumentation is available that is capable of separately monitoring several of the acoustic-emission segments. For instance, the expulsion count, phase-transformation count, and cracking count can be monitored and recorded for each resistance spot weld, giving a permanent record of quality.

Acoustic emission surveillance of spot-welding can take two forms, namely monitoring and control. The purpose of on-line monitoring is to identify and segregate unacceptable welds for quality evaluation purposes. On-line control, on the other hand, utilizes the acoustic emission instrumentation to complete a feedback loop between the process and the source of welding current by automatically adjusting one or two process variables to compensate for deteriorating welding conditions.

Monitoring During Proof Testing. AET methods have been applied to welded pressure vessels and other welded structures during proof testing, such as pressurization. A sound vessel stops emitting signals when the load is reduced, and does not emit further bursts until the previous load has been exceeded. A growing crack emits an increasing number as it is loaded. Location of suspect areas in such structures is a very well established AET technique. Locating systems have been developed, some providing sophisticated analysis of the signal data collected.

RECORDING OF RESULTS

With any type of inspection, defective areas in a weldment must be identified in some manner to assure that they will be located and repaired properly. Many identification methods are available. In addition to logging the results by type, size, and location, the defective area should be marked directly on the weldment. The following rules should be followed:

(1) Marking should be positive, clear, and in a color contrasting with the metal.

(2) Appropriate personnel should be familiar with the marking system.

(3) The color used for marking defects should not be used on the weldment for any other purpose.

(4) The marking material should withstand exposure to handling or further processing until the defective area can be repaired, but should not damage the weldment during subsequent processing or repair.

(5) If service conditions dictate, all marks should be removed after repair and final inspection.

After repair is completed, the same inspection should be repeated and the results carefully recorded.

DESTRUCTIVE TESTING

GENERAL DESCRIPTION

Destructive tests of welds are designed to determine specific properties, such as chemical, mechanical, and metallurgical properties, and to locate discontinuities such as cracks, porosity, incomplete fusion, inadequate joint penetration, and entrapped slag. They are normally performed on sample weldments made with procedures similar to those to be used in production. Destructive tests are conducted for the qualification of welding procedures, welders, and welding operators, and for production quality control. The tests generally specified are those that will provide reasonable assurance of the dependability of the weldment in service.

The ideal test would, of course, be one that exactly duplicates the service conditions. However, the difficulty of applying such tests is that duplication of service conditions normally would require extensive testing or long exposure to those conditions. Thus, certain standard tests are alternatively used. Details of common destructive tests of welded joints and weld metals can be found in Volume 1, 7th Edition of this Handbook; *AWS B4.0, Standard Methods for Mechanical Testing of Welds* (latest edition); or the applicable AWS filler metal specification (A5.XX series).

CHEMICAL ANALYSIS

Chemical analyses can be used to determine the compositions of solid welding wires, deposited weld metals, and base metals to verify code or specification requirements. Prior to sampling, the surface should be cleaned to bright metal. During removal, care must be taken to prevent sample contamination because contamination will give false results.

The laboratory performing the analysis may use one or more analytical techniques depending on the metal and the range of the elements in the metal. Regardless of the analytical technique, all procedures should be verified by analyzing reference standards available from the National Bureau of Standards and other sources.

MECHANICAL TESTING

Mechanical tests for welds are similar to those applied to base metals. Tension, bend, impact, and hardness tests are widely used for determining the mechanical properties of weld metal and welded joints. There are other tests used to determine special properties for design purposes, such as fatigue, fracture toughness, corrosion, and creep-rupture tests.

Tension Testing

Tensile properties are generally the bases for design of weldments, and those of the base metal and the weld metal must conform to design requirements. Normally, the weld metal used for butt joints should have tensile properties equal to or somewhat higher than those of the base metal. In the case of fillet, plug, and spot welds, the tensile-shear strength of the weld metal or heat-affected zone is usually the significant factor. Testing is conducted in accordance with *AWS B4.0, Standard Methods for Mechanical Testing of Welds* or the applicable code or specification.

Tension test specimens are taken either longitudinally or transversely to a butt joint. Longitudinal specimens composed only of weld metal are usually round, and are referred to as *all-weld-metal tensile tests.* Properties determined are tensile and yield strength, elongation, and reduction of area.

Transverse specimens include the weld metal, heat-affected zone, and base metal in the test section. These specimens are usually rectangular, but may be round. Normally, only the tensile strength is determined and the location of fracture noted. The test results determine the adequacy of the overall weld joint strength.

Full-section transverse tension specimens are used to test circumferential butt joints in pipe or tubing not exceeding 2 inches in diameter. With larger sizes, rectangular transverse test specimens are cut from the pipe wall.

In some cases, a tension specimen is removed with the weld runnng the length of and centered in the reduced section. Base metal and weld metal are simultaneously exposed to the same load conditions and strain. Both zones elongate until the specimen fractures. The point of fracture initiation indicates the zone of least ductility but not necessarily of least strength. Calculated tensile and yield strengths are composite values.

Bend Testing

Bend tests are used to evaluate both soundness and ductility of welded joints. Hence, they are used extensively in welding procedure and welder performance qualification tests. Most codes and specifications require the specimens to be bent 180 degrees into a "U" shape around a standard plunger having a radius of twice the thickness of the test specimen. This test elongates the outer fibers of the bend specimen by about 20 percent. Specimens of higher strength base metals (over 50,000 psi) are bent around a larger radius because of their inherently lower ductility.

Various welding specifications may require testing of transverse face, root, or side bend specimens or longitudinal face or root bend specimens for qualification of welding procedures or welders.

Impact Testing

Impact tests are designed to evaluate the notch toughness of a metal at various temperatures. The results are used to evaluate the suitability of the metal for service at room or lower temperatures.

One type of test is the Charpy V-notch impact test, which uses a specimen in the form of a notched beam. The notch may be in the base metal, the weld metal, or the heat-affected zone. The specimen is cooled to the desired test temperature and then quickly placed on two anvils with the notch centered between them. The specimen is struck at a point opposite the notch by the tip of a swinging pendulum. The amount of energy required to fracture the specimen represents the notch toughness of the metal at the test temperature. In addition to the amount of energy required for fracture, some codes require a measurement of lateral expansion for compliance.

When the test results are plotted as a function of the test temperature, the resulting curve is used to evaluate the expected performance of the base or weld metal in service. With some steels and other metals, the required energy drops rapidly within a narrow temperature range, and the fracture changes from ductile to brittle behavior. This transition temperature range varies with chemical composition, welding procedure, and postweld heat treatment.

Hardness Testing

There are three static hardness tests commonly used to measure the effects of processing on metals and for quality control in production. They are the Brinell, Rockwell, and Vickers hardness tests that are based on the size of an indentation made with a particular indenter design under a specified load. The choice of method depends on the size and finish of the area to be tested, the composition and thicknesses of the test object, and the end use of the object.

In the case of carbon and alloy steels, there is an approximate relationship between Brinell, Rockwell, and Vickers hardness numbers, and between hardness number and tensile strength. These data are presented in tabular form in several Handbooks.[16] With Brinell hardness testing, the approximate tensile strength of carbon and low alloy steels is 500 times the hardness number. The Brinell test is commonly used for quality control in industry but the impression is relatively large compared to the other tests. Therefore, it can only be used for obtaining hardness on a relatively large area, and where the impression on the surface is not objectionable. Where marking will affect the end use of the product, Rockwell or Vickers tests can be applied with smaller surface indentions. Because these methods can be applied to small or narrow areas, they can be used for survey work to locate variations in hardness across a weld.

For the welding inspector, the most important difference involves the size of the indenter or penetrator, which may range from a 10 mm diameter ball to a small diamond tip. For reasonably homogenous base metal or weld metal, the size of the indentation has little effect on the test results. However, with a narrow heat-affected zone or weld bead that contains several metallurgically differing zones, the size of the indentation is of major importance. A small indenter may detect narrow areas of different hardness while the 10 mm ball produces an average value for the entire zone. Any test that includes heat-affected zone hardness should specify and record the size of the indenter employed.

In hardness testing, specimen preparation is important for reliable results. The surface should be flat, reasonably free from scratches, and must be normal to the applied load for uniform indentations. With thin, soft metals, the testing method must produce a shallow indentation that is not restricted by the anvil of the testing machine. This can be accomplished, particularly with a Rockwell testing machine, by the use of a small indenter and a light load.

Brinell Hardness. The Brinell hardness test consists of impressing a hardened steel ball into the test surface using a specified load for a definite time. Following this, the diameter of the impression is accurately measured, and converted to a hardness number from a table. Stationary machines impress a 10 mm ball into the test object. The load for steel is 3000 kg and for softer metals, 500 or 1500 kg.

Two diameters of the impression are measured at 90 degrees by the eye using a special Brinell microscope graduated in tenths of millimeters. The mean diameter is used to determine the Brinell hardness number from a table. Testing should be done to the requirements of *ASTM E10, Standard Test Method for Brinell Hardness of Metallic Materials.*

Rockwell Hardness. The Rockwell hardness test measures the depth of residual penetration made by a small hardened steel ball or a diamond cone. The test is performed by applying a minor load of 10 kg to seat the penetrator in the surface of the specimen and hold it in position. The machine dial is turned to a "set" point, and a major load is applied. After the pointer comes to rest, the major load is released, with minor load remaining. The Rockwell hardness number is read directly on the dial. Hardened steel balls of 1/8- or 1/16-in. diameter are used for soft metals, and a cone-shaped diamond penetrator is used for hard metals. Testing is conducted in accordance with *ASTM E18, Standard Test Method for Rockwell Hardness and Rockwell Superficial Hardness of Metallic Materials.*

---

16. For example, *The Metals Handbook, Vol. 11, Nondestructive Testing and Quality Control,* 8th Ed., Metals Park, OH: American Society for Metals.

Vickers Hardness. The Vickers hardness test is generally used as a laboratory tool to determine hardness of various metallurgically different areas on a cross section through a welded joint. The test consists of impressing a pyramid-shaped penetrator under a predetermined load into a polished surface of a specimen. The Vickers number is related to the impressed load and the surface area of the resulting indentation. The Vickers indenter is a diamond and, therefore, is effective for testing hard metals. The selected load depends on the hardness of the metal to be tested. A reading is obtained by measuring the diagonals of the impression. The hardness value is obtained from a chart based on the mean diagonal value. Testing is conducted as described in *ASTM E92, Standard Test Method for Vickers Hardness of Metallic Materials.*

METALLOGRAPHIC EXAMINATION

Metallographic tests can be used to determine:

(1) The soundness of the joint

(2) The distribution of nonmetallic inclusions in the joint

(3) The number of weld passes in a weld

(4) The metallurgical structure in the weld metal and heat-affected zone

(5) The extent of the heat-affected zone

(6) The location and depth of joint penetration of a weld

Macroscopic examination involves visual examination of as-polished or polished-and-etched surfaces by the unaided eye or at magnifications up to 10 power. Etching of the surface can reveal gross structure and weld bead configuration. Pores, cracks, and inclusions are better observed on an as-polished surface. Microscopic examination is carried out by observing a highly polished and etched surface at magnifications of about 50 times size or higher.

Samples are obtained by sectioning test welds or production control welds, including run-off tabs. They can be prepared by cutting, machining, or grinding to reveal the desired surface, and can be subjected to further preparation as needed to reveal the desired structure.

Macroscopic Examination

Macroetching is used to reveal the heterogeneity of metals and alloys. Typical applications of macroetching in the welding of metals are the study of weld structure; measurement of joint penetration; dilution of filler metal by base metal; and presence of flux, porosity, and cracks in weld and heat-affected zones. When macroscopic examination is used as an inspection procedure, sampling should be done in an early stage of manufacturing so that work is stopped if the weld proves faulty.

Sample preparation need not be elaborate. Any method of preparing a smooth surface with a minimum amount of cold work is satisfactory. Cross sections may be faced on a lathe or a shaper. The usual procedure is to take a roughing cut, followed by a finish cut with sharp tools. This should provide a smooth surface and remove cold work from prior operations. Grinding is usually conducted in the same manner, using free-cutting wheels and light finishing cuts. When fine detail is required, the specimen should be polished through a series of metallographic papers.

After surface preparation, the sample is cleaned carefully with suitable solvents. Any grease, oil, or other residue will produce uneven etching. Once cleaned, the sample surface should not be touched or contaminated in any way.

Recommended solutions and procedures for macroetching are given in *ASTM E340, Standard Method for Macroetching Metals and Alloys.* Caution must be observed in handling chemicals and mixing solutions. Many of the etchants are strong acids that require special handling and storage.[17] In all cases, the various chemicals should be added slowly to the water or other solvent while stirring.

Microscopic Examination

In examining for exceedingly small discontinuities or for metallurgical structure at high

17. The users of chemical etchants must be familiar with the proper handling and storage of the chemicals and the appropriate safety precautions. Refer to Sax, N.I., *Dangerous Properties of Industrial Materials,* latest Ed., New York: Van Nostrand Reinhold Co.

magnification, specimens may be cut from the actual weldment or from welded test samples. The samples are given a highly polished, mirror-like surface and then etched for examination with a metallograph to reveal the microstructure of the base metal, heat-affected zone, fusion zone, and weld metal. Procedures for selection, cutting, mounting, and polishing metallographic specimens are given in *ASTM E3, Standard Methods of Preparation of Metallographic Specimens.* Recommended chemical solutions for etching various metals and alloys are given in *ASTM E407, Standard Methods for Microetching Metals and Alloys.* Safety precautions in handling etching chemicals are also covered in this standard.

PROOF TESTING

PURPOSE

Many welded components are proof tested during or subsequent to fabrication. The purposes of these tests include:

(1) Assurance of safe operation
(2) Detection of any weakness of the design
(3) Exposure of quality deficiencies
(4) Prevention of in-service failures

Proof testing is achieved by applying one or more tests that exceed actual service requirements by a predetermined factor of safety. This may involve overloading the component or testing for leaks, or both.

LOAD TESTING

Welded components can be proof tested by applying specific loads without failure or permanent deformation. Such tests are usually designed to subject the parts to stresses exceeding those anticipated during service. However, the stresses are maintained below or at the minimum specified yield strength of the lower strength materials. Many load test requirements and application details are mandated by codes, specifications, and contractual documents that apply to individual product forms.

Structural members are often proof tested by demonstrating their ability to carry loads equal to or larger than any anticipated service conditions. This can be accomplished by statically loading with a testing machine, sand bags, or scrap iron, or by dynamically loading with special testing equipment. Acceptance is based on freedom from cracking or objectionable permanent deformation.

HYDROSTATIC TESTING

Closed containers are usually proof tested by filling them with water and applying a predetermined test pressure. For components built in accordance with the *ASME Boiler and Pressure Vessel Code,* this pressure is 150 percent of design pressure. For other components, the test pressure may be based upon a percentage of the minimum yield strength. After a fixed holding time, the container is inspected for soundness by visually checking for leakage, or by closing off the vessel and monitoring the hydrostatic pressure. Visual inspection can be enhanced by an indicating system applied to the outside of the vessel, such as light blue chalk that turns dark blue in the presence of a small amount of water. To increase sensitivity, an enhancing material, such as water-soluble colored or fluorescent dyes can be added to the test water for detection of small leaks by developers or ultraviolet light.

Open containers (e.g., storage tanks) may also be hydrostatically tested by filling them with water or partially submerging them into water (e.g., ship barges). The hydrostatic pressure exerted against any boundary is governed by the head of water.

Hydrostatic testing is a relatively safe operation in that water is practically noncompressible and, therefore, stores little energy. A small leak results in a meaningful pressure drop that limits the driving force needed to propagate the leak. However, three questions should be considered to conduct safe hydrostatic tests:

(1) Are the foundation and the support structure strong enough to hold the water-filled container? This is of special importance if the containers are designed to hold a gas or a lightweight liquid.

(2) Are there any pockets where energy can build up in the form of compressed air?

(3) Is the metal at a temperature where it has adequate notch toughness to assure that a relatively small leak or discontinuity will not propagate into a catastrophic brittle fracture?

PNEUMATIC TESTING

Pneumatic testing is similar to hydrostatic testing except that compressed air is used to pressurize a closed vessel. This type of test is primarily used for small units that can be submerged in water during the test. The water provides a convenient leak indicator and an effective energy absorber in the event that the container fails.

Other applications include units mounted on foundations that are not able to support the weight associated with hydrostatic tests, and vessels where water or other liquid may be harmful or cannot be adequately removed. An example of the latter is a plate-fin heat exchanger designed for cryogenic service.

Acceptance of pneumatic tests is based on freedom from leakage. Small leaks are seldom detected without some indicating devices. If units cannot be submerged in water, spraying with a soap or detergent solution and checking for bubbles is an effective procedure. This procedure is called a soap bubble test. For special applications, sound detection devices are available that report and locate all but the smallest air leaks.

During any pneumatic test, large amounts of energy may be stored in compressed air or gas in a large volume or under high pressure, or both. A small leak or rupture can easily grow into a catastrophic failure, and can endanger adjacent life and property. Extensive safety planning is needed by proper quality control during fabrication, by thorough monitoring during testing, and by observing the testing operation from a safe distance. All pneumatic testing should be done to a written test procedure for the product to be tested.

SPIN TESTING

Welded components that rotate in service can be proof tested by spinning them at speeds above the design values to develop desired stresses from centrifugal forces. Visual and other nondestructive testing plus dimensional measurements are employed to determine the acceptability of the parts. Spin testing must be done in a safety enclosure in case the component should rupture.

LEAK TESTING

When freedom from leakage is of primary importance and a high pressure test is not desirable or possible, a number of low-pressure leak testing techniques are available. All are based upon the principal of filling a container with a product that has a low viscosity and has the ability to penetrate through very small openings. For instance, if a container is designed to hold water or oil, a low-pressure pneumatic test may be an acceptable leak test.

To increase the effectiveness and accuracy of such leak tests, tracer gases are used. These include freon for standard tests and helium for critical applications. Leak detection instruments (calibrated sniffers) are used to detect the presence of the tracer gas escaping through any leaks in the vessel.

VACUUM BOX TESTING

Pressure leak testing requires the ability to pressurize one side and inspect from the other side of the component. This represents a limitation for components that cannot be pressurized. However, a vacuum box test can be used. This involves coating a test area with a soap or detergent film and placing a gasketed transparent box over the area to be inspected. After pulling a partial vacuum of not less than 2 psi in the box, the inspector looks for any bubble formations that indicate the presence of a leak. For critical service, such as LNG tanks, vacuum levels as high as 8 psi may be selected.

MECHANICAL STRESS RELIEVING

Proof-testing operations can reduce the residual stresses associated with welding. In the as-welded condition, any weldment will have peak residual stresses equal to or slightly below the yield strength of the metal. When proof testing the weldments, the yield strength may be exceeded in highly stressed areas, and the metal will plastically deform. After the load is removed, the peak stresses in those areas will be lower than the original stresses. This result tends to improve product reliability.

BRAZED JOINTS

GENERAL CONSIDERATIONS

Inspection of a completed assembly or subassembly is the last step in the brazing operation, and is essential for assuring satisfactory and uniform quality of a brazed unit. The design of a brazement is extremely important to the inspection operation and, where practical, should be such that completed joints can be easily and adequately examined. An intelligent choice of brazing processes, brazing filler metal, joint design, and cleaning methods will also assist in the inspection process. The inspection method chosen to evaluate a final brazed component should depend on the service requirements. In many cases, the inspection methods are specified by the user or by regulatory codes.

Testing and inspection of brazed joints can be conducted on procedure qualification test joints or on a finished brazed assembly, and the tests may be either destructive or nondestructive. Preproduction and workmanship samples are often used for comparison purposes during production. These may be sample specimens made during the development of the brazing procedure or samples taken from actual production. In any case, they should show the minimum acceptable production quality of the brazed joint.

TESTING METHODS

Brazed joints and completed brazements can be tested both nondestructively and destructively. Nondestructive test methods that may be used are visual examination, liquid penetrant, radiographic, and ultrasonic as well as proof and leak testing. Destructive testing methods include metallographic examination and peel, tension, shear, and torsion tests. All of these testing methods are discussed in the *Welding Handbook*, Vol. 2, 7th Edition.

Metric Conversion Factors

°C = 0.56 (°F − 32)
1 footcandle = 10.76 lm/m^2
1 A/in. = 0.039 A/mm
1 in. = 25.4 mm
1 psi = 6.89 kPa
1 μW/cm^2 = 6.45 μW/in.^{2}

SUPPLEMENTARY READING LIST

ASME Boiler and Pressure Vessel Code, Section V, Nondestructive Examination, New York: American Society of Mechanical Engineers (latest Ed.).

ASNT No. SNT-TC-1A, *Personnel Qualification and Certification in Nondestructive Testing,* Columbus, OH: The American Society for Nondestructive Testing.

AWS B1.0, Guide for the Nondestructive Inspection of Welds, Miami, FL: American Welding Society (latest Ed.).

AWS D1.1, Structural Welding Code—Steel, Miami, FL: American Welding Society (latest Ed.).

Metals Handbook, 8th ed., Vol. 11, *Nondestructive Inspection and Quality Control,* Metals Park, OH: American Society for Metals, 1976.

Nichols, R.W., ed. *Non-Destructive Examination in Relation to Structural Integrity,* London: Applied Science Publishers, Ltd., 1979.

Nondestructive Testing Handbook, Vol. 1, *Leak Testing,* Vol. 2, *Liquid Penetrant Tests,* Columbus, OH: Society for Nondestructive Testing, 1982.

Stout, R.D., *Hardness as an Index of the Weldability and Service Performance of Steel Weldments,* New York: Welding Research Council Bulletin 189, Nov. 1973.

Welding Inspection, 2nd Ed., Miami, FL: American Welding Society, 1980.

10
Safe Practices

General Welding Safety *380*
Fumes and Gases *387*
Handling of Compressed Gases *397*
Electrical Safety *402*
Processes *405*
Supplementary Reading List *415*

Chapter Committee

G.R. SPIES, *Chairman*
The B.O.C. Group, Inc.

G.C. BARNES
Allegheny International

K.L. BROWN
Lincoln Electric Company

G.N. CRUMP
Hobart Brothers Company

O.J. FISHER
Babcock and Wilcox Company

J. KUJAWA
Miller Electric Company

K.A. LYTTLE
Union Carbide Corporation

Welding Handbook Committee Member

A.F. MANZ
Union Carbide Corporation

10
Safe Practices

GENERAL WELDING SAFETY

Safety is an important consideration in all welding, cutting, and related work. An activity is not satisfactorily completed if someone is injured. The hazards that may be encountered with most welding, cutting, and related processes, and also the practices that will minimize personal injury and property damage, are covered in this chapter. In addition, the user should always read and understand the manufacturer's instructions on safe practices for the materials and equipment before work proceeds.

MANAGEMENT SUPPORT

The most important component of an effective safety and health program is management support and direction. Management must clearly state its objectives and demonstrate its commitment to safety and health by consistent execution of safe practices.

Management must designate approved areas where welding and cutting operations may be carried on safely. When these operations must be done in other than designated areas, management must assure that proper procedures to protect personnel and property are established and followed.

Management must be certain that only approved welding, cutting, and allied equipment is used. Such equipment includes torches, regulators, welding machines, electrode holders, and personal protective devices. Adequate supervision must be provided to assure that all equipment is properly used and maintained.

TRAINING

Thorough and effective training is a key aspect of a safety program. Welders[1] and other equipment operators perform most safely when they are properly trained in the subject. Proper training includes the safe use of the equipment and the process, and also the safety rules that must be followed for the protection of all concerned. Personnel need to know and understand the rules and consequences of disobeying them. For example, welders must be trained to position themselves while welding, cutting, or performing related work so that their heads are not in the gases and fumes.[2] This helps to minimize exposure to fumes and gases.

Certain AWS specifications call for the use of precautionary labels on packages of welding and brazing consumables. These labels concerning the safe use of the products should be read and followed by those persons using the particular consumable (welding electrode, brazing filler

1. The term *welder* is intended to include welding and cutting operators, brazers, and solderers.

2. Fume is the cloud-like plume containing minute solid particles arising during the melting of a metal. In distinction to a gas, fumes are metallic vapors that have condensed to solid and are often associated with a chemical reaction, such as oxidation.

metal, flux). A typical label is the following one used on packages of arc welding filler metals.

> *WARNING: Protect yourself and others. Read and understand this label. FUMES AND GASES can be dangerous to your health. ARC RAYS can injure eyes and burn skin. ELECTRIC SHOCK can kill.*
>
> - *Read and understand the manufacturer's instructions and your employer's safety practices.*
> - *Keep your head out of the fumes.*
> - *Use enough ventilation, exhaust at the arc, or both, to keep fumes and gases from your breathing zone, and the general area.*
> - *Wear correct eye, ear, and body protection.*
> - *Do not touch live electrical parts.*
> - *See American National Standard Z49.1 Safety in Welding and Cutting published by the American Welding Society, 550 N.W. LeJeune Rd., Miami, Florida 33126; OSHA Safety and Health Standards, 29 CFR 1910, available from the Government Printing Office, Washington, D.C. 20402.*
>
> *DO NOT REMOVE THIS LABEL*

The proper use and maintenance of the equipment must also be taught. For example, defective or worn electrical insulation cannot be tolerated in arc welding or cutting, nor worn or defective hoses in gas welding or brazing. Proper training in equipment operation is fundamental to safe operation.

All persons must be trained to recognize safety hazards. If they are to work in an unfamiliar situation or environment, they must be thoroughly briefed on the potential hazards involved. For example, consider a person who must work in a confined space. If the ventilation is poor and an air-supplied helmet is required, the need to use such equipment must be thoroughly explained to the welder. The consequences of improperly using the equipment must be thoroughly covered. When a welder feels that the safety precautions for a given task are not adequate, they should be questioned before proceeding.

GENERAL HOUSEKEEPING

Good housekeeping is essential to avoid injuries. A welder's vision may be restricted by necessary eye protection. Persons passing a welding station while work is in progress must shield their eyes from the flame or arc radiation. The limited vision of the welder and passers-by makes them vulnerable to tripping over objects on the floor. Therefore, welders and supervisors must always make sure that the area is clear of tripping hazards. Management also has the responsibility to lay out the production area so that gas hoses, cables, mechanical assemblies, and other equipment do not cross walkways or interfere with the routine performance of tasks.

When work is above ground or floor level, safety rails or lines must be provided to prevent falls as a result of restricted vision from eye protection. Safety lines and harnesses can be helpful to restrict workers to a safe area, and to catch them in case of a fall.

Unexpected events, such as fire and explosions, do occur in industrial environments. It is important that escape routes be known and kept clear so that orderly, rapid, and safe evacuation of a problem area can take place. Storage of goods and equipment in evacuation routes should be avoided. If an evacuation route must be temporarily blocked, employees who would normally use that route must be trained to use an alternate route.

PROTECTION IN THE GENERAL AREA

Equipment, machines, cables, hoses, and other apparatus should always be placed so that they do not present a hazard to personnel in passageways, on ladders, or on stairways. Warning signs should be posted to designate welding areas, and to specify that eye protection should be worn.

Protective Screens

Workers or other persons adjacent to welding and cutting areas must be protected from radiant energy and hot spatter by flame-resistant screens or shields, or by suitable eye and face protection and protective clothing. Appropriate radiation-protective, semi-transparent materials are permissible. Where operations permit, work stations should be separated by noncombustible screens or shields, as shown in Fig. 10.1. Booths and screens should permit circulation of air at floor level as well as above the screen.

Wall Reflectivity

Where arc welding or cutting is regularly carried on adjacent to painted walls, the walls should be painted with a finish having low reflectivity to ultraviolet radiation.[3] Finishes formulated with certain pigments, such as titanium dioxide or zinc oxide, have low reflectivity to ultraviolet radiation. Color pigments may be added if they do not increase reflectivity. Pigments based on powdered or flaked metals are not recommended because of their high reflectivity to ultraviolet radiation.

3. For further guidance, see *Ultraviolet Reflectance of Paint,* American Welding Society, 1976.

PUBLIC DEMONSTRATIONS

It is the responsibility of the persons putting on a public demonstration or exhibition to assure the safety of observers and the general public. Observers are not likely to have the necessary protective equipment to enable them to observe demonstrations safely. For exhibits involving observation of arc or gas welding processes other than submerged arc welding, appropriate eye protection for both observers and passers-by is mandatory. Fume exposure must also be minimized by appropriate ventilation. Electric cables and hoses must be routed to avoid audience exposure to possible electric shock or tripping hazards. Protection must be provided against fires from fuels, combustibles, and overheated apparatus and wiring. Fire extinguishers must be on hand in case of fire. Combustible materials must be shielded from flames, sparks, and molten metal.

Fig. 10.1—Protective screens between work stations

Safety precautions at public events should be passive types. Passive safety devices allow untrained individuals to safely observe demonstrations and exhibitions. That is, they should not require the audience to participate in the use of safety equipment. For example, a protective, moveable transparent screen allows an audience to observe a welding operation when the screen is in place. After welding is completed, the screen can be moved to allow the audience to observe the completed weld. Additional information is given in *ANSI/ASC Z49.1, Safety in Welding and Cutting* (available from the American Welding Society).

FIRE

With most welding, cutting, and allied processes, a high-temperature heat source is always present. Open flames, electric arcs, hot metal, sparks, and spatter are ready sources of ignition. Many fires are started by sparks, which can travel horizontally up to 35 feet from their source and fall much greater distances. Sparks can pass through or lodge in cracks, holes, and other small openings in floors and walls.

The risk of fire is greatly increased by failure to remove combustibles from the work area, or by welding or cutting too close to combustibles that have not been shielded. Materials most commonly ignited are combustible floors, roofs, partitions, and building contents including trash, wood, paper, textiles, plastics, chemicals, and flammable liquids and gases. Outdoors, the most common sources of combustion are dry grass and brush.

The best protection against fire is to perform welding and cutting in specially designated areas or enclosures of noncombustible construction that are kept free of combustibles. Combustibles should always be removed from the work area or shielded from the operation. Fuel gas is a common flammable often found in cutting and welding areas. Special attention should be given to fuel gas cylinders, hoses, and apparatus to prevent gas leakage.

Combustibles that cannot be removed from the area should be covered with tight fitting, flame resistant material. These include combustible walls and ceilings. Floors should be free of combustible materials for a radius of 35 feet around the work area. All doorways, windows, cracks, and other openings should be covered with a flame-resistant material or, if possible, the work area should be enclosed with portable flame-resistant screens.

If welding or cutting is to be performed on or adjacent to a metal wall, ceiling, or partition, combustibles on the other side must be moved to a safe location. If this cannot be done, a fire watcher should be stationed where the combustibles are located. Welding or cutting should not be performed on material having a combustible coating or internal structure, as in walls or ceilings. Hot scrap or slag must not be placed in containers holding combustible materials. Suitable fire extinguishers should be available nearby. A thorough examination for evidence of fire should be made before leaving the work area. Fire inspection should be continued for at least 30 minutes after the operation is completed.

Welding, brazing, and cutting should not be performed on combustible staging, that may readily be ignited by heat from the operation. Welders must be alert for traveling vapors of flammable liquids. Vapors are often heavier than air and can travel along floors and in depressions for distances of several hundred feet. Light vapors can travel along ceilings to adjacent rooms.

EXPLOSION

Flammable gases, vapors, and dusts, when mixed with air or oxygen in certain proportions, present danger of explosion as well as fire. To avoid such explosions, welding, brazing, soldering, or cutting must not be done in atmospheres containing flammable gases, vapors, or dusts. Such flammables must be kept in leak-tight containers or be well removed from the work area. Heat may cause low-volatile materials to produce flammable vapors.

Hollow containers must be vented before applying heat. Heat must not be applied to a container that has held an unknown material, a combustible substance, or a substance that may form flammable vapors on heating.[4] The container must first be thoroughly cleaned or inerted. Heat should never be applied to a workpiece covered by an unknown substance or to a substance that may form flammable or toxic vapors on heating.

Explosions may occur from simple expansion of gas trapped in pockets, such as casting porosity. Adequate eye and body protection should be worn when operations involve these risks.

BURNS

Burns of the eye or body are serious hazards of welding, brazing, soldering, and cutting. Eye, face, and body protection for the operator and others in the work area are required to prevent burns from ultraviolet and infrared radiation, sparks, and spatter.

Eye and Face Protection

Arc Welding and Cutting. Welding helmets or handshields containing appropriate filter lenses and cover plates must be used by operators and nearby personnel when viewing the arc.[5] Suggested shade numbers of filter plates for various welding, brazing, soldering, and thermal cutting operations are given in Table 10.1.

Safety spectacles, goggles, or other suitable eye protection must also be worn during welding and cutting operations. They must have side shields when there is danger of exposure to injurious rays or flying particles from grinding or chipping operations. Spectacles and goggles may have clear or colored lenses, depending on the intensity of the radiation that may come from adjacent welding or cutting operations when the welding helmet is raised or removed. Number 2 filter lenses are recommended for general purpose protection.

4. Additional information is given in *AWS F4.1-80, Recommended Safe Practices for the Preparation for Welding and Cutting of Containers and Piping That Have Held Hazardous Substances.*

5. Standards for welding helmets, handshields, face shields, goggles, and spectacles are given in *ANSI Publication Z87.1, Practice for Occupational and Educational Eye and Face Protection,* latest edition, American National Standards Institute.

Oxyfuel Gas Welding and Cutting, Submerged Arc Welding. Safety goggles with filter lenses (Table 10.1) and side shields must be worn while performing oxyfuel gas welding and cutting, and submerged arc welding. During submerged arc welding, the arc is covered by flux and not readily visible; hence, an arc welding helmet is not needed.

Torch Brazing and Soldering. Safety spectacles with or without side shields and appropriate filter lenses are recommended for torch brazing and soldering. As with oxyfuel gas welding and cutting, a bright yellow flame may be visible during torch brazing. A filter similar to that used with those processes may be used for torch brazing.

Resistance Welding and Brazing. Operators and helpers must wear safety spectacles, goggles, or use a face shield to protect their eyes from spatter. Filter lenses are not necessary but may be used for comfort.

Protective Clothing

Sturdy shoes or boots and heavy clothing similar to that in Fig. 10.2 should be worn to protect the whole body from flying sparks, spatter, and radiation burns. Woolen clothing is preferable to cotton because it is not so readily ignited. Cotton clothing, if used, should be chemically treated to reduce its combustibility. Clothing treated with nondurable flame retardants must be retreated after each washing or cleaning. Clothing or shoes of synthetic or plastic materials, which can melt and cause severe burns, are not recommended. Outer clothing should be kept reasonably free of oil and grease, especially in an oxygen-rich atmosphere.

Cuffless pants and covered pockets are recommended to avoid spatter or spark entrapment. Pockets should be emptied of flammable or readily ignitable material before welding because they may be ignited by sparks or weld spatter and result in severe burns. Pants should be worn outside of shoes. Protection of the hair with a cap is also recommended. Flammable hair preparations should not be used.

Durable gloves of leather or other suitable material should always be worn. Gloves not only protect the hands from burns and abrasion, but

Table 10.1
Suggested viewing filter plates

Operation	Plate thickness, in.	Plate thickness, mm	Welding current, A	Lowest shade number	Comfort shade number[a]
Shielded metal arc welding	—	—	Under 60	7	—
			60-160	8	10
			160-250	10	12
			250-550	11	14
Gas metal arc and flux cored arc welding	—	—	Under 60	7	—
			60-160	10	11
			160-250	10	12
			250-500	10	14
Gas tungsten arc welding	—	—	Under 50	8	10
			50-150	8	12
			150-500	10	14
Plasma arc welding	—	—	Under 20	6	6-8
			20-100	8	10
			100-400	10	12
			400-800	11	14
Oxyfuel gas welding (steel)[b]	Under 1/8	3.2	—	—	4, 5
	1/8-1/2	3.2-12.7	—	—	5, 6
	Over 1/2	12.7	—	—	6, 8
Plasma arc cutting[c]	—	—	Under 300	8	9
			300-400	9	12
			400-800	10	14
Air-carbon arc cutting	—	—	Under 500	10	12
			500-1000	11	14
Oxyfuel gas cutting (steel)[b]	Under 1	25	—	—	3, 4
	1-6	25-150	—	—	4, 5
	Over 6	150	—	—	5, 6
Torch brazing	—	—	—	—	3, 4
Torch soldering	—	—	—	—	2

a. To select the best shade for the application, first select a dark shade. If it is difficult to see the operation properly, select successively lighter shades until the operation is sufficiently visible for good control. However, do not go below the lowest recommended number, where given.
b. With oxyfuel gas welding or cutting, the flame emits strong yellow light. A filter plate that absorbs yellow or sodium wave lengths of visible light should be used for good visibility.
c. The suggested filters are for applications where the arc is clearly visible. Lighter shades may be used where the arc is hidden by the work or submerged in water.

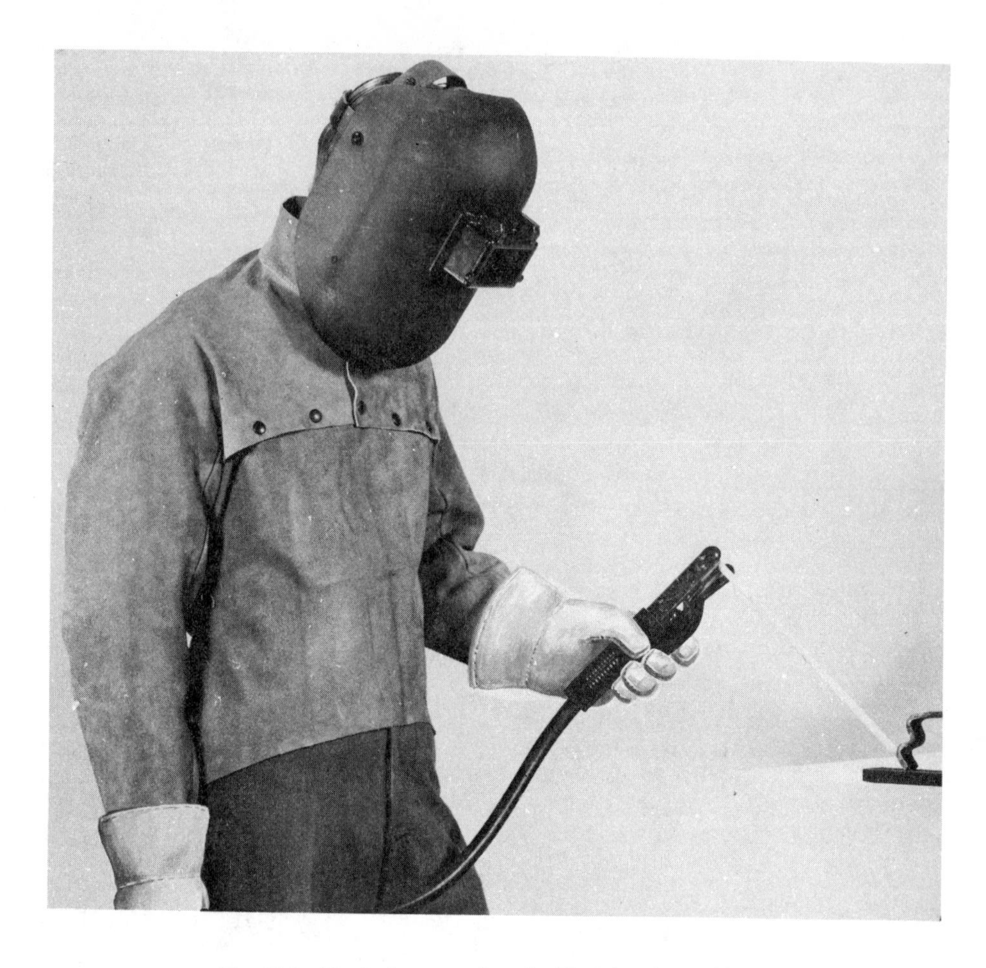

Fig. 10.2—Typical protective clothing for arc welding

also provide insulation from electrical shock. A variety of special protective clothing is also available for welders. Aprons, leggings, suits, capes, sleeves, and caps, all of durable materials, should be worn when welding overhead or when special circumstances warrant additional protection of the body.

Sparks or hot spatter in the ears can be particularly painful and serious. Properly fitted, flame-resistant ear plugs should be worn whenever operations pose such risks.

NOISE

Excessive noise, particularly continuously high levels, can damage hearing. It can cause either temporary or permanent hearing loss. U.S. Department of Labor Occupational Safety and Health Administration regulations describe allowable noise exposure levels. Requirements of these regulations may be found in General Industry Standards, 29 CFR 1910.95.

In welding, cutting, and allied operations, noise may be generated by the process or the equipment, or both.[6] Processes that tend to have high noise levels are air carbon arc and plasma arc cutting. Engine-driven generators sometimes emit a high noise level.

6. Additional information is presented in *Arc Welding and Cutting Noise,* American Welding Society, 1979.

The most direct way to control excessive noise is to reduce the intensity at the source, or block it with appropriate sound absorbing materials. When control methods fail to bring noise within allowable limits, personal protective devices such as ear muffs or ear plugs should be employed.

HOT WORK PERMIT SYSTEM

When welding, cutting, or similar hot working operations are to be performed in areas not normally assigned for such operations, a hot work permit system should be used. The purpose of the hot work permit system is to assure that fire protection provisions are observed. Permit systems vary from simple verbal procedures to formal paperwork, depending upon site circumstances.

The essence of a permit system is that no work can be performed until formal approval has been given by a responsible official. The responsible official must establish procedures that properly protect all fire risks before approval is given. Hot work permits should be issued only for the period for which the approving official is responsible. Generally, this would be the duty shift of that official. Work during a succeeding shift would have to receive a permit from the responsible official on that shift.

Whenever hot work is to be performed in areas containing combustibles and there is danger of fire from undetected smoldering combustibles, a firewatcher should be posted. The firewatcher's duty is to maintain surveillance of the area for a period of sufficient duration to detect any fires that might have been set. Another duty of the firewatcher is to maintain fire surveillance in areas not readily visible to the operator. An example is the opposite side of a metal wall where a fire could start in combustibles close to the wall.

Integral with the hot work permit and firewatcher systems is, of course, the assurance that adequate fire extinguisher provisions are maintained.

MACHINERY GUARDS

Welders and other workers must be protected from injury by machinery and equipment that they are operating or by other machinery operating in the work area. Moving components and drive belts must be covered by guards to prevent physical contact.

When repairing machinery by welding or brazing, the power to the machinery must be disconnected, locked out, and tagged to prevent inadvertent operation and injury. Rotating and automatic welding machines and welding robots must be equipped with appropriate guards or sensing devices to prevent operation when someone is in the danger area.

Pinch points on welding and other mechanical equipment can also result in serious injury. Examples of such equipment are resistance welding machines, robots, automatic arc welding machines, jigs, and fixtures. To avoid injury with such equipment, the machine should be equipped so that the operators hands must both be at safe locations when the machine is actuated, or the pinch points must be suitably guarded mechanically. Metalworking equipment should not be located where a welder could accidently fall into or against it while welding. During maintenance of the equipment, pinch points should be blocked to prevent them from closing in case of equipment failure. In very hazardous situations, an observer should be stationed to prevent someone from turning the power on until the repair is completed.

FUMES AND GASES

Protection of operators and other persons from overexposure to fumes and gases produced during welding, brazing, and cutting is an important safety concern. Excess exposure is exposure that exceeds the permissible limits specified by a government agency, such as the U.S. Department of Labor, Occupational Safety and Health Administration (OSHA), Regulations 29 CFR

1910.1000, or by other recognized authority, such as the American Conference of Governmental Industrial Hygienists (ACGIH) in their publication, *Threshold Limit Values for Chemical Substances and Physical Agents in the Workroom Environment.* Persons with special health problems may have unusual sensitivity that requires more stringent protection.

Fumes and gases are usually a greater concern in arc welding than in oxyfuel gas welding, cutting, or brazing because of the higher temperatures involved, the greater variety of materials, and the larger number of process variables.

Protection from excess exposure is usually accomplished by ventilation. Where exposure would exceed permissible limits with available ventilation, respiratory protection must be used. Protection must be provided not only for the welders and cutting operators but also for other persons in the area.

ARC WELDING

Nature and Sources

Fumes and gases from arc welding and cutting cannot be classified simply. The composition and quantity depend upon the base metal composition; the process and consumables used; coatings on the work, such as paint, galvanizing, or plating; contaminants in the atmosphere, such as halogenated hydrocarbon vapors from cleaning and degreasing activities; and other factors.

In welding and cutting, the composition of the fume usually differs from the composition of the electrode or consumables. Reasonably expected fume constituents from normal operations include products of volatilization, reaction, or oxidation of consumables, base metals, coatings, and atmospheric contaminants. Reasonably expected gaseous products include carbon monoxide, carbon dioxide, fluorides, nitrogen oxides, and ozone.

The quantity and chemical composition of air contaminants change substantially with the process used and with the wide range of variables inherent in each process. During arc welding, the arc energy and temperature depend on the process as well as the welding variables. Therefore, fumes and gases are generated in varying degrees in different welding operations.

Welding fume is a product of vaporization, oxidation, and condensation of components in the consumable and, to some degree, in the base metal. It is the electrode, rather than the base metal, that is the major source of fume. However, significant fume contributions can originate from the base metal if it contains alloying elements or a coating that is volatile at elevated temperatures.

Various gases are generated during welding. Some are a product of the decomposition of fluxes and electrode coatings. Others are formed by the action of arc heat or ultraviolet radiation emitted by the arc on atmospheric constituents and contaminants. Potentially hazardous gases include carbon monoxide, oxides of nitrogen, ozone, and phosgene or other decomposition products of chlorinated hydrocarbons.

Helium and argon, although chemically inert and nontoxic, are simple asphyxiants, and could dilute the atmospheric oxygen concentration to potentially harmful low levels. Carbon dioxide (CO_2) and nitrogen can also cause asphyxiation.

Ozone may be generated by ultraviolet radiation from welding arcs. This is particularly true with gas shielded arcs, especially when argon is used. Photochemical reactions between ultraviolet radiation and chlorinated hydrocarbons result in the production of phosgene and other decomposition products.

Thermal effects are responsible for the formation of oxides of nitrogen from atmospheric nitrogen. For this reason, nitrogen oxides may be produced by a welding arc or other high temperature heat sources. The thermal decomposition of carbon dioxide and inorganic carbonate compounds by an arc results in the formation of carbon monoxide. The levels can be significant when using a CO_2 shielded welding process.

Reliable estimates of fume composition cannot be made without considering the nature of the welding process and system being examined. For example, aluminum and titanium are normally arc welded in an atmosphere of argon or helium, or mixtures of the two gases. The arc creates relatively little fume but an intense radiation that can produce ozone. Similarly, gas shielded arc welding of steels also creates a relatively low fume level.

Arc welding of steel in oxidizing environments generates considerable fumes, and can produce carbon monoxide and oxides of nitrogen. These fumes generally are composed of discreet particles of amorphous slags containing iron, manganese, silicon, and other metallic constituents, depending on the alloy system involved. Chromium and nickel compounds are found in fumes when stainless steels are arc welded.

Some coated and flux cored electrodes are formulated with fluorides. The fumes associated with those electrodes can contain significantly more fluorides than oxides.

Factors Affecting Generation Rates

The rate of generation of fumes and gases during arc welding of steel depends on numerous variables. Among these are:

(1) Welding current
(2) Arc voltage (arc length)
(3) Type of metal transfer or welding process
(4) Shielding gas

These variables are interdependent and can have a substantial effect on total fume generation.

Welding current. In general, fume generation rate increases with increased welding current. The increase, however, varies with the process and electrode type. Certain covered, flux-cored, and solid wire electrodes exhibit a non-proportional increase in fume generation rate with increasing current.

Studies have shown that fume generation rates with covered welding electrodes are proportional to the welding current raised to a power.[7] The exponent is 2.24 for E6010 electrodes and 1.54 for E7018 electrodes. Similar trends were reported in other studies.

Flux cored and solid electrode fume generation rates are more complexly related to current. Welding current levels affect the type of metal droplet transfer. As a result, fume generation rate can decrease with increasing current until some minimum is reached. Then, it will increase in a relatively proportional fashion.

An increase in current can increase ultraviolet radition from the arc. Therefore, the generation of gases formed photochemically by this radiation can be expected to increase as welding current is increased. One such gas is ozone. Measurements of ozone concentration during gas metal arc and gas tungsten arc welding have shown such behavior.

Arc Voltage (Arc Length). Arc voltage and arc length are closely related. For a given arc length there is a corresponding arc voltage, mostly dependent upon the type of electrode, welding process, and power supply. In general, increasing arc voltage (arc length) increases the fume generation rate for all open arc welding processes. The levels differ somewhat for each process or electrode type.

Type of Transfer. When gas metal arc welding steel using a solid wire electrode, the mode of metal transfer depends upon the current level. At low welding currents, short-circuiting transfer takes place, where droplets are deposited during short circuits between the electrode and molten weld pool. As the current is increased, metal transfer changes to globular type where large globules of metal are projected across the arc into the weld pool. At high currents, transfer changes to a spray mode where fine metal droplets are propelled rapidly across the arc. The fume generation rate appears to follow a transition also. It is relatively high during short-circuiting transfer because of arc turbulence. As the transition current is approached in an argon-rich shielding gas, the fume rate decreases and then increases again as spray transfer is achieved. In the spray region, the rate of fume generation is proportional to welding current.

For other welding processes, the type of metal transfer does not vary substantially with current. In these cases, fume generation follows the relationship for welding current changes.

Shielding Gas. When gas metal arc welding or flux cored arc welding with certain electrodes, shielding gas must be used. The type of shielding gas affects both the composition of the fume and the rate of generation. It also affects the kind of gases found in the welding environment. For example, the fume generation rate is higher with CO_2 shielding than with argon-rich shielding. The rate of fume formation with argon-oxygen

7. Refer to *Fumes and Gases in the Welding Environment,* American Welding Society, 1979.

or argon-CO_2 mixtures increases with the oxidizing potential of the mixture.

For welding processes where inert gas shielding used, such as gas tungsten arc welding, the fume generation rate varies with the type of gas or gas mixture. For example, there can be more fume with helium than with argon shielding.

By-product gases also vary with shielding gas composition. The rate of formation of ozone depends upon the wave lengths and intensity of the ultraviolet rays generated in the arc; ozone is more commonly found with argon-rich gases than with carbon dioxide. Oxides of nitrogen are present in the vicinity of any open arc process. Carbon monoxide is commonly found around CO_2 shielded arcs.

Welding Process. Studies of the relative fume generation rates of processes for welding on mild steel have shown definite trends. Considering the ratio of the weight of fumes generated per weight of metal deposited, covered welding electrodes and self-shielded, flux cored electrodes produce the most fume. Gas-shielded flux cored electrodes are next, following by solid wire electrodes. The submerged arc process consistently produces the least fumes of any arc welding process because of the flux cover.

Consumables. With a specific process, the fume rate depends upon the composition of the consumables. Some components of the covering of shielded metal arc welding electrodes and the core of a flux cored electrode are designed to decompose to form protective gases during welding. Hence, they will generate relatively high fume levels.

Because the constituents of many covered and flux cored electrodes are proprietary, the fume generation rates of similar electrodes produced by different manufacturers can vary substantially. The only reliable method for comparing filler metals, even within the same AWS Classification, is actual product testing to determine specific fume generation characteristics.

OXYFUEL GAS WELDING AND CUTTING

The temperatures encountered in oxyfuel gas welding and cutting are lower than those found in electric arc processes. Consequently, the quantity of fumes generated is normally lower. The gases formed are reaction products of fuel gas combustion and of chemical reactions between the gases and other materials present. Fumes generated are the reaction products of the base metals, coatings, filler metals, fluxes, and the gases being used. In oxyfuel gas cutting of steel, the fumes produced are largely oxides of iron.

Fume constituents of greater hazard may be expected when coatings such as galvanizing, paint primers, or cadmium plating are present. Gases of greatest concern include oxides of nitrogen, carbon monoxide, and carbon dioxide. Oxides of nitrogen may be present in especially large amounts when oxyfuel gas cutting stainless steels using either the chemical flux or the iron powder process.

EXPOSURE FACTORS

Position of the Head

The single most important factor influencing exposure to fume is the position of the welder's head with respect to the fume plume. When the head is in such a position that the fume envelops the face or helmet, exposure levels can be very high. Therefore, welders must be trained to keep their heads to one side of the fume plume. In some cases, the work can be positioned so the fume plume rises to one side.

Type of Ventilation

Ventilation has a significant influence on the amount of fumes in the work area, and hence the welder's exposure. Ventilation may be local, where the fumes are extracted near the point of welding, or general, where the shop air is changed or filtered. The appropriate type will depend on the welding process, the material being welded, and other shop conditions. Adequate ventilation is necessary to keep the welder's exposure to fumes and gases within safe limits.

Shop Size

The size of the welding or cutting enclosure is important. It affects the background fume level. Fume exposure inside a tank, pressure vessel, or other confined space will certainly be higher than in a high-bay fabrication area.

Number of Welders

The number of welders, the welding process, and the duration of operation in a particular area determine the background fume level in the general work area.

Design of Welding Helmet

The extent to which the helmet curves under the chin towards the chest affects the amount of fume exposure. Close-fitting helmets can be effective in reducing exposure.

Base Metal and Surface Condition

The type of base metal being welded influences both the constituents and the amount of fume generated. Surface contaminants or coatings may contribute significantly to the hazard potential of the fume. Paints containing lead and platings containing cadmium generate dangerous fumes during welding. Galvanized material evolves zinc fume.

VENTILATION

Adequate ventilation is the key to control of fumes and gases in the welding environment.[8] Natural, mechanical, or respirator ventilation must be provided for all welding, cutting, brazing, and related operations. The ventilation must ensure that hazardous concentrations of airborne contaminants are maintained at recommended levels. These levels must be no higher than the allowable levels specified by the U.S. Occupational Safety and Health Administration or other applicable authorities.

Many ventilation methods are available. They range from natural convection to localized devices, such as air-ventilated welding helmets. Examples of ventilation include:

(1) Natural
(2) General area mechanical ventilation
(3) Overhead exhaust hoods
(4) Portable local exhaust devices
(5) Downdraft tables
(6) Cross draft tables
(7) Extractors built into the welding equipment
(8) Air-ventilated helmets

The bulk of fumes generated during welding and cutting consists of small particles that remain suspended in the atmosphere for a considerable time. As a result, fume concentration in a closed area can build up over time, as can the concentration of any gases evolved or used in the welding process. The particles eventually settle on the walls and floor, but the settling rate is low compared to the generation rate of the welding or cutting processes. Therefore, fume concentration must be controlled by ventilation.

General Ventilation

In most cases, general ventilation is more effective in protecting personnel in adjacent areas than in protecting the welder. Such ventilation may occur naturally outdoors or when the shop doors and windows are open. It is acceptable when precautions are taken to keep the welder's breathing zone away from the fume plume, and when sampling of the atmosphere shows that concentrations of contaminants do not exceed recommended limits. Natural ventilation often meets these criteria when all of the following conditions are present:

(1) Space of more than 10,000 ft^3 per welder is provided.
(2) Ceiling height is more than 16 ft.
(3) Welding is not done in a confined area.
(4) The general welding area is free of partitions, balconies, or other structural barriers that significantly obstruct cross ventilation. The general area refers to a building or a room in a building, not a welding booth or screened area that is used to provide protection from welding radiation.
(5) Toxic materials with low permissible exposure limits are not present deliberately.

When natural ventilation is not sufficient, fans may be used to force and direct the required amount of air through a building or work room.

The effectiveness of general ventilation, natural or forced, is dependent upon the design of the system. The locations where fresh air is introduced and contaminated air is exhausted must be positioned to carry away the welding

8. Additional information is available in *Welding Fume Control with Mechanical Ventilation,* 2nd Ed., Fireman's Fund Insurance Companies, 1980. (Available from the American Welding Society.)

fumes and gases, and not concentrate them in dead zones. In some cases, the fresh air supply may be located so that incoming air provides the required protection for the welders as well as for personnel in the general area.

Air movement should always be from either side across the welder. This makes it easier for the welder to keep out of the plume and to avoid fumes and gases from entering the welding helmet. Air should never blow toward the face or back of the welder because it may force the fume behind the helmet.

General mechanical ventilation may be necessary to maintain the background level of airborne contaminants within acceptable limits. It is not usually satisfactory for health hazard control, but is often used to supplement local ventilation.

Local Ventilation

General ventilation may control contamination levels in the general area, but in many cases, it will not provide the local control needed to protect the welder. Local exhaust ventilation is an effective way of providing this protection. It provides an efficient and economical means of fume control, and may be applied by one of the following methods:

(1) *A fixed open or enclosing hood.* It consists of a top and at least two sides, that surrounds the welding or cutting operation. It must have sufficient air flow and velocity to keep contaminant levels at or below recommended limits.

(2) *A movable hood with a flexible duct.* The hood is positioned by the welder as close as practicable to the point of welding, as illustrated in Fig. 10.3. It should have sufficient air flow to

Fig. 10.3—An adjustable hood for local exhaust ventilation

produce a velocity of 100 ft/min in the zone of welding. Air flow requirements range from 150 ft^3/min, when the hood is positioned at 4 to 6 inches from the weld, to 600 ft^3/min at 10 to 12 inches. This is particularly applicable for bench work, but may be used for any location, provided the hood is moved as required. An air velocity of 100 ft/min will not disturb the torch gas shield during gas shielded arc welding, provided adequate gas flow rates are used. Higher air velocities may seriously disturb the gas shield and render it ineffective.

(3) *Crossdraft or downdraft table.* A crossdraft table is a welding bench with the exhaust hood placed to draw the air horizontally across the table. The welder should face in a direction perpendicular to the air flow. A downdraft table has a grill to support the work above an exhaust hood that draws the air downward and away from the welder's head.

(4) *Gun-mounted fume removal equipment.* This equipment extracts the fumes at the point of welding. The exhaust rate must be set so that it does not interfere with the shielding gas pattern provided by the welding process.

(5) *A water table, used for oxyfuel gas and plasma arc cutting operations.* This unit is simply a cutting table filled with water near to or contacting the bottom surface of the work. Much fume emerging from the cut is captured in the water.

Where permissible, air cleaners that have high efficiencies in the collection of submicron particles may be used to recirculate a portion of ventilated air that would otherwise be exhausted. Some air cleaners do not remove gases; therefore, adequate monitoring must be done to keep concentrations of harmful gases within safe limits.

Respiratory Protective Equipment

Where natural or mechanical ventilation is not adequate or where very toxic materials require a supplement to ventilation, respiratory protective equipment must be used.[9] Airline respirators or face masks that give protection against all contaminants are generally preferred. Air-supplied welding helmets are also available commercially. Filter-type respirators, approved by the U.S. Bureau of Mines for metal fume, give adequate protection against particulate contaminants that are less toxic than lead, provided they are used and maintained correctly. Their general use is not recommended, however, because of the difficulty in assuring proper use and maintenance. They will not protect against mercury vapor, carbon monoxide, or nitrogen dioxide. For these hazards an airline respirator, hose mask, or gas mask is required.

SPECIAL VENTILATION SITUATIONS

Welding in Confined Spaces

Special consideration must be given to the safety and health of welders and other workers in confined spaces.[10] Gas cylinders must be located outside of the confined space to avoid possible contamination of the space with leaking gases or volatiles. Welding power sources should also be located outside to avoid danger of engine exhaust and electric shock.

A means for removing persons quickly in case of emergency must be provided. Safety belts and lifelines, when used, should be attached to the worker's body in a manner that avoids the possibility of the person becoming jammed in the exit. A trained helper should be stationed outside the confined space with a preplanned rescue procedure to be put into effect in case of emergency.

In addition to keeping airborne contaminants in breathing atmospheres at or below recommended limits, ventilation in confined spaces must also (1) assure adequate oxygen for life support but not an oxygen-enriched atmosphere, and (2) prevent accumulation of flammable mixtures. Asphyxiation can quickly result in unconsciousness and death without warning. A confined space must not be entered unless (1) it is well ventilated, or (2) the welder is properly trained to work in such spaces and is wearing an

---

9. Refer to *ANSI Publication Z88.2, Practices for Respiratory Protection,* latest edition, for additional information.

10. See *ANSI Publication Z117.1, Safety Requirements for Working in Tanks and Other Confined Spaces,* latest edition, for further precautions.

approved air-supplied breathing apparatus.[11] A similarly equipped second person must be present. Confined spaces should be tested for (1) toxic or flammable gases and vapors, and (2) adequate or excess oxygen before entering, using instruments approved by the U.S. Bureau of Mines. Heavier-than-air gases, such as argon, MPS, propane, and carbon dioxide, may accumulate in pits, tank bottoms, low areas, and near floors. Lighter-than-air gases, such as helium and hydrogen, may accumulate in tank tops, high areas, and near ceilings. The precautions for confined spaces also apply to these areas. If practical, a continuous monitoring system with audible alarms should be used for confined space work.

Oxygen-enriched atmospheres can pose great danger to occupants of confined areas. They are especially hazardous at oxygen concentrations above 25 percent. Materials that burn normally in air may flare up violently in an oxygen-enriched atmosphere. Clothing may burn fiercely; oil or grease soaked clothing or rags may catch fire spontaneously; paper may flare into flame. Very severe and fatal burns can result.

Protection in confined spaces must be provided not only for welders but also for other persons in the enclosure. Only clean, respirable air must be used for ventilation. Oxygen, other gases, or mixtures of gases must never be used for ventilation.

When welding, cutting, or related processes are performed in confined areas immediately dangerous to life and health, positive pressure self-contained breathing apparatus must be used. It must have an emergency air supply of at least five minutes duration in the event that the main source fails.

Welding of Containers

Welding or cutting on the outside or inside of containers or vessels that have held dangerous substances presents special hazards. Flammable or toxic vapors may be present, or may be generated by the applied heat. The immediate area outside and inside the container should be cleared of all obstacles and hazardous materials.[12] When repairing a container in place, care must be taken to prevent entry of hazardous substances released from the floor or the soil beneath the container. The required personal and fire protection equipment must be available, serviceable, and in position for immediate use.

When welding or cutting inside of vessels that have held dangerous materials, the precautions for confined spaces must also be observed.

Gases generated during welding should be discharged in a safe and environmentally acceptable manner. Provisions must be made to prevent pressure buildup inside containers. Testing for gases, fume, and vapors should be conducted periodically to ensure that recommended limits are maintained during welding.

An alternative method of providing safe welding of containers is to fill them with an inert medium such as water, inert gas, or sand. When using water, the level should be kept to within a few inches of the point where the welding is to be done. The space above the water should be vented to allow the heated air to escape. With inert gas, the responsible individual needs to know the percentage of inert gas that must be present in the tank to prevent fire or explosion, and how to safely produce and maintain a safe atmosphere during welding.

Highly Toxic Materials

Certain materials, which are sometimes present in consumables, base metals, coatings, or atmospheres for welding or cutting operations, have permissible exposure limits of 1.0 mg/m^3 or less. Among these materials are the following metals and their compounds:

(1) Antimony
(2) Arsenic
(3) Barium
(4) Beryllium
(5) Cadmium
(6) Chromium
(7) Cobalt

11. Air-supplied respirators or hose masks approved by the U.S. Bureau of Mines or other recognized agency.

12. Refer to *AWS F4.1-80, Recommended Safe Practices for the Preparation for Welding and Cutting Containers and Piping That Have Held Hazardous Substances,* for more complete procedures.

(8) Copper
(9) Lead
(10) Manganese
(11) Mercury
(12) Nickel
(13) Selenium
(14) Silver
(15) Vanadium

One or more of these materials may evolve as fume during welding or cutting. Typical examples are given in Table 10.2.

Table 10.2
Possible toxic materials evolved during welding or thermal cutting

Base or filler metal	Evolved metals or their compounds
Carbon and low alloy steels	Chromium, manganese, vanadium
Stainless steels	Chromium, nickel
Manganese steels and hardfacing materials	Chromium, cobalt, manganese, nickel, vanadium
High copper alloys	Beryllium, chromium, copper, lead, nickel
Coated or plated steel or copper	Cadmium[a], chromium, copper, lead, nickel, silver

a. When cadmium is present in a filler metal, a warning label must be affixed to the container or coil. Refer to *ANSI/ASC Z49.1, Safety in Welding and Cutting.*

Manufacturer's Material Safety Data Sheets should be consulted to determine if any of these materials are present in welding rods or fluxes being supplied. The appropriate Material Safety Data Sheets should be requested from suppliers when there is a possibility that such materials may be present. However, welding rods or fluxes are not the only source of these materials. They may also be present in base metals, coatings, or other sources in the work area. Radioactive materials under Nuclear Regulatory Commission jurisdiction require special considerations.

Persons should not consume food in areas where fumes that contain materials with very low allowable exposure limits may be generated. They should also practice good personal hygiene to prevent ingestion of these contaminants.

When toxic materials are encountered as designated constituents in welding, brazing, or cutting operations, special ventilation precautions must be taken to assure that the levels of these contaminants in the atmosphere are at or below the limits allowed for human exposure. All persons in the immediate vicinity of welding or cutting operations involving these materials must be similarly protected. Unless atmospheric tests under the most adverse conditions establish that exposure is within acceptable concentrations, the following precautions must be observed.

Confined Spaces. Whenever any of these materials are encountered in confined space operations, local exhaust ventilation and respiratory protection must be used.

Indoors. When any of these materials are encountered in indoor operations, local exhaust mechanical ventilation must be used. When beryllium is encountered, respiratory protection in addition to local exhaust ventilation is mandatory.

Outdoors. Whenever any of these materials are encountered in outdoor operations, respiratory protection approved by the Mine Safety and Health Association (MSHA), the National Institute of Occupational Safety and Health (NIOSH), or other approving authority may be required.

Fluorine Compounds

Fumes and gases from fluorine compounds can be dangerous to health, and can burn the eyes and skin on contact. When welding or cutting in confined spaces involves fluxes, coatings, or other material containing fluorine compounds, local mechanical ventilation or respiratory protection must be provided.

When welding or cutting such materials in open spaces, the need for local exhaust ventilation or respiratory protection will depend upon the circumstances. Such protection is not necessary when air samples taken in breathing zones indicate that the fluorides are within allowable limits. However, local exhaust ventilation is desirable for fixed-location production welding and all production welding on stainless steels when filler metals or fluxes containing fluorides are used.

Zinc

Fumes containing zinc compounds may produce symptoms of nausea, dizziness, or fever (sometimes called *metal fume fever*). Welding or cutting where zinc may be present in consumables, base metals, or coatings should be done as described for fluorine compounds.

Cleaning Compounds

Cleaning compounds often require special ventilation precautions because of their possible toxic or flammable properties. Manufacturers' instructions should be followed before welding or cutting is done with cleaned consumables or on cleaned based metal.

Chlorinated Hydrocarbons

Degreasing or cleaning operations involving chlorinated hydrocarbons must be so located that vapors from these operations do not enter the atmosphere surrounding molten weld metal or the welding arc. A reaction product having an objectionable, irritating odor, and containing highly toxic phosgene gas is produced when such vapors enter the atmosphere of arc welding operations. Low levels of exposure can produce feelings of nausea, dizziness, and weakness. Heavy exposures may produce serious health impairment.

Cutting of Stainless Steel

Oxyfuel gas, gas shielded arc, or plasma arc cutting of stainless steel should be done using local mechanical ventilation to remove the fumes generated. Plasma arc cutting may be done under water to capture the fumes.

MEASUREMENT OF EXPOSURE

The American Conference of Governmental Industrial Hygienists (ACGIH) and the U.S. Department of Labor, Occupational Health and Safety Administration (OSHA) have established allowable limits of airborne contaminants. They are called threshold limit values (TLV)[13] or permissible exposure limits (PEL).

The TLV is the concentration of an airborne substance to which most workers may be repeatedly exposed, day after day, without adverse effect. In adapting these to the working environment, a TLV-TWA (Threshold Limit Value-Time Weighted Average) quantity is defined. TLV-TWA is the time-weighted average concentration for a normal 8-hour workday or 40-hour workweek to which nearly all workers may be repeatedly exposed without adverse effect. TLV-TWA values should be used as guides in the control of health hazards, and should not be interpreted as sharp lines between safe and dangerous concentrations.

TLV's are revised annually as necessary. They may or may not correspond to OSHA permissible exposure limits (PEL) for the same materials. In many cases, current ACGIH values for welding materials are more stringent than OSHA levels.

The only way to assure that airborne contaminant levels are within the allowable limits is to take air samples at the breathing zones of the personnel involved. An operator's actual on-the-job exposure to welding fume should be measured following the guidelines provided in ANSI/AWS F1.1-78, *Method for Sampling Airborne Particulates Generated by Welding and Allied Processes.* This document describes how to obtain an accurate breathing zone sample of welding fume for a particular welding operation. Both the amount of fume and the composition of the fume can be determined in a single test using this method. Multiple samples are recommended for increased accuracy. When a helmet is worn, the sample should be collected inside the helmet in the welder's breathing zone.

---

13. TLV is a registered trademark of the American Conference of Governmental Industrial Hygienists.

HANDLING OF COMPRESSED GASES

GAS CYLINDERS AND CONTAINERS

Gases used in the welding and cutting industry are packaged in containers, usually called cylinders or tanks.[14] Only cylinders designed and maintained in accordance with U.S. Department of Transportation (DOT) Specifications may be used in the United States. The use of other cylinders may be extremely dangerous and is illegal. Cylinders requiring periodic retest under DOT regulations may not be filled unless the retest is current.

Filling

Cylinders may be filled only with the permission of the owner. Cylinders should be filled only by recognized gas suppliers or those with the proper training and facilities to do so. Filling one cylinder from another is dangerous and should not be attempted by anyone not qualified to do so. Combustible or incompatible combinations of gases must never be mixed in cylinders.

Usage and Storage

Welding must not be performed on gas cylinders. Cylinders must not be allowed to become part of an electrical circuit because arcing may result. When used in conjunction with arc welding, cylinders must not be grounded. To avoid arcing or interference with valve operation, electrode holders, welding torches, cables, hoses, and tools should not be stored on gas cylinders. Arc-damaged gas cylinders may rupture and result in injury or death.

Cylinders must not be used as work rests or rollers. They should be protected from bumps, falls, falling objects, and the weather. Cylinders should be kept clear of passageways where they might be struck by vehicles and should not be dropped. They should be kept in areas where temperatures do not fall below -20° F or exceed 130° F. Any of these exposures, misuses, or abuses could damage them to the extent that they might fail with serious consequences.

Cylinders must not be hoisted using ordinary slings or chains. A proper cradle or cradle sling that securely retains the cylinder should be used. Electromagnets should not be used to handle cylinders.

Cylinders must always be secured by the user against falling in both use and storage. Acetylene and liquefied gas cylinders should always be stored and used in the upright position. Other cylinders are preferably stored and used in the upright position, but this is not essential in all circumstances.

Before using gas from a cylinder, the contents should be identified by the label thereon. Contents must never be identified by any other means such as cylinder color, banding, or shape. These may vary among manufacturers, geographical area, or product line and could be completely misleading. The label on the cylinder is the only proper notice of the contents of that cylinder. If a label is not on a cylinder, the contents should not be used and the cylinder should be returned to the supplier.

A valve protection cap is provided on many cylinders to protect the safety device and the cylinder valve. This cap should always be in place except when the cylinder is in use. The cylinder should never be lifted manually or by hoist with the valve protection cap. The threads that secure these valve protection caps are intended only for that purpose, and may not be capable of supporting full cylinder weight. The caps should always be threaded completely onto the cylinders and hand tightened.

Gas cylinders and other containers must be stored in accordance with all state and local regulations and the appropriate standards of OSHA and the National Fire Protection Association. Safe handling and storage procedures are discussed in the *Handbook of Compressed Gases* published by the Compressed Gas Association.

Withdrawl of Gas

Many gases in high-pressure cylinders are filled to pressures of 2000 psig or more. Unless

14. For additional information on compressed gases, refer to the Compressed Gas Association, *Handbook of Compressed Gases,* 2nd Ed., New York: Van Nostrand Reinhold, 1981.

the equipment to be used with the gas is designed for use with full cylinder pressure, an approved pressure-reducing regulator must be used to withdraw gas from a cylinder or manifold. Simple needle valves should never be used. A pressure-relief or safety valve, rated to function at less than the maximum allowable pressure of the welding equipment, should also be employed. Its function is to prevent failure of the equipment at pressures in excess of working limits if the regulator should fail in service.

Valves on cylinders containing high pressure gas, particularly oxygen, should always be opened slowly to avoid the high temperature of adiabatic recompression, which can occur if the valves are opened rapidly. In the case of oxygen, the intense heat can ignite the valve seat which, in turn, may cause the metal to melt or burn. The cylinder valve outlet should point away from the operator and other persons when opening the valve to avoid injury should a fire occur.

Prior to connecting a gas cylinder to a pressure regulator or a manifold, the valve outlet should be cleaned of dirt, moisture, and other foreign matter by first wiping it with a clean, oil-free cloth. Then, the valve should be opened momentarily and closed immediately. This is known as *cracking* the cylinder valve. Fuel gas cylinders must never be cracked near sources of ignition (i.e., sparks and flames), while smoking, or in confined spaces.

A regulator should be drained of gas pressure prior to connecting it to a cylinder and also after closing the cylinder valve upon shutdown of operation. The outlet threads on cylinder valves are standardized for specific gases so that only regulators or manifolds with similar threads can be attached.[15] Standard valve thread connections for gases normally used for welding, brazing, and allied processes are given in Table 10.3.

It is preferable not to open valves on low pressure fuel gas cylinders more than one turn. This usually provides adequate flow and permits rapid closure of the cylinder valves in an emergency. High pressure cylinder valves, on the other hand, usually must be opened fully to backseat the packing and prevent packing leaks during use.

Table 10.3
Compressed Gas Association standard and alternate valve thread connections for compressed gas cylinders

Gas	Connection number
Acetylene	510 or 300
Argon	580
Butane	510
Carbon dioxide (CO_2)	320
Helium	580
Hydrogen	350
Methylacetylene-propadiene (MPS)	510
Nitrogen	580 or 555
Oxygen	540
Propane	510
Propylene	510

The cylinder valve should be closed after each use of a cylinder and when an empty cylinder is to be returned to the supplier. This prevents loss of product through leaks that might develop and go undetected while the cylinder is unattended, and also avoids hazards that might be caused by leaks. It also prevents backflow of contaminants into the cylinder. It is advisable to return cylinders with about 25 psi of contents remaining. This prevents possible contamination by the atmosphere during shipment.

Pressure Relief Devices

Pressure relief devices on cylinders should not be tampered with. These are intended to provide protection in the event the cylinder is subjected to a hostile environment, usually fire or other source of heat. Such environments may raise the pressure within cylinders. To prevent

15. Refer to *ANSI/CGA Publication V-1, Compressed Gas Cylinder Valve Outlet and Inlet Connections,* Compressed Gas Association.

cylinder pressures from exceeding safe limits, the safety devices are designed to relieve the contents.

CRYOGENIC CYLINDERS AND TANKS

Cryogenic cylinders and tanks are used to store at very low temperatures those liquids that change to gases at normal conditions of temperature and pressure. So-called cryogenic liquids for commercial purposes usually include oxygen, nitrogen, and argon, although other gases may be handled as cryogenic liquids.

Cyrogenic cylinders and tanks for storing liquids are usually double-walled vessels that are evacuated and insulated between the walls. They are designed to contain liquids at very cold temperatures with a minimum of heat gain. Heat gain from the atmosphere causes evaporation of the product. Liquid containers hold a greater amount of product for a given volume than high pressure gas cylinders. For safety, liquid containers must be handled carefully. They must always be maintained in an upright position. Whenever they are moved, a cylinder handling truck designed for this purpose must be used. They should not be rolled on a bottom edge, as is often done with high pressure cylinders.

Failure to properly handle liquid cylinders can result in rupture of either the inner or outer cylinder wall with consequent loss of vacuum and rapid rise of internal pressure. This will result in activation of the cylinder's protective devices and loss of the cylinder contents. The cylinder protection devices must never be tampered with nor deactivated. Overpressurization could result in explosive failure.

Damage to the internal walls or fittings of a liquid container is often evidenced by visible frosting on the exterior of the container. Whenever frosting appears on the exterior of a liquid container, personnel should be kept clear and the gas supplier notified. In general, people should stay clear of the container until the frost disappears, usually indicating that the contents have dissipated and any dangerous internal pressure has been relieved.

Although the contents of liquid containers are predominantly in liquid form at a very cold temperature, the product withdrawn from these containers should be a gas at room temperature. The conversion takes place within a vaporizer system that evaporates the liquid and warms the gas to atmospheric temperature.

In some cases, the user may want to withdraw liquid from the container. When liquid is withdrawn, protective clothing should be worn to prevent bodily contact with the cold product. Loose-fitting insulated gloves and an adequate face shield are essential. Contact of liquid with the skin will cause burns similar to heat burns, and prolonged contact will cause severe frostbite. Injuries of this nature should be treated in the same manner as injuries from exposure to low temperatures.

The properties of many materials are drastically different at the low temperatures of cryogenic liquids, compared to their normal properties. Many metals, including carbon steel, and most elastomers, such as rubber, become extremely brittle. When cryogenic liquids are to be withdrawn from cylinders, all materials in the transfer line must have satisfactory properties at the low temperatures.

Liquid oxygen may react with explosive violence on contact with asphalt or similar bituminous materials. Therefore, liquid oxygen tanks must never be mounted on such surfaces, and liquid oxygen must not be allowed to contact them. Liquid oxygen tanks must always be installed on concrete pads.

REGULATORS

Pressure reducing regulators[16] must be used only for the gas and pressure given on the label. They should not be used with other gases or at other pressures even though the cylinder valve outlet threads may be the same. The threaded connections to the regulator must not be forced. Improper fit of threads indicates an improper regulator or hose for the gas.

Use of adaptors to change the cylinder connection thread is not recommended because of the danger of using an incorrect regulator or

---

16. Gas regulators should meet the requirements of the Compressed Gas Association Publication *E-4, Standard for Gas Regulators for Welding and Cutting,* and other code regulations.

of contaminating the regulator. For example, gases that are oil-contaminated can deposit an oily film on the internal parts of the regulator. This film can contaminate oil-free gas or result in fire or explosion in the case of oxygen.

The threads and connection glands of regulators should be inspected before use for dirt or damage. If a hose or cylinder connection leaks, it should not be forced with excessive torque. Damaged regulators and components should be repaired by properly trained mechanics or returned to the manufacturer for repair.

A suitable valve or flowmeter should be used to control gas flow from a regulator. The internal pressure in a regulator should be drained before it is connected to or removed from a gas cylinder or manifold.

MANIFOLDS

A manifold is used when a gas is needed without interruption or at a high delivery rate, or both, for one or more operations in a plant. A manifold system must be suitable for the specific gas and operating pressure, and be leak tight. The components should be approved for such purpose, and used only for the gas and pressure for which they are approved. Oxygen and fuel gas manifolds must meet specific design and safety requirements.[17]

Piping and fittings for acetylene and methylacetylene-propadiene (MPS) manifolds must not be unalloyed copper. These fuel gases react with copper under certain conditions to form unstable copper acetylide. This compound may detonate under shock or heat. However, copper alloys containing less than 70 percent copper may be used in systems for these fuel gases.

Piping systems must contain an appropriate over-pressure relief valve. Each fuel gas cylinder lead should incorporate a backflow check valve and flash arrester. Backflow check valves must also be installed in each line at each station outlet where both fuel gas and oxygen are provided for a welding, cutting, or preheating torch.

17. Refer to *ANSI/NFPA 51, Oxygen-Fuel Gas Systems for Welding, Cutting, and Allied Processes,* National Fire Protection Association, latest edition, for information on manifold and piping systems.

OXYGEN

Oxygen is nonflammable but it supports the combustion of flammable materials. It can initiate combustion, and vigorously accelerates it. Therefore, oxygen cylinders and liquid containers should not be stored in the vicinity of combustibles or with cylinders of fuel gas. Oxygen should never be used as a substitute for compressed air. Oxygen should always be referred to by name, never as *air*.

Oil, grease, and combustible dusts may spontaneously ignite on contact with oxygen. Hence, all systems and apparatus for oxygen service must be kept free of these materials. Valves, piping, or system components that have not been expressly manufactured for oxygen service must be cleaned and approved for this service before use.[18]

Apparatus that has been manufactured expressly for oxygen service, and is usually so labeled, must be kept in the clean condition as originally received. Oxygen valves, regulators, and apparatus must never be lubricated with oil. If lubrication is required, it must be done only with lubricants that are compatible with oxygen.

Oxygen must never be used to power compressed air tools. These are almost always oil lubricated. Similarly, oxygen must not be used to blow dirt from work and clothing because they are often contaminated with oil or grease.

Only clean clothing should be worn when working with oxygen systems. Similarly, oxygen must not be used to ventilate confined spaces. Severe burns may result from ignition of clothing or the hair in an oxygen-rich atmosphere.

FUEL GASES

Fuel gases commonly used in oxyfuel gas welding (OFW) and cutting (OFC) are acetylene, methylacetylene-propadiene (MPS), natural gas, propane, and propylene. Hydrogen is used in a few applications. These gases should always be referred to by name. Gasoline is sometimes used as fuel for oxygen cutting. It vaporizes in the torch.

18. Refer to Compressed Gas Association Publication *G4.1, Cleaning Equipment for Oxygen Service.*

Acetylene in cylinders is dissolved in a solvent so that it can be safely stored under pressure. In the free state, acetylene should never be used at pressures over 15 psig (30 psia). Acetylene at pressures above 15 psig may dissociate with explosive violence.

Acetylene and methylacetylene-propadiene (MPS) should never be used in contact with silver or mercury, or as stated previously, copper. These gases react with the former two metals in the same manner as with copper to form unstable compounds that may detonate under shock or heat. As mentioned before, the valves on fuel gas cylinders should never be cracked to clean the outlet near possible sources of ignition or in confined spaces.

When used for a brazing furnace atmosphere, fuel gases must be burned or vented to a safe location. Prior to filling a furnace or retort with fuel gas, the equipment must first be purged with a nonflammable gas, such as nitrogen or argon, to prevent formation of an explosive air-fuel mixture.

Special attention must be given when using hydrogen. Flames of hydrogen may be difficult to see and result in serious burns or fires.

FUEL GAS FIRES

The best procedures for avoiding fire from a fuel, gas or liquid, is to keep it contained within the system, that is, prevent leaks. All fuel systems should be checked carefully for leaks upon assembly and at frequent intervals thereafter. Fuel gas cylinder should be examined for leaks, especially at fuse plugs, safety devices, and valve packings. One common source of fire in welding and cutting is ignition of leaking fuel by flying sparks or spatter.

In the event of a fuel fire, an effective means for controlling the fire is to shut off the fuel valve, if accessible. Fuel gas valves should not be opened beyond the point necessary to provide adequate flow and yet permit quick shutoff in an emergency. In most cases, this is less than one turn.

Most fuel gases in cylinders are in liquid form or dissolved in liquids. Therefore, the cylinders should always be used in the upright position to prevent liquid surges into the system.

A fuel gas cylinder can develop a leak and sometimes result in a fire. In case of fire, the fire alarm should be sounded, and trained fire personnel should be summoned immediately. A small fire in the vicinity of a cylinder valve or a safety device should be extinguished, if possible, by closing the valve or by the use of water, wet cloths, or fire extinguishers. If the leak cannot be stopped, the cylinder should be removed by trained fire personnel to a safe outdoor location, and the supplier notified. A warning sign should be posted, and no smoking or other ignition sources should be permitted in the area.

In the case of a large fire at a fuel gas cylinder, the fire alarm should be actuated, and all personnel should be evacuated from the area. The cylinder should be kept wet by fire personnel with a heavy stream of water to keep it cool. It is usually better to allow the fire to continue to burn and consume the issuing gas rather than attempt to extinguish the flame. If the fire is extinguished, there is danger that the escaping gas may reignite with explosive violence.

SHIELDING GASES

Argon, helium, carbon dioxide (CO_2), and nitrogen are used as shielding gases for welding. All, except carbon dioxide, are used as brazing atmospheres. They are odorless and colorless, and can displace air needed for breathing.

Confined spaces filled with these gases must be well ventilated before personnel enter them. If there is any question, either the space should be checked for adequate air concentration with an oxygen analyzer, or an air-supplied respirator should be worn by anyone entering the space. Containers of these gases should not be placed in confined spaces.

ELECTRICAL SAFETY

ELECTRIC SHOCK

Electric shock can cause sudden death. Injuries and fatalities from electric shock in welding and cutting operations can occur if proper precautionary measures are not followed. Most welding and cutting operations employ some type of electrical equipment. For example, automatic oxyfuel gas cutting machines use electric motor drives, controls, and systems.

Some electrical accidents may not be avoidable, such as those caused by nature. However, the majority are avoidable, including those caused by lack of proper training.

Shock Mechanism

Electric shock occurs when an electric current of sufficient magnitude to create an adverse effect passes through the body. The severity of the shock depends mainly on the amount of current, the duration of flow, the path of flow, and the state of health of the person. The current is caused to flow by the applied voltage. The amount of current depends upon the applied voltage and the resistance of the body path. Frequency of the current may also be a factor when alternating current is involved.

Shock currents greater than about 6 milliamperes (mA) are considered primary because they are capable of causing direct physiological harm. Steady state currents between 0.5 and 6 mA are considered secondary shock currents. Secondary shock currents are defined as those capable of causing involuntary muscular reactions without normally causing direct physiological harm. The 0.5 mA level is called the perception threshold because it is the point at which most people just begin to feel the tingle from the current. The level of current sensation varies with the weight of the individual and to some extent between men and women.

Shock Sources

Most electrical equipment can be a shock hazard if improperly used or maintained. Shock can occur from lightning-induced voltage surges in power distribution systems. Even earth grounds can attain high potential relative to true ground during severe transient phenomenon. Such circumstances, however, are rare.

In welding and cutting work, most electrical equipment is powered from ac sources of between 115 and 575 volts, or by engine-driven generators. Most welding is done with less than 100 arc volts. (Fatalities have resulted with equipment operating at less than 80 volts.) Some arc cutting methods operate at over 400 V, and electron beam welding machines at up to about 150 kV. Most electric shock in the welding industry occurs as the result of accidental contact with bare or poorly insulated conductors operating at such voltages. Therefore, welders must take precautions against contacting bare elements in the welding circuit, as well as those in the primary circuits.

Electrical resistance is usually reduced in the presence of water or moisture. Electrical hazards are often more severe under such circumstances. Welders must take special precautions to avoid shock when working under damp or wet conditions, including perspiration.

EQUIPMENT SELECTION

Electric shock hazards are minimized by proper equipment installation and maintenance, good operator practices, proper operator clothing and body protection, and the use of equipment designed for the job and situation. Equipment should meet applicable NEMA or ANSI standards, such as *ANSI/UL 551, Safety Standard for Transformer Type Arc Welding Machines,* latest edition.

If a significant amount of welding and cutting work is to be done under electrically hazardous conditions, automatic machine controls that reduce the no-load (open circuit) voltage to a safe level are recommended. When special welding and cutting processes require open-circuit voltages higher than those specified in *ANSI/NEMA Publication EW-1, Electrical Arc Welding Apparatus,* insulation and operating procedures that are adequate to protect the welder from these higher voltages must be provided.

PERSONNEL TRAINING

A good safety training program is essential. Operators must be fully instructed in electrical safety by a competent person before being allowed to commence operations. As a minimum, this training should include the points covered in *ANSI/ASC Z49.1, Safety in Welding and Cutting.* Persons should not be allowed to operate electrical equipment until they have been properly trained.

INSTALLATION

Equipment should be installed in a clean, dry area. When this is not possible, it should be adequately guarded from dirt and moisture. Installation must be done to the requirements of *ANSI/NFPA 70, National Electric Code,* and local codes. This includes necessary disconnects, fusing, and type of incoming power lines.

Terminals for welding leads and power cables must be shielded from accidental contact by personnel or by metal objects, such as vehicles and crane hooks. Connections between welding leads and power supplies may be guarded using (1) dead front construction and receptacles for plug connections, (2) terminals located in a recessed opening or under a non-removable hinged cover, (3) insulating sleeves, or (4) other equivalent mechanical means.

GROUNDING

The workpiece being welded and the frame or chassis of all electrically powered machines must be connected to a good electrical ground. Grounding can be done by locating the workpiece or machine on a grounded metal floor or platen, or by connecting it to a properly grounded building frame or other satisfactory ground. Chains, wire ropes, cranes, hoists, and elevators must not be used as grounding connectors nor to carry welding current.

The work lead is not the grounding lead. The former connects the work terminal on the power source to the workpiece. A separate lead is required to ground the workpiece or power source work terminal.

Care should be taken to avoid double grounding. Otherwise, the welding current may flow through a connection intended only for safety ground, and may be of higher magnitude than the grounding conductor can safely carry. Special radio-frequency grounding may be necessary for arc welding machines equipped with high-frequency arc initiating devices.[19]

Connections for portable control devices, such as push buttons to be carried by the operator, must not be connected to circuits with operating voltages above about 120 V. Exposed metal parts of portable control devices operating on circuits above 50 V must be grounded by a grounding conductor in the control cable. Controls using intrinsically safe voltages below 30 V are recommended.

CABLES AND CONNECTIONS

Electrical connections must be tight, and checked periodically for tightness. Magnetic work clamps must be free of adherent metal particles and spatter on contact surfaces. Coiled welding leads should be spread out before use to avoid overheating and damage to the insulation. Jobs alternately requiring long and short leads should be equipped with insulated cable connectors so that idle lengths can be disconnected when not needed.

Equipment, cables, fuses, plugs, and receptacles must be used within their current carrying and duty cycle capacities. Operation of apparatus above its current rating or duty cycle leads to overheating and rapid deterioration of insulation and other parts. Actual welding current may be higher than that shown by indicators on the welding machine if welding is done with short leads or low voltage, or both. High currents are likely with general purpose welding machines when they are used with processes that use low arc voltage, such as gas tungsten arc welding.

Welding leads should be the flexible type of cable designed especially for the rigors of welding service. Insulation on cables used with high voltages or high-frequency oscillators must provide adequate protection. Cable insulation must be

---

19. See *Recommended Installation and Test Procedures for High Frequency Stabilized Arc Welders,* National Electrical Manufacturers Association, 1980.

kept in good condition, and cables repaired or replaced promptly when necessary.

OPERATIONS

Welders should not allow the metal parts of electrodes, electrode holders, or torches to touch their bare skin or any wet covering of the body. Dry gloves in good condition must always be worn. The insulation on electrode holders must be kept in good repair. Electrode holders must not be cooled by immersion in water. If water-cooled guns or holders are used, they should be free of water leaks and condensation that would adversely affect the welder's safety. With gas tungsten arc welding, the power to the electrode holder should always be de-energized while the electrode is being changed. Welders should not drape or coil the welding leads around their bodies.

When welding or arc cutting is to be done under damp or wet conditions including heavy perspiration, the welder must wear dry gloves and clothing in good condition to prevent electric shock. The welder should be protected from electrically conductive surfaces, including the earth, by rubber-soled shoes as a minimum, and preferably by an insulating layer such as a rubber mat or wooden board. Similar precautions against accidental contact with bare conducting surfaces must be taken when the welder is required to work in a cramped kneeling, sitting, or lying position. Rings and jewelry should be removed before welding to decrease the possibility of electric shock.

A welding circuit must be de-energized to avoid electric shock while the electrode, torch, or gun is being changed or adjusted. One exception concerns shielded metal arc welding electrodes. When the circuit is energized, electrodes must be changed with dry welding gloves, not with bare hands. In any case, de-energization of the circuit is desirable for optimum safety.

When the welder has completed the work or has occasion to leave the work station for an appreciable time, the welding circuit should be de-energized by turning off the welding machine. Similarly, when the machine is to be moved, the input power supply should be electrically disconnected. When equipment is not in use, exposed electrodes should be removed from the holders to eliminate the danger of accidental electrical contact with persons or conducting objects. Also, welding guns of semiautomatic welding equipment should be placed so that the gun switch cannot be operated accidentally.

MULTIPLE ARC WELDING OPERATIONS

There can be increased danger of electrical shock when several welders are welding on a large metal structure, such as a building frame or ship, and are using the structure in the return welding circuit. Proper electrical contact must exist at all joints in the structure. Sparking or heating at any point makes the structure unsuitable as a return circuit.

Where two or more welders are working on the same structure and one is likely to touch simultaneously the exposed parts of more than one electrode holder, the welding machines must be connected to minimize shock hazard. It is preferable that all dc welding machines be connected with the same polarity. A test lamp or voltmeter can be used to determine whether the polarities are matched. It is preferable to connect all single-phase ac welding machines to the same phase of the supply circuit with the same instantaneous polarity.

In some cases, the preferable connections may not be possible. Welding may require the use of both dc polarities, or supply circuit limitations may require distribution of ac welding machines among the phases of the supply circuit. In such cases, no-load voltage between electrode holders or welding guns may be twice the normal voltage. The welders and other area personnel must be instructed to avoid simultaneous contact with more than one electrode holder, welding gun, or installed electrode.

MODIFICATION AND MAINTENANCE

Commutators on rotating welding machines should be kept clean to prevent excessive arcing. Rectifier type welding machines should be inspected frequently for accumulations of dust or lint that interfere with ventilation. Electrical coil ventilating ducts should be similarly inspected. It is good practice to occasionally blow out the welding machine with clean, dry compressed air at low pressure using adequate safety precautions. Air filters in the ventilating systems of electrical components are not recommended unless provided by the manufacturer of the welding machine. The reduction of air flow resulting from the use of an air filter on equipment not so designed can subject internal components to an overheating condition and subsequent failure. Machines that have become wet should be thoroughly dried and properly retested before being used.

PREVENTION OF FIRES

Fires resulting from electric welding equipment are generally caused by overheating of electrical components, flying sparks or spatter from the welding or cutting operation, or the mishandling of fuel in engine-driven equipment. Most precautions against electrical shock are also applicable to the prevention of fires caused by overheating of equipment. Avoidance of fire from sparks and spatter was covered previously.

The fuel systems of engine driven-equipment must be in good condition. Leaks must be repaired promptly. Engine-driven machines must be turned off before refueling, and any fuel spills should be wiped up before the engine is restarted. Otherwise, the ignition system, electrical controls, or engine heat may start a fire.

PROCESSES

OXYFUEL GAS WELDING AND CUTTING

Torches

Only approved welding and cutting torches should be used.[20] They should be kept in good working order, and serviced at regular intervals by the manufacturer or qualified technicians. A torch must be used only with the fuel gas for which it is designed. The fuel gas and oxygen pressures should be those recommended by the torch manufacturer.

The procedures recommended by the manufacturer should be follwed when lighting and extinguishing the torch. The torch should be lighted only with a friction lighter, pilot light, or similar ignition source. Matches or cigarette lighters should not be used.

Hoses

Hoses used should be only those specified for oxyfuel gas welding and cutting systems. Generally, these hoses are manufactured in accordance with *Specification IP-7 for Rubber Welding Hose,* Compressed Gas Association and Rubber Manufacturers Association. Fuel gas hose is usually red with left-hand threaded fittings; green hose with right-hand threaded fittings is generally used for oxygen.[21] Hoses should be free of oil and grease, and in good condition. When parallel lengths are taped together for convenience, no more than 4 in. of any 12 in. section of hose should be covered.

Only proper ferrules and clamps should be used to secure hose to fittings. Long runs of hose should be avoided. Excess hose should be coiled to prevent kinks and tangles, but it should not be

20. Oxyfuel gas torches should meet the requirements of Compressed Gas Association *Publication E-5, Torch Standard for Welding and Cutting,* and appropriate government regulations.

21. Hose connections must comply with Compressed Gas Association *Publication E-1, Standard Connections for Regulators, Torches and Fitted Hose for Welding and Cutting Equipment.*

wrapped around cylinders or cylinder carts while in use.

Backfire and Flashback

A backfire during welding or cutting is a momentary flame outage and reignition of the torch flame, accompanied by a pop or bang depending upon the size of the tip. There is a momentary retrogression of the flame back into the tip and torch. In the case of severe backfire, the flame may burn back into the hoses. Occasionally, backfire flames burn through the hose (especially oxygen hose) and can cause injury.

A flashback is an occurrence initiated by a backfire where the flame continues to burn inside the equipment instead of being re-established at tip. Flashbacks result in very rapid internal heating of the equipment, and can quickly destroy it. The flashback is usually recognized by a whistling or squealing sound. The equipment will heat up rapidly and sparks may be issuing from the tip. The flashback should be extinguished by turning off the torch valves as quickly as possible. Different manufacturers may recommend shutting off either the fuel or oxygen first, but the most important concern is to get both valves closed quickly.

Backfires and flashbacks should not ordinarily be a concern if the apparatus is operated in accordance with the manufacturer's instructions. Generally, they occur from allowing a tip to become overheated, forcing the tip into the work, or providing insufficient gas flow for the size of the tip in use. If frequent backfiring or flashbacks are experienced, the cause should be investigated. There is probably something wrong with the equipment.

To prevent backfires or flashbacks, hose lines should be purged before lighting the oxyfuel gas torch. Purging flushes out any combustible oxygen-fuel gas or air-fuel gas mixtures in the hoses. It is done by first opening either the fuel or oxygen valve on the torch and allowing that gas to flow for several seconds to clear that hose of any possible gas mixtures. That valve is then closed, and the other valve is opened to allow the other gas to flow for a similar period of time. Purging should always be done before any welding or cutting tip is lighted. The purge stream must not be directed towards any flame or source of ignition. Torches should not be purged in confined spaces because of possible explosion of accumulated gases.

Hose Line Safety Devices

Reverse-flow check valves and flashback arresters for hose line service are available. These devices can prevent backflow of gases and flashbacks into hoses provided they are operating properly. They must be used strictly in accordance with the manufacturer's instructions, and maintained regularly in accordance with the manufacturer's recommendations.

Shutdown Procedures

When oxyfuel gas operations are completed, the equipment should always be completely shut down and the gas presssures drained from the system. Cylinder or supply valves must be closed. The equipment should not be left unattended until the shutdown has been completed. Oxyfuel gas cylinders and supplies must always be stored in well ventilated spaces, and never in confined areas and unventilated cabinets. Even small gas leaks in confined spaces can result in development of explosive mixtures that might be set off with disastrous results. For the same reason, oxyfuel gas cylinders should never be transported in enclosed vehicles, particularly not in closed vans or the trunks of private automobiles.

ARC WELDING AND CUTTING

The potential hazards of arc welding are fumes and gases, electric shock, infrared and ultraviolet radiation, burns, fire, explosion, and sometimes, noise. These hazards have been described previously in this chapter. Precautions given to avoid injury or death with these processes should be followed.

Precautionary labels on all equipment and materials should be read, understood, and followed. The manufacturer should be consulted when questions arise concerning safe use of the product or equipment.

The safety precautions for arc welding also apply to arc cutting. In addition, noise during arc cutting operations can be high; prolonged exposure could lead to hearing damage. Where necessessary, ear protection should be provided for the operator and others in the area.

Two common accessories are available for mechanized plasma arc cutting machines to aid in fume and noise control. One is a water table. It is simply a cutting table filled with water to the bottom surface of the plate or above the plate. In the latter case, cutting is done under water using a special torch, to minimize noise and reduce radiation. The high-speed gases emerging from the plasma jet produce turbulence in the water. This action traps almost all of the fume particles in the water.

The second accessory is a water muffler to reduce noise. The muffler is a nozzle, attached to a special torch body, that produces a curtain of water around the front of the torch. It is always used in conjunction with a water table. The combination of a water curtain at the top of the plate and a water table contacting the plate bottom encloses the arc in a noise-reducing shield. The noise output is reduced by roughly 20 dB. This equipment should not be confused with cutting variations using water injection or water shielding.

RESISTANCE WELDING

The main hazards that may arise with resistance welding processes and equipment are as follows:

(1) Electric shock from contact with high voltage terminals or components

(2) Ejection of small particles of molten metal from the weld

(3) Crushing of some part of the body between the electrodes or other moving components of the machine

Mechanical Considerations

Guarding. Initiating devices on resistance welding equipment, such as push buttons and switches, should be positioned or guarded to prevent the operator from inadvertently activating them.

With some machines, the operator's hands can be expected to pass under the point of operation during loading and unloading. These machines should be effectively guarded by proximity-sensing devices, latches, blocks, barriers, dual hand controls, or similar accessories that prevent (1) the hands from passing under the point of operation or (2) the ram from moving while the hands are under the point of operation.

Static Safety Devices. Press, flash, and upset welding machines should have static safety devices, such as pins, blocks, or latches, to prevent movement of the platen or head during maintenance or setup for welding.

Portable Welding Machines. The support system of suspended portable welding gun equipment, with the exception of the gun assembly, should be capable of withstanding the total mechanical shock load in the event of failure of any component of the system. Devices, such as cables, chains, or clamps, are considered satisfactory.

Guarding should be provided around the mounting and actuating mechanism of the movable arm of a welding gun if it can cause injury to the operator's hands. If suitable guarding cannot be achieved, the gun should have two handles and operating switches that must be actuated to energize the machine.

Stop Buttons. One or more emergency stop buttons should be provided on all welding machines, with a minimum of one at each operator position.

Guards. Eye protection against expelled metal particles must be provided by a guard of suitable fire-resistant material or by the use of approved personal protective eye wear. For flash welding equipment, flash guards of suitable fire-resistant material must be provided to control flying sparks and molten metal.

Electrical Considerations

All external weld initiating control circuits should operate at a maximum of about 120 V for stationary equipment and about 36 V for portable equipment.

Resistance welding equipment containing high voltage capacitors must have adequate electrical insulation and be completely enclosed. All enclosure doors must be provided with suitable interlocks that are wired into the control circuit. The interlocks must effectively interrupt power and discharge all high voltage capacitors when the door or panel is open. In addition, a manually operated switch or suitable positive device should be provided to assure complete discharge of all high voltage capacitors. The doors or panels must be kept locked except during maintenance.

Grounding. The welding transformer secondary should be grounded by one of the following methods:

(1) Permanent grounding of the welding secondary circuit

(2) Connection of a grounding reactor across the secondary winding with a reactor tap to ground

As an alternative on stationary machines, an isolation contactor may be used to open all of the primary lines.

The grounding of one side of the secondary windings on multiple spot welding machines can cause undesirable transient currents to flow between their transformers when they are either connected to different primary phases or have different secondary voltages, or both. A similar condition can also exist with portable spot welding guns when several units are used to weld the same assembly or another one that is nearby. Such situations require use of a grounding reactor or isolation contactor.

Installation

All equipment should be installed in conformance with the *ANSI/NFPA 70, National Electric Code* (latest edition). The equipment should be installed by qualified personnel under the direction of a competent technical supervisor. Prior to production use, the equipment should be inspected by competent safety personnel to ensure that it is safe to operate.

BRAZING AND SOLDERING

Hazards encountered with brazing and soldering operations are similar to those associated with welding and cutting processes. Brazing and soldering operations may require temperatures where some elements in the filler metal vaporize. Personnel and property must be protected against hot materials, gases, fumes, electrical shock, radiation, and chemicals.

It is essential that adequate ventilation be provided so that personnel do not inhale gases and fumes generated during brazing or soldering. Some filler metals and base metals contain toxic materials such as cadmium, beryllium, zinc, mercury, or lead that vaporize during brazing. Fluxes contain chemical compounds of fluorine, chlorine, and boron that are harmful if they are inhaled or contact the eyes or skin. Ventilation to avoid these hazards was described previously.

Brazing Atmospheres

Flammable gases are used as atmospheres for some furnace brazing operations. These include combusted fuel gas, hydrogen, and dissociated ammonia. Prior to introducing the atmosphere, the furnace or retort must be purged of air by safe procedures recommended by the manufacturer.

In addition, adequate area ventilation must be provided to exhaust and discharge to a safe place all explosive or toxic gases that may emanate from furnace purging and brazing operations. Local environmental regulations should be consulted when designing the exhaust system.

Dip Brazing and Soldering

In dip brazing and soldering, the parts to be immersed in the bath must be completely dry. The presence of moisture on the parts will cause an instantaneous generation of steam that may expel the contents of the dip pot with explosive force and create a serious burn hazard. Predrying of the parts will prevent this problem. If supplementary flux is necessary, it must be adequately dried to remove not only moisture but also water of hydration to avoid explosion hazards.

Solder Flux

Some fluxes, such as the rosin, petrolatum, and reaction types, give off considerable smoke depending on the soldering temperature and the duration of heating. The American Conference of Governmental Industrial Hygienists has currently established a threshold limit valve for pyrolysis products of rosin core solder of 0.1 mg/m^3 aliphatic aldehydes, measured as formaldehyde.

Other fluxes give off fumes that are harmful if breathed in any but small quantities. Prolonged inhalation of halides and some of the newer organic fluxes should be avoided. The aniline type fluxes and some of the amines also evolve fumes that are harmful, and can cause dermatitis. Fluorine in flux can be dangerous to health, cause burns, and be fatal if ingested.

HIGH FREQUENCY WELDING

High frequency generators are electrical devices and require all usual safety precautions in handling and repairing such equipment. Voltages are in the range from 400 to 20,000 V and are lethal. These voltages may be either low or high frequency. Proper care and safety precautions should be taken while working on high frequency generators and their control systems. Units must be equipped with safety interlocks on access doors and automatic safety grounding devices to prevent operation of the equipment when access doors are open. The equipment should not be operated with panels or high voltage covers removed or with interlocks and grounding devices blocked.

The output high-frequency primary leads should be encased in metal ducting and should not be operated in the open. Induction coils and contact systems should always be properly grounded for operator protection. High frequency currents are more difficult to ground than low frequency currents, and grounding lines must be kept short and direct to minimize inductive impedance. The magnetic field from the output system must not induce heat in adjacent metallic sections and cause fires or burns.

Injuries from high frequency power, especially at the upper range of welding frequencies, tend to produce severe local surface tissue damage. However, they are not likely to be fatal because current flow is shallow.

ELECTRON BEAM WELDING

The primary hazards associated with electron beam welding equipment are electric shock, x-radiation, fumes and gases, and damaging visible radiation. Precautionary measures must be taken at all times to assure that proper protective procedures are always observed. *AWS F2.1, Recommended Safe Practices for Electron Beam Welding and Cutting* and *ANSI/ASC Z49.1, Safety in Welding and Cutting* (latest editions) give the general safety requirements that should be strictly adhered to at all times.

Electric Shock

Every electron beam welding machine operates at voltages above 20 kV. These voltages can cause fatal injury, regardless of whether the machine is referred to as a *low voltage* or a *high voltage machine.* The manufacturers of electron beam welding equipment, in meeting various regulatory requirements, produce machines that are well-insulated against high voltage. However, precautions should be exercised with all systems when high voltage is present. The manufacturer's instructions should be followed for operation and maintenance of the equipment.

X-radiation

The X-rays generated by an electron beam welding machine are produced when electrons, traveling at high velocity, collide with matter. The majority of X-rays are produced when the electron beam impinges upon the workpiece. Substantial amounts are also produced when the beam strikes gas molecules or metal vapor in the gun column and work chamber. Underwriters Laboratories and OSHA regulations have established firm rules for permissible x-radiation exposure levels, and producers and users of equipment must observe these rules.

Generally, the steel walls of the chamber are adequate protection in systems up to 60 kV, assuming proper design. High-voltage machines utilize lead lining to block x-ray emission beyond the chamber walls. Leaded glass windows are employed in both high and low voltage electron beam systems. Generally, the shielded vacuum chamber walls provide adequate protection for the operator.

In the case of nonvacuum systems, a radiation enclosure must be provided to assure the safety of the operator and other persons in the area. Thick walls of high-density concrete or other similar material may be employed in place of lead, especially for large radiation enclosures on nonvacuum installations. In addition, special safety precautions should be imposed to prevent personnel from accidentally entering or being trapped inside the enclosure when equipment is in operation.

A complete x-ray radiation survey of the electron beam equipment should always be made at the time of installation and at regular intervals thereafter. This should be done by personnel trained to make a proper radiation survey to assure initial and continued compliance with all radiation regulations and standards applicable to the site where the equipment is installed.

Fumes and Gases

It is unlikely that the very small amount of air in a high vacuum electron beam chamber would be sufficient to produce ozone and oxides of nitrogen in harmful concentrations. However, nonvacuum and medium vacuum electron beam systems are capable of producing these by-products, as well as other types of airborne contaminants, in concentrations above acceptable levels.

Adequate area ventilation must be employed to reduce concentrations of any airborne contaminants around the equipment below permissible exposure limits, and proper exhausting techniques should be employed to maintain residual concentrations in the enclosure below those limits.

Visible Radiation

Direct viewing of intense radiation emitted by molten weld metal can be harmful to eyesight. Viewing of the welding operation should be done through a filter lens commonly used for arc welding.

LASER BEAM WELDING AND CUTTING

The basic hazards associated with laser operation are:

(1) Eye damage from the beam including burns of the cornea or retina, or both

(2) Skin burns from the beam

(3) Respiratory system damage from hazardous materials evolved during operation

(4) Electrical shock

(5) Chemical hazards

(6) Contact with cryogenic coolants

Laser manufacturers are required to qualify their equipment with the U.S. Bureau of Radiological Health (BRH). Electrical components should be in compliance with NEMA standards. User action is governed by OSHA requirements. In all cases, *American National Standard Z136.1, Safe Use of Lasers* (latest edition), should be followed.

Eye Protection

Eye injury is readily caused by laser beams. With laser beams operating at visible or near infrared wavelengths, even a five milliwatt beam can inflict retinal damage. Safety glasses are available that are substantially transparent to visible light but are opaque to specific laser beam outputs. Selective filters for ruby, Nd-YAG, and other laser systems are available. Glasses appropriate to the specific laser system must be used. At longer infrared wavelengths (such as that of a CO_2 laser), ordinarily transparent materials such as glass are opaque. Clear safety glasses with side shields may be used with these systems, and the only light reaching the eye will be from incandescence of the workpiece. Nevertheless, extreme brilliance can result if a plasma is generated at high power, and filter lens should then be used for viewing the operation.

Burns

Laser burns can be deep and very slow to heal. Exposure must be avoided by appropriate enclosure of the beam or by methods that prevent beam operation unless the beam path is unobstructed. This is particularly important for nonvisible beams that provide no external evidence of their existance unless intercepted by a solid.

Electric Shock

High voltages as well as large capacitor storage devices are associated with lasers. Therefore, the possibility for lethal electrical shock is ever present. Electrical system enclosures should have appropriate interlocks on all access doors and provisions for discharging capacitor banks before entry. The equipment should be appropriately grounded.

Fumes and Gases

Hazardous products may be generated from interaction of the beam and the workpiece. For example, plastic materials used for "burn patterns" to identify beam shape and distribution in high power CO_2 laser systems can generate highly toxic vapors if irradiated in an oxygen-lean atmosphere.

In deep penetration welding, fine metal fume can arise from the joint. Also, intense plasma generation can produce ozone. Consequently, adequate ventilation and exhaust provisions for laser work areas are necessary.

FRICTION WELDING

Friction welding machines are similar to machine tool lathes in that one workpiece is rotated by a drive system. They are also similar to hydraulic presses in that one workpiece is forced against the other. Therefore, safe practices for lathes and power presses should be used as guides for the design and operation of friction welding machines.

Machines should be equipped with appropriate mechanical guards and shields as well as two-hand operating switches and electrical interlocks. These devices should be designed to prevent operation of the machine when the work area, rotating drive, or force system are accessible to the operator or others.

Operating personnel should wear appropriate eye protection and safety apparel commonly used with machine tool operations. In any case, applicable OSHA standards should be strictly observed.

EXPLOSION WELDING

Explosives and explosive devices are a part of explosion welding. Such materials and devices are inherently dangerous, but there are safe methods for handling them. However, if the materials are misused, they can kill or injure, and destroy or damage property.

Explosive materials should be handled and used only by trained personnel who are experienced in that field. Handling and safety procedures must comply with all applicable federal, state, and local regulations. Federal jurisdiction on the sale, transport, storage, and use of explosives is through the U.S. Bureau of Alcohol, Tobacco, and Firearms; the Hazardous Materials Regulation Board of the U.S. Department of Transportation; the Occupational Safety and Health Administration; and the Environmental Protection Agency. Many states and local governments require a blasting license or permit, and some cities have special explosive requirements.

The Institute of Makers of Explosives provides educational publications to promote the safe handling, storage, and use of explosives. The National Fire Protection Association provides recommendations for safe manufacture, storage, handling, and use of explosives.[22]

ULTRASONIC WELDING

With high-power ultrasonic equipment, high voltages are present in the frequency converter, the welding head, and the coaxial cable connecting these components. Consequently, the equipment should not be operated with the panel doors open or housing covers removed. Door interlocks are usually installed to prevent introduction of power to the equipment when the high voltage circuitry is exposed. The cables are shielded fully and present no hazard when properly connected and maintained.

Because of hazards associated with application of clamping force, the operator should not place hands or arms in the vicinity of the welding tip when the equipment is energized. For manual operation, the equipment should be activated by dual palm buttons that meet the requirements of the Occupational Safety and Health Administration (OSHA). Both buttons must be pressed simultaneously to actuate a weld cycle, and both must be released before the next cycle is initiated. For automated systems in which the weld cycle is sequenced with other operations, guards should be installed for operator protection. Such hazards can be further minimized by setting the welding stroke to the minimum that is compatible with workpiece clearance.

THERMIT WELDING

The presence of moisture in a thermit mix, in the crucible, or on the workpieces can lead to rapid formation of steam when the chemical reaction takes place. This may cause ejection of molten metal from the crucible. Therefore, the

22. See *ANSI/NFPA 495, Manufacture, Transportation, Storage and Use of Explosive Materials,* latest edition.

thermit mix should be stored in a dry place, the crucible should be dry, and moisture should not be allowed to enter the system before or during welding.

The work area should be free of combustible materials that may be ignited by sparks or small particles of molten metal. The area should be well ventilated to avoid the buildup of fumes and gases from the reaction. Starting powders and rods should be protected against accidental ignition.

Personnel should wear appropriate protection against hot particles or sparks. This includes full face shields with filter lenses for eye protection and headgear. Safety boots are recommended to protect the feet from hot sparks. Clothing should not have pockets or cuffs that might catch hot particles.

Preheating should be done using the safety precautions applicable to oxyfuel gas equipment and operations.

THERMAL SPRAYING

The potential hazards to the health and safety of personnel involved in thermal spraying operations and to persons in the immediate vicinity are as follows:

(1) Electrical shock
(2) Fire
(3) Fumes and gases
(4) Dust
(5) Arc radiation
(6) Noise

These hazards are not unique to thermal spraying methods. For example, flame spraying has hazards similar to those associated with the oxyfuel gas welding and cutting processes. Likewise, arc spraying and plasma spraying are similar in many respects to gas metal arc and plasma arc welding, respectively. Safe practices for these processes should be followed when thermal spraying with similar equipment. However, thermal spraying does generate dust and fumes to a greater degree.[23]

23. Additional information may be found in *AWS C2.1, Recommended Safe Practices for Thermal Spraying,* Miami: American Welding Society, 1973.

Fire Prevention

Airborne finely divided solids, especially metal dusts, must be treated as explosives. To minimize danger from dust explosions, spray booths must have adequate ventilation.

A wet collector of the water-wash type is recommended to collect the spray dust. Bag or filter type collectors are not recommended. Good housekeeping in the work area should be maintained to avoid accumulation of metal dusts, particularly on rafters, tops of booths, and in floor cracks.

Paper, wood, oily rags, and other combustibles in the spraying area can cause a fire, and should be removed before the equipment is operated.

Protection of Personnel

The general requirements for the protection of thermal spray operators are the same as for welders.

Eye Protection. Helmets, hand shields, face shields, or goggles should be used to protect the eyes, face, and neck during all thermal spraying operations. Safety goggles should be worn at all times. Table 10.4 is a guide for the selection of the proper filter shade number for viewing a specific operation.

Table 10.4
Recommended eye filter plates for thermal spraying operations

Operation	Filter shade numbers
Wire flame spraying (except molybdenum)	2 to 4
Wire flame spraying of molybdenum	3 to 6
Flame spraying of metal powder	3 to 6
Flame spraying of exothermics or ceramics	4 to 8
Plasma and arc spraying	9 to 12
Fusing operations	4 to 6

Respiratory Protection. Most thermal spraying operations require that respiratory protective devices be used by the operator. The nature, type, and magnitude of the fume and gas exposure determine which respiratory protective device should be used. All devices selected for use should be of a type approved by the U.S. Bureau of Mines, National Institute for Occupational Safety and Health, or other approving authority for the purpose intended.

Ear Protection. Ear protectors or properly fitted soft rubber ear plugs should be worn to protect the operator from the high-intensity noise from the process. Federal, state, and local codes should be checked for noise protection requirements.

Protective Clothing. Appropriate protective clothing required for a thermal spraying operation will vary with the size, nature, and location of the work to be performed. When working in confined spaces, flame-resistant clothing and gauntlets should be worn. Clothing should be fastened tightly around the wrists and ankles to keep dusts from contacting the skin.

The intense ultraviolet radiation of plasma and electric arc spraying can cause skin burns through normal clothing. Protection against radiation during arc spraying is practically the same as that for normal arc welding at equivalent current levels.

ADHESIVE BONDING

Adequate safety precautions must be observed with adhesives. Corrosive materials, flammable liquids, and toxic substances are commonly used in adhesive bonding. Therefore, manufacturing operations should be carefully supervised to ensure that proper safety procedures, protective devices, and protective clothing are being used. All federal, state, and local regulations should be complied with, including OSHA Regulation 29CRF 1900.1000, *Air Contaminants.*

General Requirements

Flammable Materials. All flammable materials, such as solvents, should be stored in tightly sealed drums and issued in suitably labeled safety cans to prevent fires during storage and use. Solvents and flammable liquids should not be used in poorly ventilated, confined areas. When solvents are used in trays, safety lids should be provided. Flames, sparks, or spark-producing equipment must not be permitted in the area where flammable materials are being handled. Fire extinguishers should be readily available.

Toxic Materials. Severe allergic reactions can result from direct contact, inhalation, or ingestion of phenolics and epoxies as well as most catalysts and accelerators. The eyes or skin may become sensitized over a long period of time even though no signs of irritation are visible. Once workers are sensitized to a particular type of adhesive, they may no longer be able to work near it because of allergic reactions. Careless handling of adhesives by production workers may expose others to toxic materials if proper safety rules are not observed. For example, coworkers may touch tools, door knobs, light switches, or other objects contaminated by careless workers.

For the normal individual, proper handling methods that eliminate skin contact with an adhesive should be sufficient. It is mandatory that protective equipment, barrier creams, or both be used to avoid skin contact with certain types of formulations.

Factors to be considered in determining the extent of precautionary measures to be taken include:

(1) The frequency and duration of exposure

(2) The degree of hazard associated with a specific adhesive

(3) The solvent or curing agent used

(4) The temperature at which the operations are performed

(5) The potential evaporation surface area exposed at the work station

All these elements should be evaluated in terms of the individual operation.

Precautionary Procedures

A number of measures are recommended in the handling and use of adhesives and auxiliary materials.

Personal Hygiene. Personnel should be instructed in proper procedures to prevent skin contact with solvents, curing agents, and uncured base adhesives. Showers, wash bowls, mild soaps, clean towels, refatting creams, and protective equipment should be provided.

Curing agents should be removed from the hands with soap and water. Resins should be removed with soap and water, alcohol, or a suitable solvent. Any solvent should be used sparingly and be followed by washing with soap and water. In case of allergic reaction or burning, prompt medical aid should be obtained.

Work Area. Areas in which adhesives are handled should be separated from other operations. These areas should contain the following facilities in addition to the proper fire equipment:

(1) A sink with running water
(2) An eye shower or rinse fountain
(3) First aid kit
(4) Ventilating facilities

Ovens, presses, and other curing equipment should be individually vented to remove fume. Vent hoods should be provided at mixing and application stations.

Protective Devices. Plastic or rubber gloves should be worn at all times when working with potentially toxic adhesives. Contaminated gloves must not contact objects that others may touch with their bare hands. Those gloves should be discarded or cleaned using procedures that remove the particular adhesive. Cleaning may require solvents, soap and water, or both. Hands, arms, face, and neck should be coated with a commercial barrier ointment or cream. This type of material may provide short-term protection and facilitate removal of adhesive components by washing.

Full face shields should be worn for eye protection whenever the possibility of splashing exists, otherwise glasses or goggles should be worn. In case of irritation, the eyes should be immediately flushed with water and then promptly treated by a physician.

Protective clothing should be worn at all times by those who work with the adhesives. Shop coats, aprons, or coveralls may be suitable, and they should be cleaned before reuse.

Metric Conversion Factors

1 in. = 25.4 mm
1 ft = 0.305 m
1 ft^3 = 0.028 m^3
1 ft/min = 5.08 mm/s

SUPPLEMENTARY READING LIST

ANSI/ASC Z49.1-1983, *Safety in Welding and Cutting*. Miami: American Welding Society.

ANSI/NFPA 51-1983, *Oxygen-Fuel Gas Systems for Welding, Cutting and Allied Processes,* Quincy, MA: National Fire Protection Association.

ANSI/NFPA 51B-1977, *Cutting and Welding Processes,* Quincy, MA: National Fire Protection Association.

Arc Welding and Cutting Noise, Miami: American Welding Society, 1979.

Balchin, N.C., *Health and Safety in Welding and Allied Processes,* 3rd Ed., England: The Welding Institute, 1983.

Barthold, L.O., et al., *Electrostatic effects of overhead transmission lines, Part I - Hazards and Effects,* IEEE Transactions, Power Apparatus and Systems, Vol. PAS-91, 1972: 422-444.

Compressed Gas Association, Inc., *Handbook of Compressed Gases,* 2nd Ed., New York: Van Nostrand Reinhold Co., 1981.

Dalziel, Charles F., *Effects of electric current on man,* ASEE Journal, 1973, June: 18-23.

Effects of Welding on Health I, II, III, and IV, Miami: American Welding Society, 1979, 1981, 1983.

Fumes and Gases in the Welding Environment, Miami: American Welding Society, 1979.

Handling Acetylene Cylinders in Fire Situations, SB-4, New York: Compressed Gas Association, 1972.

Metals Handbook, Vol. 4, *Heat Treating,* 9th Ed., Metals Park, OH: American Society for Metals, 1981: 389-416.

Recommended Safe Practices for Electron Beam Welding and Cutting, AWS F2.1-78, Miami, American Welding Society, 1978.

Recommended Safe Practices for the Preparation for Welding and Cutting of Containers That Have Held Hazardous Substances, AWS F4.1-80, Miami: American Welding Society, 1980.

Recommended Safe Practices for Thermal Spraying, AWS C2.1-73, Miami, American Welding Society, 1973.

Safe Handling of Compressed Gases in Containers, P-1, New York: Compressed Gas Association, 1974.

The Facts About Fume, England: The Welding Institute, 1976.

The Welding Enviornment, Miami: American Welding Society, 1973.

Ultraviolet Reflectance of Paint, Miami: American Welding Society, 1976.

Welding Fume Control with Mechanical Ventilation, 2nd Ed., San Francisco: Fireman's Fund Insurance Companies, 1981.

Welding Handbook
Index of Major Subjects

	Seventh Edition Volume	Sixth Edition Section	Chapter
A			
Adhesive bonding of metals	3		11
Air carbon arc cutting	2		13
Aircraft		5	91
Alternating current power sources	2		1
Aluminum and aluminum alloys, welding of	4		8
Aluminum structure design	5		1
Applied liners		5	93
Arc characteristics	1		2
Arc cutting	2		13
Arc stud welding	2		8
Arc welding power sources	2		1
Atmosphere, brazing	2		11
Atomic hydrogen welding	3		13
Austenitic manganese steel, welding of	4		4
Austenitic (Cr-Ni) stainless steels welding of	4		2
Automatic brazing	5		5
Automatic welding	5		5
Automotive products		5	90
B			
Beryllium, welding of	4		11
Boilers		5	84

	Seventh Edition Volume	Sixth Edition Section	Chapter
Bonding, adhesive	3		11
Brass	4		7
Braze welding	2		11
Brazed joints, discontinuities in	5		8
Brazed joints, inspection of	5		9
Brazer performance qualification	5		7
Brazing	2		11
Brazing, diffusion	3		10
Brazing automation	5		5
Brazing fixtures	5		4
Brazing metallurgy	1		4
Brazing procedure specifications	5		7
Brazing safety	5		10
Brazing symbols	5		2
Bridges		5	82
Bronze	4		7
Buildings		5	81

C

	Seventh Edition Volume	Sixth Edition Section	Chapter
Capacitor discharge stud welding	2		8
Carbon arc welding	3		13
Carbon steels, welding of	4		1
Cast iron, welding of	4		5
Cast steel, welding of	4		1
Chromium (4 to 10%)-molybdenum stainless steels, welding of	4		2
Chromium stainless steels, welding of	4		1
Clad steel liners		5	93
Clad steel, welding of	4		12
Cobalt alloys	4		6
Codes for welding	5		6
Cold welding	3		13
Columbium, welding of	4		10
Compressed gases, handling of	5		10

	Seventh Edition Volume	Sixth Edition Section	Chapter
Cooling rates	1		3
Copper and copper alloys, welding of	4		7
Corrosion of welded joints	1		5
Cost estimating	5		3
Cutting, air carbon arc	2		13
Cutting, arc	2		13
Cutting, laser beam	3		6
Cutting, metal powder	2		13
Cutting, oxyfuel gas	2		13
Cutting, oxygen	2		13
Cutting, plasma arc	2		13
Cutting, safe practices in	5		10
Cutting processes, survey of	1		1

D

	Seventh Edition Volume	Sixth Edition Section	Chapter
Definitions, terms and	1		App. A
Design of aluminum structures	5		1
Design of welded joints and weldments	5		1
Destructive testing	5		9
Die steels	4		3
Diffusion brazing	3		10
Diffusion welding	3		10
Direct current power sources	2		1
Discontinuities, significance of	1		5
	5		8
Discontinuities in brazes	5		8
Discontinuities in welds	5		8
Dissimilar welds	4		12
Distortion	1		6
Distortion, control of	1		6
Distortion, correction of	1		6
Distortion, types of	1		6

	Seventh Edition Volume	Sixth Edition Section	Chapter
E			
Economics	5		2
Electrical safety	5		10
Electrogas welding	2		7
Electron beam welding	3		5
Electroslag welding	2		7
Elevated temperature behavior of welded joints	1		5
Energy sources for welding	1		2
Equipment, resistance welding	3		3
Estimating costs	5		3
Explosion welding	3		8
F			
Fatigue properties of welded joints	1		5
Ferritic stainless steels	4		2
Filler metals	4		1-12
Filler metals, brazing	2		11
Filler metals, soldering	2		12
Filler metals, surfacing	2		14
Fixtures	5		4
Flash welding	3		2
Flux cored arc welding	2		5
Flux cutting, chemical	2		13
Fluxes, brazing	2		11
Fluxes, soldering	2		12
Forge welding	3		13
Fracture mechanics	1		5
	5		8
Fracture toughness	1		5
Friction welding	3		7
Fuel gases, characteristics of	2		10
Fumes and gases	5		10
Fusion weld discontinuities	5		8

	Seventh Edition Volume	Sixth Edition Section	Chapter
G			
Galvanized steel	4		1
Gases, physical properties	1		2
Gas metal arc spot welding	2		4
Gas metal arc welding	2		4
Gas tungsten arc welding	2		3
Gas welding, oxyfuel	2		10
Gold, welding of	4		11
H			
Hafnium, welding of	4		10
Hardfacing	2		14
Heat flow in welding	1		3
High frequency welding	3		4
Hot pressure welding	3		13
I			
Industrial piping		5	85
Inspection, brazing	2		11
Inspection of welding	5		9
Iron, welding of	4		1
J			
Jewelry, welding of	4		11
Joining processes, survey of	1		1
L			
Laser beam cutting	3		6
Laser beam welding	3		6

	Seventh Edition Volume	Sixth Edition Section	Chapter
Launch vehicles		5	92
Lead, welding of	4		11
Liners, applied		5	93
Liners, clad steel		5	93
Liquid penetrant testing	5		9
Low alloy steels, welding of	4		1

M

Magnesium and magnesium alloys, welding of	4		9
Magnetic fields, influence on arcs	1		2
Magnetic particle testing	5		9
Maraging steels	4		4
Martensitic stainless steels	4		2
Mechanical properties	1		5
Mechanical testing	1		5
	5		9
Mechanical treatments of weldments	1		6
Melting rates (electrode)	1		2
Metal powder cutting	2		13
Metal transfer	1		2
Metallurgy, general	1		4
Metallurgy of brazing and soldering	1		4
Metallurgy of surfacing alloys	2		14
Metallurgy of welding	1		4
Metals, physical properties	1		2
Metric practice guide	1		App. B
Molybdenum, welding of	4		11

N

Narrow gap welding	2		4
NEMA power source requirements	2		1
Nickel and nickel alloys, welding of	4		6

	Seventh Edition Volume	Sixth Edition Section	Chapter
Nondestructive examination (testing)	5		9
Nondestructive examination symbols	5		2
Nuclear power plants		5	87

O

Oxyfuel gas cutting	2		13
Oxyfuel gas welding	2		10
Oxygen cutting	2		13

P

Palladium, welding of	4		11
Percussion welding	3		2
Performance qualification	5		7
Physical properties of metals and gases	1		2
Physics of welding	1		2
Pipelines, transmission		5	86
Piping, industrial		5	85
Plasma arc cutting	2		13
Plasma arc welding	2		9
Platinum, welding of	4		11
Positioners	5		4
Power sources, arc welding	2		1
Power sources, special	2		1
Precious metals, welding of	4		11
Precipitation-hardening steels	4		2
Precoated steels	4		1
Pressure gas welding	3		13
Pressure vessels		5	84
Procedure qualification	5		7
Process safety	5		10
Processes, brazing	2		11
Projection welding	3		1

	Seventh Edition Volume	Sixth Edition Section	Chapter
Proof testing	5		9
Properties, mechanical	5		1

Q

Qualification	5		7
Quenched and tempered steels	4		1

R

Radiographic testing	5		9
Railroads		5	89
Reactive metals, welding of	4		10
Refractory metals, welding of	4		10, 11
Residual stresses	1		6
Residual stresses, causes of	1		6
Residual stresses, effects of	1		6
Residual stresses, measurement of	1		6
Residual stresses, reduction of	1		6
Resistance weld discontinuities	5		8
Resistance welding electrodes	3		3
Resistance welding equipment	3		3
Robotic welding	5		5
Roll welding	3		13

S

Safe practices	5		10
Safe practices, brazing	2		11
Seam welding	3		1
Shielded metal arc welding	2		2
Ships		5	88
Silver, welding of	4		11
Soldering	2		12
Soldering metallurgy	1		4

	Seventh Edition Volume	Sixth Edition Section	Chapter
Soldering safety	5		10
Solders	2		12
Solid-state circuitry, power source	2		1
Solid-state weld discontinuities	5		8
Solidification rates	1		3
Specifications for welding	5		6
Spot welding	3		1
Spot welding, gas metal arc	2		4
Spraying, thermal	3		12
Standards for welding	5		6
Steel, welding of	4		1
Storage tanks, field-welded		5	83
Stresses, residual	1		6
Structural tubular connections	5		1
Stud welding	2		8
Submerged arc welding	2		6
Surfacing	2		14
Surfacing methods	2		14
Survey of joining and cutting processes	1		1
Symbols for welding, brazing, and nondestructive examination	5		2

T

	Seventh Edition Volume	Sixth Edition Section	Chapter
Tantalum, welding of	4		10
Tensile properties of welded joints	1		5
Terms and definitions	1		App. A
Testing, mechanical	1		5
Testing, weldability	1		4
Testing of welded joints	1		5
Thermal spray operator qualification	5		7
Thermal spraying	3		12
Thermal treatments of weldments	1		6
Thermit welding	3		13
Titanium and titanium alloys, welding of	4		10

	Seventh Edition Volume	Sixth Edition Section	Chapter
Tool steels	4		3
Toughness, fracture	1		5
Transmission pipelines		5	86
Tubular connections, structural	5		1
Tungsten, welding of	4		11
Turning rolls	5		4
Turntables	5		4

U

Ultra high strength steels	4		4
Ultrasonic testing	5		9
Ultrasonic welding	3		9
Upset welding	3		2
Uranium, welding of	4		11

V

Ventilation	5		10
Visual inspection	5		9

W

Weld discontinuities, significance of	1		5
	5		8
Weld distortion	1		6
Weld quality	5		8
Weld thermal cycles (typical)	1		3
Weldability testing	1		4
Welded joints, corrosion testing	1		5
Welded joints, design of	5		1
Welded joints, elevated temperature behavior	1		5
Welded joints, fatigue properties	1		5

	Seventh Edition Volume	Sixth Edition Section	Chapter
Welded joints, performance of	5		1
Welded joints, tensile properties	1		5
Welded joints, testing of	1		5
Welder performance qualification	5		7
Welding, filler metals for	4		1-12
Welding, inspection of	5		9
Welding, physics of	1		2
Welding, safe practices in	5		10
Welding applications			
Aircraft		5	91
Applied liners		5	93
Automotive products		5	90
Boilers		5	84
Bridges		5	82
Buildings		5	81
Clad steel liners		5	93
Industrial piping		5	85
Jewelry	4		11
Launch vehicles		5	92
Nuclear power plants		5	87
Pressure vessels		5	84
Railroads		5	89
Ships		5	88
Storage tanks		5	83
Transmission pipelines		5	86
Welding automation	5		5
Welding codes	5		6
Welding costs	5		2
Welding fixtures	5		4
Welding inspector(s)	5		9
Welding inspector qualification	5		7
Welding metallurgy	1		4
Welding of metals			
Aluminum and aluminum alloys	4		8
Austenitic manganese steel	4		4

	Seventh Edition Volume	Sixth Edition Section	Chapter
Austenitic (Cr-Ni) stainless steels	4		2
Beryllium	4		11
Carbon steels	4		1
Cast iron	4		5
Cast steel	4		1
Chromium (4 to 10%)-molybdenum steels	4		1
Chromium stainless steels	4		2
Cobalt alloys	4		6
Columbium	4		11
Copper and copper alloys	4		7
Gold	4		11
Hafnium	4		10
Iron	4		1
Lead	4		11
Low alloy steels	4		1
Magnesium and magnesium alloys	4		9
Molybdenum	4		11
Nickel and high nickel alloys	4		6
Palladium	4		11
Platinum	4		11
Precious metals	4		11
Reactive metals	4		10
Refractory metals	4		10, 11
Silver	4		11
Steel	4		1
Tantalum	4		10
Titanium and titanium alloys	4		10
Tungsten	4		11
Uranium	4		11
Wrought iron	4		1
Zinc-coated steel	4		1
Zirconium	4		10
Welding procedure specifications	5		7
Welding processes			
Atomic hydrogen welding	3		13

	Seventh Edition Volume	Sixth Edition Section	Chapter
Bare metal arc welding	3		13
Brazing	2		11
Carbon arc welding	3		13
Cold welding	3		13
Diffusion brazing	3		10
Diffusion welding	3		10
Electrogas welding	2		7
Electron beam welding	3		5
Electroslag welding	2		7
Explosion welding	3		8
Flash welding	3		2
Flash welding machines	3		3
Flux cored arc welding	2		5
Forge welding	3		13
Friction welding	3		7
Gas metal arc welding	2		4
Gas tungsten arc welding	2		3
Gas welding	2		10
High frequency welding	3		4
Hot pressure welding	3		13
Laser beam cutting	3		6
Laser beam welding	3		6
Oxyfuel gas welding	2		10
Percussion welding	3		2
Plasma arc welding	2		9
Projection welding	3		1
Projection welding machines	3		3
Seam welding, resistance	3		1
Seam welding machines	3		3
Shielded metal arc welding	2		2
Spot welding, resistance	3		1
Spot welding machines	3		3
Stud welding	2		8
Submerged arc welding	2		6
Thermit welding	3		13

	Seventh Edition Volume	Sixth Edition Section	Chapter
Ultrasonic welding	3		9
Upset welding	3		2
Upset welding machines	3		3
Welding safety	5		10
Welding specifications	5		6
Welding standards	5		6
Welding symbols	5		2
Weldment design	5		1
Wrought iron, weldability of	4		1

Z

	Seventh Edition Volume	Sixth Edition Section	Chapter
Zinc-coated steel (galvanized)	4		1
Zirconium, weldability of	4		10

INDEX

AAR. *See* Association of American Railroads.
AASHTO. *See* American Association of State Highway and Transportation Officials.
ABS. *See* American Bureau of Shipping.
Acceptable weld, definition, 278
Acoustic emission testing, 321, 324. *See also* Nondestructible testing.
  acoustic emissions, location of, 366
  applications, 366-70
  general description, 366
  monitoring continuous welding operations, 366-68
  monitoring resistance spot welding, 368-69
  proof testing, during, 369-70
  recording of results, 370
Aerospace Material Specifications, 239-43
Airborne contaminants
  permissible exposure limits, 399
  threshold limit values, 396
AISC. *See* American Institute of Steel Construction
Aluminum structures
  butt joints, 81
  butt joints, minimum strength of, 90
  designing for welding, 81
  fillet welds, minimum size, 86
  fillet welds, shear strength of, 86, 87-88
  joint design, 83-84
  lap joints, 82
  service temperature, effects of 89-90
  T-joints, 82-83
  weld joints, 81-83
  welding, effects of, 84-85
  welding stresses, reduction of, 86-87
  welds, fatigue strength of, 86, 89
  welds, strength distribution across, 84
  welds, stress distribution in, 85-86
American Association of State Highway and Transportation Officials, 218, 219
American Bureau of Shipping, 218, 219-21
American Institute of Steel Construction, 218, 221
American National Standards Institute, 217, 218, 221-22, 237
American Petroleum Institute, 218, 222-23
American Railway Engineering Association, 218, 223
American Society for Nondestructive Testing, 271
American Society for Testing and Materials, 218, 225-26
American Society of Mechanical Engineers, 218, 223-25
American Water Works Association, 218, 226
American Welding Society, 217, 218, 226-31
  filler metal specifications, 227-28
  mechanical testing standards, 228-29
  qualification standards, 228
  recommended practices, 229-30
  Standard for Qualification and Certification of Welding Inspectors, 271, 273
  symbols for welding and nondestructive testing, 227
  welding terms and definitions, 227
AMS. *See* Aerospace Material Specifications.
ANSI. *See* American National Standards Institute.
API. *See* American Petroleum Institute.

Arc blow, 153
Arc strikes, 292, 302
  definition of, 278
Arc welding automation. *See* Automation, arc welding.
Arc welding economics, *See* Economics of arc welding.
AREA. *See* American Railway Engineering Association.
ASME. *See* American Society of Mechanical Engineers.
Asphyxiation, 393
Association Francaise de Normalisation, 217
Association of American Railroads, 218, 231
ASTM. *See* American Society for Testing and Materials.
ASTM Standards, 225-26
Automatic brazing. *See* Automation, brazing.
Automatic welding, 182-84. *See also* Automation, welding.
Automatic welding, flexible, 184-85. *See also* Robots, welding; Robots, arc welding.
Automatic welding controller(s), 184
Automation. *See also* Automation, arc welding; Automation, brazing; Automation, resistance welding; Automation, welding; Robots, welding; Robots, arc welding.
  arc welding, 178-85
  brazing, 208-10
  design considerations, 211
  equipment reliability, 212
  extent of, 186-87
  factory integration, 210-11
  fixturing and positioning, 211
  inspection and quality control, 212
  manufacturing considerations, 211-12
  meaning of, 178
  problems of, 210-12
  resistance welding, 193-200
  risk factors, 210-12
  supplementary reading list, 213
  system integration, 211-12
  use of, 178
  vendor assistance, 212
Automation, arc welding. *See also* Automation.
  fumes and gases, handling, 205
  robots, arc welding, 200-203. *See also* Robots, welding.
  seam tracking systems, 205-8
  types, 200
  welding gun cleaners, 205
Automation, brazing. *See also* Automation.
  advantages of, 210
  equipment, 209-10
  filler metal application, 210
  flux application, 210
  fundamentals, 208-9
  heating time, 209-10
  temperature control, 209
  torch heating, 209
Automation, reistance welding. *See also* Automation.
  accessory equipment, 199
  adaptive process control, 199
  controls, 197-99
  interface between components, 198-99
  material handling, 193
  monitoring devices, 198, 199
  production rates, 195-96
  production requirements, 193-96
  programming, 199-200
  quality control, 196
  resistance welding equipment, 197
  spot welding guns, 197
  welding electrodes, 197
Automation, welding. *See also* Automation.
  designing for, 188
  environment for, 189-90
  extent of, 186-87
  feasibility testing, 188-89
  fixturing, 188
  flexibility and expansion, 192
  floor space, 186
  fundamentals, 178-85
  governing factors, 186-87
  inventory turn over, 187
  investment, 187
  maintenance of, 192-93
  manpower for, 187
  material handling, 192
  part tolerances and fit-up, 188
  planning, 187-93
  procurement scheduling, 190
  product quality, 186
  production considerations, 192-93
  production levels, 186
  safety, 188
  shop acceptance, 192

space allocation for, 192
specifications for, 189
training and education, 190-92
utilities for, 189
welding process selection, 188
AWS. *See* American Welding Society.
AWWA. *See* American Water Works Association.

Bend tesing of welds, 371
Boiler and Pressure Vessel Code, ASME, 223-24, 226, 237-38, 268, 314, 364
Brazed joints, discontinuities in. *See* Discontinuities in brazed and soldered joints.
Brazed joints, inspection of, 376
Brazement design, 1n
Brazer performance qualification. *See also* Welder performance qualification.
acceptance criteria, 268
purpose, 268
qualification variables, 268
specimen testing, 268, 270
test record, 268, 271, 272
workmanship test brazements, 268, 270
Brazing automation. *See* Automation, brazing.
Brazing cost estimating, 123-24
Brazing economics. *See* Economics of brazing and soldering.
Brazing fixtures. *See* Fixtures, brazing.
Brazing procedure specifications, 253-55. *See also* Procedure specifications; Welding procedure specifications.
qualification of, 259, 261
Brazing symbols
applications, 113-15
brazing processes, letter designations for, 113
double-flare-bevel-groove and fillets, 115
flare-bevel-groove and fillet, 115
lap joint with fillet, 114
scarf joint, 114
square groove, 115
T-joint, 115
Brinell hardness test, 372
British Standards Association, 217
Brittle fracture of welds, 313-14
Building standards, 221
Burn prevention, 384-86
Burn-through, definition of, 278

Canadian Standards Association, 217, 218, 231-32
Carbon monoxide, 388
Carriages, side beam, 181-82
Carriages, welding machine, 179-80
Certification. *See* Qualification and certification.
CGA. *See* Compressed Gas Association.
Charpy V-notch impact test, 371-72
Chemical analysis, 370
Chlorinated hydrocarbons, ventilation of, 396
Codes. *See* Standards.
Compressed Gas Association, 218, 232
Compressed gas cylinders. *See also* Compressed gases, handling of.
filling, 397
gas withdrawal, 397-98
storage, 397
use, 397
valve(s), 398
valve threads, standard, 398
Compressed gases, handling of. *See also* Compressed gas cylinders.
acetylene, 401
argon, 401
carbon dioxide, 401
cryogenic cylinders and tanks, 399
fuel gases, 400-401
helium, 401
inert gases, 401
manifolds, 400
methylacetylene-propadiene (MPS), 401
nitrogen, 401
oxygen, 400
pressure relief devices, 398-99
regulators, 398, 399-400
shielding gases, 401
Concavity, weld root, 352
Confined spaces, 395-96
welding in, 393-94
working in, 381
Containers, safe practices for, 383-84
Containers, welding on, 394
Corrosion properties of metals, 12-13
Cost estimating. *See also* Welding cost estimating.
direct labor, 123
direct materials, 123
elements of, 122-23
factory overhead, 123

manufacturing costs, 123
purposes of, 122
small tools, jigs, and fixtures, 123
supplementary reading list, 148
Cracks, 287, 304, 305, 349, 351-52
base metal, 296, 301-2
cold, 289
crater, 280, 290, 301
definition, 278
delayed, 289
hot, 278, 279
hydrogen induced, 279, 301-2
longitudinal, 280, 300-301
significance, 302
surface, 333
throat, 280, 289
toe, 280, 290
transverse, 280, 300
underbead, 280, 290
weld metal, 295-96, 300-301
Creep failure of welds, 314
Cryogenic cylinders and tanks for liquids, 399
Cryogenic liquid transfer lines, 399
CSA. *See* Canadian Standards Association.

Defect, definition of, 278, 321
Defective weld, definition of, 278
Delamination, 280
Design considerations
bending, 22-25
compressive loading, 20-21, 22
deflections of beams, 23
design approach, 17
design formulas, 18-20
designing for rigidity, 18
designing for strength, 18
diagonal bracing, 28
forces, transfer of, 28-31
loading, types of, 20-28
shear, 25-26
shear stress in torsion, 27-28
structural safety, 17-18
tension loading, 20
torsion, 26-28
torsional resistance, 26-27
welded design, basis for, 18
welding process, influence of, 31-32
Design for welding 1-91. *See also* Aluminum structures; Design considerations; Fillet welds; Properties of metals; Steel welds, sizing of; Welded joints, design of; Weldment design program.
designer(s) of weldments, 2-3
objectives, 1
supplementary reading list, 91
Design of brazements, 1n
Design of welded joints. *See* Welded joints, design of.
Destructive testing. *See also* Mechanical testing; Nondestructive testing; Proof testing.
chemical analysis, 370
chemical solutions for etching, 374
general description, 370
macroetching procedures, 373
macroscopic examinations, 373
metallographic examinations, 373-74
microscopic examinations, 37374
Deutsches Institute fuer Normung, 217
DIN, 217
Discontinuities. (*See also* the specific type of discontinuity.)
acceptance standards, 308, 311
in brazed joints, 306-8, 309-10
definition of, 278, 321
discontinuity-material relationships, 311-14
effects on mechanical properties, 311-14
in flash welds, 303-6
in friction welds, 303-6
fusion type discontinuity, 287
in fusion welds, 278-93
in resistance welds, 303-6
significance of, 308, 311-15
in soldered joints, 306-8
in solid state welds, 303-6
in upset welds, 303-6
with various welding processes, 287
Discontinuities in brazed and soldered joints
acceptance limits, 307
base metal erosion, 307
cracks, 307
flux entrapment, 307
lack of fill, 307
noncontinuous fillets, 307
poor surface appearance, 307
typical discontinuities and causes, 307-8
Discontinuities in fusion welds. (*See also* specific types of discontinuity.)
base metal flaws, 290-92, 301-2

causes, 293ff
classification of, 278-79
common types, 280-86
cracks, 289-90, 300-302
dimensional discrepancies, 292-93
excessive reinforcement, 312
inadequate joint penetration, 288-89, 299-300
inadequate joint properties, 302
incomplete fusion, 288, 299
joint misalignment, 293
joint preparation, incorrect, 293
location and occurrence, 279-88
material related, 279
overlap, 289
planar types, 279
porosity, 288, 293-97
process related, 278, 287
remedies for, 293ff
slag inclusions, 288, 298, 312-13
surface irregularities, 290, 2911, 302
surface pores, 290
tungsten inclusions, 288
undercut, 289
underfill, 289
warpage, 293
weld joint properties, inadequate, 302
weld profile, incorrect, 293, 294
weld size, incorrect, 293
Discontinuities in resistance and solid state welds
anisotropic mechanical behavior, 304
center discontinuities, 303-304, 305
cracking, 304, 305
die burns, 304
flash welds, 303ff
flat spots, 304, 306
friction welds, 303ff
inclusions, 305
misalignment, 303
upset geometry, 303
upset welds, 303ff
welding processes, 303
Dissolved gases in weld metal, 296-97. *See also* Porosity.
Distortions, 16
Duty cycle, 127-28

Eddy current testing, 321, 324. *See also* Nondestructive testing.
applications, 354
cracks, effects of, 355
demagnetization, 357
description, 354
discontinuities, detectable, 359
eddy current(s), 354-55
eddy current penetration, 358
eddy currents, properties of, 356
electromagnetic properties of coils, 355-56
electromagnetic testing, 357-59
equipment calibration, 359
magnetic saturation, ac, 357
skin effects, 357
Economic considerations for positioners, 159-60
Economic equipment selection, mathematical analysis for, 140-41
Economics of arc welding
manufacturing suggestions, 135
welding costs, minimization of, 134-35
welding fabrication suggestions, 134-35
welding procedure suggestions, 135
weldment design suggestions, 134
Economics of brazing and soldering
assembly and fixturing costs, 142
inspection costs, 143
joint designs, 142
process costs, 142-43
surface preparation costs, 142
Economics of new joining processes, 139-41
Economics of oxyfuel gas cutting, 143-47
Economics of plasma arc cutting, 143-47
Economics of thermal cutting, 143-47
Economics of welding automation and robotics, 135-39. *See also* Economics of arc welding; Welding cost estimating.
benefits of, 136
combined operations, 138
costs, 136
deposition efficiency, 137
equipment and tooling selection, 138-39
improvement factor, overall, 138
installation costs, 138
material handling, 137
operator factor, 137-38
part accuracy, 137
part geometry, 136
production estimates, 136
return on investment, percentage, 138
safety, 137
weld joint design, 138

welding site and environment, 136-37
Electrical safety, 402-5
cables and connectors, 403-4
electric shock, 402, 409, 410
electrical grounding, 403
equipment installation, 403, 408
equipment selection, 402
fire prevention, 405
grounding, electrical, 408
maintenance of welding machines, 404-5
personnel training, 403
resistance welding, 407-8
welding leads, 403-4
in welding operations, 404
welding under damp conditions, 404
Explosion prevention, 383-84
Eye protection, 322, 384, 385
for electron beam welding, 410
for laser beams, 410
for resistance welding, 407
for thermal spraying, 412

Face protection, 384
Fatigue strength
of aluminum welds, 86, 89
of steel welds, 51-58
Federal standards, 232-37
Fillet welds, *See also* Steel welds, sizing of; Tubular connections, structural; Welded joints, design of.
applications, 39
double, 36, 41, 42
effective throat, 39
intermittent, 39, 56
load(s) for steel, unit, 50
measurement gage, 335
minimum size for aluminum, 86
minimum size for steel, 51
shear strength, aluminum, 86, 87-88
single, 36, 41
size, 39-40
skewed, 61
strength of, 39
in tubular connections, 68, 78
Fire prevention, 383
electrical, 405
for thermal spraying, 412
Fissure(s), 280, 290
definition of, 277
Fitness for purpose, 276, 277
Fixtures, 204. *See also* Fixtures, brazing; Fixtures, welding.
applications, 151
definition, 150
design features, 150-51
supplementary reading list, 175
Fixtures, brazing. *See also* Fixtures.
assembly for brazing, 153-54, 157
for close-tolerance assemblies, 158
design, 156-58
for dip brazing, 156, 159
fixture materials, 154-56
for furnace brazing, 156, 158
general considerations, 153
for induction brazing, 156, 158-59
self-fixturing, 154, 157
for short runs, 156
Fixtures, welding, 183, 188. *See also* Fixtures.
design considerations. 152-53
examples, 153ff
Flat spots in welds, 304, 306
Flaws
base metal, 290-92
definition of, 278, 321
Fluorescent penetrant testing. *See* Liquid penetrant testing.
Fluorides in fumes and gases, 389
Fluoride compounds, safe practices with, 395-96
Fracture mechanics, 10, 313, 314-15. *See also* Fracture toughness.
Fracture toughness, 9-10, 313n. *See also* Fracture mechanics.
Fuel gas fires, 401
Fumes and gases, 205, 380
air sampling, 396
arc voltage, effects of, 389
from arc welding, 388-89
asphyxiants, 388
carbon monoxide, 388
definition of fume, 380n
from electron beam welding, 410
exposure factors, 390-91
exposure measurement, 396
from fluorine compounds, 389, 395-96
fume constituents, 388-89
fume generation rate factors, 389-90
hazardous types, 388
with lazer beam welding and cutting, 410-11

metal transfer effects, 389
operator protection, 387ff
oxides of nitrogen, 388
from oxyfuel gas welding and cutting, 390
ozone, 388
permissible limits, 387-88
position of welder's head, 390
shielding gas, effects of, 389-90
shop size, effects of, 390
sources, 388-89
welding consumables, effects of, 390
welding current, effects of, 389
welding helmet design, 391
welding process, effects of, 390
Fusion welds, discontinuities in. *See* Discontinuities in fusion welds.

Goodman design stress diagram, 57
Groove welds
allowable fatigue conditions, 56
complete-joint-penetration, 37, 45, 46
double-bevel-groove, 38-39
double-V-groove, 38
effective throat, 45, 50
fatigue strength, 57-58, 59
J-groove, 39
partial-joint-penetration, 37-38, 45, 77
single-bevel-groove, 38
single-V-groove, 38
square-groove, 38
U-groove, 39
Guides. *See* Standards.

Hardness testing, 372-73
Hot work permit system, 387
Hydrogen in weld metal, 296-97
Hydrostatic testing, 374-75

Impact testing of welds, 371-72, *See also* Fracture toughness.
Inadequate joint penetration, 280, 287, 288-89, 334, 352
causes, 299
definition of, 278
remedies for, 299-300
significance of, 300
Inclusions in welds, 295, 304
Incomplete fusion, 280, 287, 288, 334, 352
causes, 299
definition of, 278
remedies for, 299
significance of, 299
Inspection, 16-17, 318-77. *See also* Nondestructive testing; Proof testing.
of brazed joints, 376
destructive testing, 370-74
inspection plan, 319-20
nondestructive testing, 321-70
plan selection, 320
production sampling, 320
proof testing, 374-76
supplementary reading list, 377
welding inspector(s), 318-19
Welding Inspector Qualification and Certification Program, AWS, 318
International Organization for Standardization, 217, 218, 222, 237
ISO. *See* International Organization for Standardization.

Japanese Standards Associations, 217
Jigs. *See* Fixtures.
JIS, 217
Joint types, 32-36, 98, 101, 104

Lamellar tearing, 45, 80-81, 279, 280, 292
Laminations, 280, 292
Lap joints, double welded, 41
Laps in base metal, 280, 292
Laser beam hazards, 410
Leak testing, 375-76
Letter designations for
brazing processes, 113
nondestructive examination methods, 116
welding processes, 101
Liquid penetrant testing, 321, 323. *See also* Nondestructive testing.
applications, 329
equipment, 329
interpretation of indications, 333-34
materials, 329-30
procedures, 330-33
Load testing, 374

Machine guards, 387
Machine welding, 179-82
Macroscopic examination, 373

Magnetic particle testing, 321, 323, *See also* Nondestructive testing.
applications, 334-35
demagnetization, 341
discontinuity indications, 341-42
discontinuity indications, interpretation of, 342-43
dry method, 339
equipment, 340-41
false indications, 342
inspection media, 339-40
limitations, 335
magnetic field orientation, 335
magnetization, circular, 336-37
magnetization, continuous, 340
magnetization, localized, 338
magnetization, longitudinal, 337
magnetization, residual, 340
magnetization, yolk method, 338
magnetizing current, amount of, 339
magnetizing current, types of, 338-39
operation, sequence of, 340
principles, 334
prod magnetization, 338
recording of results, 341
requirements, essential, 34
wet method, 339-40
Manipulators, welding head, 180-81
Manufacturers associations, 244
Manufacturer's material safety data sheets, 395
Mechanical properties of metals. *See* Metals, properties of.
Mechanical testing. *See also* Proof testing.
bend testing, 371
hardness testing, 372-73
impact testing, 371-72. *See also* Fracture toughness.
tension testing, 371
Metal fume fever, 396
Metal properties. *See* Properties of metals.
Metallographic examination, 373-74
Metals, properties of
corrosion properties, 12-13
creep rupture properties, 11
ductility, 8-9
elastic limit, 5
electrical conductivity, 12
elevated temperature properties, 10-11
endurance limit, 7, 8
fatigue life, 7, 8
fatigue limit, 7
fatigue strength, 7-8
fatigue stress range, 8
fracture toughness, 9-10, 313n
low temperature properties, 10
mechanical properties, 5-11
melting temperature, 12
modulus of elasticity, 5
modulus of elasticity of steel, 20
notch toughness, 10, 11
physical properties, 12
plastic deformation, 6
proportional limit, 6
stress-strain diagram, 6
structure-insensitive properties, 3-4
structure-sensitive properties, 3-4
tensile strength, 7
thermal conductivity, 12
thermal expansion, coefficient of, 12
yield point, 7
yield strength, 7
Young's modulus, 5
Methods. *See* Standards.
Microscopic examination, 373-74
Military specifications, 233-37

National Board of Boiler and Pressure Vessel Inspectors, 218, 237-38
National Fire Protection Association, 218, 238
Naval Publications and Forms Center, 218
NFPA. *See* National Fire Protection Association.
Nitrogen in weld metal, 296-97
Noise exposure, allowable levels, 386
Nondestructive examinations. *See* Nondestructive testing; Proof testing.
Nondestructive examination symbols
acoustic emission examination, 119
all-around examination, 118
area of examination, 119
arrow location significance, 116
arrow side examination, 116, 117
combined symbols, 117, 118
elements of, 116
length of examination, 117-18
letter designations, 116
no side significance, 117
number of examinations, 118

other side examinations, 117
radiographic examination, 117
references, 117
Nondestructive testing. *See also* Acoustic emission testing; Eddy current testing; Liquid penetrant testing; Magnetic particle testing; Radiographic testing; Ultrasonic testing; Visual inspection; Nondestructive examination symbols; Destructive testing.
definitions, 321
methods, 321, 322-24
qualification of personnel, 271
reasons for use, 321

Occupational Safety and Health Administration, 232
Operating factor, 159-60
OSHA. *See* Occupational Safety and Health Administration.
Overlap, 280, 287, 289
definition of, 278
Oxygen in weld metal, 296-97
Ozone, 388

Performance qualification, 246, 247
of brazers, 268, 270-71, 272
of welders, 261-68
PFI. *See* Pipe Fabrication Institute.
Physical properties of metals. *See* Metals, properties of.
Pipe Fabrication Institute, 218, 238-39
Pneumatic testing, 375
Porosity, 280, 287, 288, 351
causes, 293, 295-96
definition of, 278
effect on fatigue life, 312
effect on mechanical properties, 297
significance of, 297
surface, 297, 334
Positioners, 204. *See also* Positioners, headstock-tailstock; Positioners, tilting-rotating.
attachment of weldment, 172
center of gravity, 168-72
economic considerations, 159-60
general description, 159
safety, 166-68
supplementary reading list, 175
technical considerations, 168-74
tilt torque, 170
turning roll(s), 160, 161
turning roll tractive effort, 170
turntable positioners, 163
types of positioning, 160
work lead connections, 174
Positioners, headstock-tailstock. *See also* Positioners.
applications, 161-63
features and accessories, 163
general description, 160-61
Positioners, tilting-rotating. *See also* Positioners.
capacity rating, 164
center of gravity, 164n
drop-center tilting, 164, 166
functions, 164
general description, 163-64
power elevated, 166
sky hook, 166, 169
special types, 166
Positions of fillet welds, 264
Positions of groove welds, 263
Procedure specifications, 246, 248-55
brazing, 253-55
essential variables, 246
non-essential variables, 246
procedure qualification record, 246
qualification of, 256-61
resistance welding, 255
welding, 248-53
Process safety, 405-14
Proof testing. *See also* Destructive testing; Mechanical testing.
hydrostatic testing, 374-75
leak testing, 375-76
load testing, 374
pneumatic testing, 375
purpose, 374
spin testing, 375
stress relief, 376
Protective clothing, 384, 386

Qualification and certification, 246-74
of brazing procedures, 259, 261
certification, 246-47
codes and standards, 247
joining processes, 247
nondestructive testing personnel, 271
of procedure specifications, 256-61
requirements, standardization of, 273

supplementary reading list, 274
thermal-spray operators, 271
welding inspectors, 271, 273
of welding procedures, 246
Qualification of brazing procedures, 259, 261
Qualification of procedure specifications, 256-61
Qualification of resistance welding procedures, 259, 261. *See also* Welding procedure specifications; Qualification of welding procedures.
Qualification of welding procedures, 256-59. *See also* Welding procedure specifications.
preparation of sample joints, 256
procedure changes, 257, 259-60
recording results, 257
steps, 256
testing of welds, 256-57
Quality of welds. *See* Weld quality.
Quenched-and-tempered steels, welding of, 31

Radiographic testing, 321, 322. *See also* Nondestructive testing.
description, 343
discontinuities, interpretation of, 349-53
essential elements, 343
exposure techniques, 349, 350
film processing, 346
film viewing, 349
gamma ray sources, 343, 344
identification markers, 349
image quality indicators, 347-49
image quality requirements, 348
penetrameters, 347-49
radiation absorption, 346
radiation monitoring, 354
radiation protection, 353
radiographer, 346
radiographic image quality, 347
radiographs, interpretation of, 346-47, 349
radioisotopes, 343, 344
recording or viewing means, 346
safe practices, 353-54
sensitivity, 353
steel thickness limitations, 345
x-ray sources, 343, 344
Radiography. *See* Radiographic testing.
Radius of gyration, 21
Recommended practices. *See* Standards.
Resistance welding automation. *See* Automation, resistance welding.
Resistance welding controls, 197-99
Resistance welding equipment, 197
Resistance welding procedure qualification, 259, 261
Resistance welding procedure specifications, 255
Respiratory protective equipment, 393
Robotics, economics of. *See* Economics of welding automation and robotics.
Robots, arc welding. *See also* Robots, welding.
accessories, 205-8
accuracy, 202
arc voltage control, 204
axes of motion, 201-2
control interfaces, 204
features, 200-201
fixtures, control of, 204
material handling, 204
positioners, use of, 204
programming, 205
repeatability, 202
welding equipment, 203
welding equipment control, 204
welding guns, 203
welding power sources, 203
welding process control, 202-3
welding torch(es), 203
welding torch mounts, 203
welding wire feeders, 203
welding wire supply, 208
Robots, welding, 16, 184-185. *See also* Robots, arc welding; Automation, welding.
accuracy, 196
anthropomorphic, 196, 200
articulated, 185
axes, number of, 196
cylindrical coordinate, 196
jointed arm, 196, 200
load capacity, 196
maintenance, 197
programming, 185
rectilinear, 185, 196, 201
reliability, 196-97
repeatability, 196
safety, 197
spherical coordinate, 196
Rockwell hardness test, 372

SAE. *See* Society of Automotive Engineers.

Safe practices, 380-415. *See also* Fumes and gases; Ventilation; Compressed gases, handling of; Compressed gas cylinders; Electrical safety.
for arc cutting, 406-7
for arc welding, 406-7
burn prevention, 384-86
combustibles, 383
confined spaces, 381
containers, venting of, 383-84
for electron beam welding, 409-10
escape routes, 381
excessive noise, protection from, 386-87
explosion prevention, 383-84
for explosion welding, 411
eye protection, 382, 384, 385, 407, 410
face protection, 384
fire prevention, 383
fire watch, 383
flammable vapors, 383
for friction welding, 411
fuel gas fires, 401
general area, 381-82
for high frequency welding, 409
hot work permit system, 387
housekeeping, 381
for laser beam welding and cutting, 410-11
for liquid oxygen, 399
machine guards, 387
management support, 380
pinch points of equipment, 387
protective clothing, 384, 386
protective screens, 381-82
public demonstrations, 382-83
safety labels, 380-81
safety lines, 381
safety rails, 381
safety training, 380-81
supplementary reading list, 415
for thermit welding, 411-12
tripping hazards, 381
for ultrasonic welding, 411
viewing filter plates, 385
wall reflectivity, 382
warning signs, 381
Safe practices for adhesive bonding
adhesives, handling of, 413-14
flammable material storage, 413
protective clothing, 414
toxic materials, handling of, 413
Safe practices for brazing and soldering
brazing atmospheres, 408
dip brazing and soldering, 408
hazards, 408
for soldering fluxes, 408-9
ventilation, 408
Safe practices for oxyfuel gas welding and cutting
backfire prevention, 406
flashback arresters, 406
flashback prevention, 406
gas hoses, 405-6
reverse flow check valves, 406
shutdown procedures, 406
torches, 405
Safe practices for radiographic testing, 353-54
Safe practices for resistance welding
electrical considerations, 407-8
grounding, 408
guarding of machines, 407
mechanical considerations, 407
Safe practices for thermal spraying, 412-13
ear protection, 413
eye protection, 412
protective clothing, 413
respiratory protection, 413
Safety, positioner, 166-68
Safety in Welding and Cutting, ANSI/ASC Z49.1, 383, 403
Safety labels, 380-81
Safety of structures, 17-18
Safety standards, 223, 230-31, 238
Seam tracking systems, 205-8
Seams in base metal, 280, 292
Shrinkage voids, definition of, 278
Side beam carriage, 181-82
Slag, surface, 334
Slag inclusion(s), 279, 280, 287, 288, 351
causes and remedies, 298
definition of, 278
effects on mechanical properties, 298
Society of Automotive Engineers, 218, 239-44
automotive welding standards, 239
Aerospace Material Specifications, 239-43
Soldered joints, discontinuities in. *See* Discontinuities in brazed and soldered joints.
Soldering cost estimating, 123-24
Soldering economics. *See* Economics of brazing and soldering.

Spin testing, 375
Stainless steel, ventilation during cutting, 396
Standards, 215-44
American National Standards, 221-22
applications, 217-19
automotive welding, 229
for bridges, 219
for buildings, 221
filler metals, 227-28
machinery and equipment welding, 229
marine welding, 230
mechanical testing, 228-29
military, 233-237
piping and tubing, welding of, 230
products covered, 220
qualifications, 228
reinforcing steel, welding of, 230
revision of, 217
safety, 238
safety in welding and cutting, 230-31
sheet metal welding, 230
for ships, 219-21
sources, 216-17, 218
structural welding, 230
Unified Numbering System, 243-44
for welding applications, 229-31
Standards, Federal, 232-37
Steel welds, sizing of, 45-67. *See also* Fillet welds; Tubular connections, structural; Welded joints, design of; Groove welds.
allowable stresses, 45ff
cyclic loading, 51-58
fatigue stress range, 51, 56
fatigue stress ratio, 58, 59
fillet welds, 56-57
fillet welds, skewed, 61
force per unit length, 62-63
primary welds, 58, 60
in rigid structures, 58
secondary welds, 58, 60
single-V-groove welds, 57-58, 59
static loading, 45-51
stress, allowable, in fatigue, 51-56
weld as a line, treating a, 62-67
weld size, 63, 66-67
welded connections, properties of, 64-66
Stresses, allowable, in steel welds, 45ff
Structural tubular connections. *See* Tubular connections, structural.
Superintendent of Documents, 218
Surface discontinuities, 326
Surface irregularities, 302
Symbols for welding, brazing, and nondestructive examinations. *See* Welding symbols; Brazing symbols; Nondestructive exmination symbols.

Tension testing of welds, 371
Test positions for welding, 250-53
Thermal spray operator qualifications, 271
Tooling. *See* Fixtures.
Toxic material safety, 394-95
Tubular connections, structural, 68-80
fatigue behavior, 80
fillet weld details, 68, 78
general collapse, 79
groove weld designs, 68, 72-75, 77
K-connections, 68ff
lamellar tearing, 79-80
load, uneven distribution of, 76
local failure, 76
T-connections, 68ff
weld joint design, 68-76
Y-connections, 68ff
Tungsten inclusion(s), 279, 280, 287, 288
Turning rolls, 160, 161
tractive effort, 170
Turntable positioners, 163

Ultrasonic testing, 321, 323. *See also* Nondestructive testing.
advantages, 360
applications, 360
basic equipment, 360-62
calibration test blocks, 363
coupling, 363
general description, 359-60
limitations, 360
longitudinal sound waves, 262-63
operator qualifications, 364
pulse-echo method, 360-62
reporting results, 364
sound behavior in metals, 362
sound wave forms, 362-63
test frequencies, 362
test procedures, 363-64, 365
transverse sound waves, 363
weld discontinuities, evaluation of, 364ff

Undercut, 280, 287, 289, 352
  definition of, 278
Underfill, 280, 289
  definition of, 278
Unified Numbering System, 243-44
Uniform Boiler and Pressure Vessel Laws Society, 218, 224
United States Code of Federal Regulations, 232, 233

Ventilation, 388, 390
  air movement, direction of, 392
  for brazing and soldering, 408
  of chlorinated hydrocarbons, 396
  of cleaning compounds, 396
  confined spaces, fluorine compounds in, 395
  confined spaces, welding in, 393-94
  containers, welding on, 394
  crossdraft table, 393
  cutting of stainless steel, 396
  downdraft table, 393
  general ventilation, 391-92
  gun-mounted equipment, 393
  local exhaust, 392-93
  local ventilation, 392-93
  methods of, 391-93
  natural ventilation, 391
  respiratory protective equipment, 393
  of toxic materials, 394-95
  water tables for welding and cutting, 393
  of zinc fume, 396
Vickers hardness test, 373
Viewing filter plates for welding, brazing, soldering, and cutting, 385
Visible penetrant testing. *See* Liquid penetrant testing.
Visual inspection, 321-28. *See also* Nondestructive testing.
  after welding, 326-28
  dimensional accuracy, 326-28
  during welding, 326
  equipment for, 325
  of postweld heat treatment, 328
  prior to welding, 325
  weld measurement gages, 325
  weld workmanship standards, 326-27

Water tables, 393
Weld distortion, 12
Weld quality, 276-316. *See also* Inspection.
  cost considerations, 277
  definitions of terms, 278
  meaning of, 276-77
  selection of, 277
  supplementary reading list, 316
Weld reinforcement, 353
Weld size, 15
Welded joints, design of 15, 32-45. *See also* Aluminum structures; Fillet welds; Steel welds, sizing of; Tubular connections, structural.
  combined welds, 41
  corner joint(s), 43-45
  cost considerations, 41-42
  double-bevel-groove weld, 35, 38-39, 42
  double-V-groove weld, 35, 38
  effective throat, 43, 44
  fillet weld(s), 39-41
  groove welds, 34-39
  J-groove weld, 34, 35, 39
  joint efficiency, 36
  joint penetration, 36-38
  lamellar tearing, 45
  single-bevel-groove weld, 34-38, 42
  single-V-groove weld, 34, 38
  square-groove weld, 34, 38
  for tubular structures, 68-76
  types of joints, 32-36
  U-groove weld, 34, 35, 39
  weld type, selection of, 41-43
Welder performance qualification, 261-68. *See also* Brazer performance qualification.
  duration of qualification, 268
  limitation of variables, 262
  qualification records, 267-68, 269
  requirements, 262
  test specimens, 262, 265-67
  testing of welds, 267
  weld type and position limitations, 267
Welding automation. *See* Automation, welding.
Welding automation, economics of. *See* Economics of welding automation and robotics.
Welding cost estimating. *See also* Cost estimating.
  deposition efficiency, 125
  elements in welding costs, 124-27
  filler metal costs, 129-31
  flux costs, 131-33
  labor costs, 125, 128

material costs, 125
methods, 127-33
operating factor, 127-28
overhead costs, 125-27, 133
shielding gas costs, 133
standard data, 128, 128n
variables, 124
Welding design. *See* Design for welding.
Welding head manipulators, 180-81
Welding fixtures, 152-53
Welding inspector
duties, 319
requirements for, 318-19
Welding inspector qualification, 217, 273
Welding Inspector Qualification and Certification Program, AWS, 318
Welding machine carriages, 179-80
Welding procedure specifications, 16, 246, 248-53. *See also* Brazing procedure specifications; Procedure specifications. Qualification of welding procedures.
applications, 253
forms, 253, 254
prequalified welding procedures, 248
qualification of, 256-59
responsibility for, 249
subjects covered, 249
test positions for welding, 250-53
types, 248
Welding robots. *See* Robots, welding.
Welding safety, 380-87
Welding symbols
arrow, significance of, 95, 97
arrow side, 95
backing symbol, 98, 102
construction of, 98, 102
contour symbol, 98, 102
dimensions of weld, 96, 98
double-V-groove weld, 103, 105
elements, location of, 95-96
elements of, 94-95
examples of, 101-12
field weld symbol, 96-98, 101
fillet weld, 103, 107
flange weld, 106, 110
intermittent weld, 96, 99
joints, types of, 98, 101, 10
length of weld, 96, 98
melt-thru symbol, 98, 101
other side, 95
plug weld, 103-6, 108
process letter designations, 101
reference line(s), 94, 96, 98
references, 95
seam weld, 107, 112
single-flare-bevel-groove weld, 103, 106
single-flare-V-groove weld, 103, 106
single-V-groove weld, 101, 105
size of weld, 96, 98
slot weld, 103-6, 109
spacer symbol, 98, 102
spot weld, 106-7, 111
supplementary symbols, 96ff
weld symbols, 94-95
weld-all-around symbol, 96, 100
Weldment design. *See* Design for welding.
Weldment design program, 13-17
cleaning, 16-17
design factors, 14
designing the weldment, 14-16
existing design, analysis of, 13
fixtures, positioners, and robots, use of, 16
forming of parts, 14-15
inspection, 16-17
load conditions, determination of, 13-14
part preparation, 14
residual stresses and distortion, 16
subassemblies, use of, 16
weld joint design, 15. *See also* Welded joints, design of.
weld size and amount, 15
welding procedures, 16
Welds in steel. *See* Steel welds, sizing of.

X-radiation from electron beams, 409

Zinc fume, 396